Lecture Notes in Mathematics

Edited by A. Dold and B. Eckmann

885

Combinatorics and Graph Theory

Proceedings of the Symposium Held at the Indian Statistical Institute, Calcutta, February 25–29, 1980

Edited by S. B. Rao

Springer-Verlag
Berlin Heidelberg New York 1981

Editor
Siddani Bhaskara Rao
Indian Statistical Institute
203, Barrackpore Trunk Road
Calcutta – 700035, India

AMS Subject Classifications (1980): 05-02, 05-04, 05-06, 05 A15, 05 Bxx, 05 Cxx, 62 K10, 68 E10

ISBN 3-540-11151-4 Springer-Verlag Berlin Heidelberg New York
ISBN 0-387-11151-4 Springer-Verlag New York Heidelberg Berlin

Printed in Germany

Printing and binding: Beltz Offsetdruck, Hemsbach/Bergstr.
2141/3140-543210

PREFACE

This volume consists of the Proceedings of the second Symposium on Combinatorics and Graph Theory held at the Indian Statistical Institute, Calcutta from February 25 to February 29, 1980 under the auspices and with the full financial support of the Indian Statistical Institute. I would like to express my gratitude to Professor B.P. Adhikari, Director of the Institute, Professor A. Maitra, the then Chairman, Division of Theoretical Statistics and Mathematics and the faculty of the Division for giving me the opportunity to convene the symposium and for their valuable advice and suggestions throughout.

The aim of the symposium was to discuss the recent trends in Combinatorics and Graph Theory and to expose the young combinatorialists and graph theorists in India to the recent developments in these areas.

We would like to thank the special invited speakers – Professors C. Berge (France), P. Erdös (Hungary), E.V. Krishnamurthy (India), L. Lovász (Hungary), K.R. Parthasarathy (India), D.K. Ray-Chaudhuri (USA), J.J. Seidel (The Netherlands), S.S. Shrikhande (India), Vera T. Sós (Hungary) and M.N. Vartak (India) – for their thought-provoking special invited addresses, active participation in the deliberations of the symposium and for their valuable contributions to these proceedings (Professor L. Lovász was unable to attend the symposium for personal reasons, but we are glad to include his contribution as a part of the proceedings). We thank all the participants who have chaired the sessions.

All the contributed papers appearing in this volume were refereed. We thank all the referees for their help in this regard. We also thank all the participants for their keen interest in the symposium and for their interesting contributions to these proceedings. Both the special invited addresses and the contributed papers are arranged alphabetically according to the authors' names.

I would like to use this opportunity to thank most warmly all those, apart from the special invited speakers and the participants, who have contributed very much to the success of the symposium: colleagues at the Indian Statistical Institute, particularly those of the Stat-Math Division, Dean's office, Director's office, Eka press, Electrical and Engineering units, Estate office, Guest House, Library, Medical Welfare unit, Reprography, Photography and Audio Visual units, Transport and Travel Cell, and more particularly Dr.(Mrs.) Nirmala Achuthan, Dr. A. Ramachandra Rao, Dr. T.J. Rao, Mr. N. Srinivasan, Dr. K.S. Vijayan, Ph.D. students Mr. Prabir Das, Mr. T. Gangopadhyay and graduate students Mr. Atul Jain and Mr. Rahul Roy for their active participation, valuable help and suggestions in the various organizational committees.

I would like to express my gratitude to Professor J.K. Ghosh, present Chairman of the Stat-Math Division, for his advice and for providing all the facilities at his disposal. I would like to express my gratitude to Dr. A. Ramachandra Rao under whose excellent supervision the typing of the manuscript was done, including the difficult task of proof reading the entire manuscript, with the cooperation of Dr.(Mrs.) Nirmala Achuthan, Mr. Prabir Das and Mr. T. Gangopadhyay while I was away visiting the department of mathematics, The Ohio State University, Columbus (USA) during the academic year 1980-81. I am immensely grateful to them for their kind help in this regard.

I thank Mr. Arun Das for the elegant typing of these proceedings. Thanks are also due to Mr. S.K. De for the excellent diagrams.

Last not least I thank Mr. K. Kundu, Mr. S. DasGupta and Mr. P. Nandi for secretarial help.

February 19, 1981 SIDDANI BHASKARA RAO

CONTENTS

Special Invited Addresses

Contributed Papers

DIPERFECT GRAPHS

CLAUDE BERGE
University of Paris, Paris VI

1. INTRODUCTION

Gallai and Milgram have shown that in a directed graph, the minimum number of vertex-disjoint (elementary, directed) paths needed to cover the vertex-set is smaller than or equal to the stability number (i.e., maximum number of independent vertices). However, the various proofs of this result do not imply the existence of an optimal stable set S and of a partition of the vertex-set into paths $\mu_1,\mu_2,\ldots,\mu_k$ such that $|\mu_i \cap S| = 1$ for all i.

Similarly, Gallai and Roy have shown independently that in a directed graph, the maximum number of vertices in an (elementary, directed) path is greater than or equal to the chromatic number; here again, we do not know if there exists an optimal coloring $(S_1,S_2,\ldots,S_k)$ and a path μ such that $|\mu \cap S_i| = 1$ for all i.

In this paper, we give proofs and applications for the two above mentioned results; then we study the graphs α-diperfect and γ-diperfect, which suggest several conjectures similar to the perfect graph conjecture. I wish to thank Polly Underground who has suggested the proof of Proposition 1.

2. THE GALLAI-MILGRAM THEOREM AND SOME APPLICATIONS

Let $G = (X,U)$ be a directed graph with vertex set X, and a set $U \subseteq X \times X$ of arcs (or "directed edges"). A path will be defined as a sequence of vertices $x_1,x_2,\ldots,x_k$, all distinct, such that $(x_1,x_2) \in U$, $(x_2,x_3) \in U,\ldots,(x_{k-1},x_k) \in U$; every path $\tilde{\mu} = (x_1,x_2,\ldots,x_k)$ defines a set that we denote $\mu = \{x_1,x_2,\ldots,x_k\}$.

A family $M = \{\mu_1, \mu_2,\ldots\}$ is a <u>partition</u> of X if the $\tilde{\mu}_i$ are pairwise vertex-disjoint paths, and if $\bigcup \mu_i = X$.

We shall give a proof of the following

<u>Theorem (Gallai-Milgram)</u>. In a directed graph G, we have $\min_M |M| \le \alpha(G)$. The minimum is taken over all the partitions M of X into paths, and $\alpha(G)$ denotes the stability number of G (i.e., the maximum number of independent vertices).

<u>Proof</u>. For a family $M = \{\mu_1, \mu_2, \ldots\}$, the initial vertices of the paths $\tilde{\mu}_1, \tilde{\mu}_2, \ldots$ will be denoted by : $A(M) = \{a_1, a_2, \ldots\}$. We prove here a stronger statement, namely : <u>Given a partition M into paths, there exists a partition M' into paths such that $|M'| \le \alpha(G)$ and $A(M') \subseteq A(M)$.</u>

This proposition is valid for graphs with 1,2 vertices; assume that it is valid for all graphs with less than n vertices, and we show that it is also valid for a graph G with n vertices.

1. We may assume that $|M| = \alpha(G) + 1$. Because, if $|M| \le \alpha(G)$, the proposition is proved. If $|M| \ge \alpha(G) + 1$, consider the first path μ_1 in M, and consider the subgraph G_1 of G induced by $X - \mu_1$. Since $\alpha(G_1) \le \alpha(G)$, there exists a partition M_1 of G_1 with $A(M_1) \subseteq A(M)$ and $|M_1| \le \alpha(G_1) \le \alpha(G)$. Hence $M' = M_1 \cup \{\mu_1\}$ is a partition of G with $A(M') \subseteq A(M)$, $|M'| \le \alpha(G) + 1$, and we prove the result for M' instead of M.

2. Since $|A(M)| = |M| = \alpha(G) + 1$, the set $A(M) = \{a_1, a_2, \ldots\}$ is not stable, so two vertices of $A(M)$ are linked by an arc, say $(a_1, a_2) \in U$.

The path $\tilde{\mu}_1 = (a_1, b_1, c_1, \ldots)$ has more than one vertex : otherwise, the proposition is proved with the partition M' obtained from M by removing μ_1, μ_2 and adding $(a_1, a_2, b_2, c_2, \ldots)$. By the induction hypothesis, the subgraph G_1 induced by $X - \{a_1\}$ has a partition M_1 with $|M_1| \le \alpha(G)$, $A(M_1) \subseteq \{b_1, a_2, a_3, \ldots\}$. If $b_1 \in A(M_1)$, the proposition is proved with the partition M' obtained from M_1 by adding a_1 to the path of M_1 starting at b_1. If $b_1 \notin A(M_1)$ and $a_2 \in A(M_1)$, add a_1 to the path of M_1 starting at a_2. If $b_1 \notin A(M_1)$ and $a_2 \notin A(M_1)$, take $M' = M_1 \cup \{(a_1)\}$. This proves the theorem.

Note that this result shows that in every graph G there exists a partition M with $|M| = \alpha(G)$, but it does not show the existence of a stable set S of maximum size and of a partition $M = \{\mu_1, \mu_2, \ldots\}$ such that : $|S \cap \mu_i| = 1$ for all i.

Remark. A stronger result has been obtained by Las Vergnas [7]: Every quasi-strongly connected graph G contains a spanning arborescence with at most $\alpha(G)$ terminal vertices ("sinks"). For a proof, see [9].

The Gallai-Milgram Theorem for a graph G can be obtained by using the Las Vergnas Theorem for a graph obtained from G by adding a new vertex x_o, that we join with every vertex of G by an arc going out of x_o.

The applications of the Theorem of Gallai-Milgram are interesting and numerous.

Application 1. (Theorem of Rédei). If $G = (X,U)$ is a tournament (complete antisymmetric graph), there exists a path which meets each vertex exactly once.

This follows from the theorem, since we have for G :

$$\min|M| \leq \alpha(G) = 1.$$

Application 2. (Theorem of Dilworth). If G is a transitive graph (i.e., $(x,y) \in U$, $(y,z) \in U$ implies $(x,z) \in U$), then $\min|M| = \alpha(G)$.

If $\theta(G)$ denotes the least number of cliques needed to cover X, we have $\alpha(G) \leq \theta(G) = \min|M| \leq \alpha(G)$.

Application 3. (Theorem of Erdös-Szekeres). Let n,p,q be positive integers with $n > pq$, and let $\sigma = (a_1,a_2,\ldots,a_n)$ be a sequence of n distinct integers. Then there exists either a decreasing subsequence of σ with more than p integers, or an increasing subsequence of σ with more than q integers.

Let $G = (X,U)$ be the graph with $X = \{a_1,a_2,\ldots,a_n\}$, where $(a_i,a_j) \in U$ iff $i < j$ and $a_i < a_j$. If $\{\mu_1,\mu_2,\ldots,\}$ is a partition of X into $\alpha(G)$ paths,

$$\alpha(G).\max|\mu| \geq \Sigma|\mu_i| = n > pq.$$

So, we have either $\alpha(G) > p$ - and there exists a decreasing subsequence with $\alpha(G) > p$ integers - or $\max|\mu| > q$ - and there exists an increasing subsequence with more than q integers.

3. THE GALLAI-ROY THEOREM AND SOME APPLICATIONS

Now, we shall consider similar problems for the chromatic number $\gamma(G)$ of a directed graph G. First, we shall give a simple proof for the following

Theorem (Gallai-Roy). For a directed graph G,

$$\max|\mu| \geq \gamma(G).$$

Proof. Let $V \subseteq U$ be a set of arcs which meets all the circuits of G, and which is minimal with respect to that property. Then, the partial graph $H = (X, U-V)$ has no circuits, and for $x \in X$, we define

$$t(x) = \max\{|\mu| : \mu \text{ is a path in } H \text{ starting at } x\}.$$

Clearly, $(x,y) \in U-V$ implies $t(x) > t(y)$. Also, $(x,y) \in V$ implies $t(y) > t(x)$, because $U-V \cup \{(x,y)\}$ has a circuit. Thus,

$$(x,y) \in U \rightarrow t(x) \neq t(y).$$

Hence $t(x)$ defines a coloring of the vertices of G in $\max_x t(x)$ colors, and

$$\gamma(G) \leq \max_x t(x) \leq \max|\mu|.$$

This proves the theorem.

Remark. A stronger result has been obtained by Bondy [1] : If a strongly connected graph G has at least 2 vertices, the longest circuit ("directed cycle") has length $\geq \gamma(G)$.

The Gallai-Roy Theorem for a graph G can be obtained by using the Bondy Theorem for a graph obtained from G by adding a new vertex x_o, that we join in both directions with every vertex of G. Furthermore, it contains also the Theorem of Camion [3] which states that a strongly connected tournament has a circuit which meets each vertex exactly once. We remark again that the proof given above does not imply that every graph G has an optimal coloring and a path μ which meets each color exactly once. However, P. Underground, among others, has shown :

Proposition 1. A directed graph G with $\gamma(G) = 3$ has a 3-coloring (S_1,S_2,S_3) and a path $\mu = (a,b,c)$ which meets each color class S_i exactly once.

Proof. Suppose that G is a graph for which the proposition does not hold.

Let $(\bar{S}_1,\bar{S}_2,\bar{S}_3)$ be a 3-coloring of G, and let $\mu = (a,b,c)$ be a path of 3 vertices (which exists by the Gallai-Roy Theorem); since μ does not meet the 3 color classes, we have, say $a,c \in \bar{S}_1$, $b \in \bar{S}_3$. Let S_3 be the maximal stable set which contains $\bar{S}_3$, and put $S_1 = \overline{S}_1 - S_3$, $S_2 = \overline{S}_2 - S_3$. Then we have

$$a \in S_1,\ c \in S_1,\ b \in S_3.$$

A vertex x of $S_1 \cup S_2$ is necessarily, in $G_{S_1 \cup S_2}$, an isolated vertex, or a source, or a sink. Otherwise, there is a path (y,x,z) with $x \in S_1$, $y,z \in S_2$; since x is adjacent to a vertex $v \in S_3$, there is either a 3-colored path (y,x,v), or a 3-colored path (v,x,z), which is a contradiction.

Let T_1 be the sources, T_2 be the sinks and T_o be the isolated vertices in $G_{S_1 \cup S_2}$. Thus, $(T_1, T_2 \cup T_o, S_3)$ is a 3-coloring of G with no path 3-colored and we may assume :

$$a,c \in T_2 \cup T_o,\ b \in S_3.$$

Every arc between T_2 and S_3 is directed from S_3 (otherwise there would be a 3-colored path). So, $a \in T_o$, and $\mu = (a,b,c)$ meets the 3 colors of the coloring :

$$(T_1 \cup \{a\},\ T_2 \cup T_o - \{a\},\ S_3).$$

This contradiction proves the Proposition.

Remark. Among the applications of the Gallai-Roy Theorem, let us mention :

1. Rédei's Theorem. In every tournament, there exists a path which meets each vertex exactly once.

2. The Chvátal-Komlós Theorem. Let $G = (X,U)$ be a graph, let $U = U_1 + U_2 + \dots + U_q$ be a partition of the arc-set into q classes, let $p_1, p_2, \dots, p_q$ be integers such that $p_1 p_2 \dots p_q < \gamma(G)$. Then for some i, the partial graph $G_i = (X,U_i)$ contains a path with more than p_i vertices.

3. The generalized Erdös-Szekeres Theorem. Let $\sigma = (a_1, a_2, \dots)$ be a sequence of $p_1 p_2 \dots p_q + 1$ distinct integers; let $\rho_1, \rho_2, \dots, \rho_q$ be binary relations satisfying: for every $i < j$, there exists a relation ρ_k with $a_i\, \rho_k\, a_j$. Then for some k, there exists a subsequence $\sigma' = (a_{i_1}, a_{i_2}, \dots)$ of σ of length $> p_k$, such that :

$$a_{i_1}\, \rho_k\, a_{i_2},\ a_{i_2}\, \rho_k\, a_{i_3},\ \dots$$

The proofs are easy and are left to the reader.

4. THE α-DIPERFECT GRAPHS AND THE γ-DIPERFECT GRAPHS

A directed graph G is <u>α-diperfect</u> if for every optimal stable set S, there exists a partition of the vertex set into paths $\mu_1, \mu_2, \ldots$ such that $|S \cap \mu_i| = 1$ for all i, (and if every induced subgraph of G has the same property). Many important classes of graphs are α-diperfect.

<u>Proposition 2.</u> Every transitive graph, and more generally every perfect graph is α-diperfect.

<u>Proof</u>. If $G = (X,U)$ is transitive, we have

$$(x,y) \in U, \ (y,z) \in U \rightarrow (x,z) \in U.$$

From Dilworth's Theorem, it follows that G is perfect, i.e. :

$$\alpha(G_A) = \theta(G_A) \qquad (A \subseteq X)$$

In a perfect graph G, there exist $k = \alpha(G)$ cliques $C_1, C_2, \ldots, C_k$ which partition the vertex-set, and by Rédei's Theorem, each C_i is spanned by a path μ_i. So,every optimal stable set S satisfies :

$$|S \cap \mu_i| = 1 \qquad (i = 1,2,\ldots)$$

<u>Proposition 3</u>. Every symmetric graph is α-diperfect.

<u>Proof</u>. Let $G = (X,U)$ be a symmetric graph :

$$(x,y) \in U \rightarrow (y,x) \in U.$$

An optimal stable set S of G defines a graph G' obtained from G by removing the arcs going into S. By the Theorem of Gallai-Milgram, G' has a partition into $k = \alpha(G) = \alpha(G')$ paths $\mu_1, \mu_2, \ldots, \mu_k$. Clearly, the μ_i's are paths of G and satisfy $|S \cap \mu_i| = 1$.

Not every graph is α-diperfect. For instance, consider a graph G whose skeleton is isomorphic to an odd cycle C_{2p+1} with arrows directed as in Figure 1. Clearly, the set

FIGURE 1

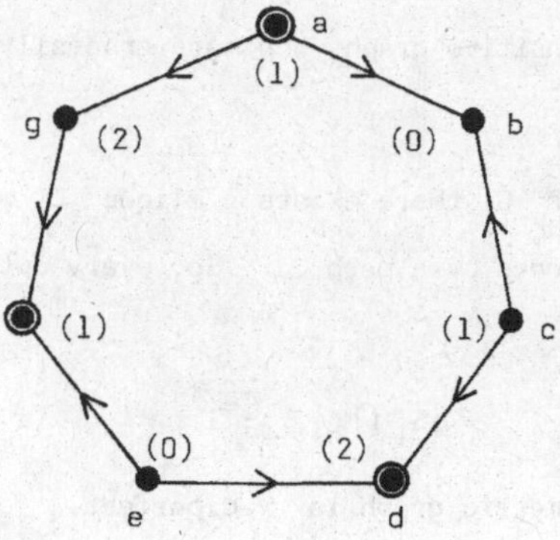

$S = \{a,d,f\}$ is an optimal stable set, but no partition into 3 paths μ_1, μ_2, μ_3 satisfies $|S \cap \mu_i| = 1$ for $i = 1,2,3$.

The graph G on Figure 2 is α-diperfect but we shall see later that it has special properties.

FIGURE 2

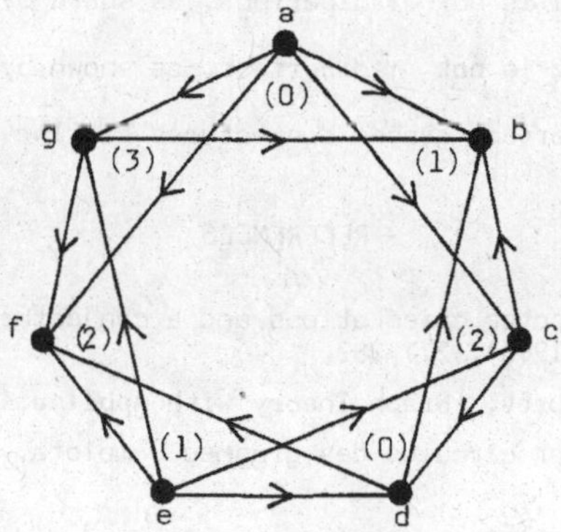

We conjecture that a graph G is α-diperfect if it has the following property : For every set A of vertices which constitute an odd cycle, the number of arcs in G_A is larger than $|A|$, and for every set B of vertices which constitute the complement of an odd cycle, the number of arcs in G_B is larger than $\binom{B}{2} - |B|$.

Clearly, the perfect graphs and the symmetric graphs fulfill the above condition.

A directed graph G is γ-diperfect if for every optimal coloring $(S_1, S_2, \ldots, S_k)$, there exists a path μ such that $|\mu \cap S_i| = 1$ for all i (and if every induced subgraph of G has the same property).

Proposition 3. Every transitive graph, and more generally, every perfect graph is γ-diperfect.

Proof. In a perfect graph G there exists a clique C with $|C| = \gamma(G)$, and by Rédei's Theorem, C is spanned by a path μ. So, every optimal coloring $(S_1,S_2,\ldots,S_k)$ satisfies

$$|S_i \cap \mu| = 1 \qquad (i = 1,2,\ldots)$$

Proposition 4. Every symmetric graph is γ-diperfect.

Proof. Let $(S_1,S_2,\ldots,S_k)$ be an optimal coloring of a symmetric graph G, and let G' be the graph obtained from G by removing the arcs going from S_j to S_i if $j > i$. By the Theorem of Gallai-Roy, G' has a path μ with $k = \gamma(G)$ vertices, therefore it meets all the S_i's.

The graph in Figure 1 is not γ-diperfect, as shown by the colors (1),(2),(3). Also, the graph in Figure 2 is not γ-diperfect, as shown by the colors (0),(1),(2),(3). We can make for the γ-diperfect graphs conjectures similar to those made above.

REFERENCES

1. J.A. Bondy, Disconnected orientations and a conjecture of Las Vergnas, J. London Math. Soc., 2, 14(1976), 277-282.
2. J.A. Bondy, U.S.R. Murty, Graph Theory with Applications, MacMillan, London, 1972.
3. P. Camion, Chemins et circuits des graphes complets, C.R. Acad. Sci. Paris,249 (1959), 2151-52.
4. V. Chvatál, and J. Komlós,
5. T. Gallai, On directed paths and circuits, Theory of Graphs (Ed. Erdös-Katona), Academic Press, New York, 1968, 115-118.
6. T. Gallai, A.N. Milgram. Verallgemeinerung eines Graphen-theoretischen Satzes von Redei, Acta Sc. Math. 21(1960), 181-186.
7. M. Las Vergnas, C.R. Acad. Sci. Paris, 1971, A. 272.
8. B. Roy, Nombre chromatique et plus longs chemins, Rev. Fr. Automat. Informat. 1(1967), 127-132.
9. A. Sache, La Théorie des Graphes, Presses Universitaires de France 1554, Paris, 1974.

SOME NEW PROBLEMS AND RESULTS IN GRAPH THEORY AND OTHER BRANCHES OF COMBINATORIAL MATHEMATICS

PAUL ERDÖS
The Hungarian Academy of Sciences
Budapest, Hungary

§ 0.

Recently I published several papers on finite and infinite combinatorial problems. I will try to make the overlap with this paper as small as possible; as a result I have to omit some of my most interesting problems, but first of all I give some references to my older papers where these questions have been discussed :

P. Erdös, Old and new problems in combinatorial analysis and graph theory, Second International Conference on Combinatorial Mathematics, New York Academy of Sciences, Vol. 319 (1979), 177-187.

P. Erdös, Problems and results on finite and infinite Combinatorial Analysis I and II, Coll. Math. Soc. J. Bólyai, 10 : Infinite and finite sets, Kenthely, Hungary (1973), 403-424, II will appear in Enseignement Math. in 1981.

P. Erdös, Some old and new problems in various branches of Combinatorics, Proc. Tenth Conference in Combinatorics, Graph Theory and Computing (1979) (Boca-Raton Conference). This paper contains extensive references to my previous papers.

P. Erdös, Combinatorial problems which I would most like to see solved, will soon appear in the new Hungarian periodical Combinatorica.

For applications of probabilistic methods to combinatorial analysis see our book, P. Erdös and J. Spencer : Probabilistic methods in Combinatorics, Acad. Press and Hung. Acad. Sci. (1974).

§ 1.

First I discuss problems connected with Ramsey's theorem and its generalisations, here I of course can not avoid overlap with previous papers. $r(n_1,\dots,n_k)$ is the smallest integer for which if one colors the edges of $K(r(n_1,\dots,n_k))$ by k colors $(K(t)$

is the complete graph of t vertices) then there is always an i, $1 \le i \le k$, so that there is a $K(n_i)$ all of whose edges are of the i-th color. Very interesting problems arise if k tends to infinity, but we will not discuss these in great detail. We just mention that it is not even known how fast $r(n_1,\dots,n_k)$ tends to infinity if all the n_i are 3. This problem goes back essentially to I. Schur who proved

$$r_k(C_3) = r_k(3,\dots,3) < e.k!\,.$$

It is not yet known if

$$r_k(C_3) < C^k \qquad \dots \qquad (1)$$

holds for all k if C is a sufficiently large absolute constant. More generally it is quite possible that there is an absolute constant C so that

$$r(n_1,\dots,n_k) < C^{n_1+\dots+n_k} \qquad \dots \qquad (2)$$

It is easy to show by induction that

$$r(n_1,\dots,n_k) < k^{\sum_{i=1}^{k}(n_i-2)} .$$

The proof or disproof of (1) and (2) seem to be very interesting questions.

Let us now restrict ourselves to $k = 2$. It is well known that

$$c_1\, n^{\frac{1}{2}}.2^{\frac{n}{2}} < r(n,n) < c_2 \binom{n}{[\frac{n}{2}]} \frac{\log\log n}{\log n} \qquad \dots \qquad (3)$$

I offered and offer 1000 rupees (or an equivalent in Swiss Francs) for a proof or disproof of

$$\lim_{n\to\infty} r(n,n)^{\frac{1}{n}} = C \qquad \dots \qquad (4)$$

I offer another 1000 rupees for the value of C. I think that perhaps the proof for the existence of C will not be difficult (though I can not do it), but I do not think the determination of C will be easy and I have no idea what its value will be.

It is known that

$$\frac{c_1 . n^2}{(\log n)^2} < r(3,n) < c_2 \frac{n^2}{\log n} . \qquad \dots \qquad (5)$$

The lower bound is due to me. The upper bound was proved very recently by Ajtai, Komlós and Szemerédi who improved the previous bound $\frac{c\, n^2 \log\log n}{\log n}$ of Graver and

Yackel.

Very likely for every k and $\varepsilon > 0$, if $n \to \infty$

$$r(k,n) > n^{k-1-\varepsilon}. \qquad \ldots \qquad (6)$$

In fact probably

$$r(k,n) > c_1 \frac{n^{k-1}}{(\log n)^{c_2}}. \qquad \ldots \qquad (6')$$

All our attempts to prove (6) and (6') - even for $k = 4$ - failed completely. It is not impossible that the difficulties are only technical. Both the upper and lower bounds of (4) are obtained by probabilistic methods and probably (6) will have to be attacked similarly.

Almost nothing is known about the local growth properties of $r(n,m)$. S. Burr and I conjectured that

$$r(n+1,n) > (1+c)\, r(n,n), \qquad \ldots \qquad (7)$$

but at the moment (7) is intractable. Faudree, Schelp, Rousseau and I needed recently a lemma stating

$$\lim_{n\to\infty} \frac{r(n+1,n) - r(n,n)}{n} = \infty. \qquad \ldots \qquad (8)$$

We could prove (8) without much difficulty, but could not prove that $r(n+1,n) - r(n,n)$ increases faster than any polynomial of n. We of course expect

$$\lim_{n\to\infty} \frac{r(n+1,n)}{r(n,n)} = C^{\frac{1}{2}}$$

where $C = \lim_{n\to\infty} r(n,n)^{\frac{1}{n}}$.

V.T. Sós and I recently needed the following results.

$$r(n+1,3) - r(n,3) \to \infty, \qquad \ldots \qquad (9)$$

and

$$r([n(1+c_1)],3) > (1+c_2)\, r(n,3). \qquad \ldots \qquad (9')$$

Both (9) and (9') must certainly be true but we could certainly not prove them. Probably

$$(r(n+1,3) - r(n,3))/n^{\frac{1}{2}} \to 0. \qquad \ldots \qquad (10)$$

All these results would easily follow if one could get a good asymptotic formula with a good error term for $r(n,3)$, but needless to say this is nowhere in sight.

One of the reasons for our inability to prove such simple results may be that we lack constructive methods for giving good lower bounds for $r(m,n)$. I offer 1000 rupees for a constructive proof of $r(n,n) > (1+c)^n$. The currently known sharpest constructive proof is due to P. Frankl who proved that

$$\lim_{n\to\infty} \frac{r(n,n)}{n^k} = \infty,$$

for every k.

Denote by $r(C_{2n+1},k)$ the largest integer t for which the edges of $K(t)$ can be colored by k colors so that there should be no monochromatic C_{2n+1}. Graham and I conjectured that

$$\lim_{n\to\infty} r(C_{2n+1},k)/r(C_3,k) = 0 \qquad \dots \quad (11)$$

(11) is open even for $n = 2$. Perhaps the proof of $r(C_5,k) < C^k$ will not be too hard. Another problem of Graham and myself states : It is well known and easy to see that the edges of $K(2^r)$ can be colored by r colors so that each color graph is bipartite and that such a decomposition does not exist for $K(2^r+1)$. Let now $f(r)$ be the smallest integer so that every coloring of the edges of $K(2^r+1)$ by r colors contains a $C_{2f(r)+1}$. Estimate $f(r)$ as well as possible.

Now I discuss the so called generalised Ramsey numbers. The systematic formulation of the problems was due to Harary and Cockayne. Let $G_1,\dots,G_k$ be k graphs, then $r(G_1,\dots,G_k)$ is the smallest integer n for which if we color the edges of $K(n)$ by k colors, then there is an i so that the edges of the i-th color contain G_i as a subgraph. Chvátal and Harary proved using the method of the proof of (3) that if G is t-chromatic, then

$$r(G,G) > (1+c)^t. \qquad \dots \quad (12)$$

After learning of (12) I conjectured that

$$\min_G r(G,G) = r(t,t). \qquad \dots \quad (13)$$

In other words, if G runs through the family of t-chromatic graphs, then the minimum of $r(G,G)$ is assumed for the complete graph $K(t)$ and further I conjecture that the minimum is assumed only for this graph. This is trivial for $t = 3$, but $t = 4$ already seems to present considerable difficulties. Let, in particular, G be the pentagonal wheel. The conjecture for $t = 4$ would follow if we could prove

$$r(G,G) > r(4,4) = 18. \qquad \dots \quad (14)$$

Perhaps (14) could be proved in cooperation with a computer, the best results so far are due to Chvátal and Schwenk and they proved that $17 \leq r(G,G) \leq 21$.

Burr recently published two excellent survey papers on the generalised Ramsey numbers, also Burr, Faudree, Rousseau, Schelp and I published several papers on this subject and several more of our papers will be published soon - here I give a short summary of some of our results and open problems.

Following some preliminary results of Bondy and myself, V. Rosta and independently Faudree and Schelp determined $r(C_n,C_m)$ for every n and m. Bondy and I conjectured

$$r(C_n,C_n,C_n) \leq 4n - 3, \qquad \dots \quad (15)$$

which is still open. For odd n, (15), if true, is best possible.

Denote by $G(n)$ a graph of n vertices. $G(n)$ is said to have edge density $\leq C$ if for every subgraph $G(m)$ of $G(n)$, we have $e(G(m)) < C.m$, where $e(G)$ denotes the number of edges of G. Burr and I conjectured that if $G(n)$ has edge density $< C$, then

$$r(G(n),G(n)) < f(C).n. \qquad \dots \quad (16)$$

The proof of this very attractive conjecture is nowhere in sight. Denote by $G_c(n)$ the graph determined by the edges of the n-dimentional cube; $G_c(n)$ has 2^n vertices and $n\,2^{n-1}$ edges. We could not decide whether for some absolute constant c_1

$$r(G_c(n),G_c(n)) < c_1.2^n \qquad \dots \quad (16')$$

is true. (16) and (16') seem to me to be two very attractive problems. Burr and I expected (16) to be true and (16') to be false.

Now I state some of the problems and results of our work with Faudree, Rousseau and Schelp. We are fairly certain that $r(K(n),C_4) < n^{2-\varepsilon}$ holds for a certain ε and all $n > n_o(\varepsilon)$, but all we could prove is that for $r \geq 5$, $r(K(n),C_r) < n^{2-\varepsilon}$.

Our most striking and original problem states the following : Denote by $\hat{r}(G_1,G_2)$ the smallest integer m for which there is a graph G of m edges so that if we color the edges of G by two colors, then either color I contains G_1 or color II contains G_2. We called $\hat{r}(G_1,G_2)$ the size Ramsey number of G_1 and G_2. Let P_n be the path of length n. Is it true that

$$\frac{\hat{r}(P_n,P_n)}{n^2} \to 0 \quad \text{but} \quad \frac{\hat{r}(P_n,P_n)}{n} \to \infty \ ? \qquad \ldots \quad (17)$$

Of course one really would like to determine $\hat{r}(P_n,P_n)$ exactly or at least to get an asymptotic formula for it; but in fact we could not make any progress with (17).

Harary recently asked the following question : Let G_n be a graph of n edges. What is the smallest possible value of $r(G_n,G_n)$? We are far from being able to give a complete solution but could prove that there is a G_n for which

$$r(G_n,G_n) < c.n^{2/3}. \qquad \ldots \quad (18)$$

Perhaps 2/3 is the best exponent, we only know that 2/3 can not be replaced by an exponent less than 3/5. We further conjectured that

$$r(G_n,3) \leq 2n + 1, \qquad \ldots \quad (19)$$

but could prove it only with 3-c instead of 2n + 1. (19), if true, is best possible.

P. Erdös and J. Spencer, Probabilistic methods in Combinatorics; Acad. Press and Hung. Acad. Sci. (1974).

S.A. Burr, Generalised Ramsey theory for graphs, in Graphs and Combinatorics, R. Bari and F. Harary (eds.), Springer Verlag, Berlin 52-75.

P. Frankl, A constructive lower bound for some Ramsey numbers, Ars Combinatoria 3(1977), 297-302.

P. Erdös and R.L. Graham, On partition theorems for finite graphs, Coll. Math. Soc. J. Bólyai 10 : Infinite and finite sets, Keszthely, Hungary (1973), 515-527.

S. Burr and P. Erdös, On the magnitude of generalised Ramsey numbers for graphs, ibid, 215-240.

V. Rosta, On a Ramsey type problem of J.A. Bondy and P. Erdös, Journal of Comb. Theory Ser. B, 15(1973), 94-120.

P. Erdős, R.J. Faudree, C.C. Rousseau and R.H. Schelp, The size Ramsey number, Periodica Math., Vol.9.

S.A. Burr and us, Ramsey minimal graphs for multiple copies, Indagationes Math., 40(1978), 187-195.

§ 2.

Now I discuss some recent problems on number theory and geometry. Let $1 \leq a_1 < \dots < a_k \leq x$. Assume $a_i + a_j \nmid a_i a_j$. Put $f(x) = \max k$. I first assumed that it will be easy to show that $f(x) = (\frac{1}{2} + o(1))x$. The odd numbers show that $f(x) > \frac{x}{2}$. I am no longer sure that my conjecture is correct. Odlyzko found with the aid of a computer that $f(1000) \geq 717$ and now I am no longer sure what happens. A related question states : Let $1 \leq b_1 < \dots < b_\ell \leq x$, $(b_i + b_j) \nmid 2b_i b_j$. Put $g(x) = \max \ell$. Is it true that $g(x) = o(x)$? Clearly if $a_i + a_j \nmid a_i a_j$ then $3x$ and $6x$ can not be both a's. I am sure that there is a sequence $u_1 < u_2 < \dots$, $\Sigma \frac{1}{u_i} < \infty$ so that the set of integers not divisible by any of the u's satisfies $(a_i + a_j) \nmid a_i a_j$ and that this sequence will give $\lim f(x)/x$. The details are not quite clear.

Silverman and I some time ago asked the following questions. Define a graph whose vertices are the integers, as follows : Join i to j if $i + j$ is a square. Is it true that this graph has infinite chromatic number? We also asked : Let $1 \leq u_1 < \dots < u_t \leq x$ and assume that $u_i + u_j$ is never a square. Put $\max t = h(x)$. $h(x) > x/3$ is trivial - take the $u_i \equiv 1 \pmod 3$. Is it true that $h(x) = (\frac{1}{3} + o(1))x$? A weaker conjecture states : Let $v_1, \dots, v_\ell$ be residue classes (mod d). Assume that $z^2 \not\equiv (v_i + v_j) \pmod d$ for $1 \leq i \leq j \leq \ell$. Is it true that $\ell \leq \frac{d}{3}$? If not - how large can ℓ be? None of these questions has been investigated very carefully and I have to ask the indulgence of the reader if the answer turns out to be trivial.

One of my oldest questions in number theory states : Is it true that the density of integers n which have two divisors $d_1 < d_2 < 2d_1$ is one, i.e., almost all integers have two divisors which are close together? I proved long ago that the density of these integers exists but could never prove that it is 1.

Denote by $d(n)$ the number of divisors of n and by $d^+(n)$ the number of integers k for which n has a divisor in $(2^k, 2^{k+1})$. If my conjecture is correct then for almost all integers n, $d^+(n) < d(n)$. I conjectured that in fact for almost all integers $d^+(n)/d(n) \to 0$. Tenenbaum and I last summer at the number theory meeting in Durham disproved this conjecture and recently Tenenbaum obtained an inequality on the density of the integers n for which $d^+(n) < C.d(n)$, but I still believe that for almost all n, $d^+(n) < d(n)$ is true.

Denote by $d_t(n)$ the number of divisors of n in $(t,2t)$ and put $D(n) = \max_t d_t(n)$. As far as I know Hooley was the first to investigate $D(n)$. He proved

$$\sum_{n=1}^{x} D(n) < x(\log x),$$

and I proved

$$\frac{1}{x} \sum_{n=1}^{x} D(n) \to \infty .$$

Hooley asked : Is it true that for every $\varepsilon > 0$

$$\sum_{n=1}^{x} D(n) = o(x(\log x)^{\varepsilon}) \text{ ?} \qquad \ldots \quad (20)$$

I several times tried to prove (20) but I was unsuccessful so far.

Now I state a few problems in Combinatorial Geometry. I published several survey papers on this subject and G. Purdy and I hope to write a book on this subject, here I only state problems which are at least partially new. Let $x_1,\ldots,x_n$ be n distinct points in k-dimensional space. Join two of the x_i if their distance is one and denote this graph by $G(x_1,\ldots,x_n)$. The chromatic number $\chi(E^{(k)})$ is defined as the upper bound of $\chi(G(x_1,\ldots,x_n))$ where n and the $\{x_i\}$ are both variable. Hadwiger and Nelson first asked for the determination of $\chi(E^{(2)})$. It is now known that $4 \leq \chi(E^{(2)}) \leq 7$ and it seems likely that $\chi(E^{(2)}) > 4$. Probably

$$\chi(E^{(k)}) > (1+c)^k, \qquad \ldots \quad (21)$$

but we are very far from being able to prove (21), the sharpest result so far is due to P. Frankl who proved $\chi(E^{(k)}) > k^{\ell}$ for every fixed ℓ if $k > k_0(\ell)$; he sharpened previous results of Larman and Rogers.

Simonovits and I define the essential chromatic number $t = \chi_e(M)$ of a metric space M as follows : t is the smallest integer so that for every $G(x_1,\dots,x_n)$ we can omit $o(n^2)$ edges from $G(x_1,\dots,x_n)$ so that the resulting graph has chromatic number $\leq t$. Simonovits and I prove $\chi_e(E^4) = 2$, $\chi_e(E^k) \geq k-2$. In fact we conjecture $\chi_e(E^k) > (1+C)^k$. $\chi_e(E^2) = \chi_e(E^3) = 1$ simply means that the number of edges of $G(x_1,\dots,x_n)$ is $o(n^2)$. Our paper on this subject will appear in Ars Combinatoria soon.

Recently the following question in elementary geometry occured to me : Is it true that for every n there are n distinct points in the plane in general position (i.e., no three on a line and no four on a circle) so that these points determine exactly $n-1$ distinct distances where further the i-th distance occurs i times. The existence of such a set is trivial for $n = 3$ and $n = 4$ (an isosceles triangle with the centre of its circumscribed circle shows this). I thought that such a set does not exist for $n = 5$ but Pomerance gave a simple example for such a system : x_1,x_2,x_3 are the vertices of an equilateral triangle, x_4 is the centre of its circumscribed circle, x_5 is the point of intersection of the perpendicular bisector of (x_3,x_4) with the circle with centre x_1 and radius $d(x_1,x_2)$ where $d(x_1,x_2)$ is the distance of x_1 to x_2. It is easy to see that these points are in general position and the i-th distance occurs i times. I believe no such system exists for $n \geq 6$. Perhaps, if we also require that no circle whose centre is one of our points should contain three of our points, then such a system can not exist for $n = 5$.

P. Erdös, On some problems of elementary geometry. Annali Math. Pure Apl., 103(1975), 99-108.

D.E. Larman and C.A. Rogers, The realisation of distances within sets in Euclidean space, Mathematica, 19(1972), 1-24.

P. Erdös, Combinatorial problems in geometry and number theory, Proc. Symp. Pure Math., Amer. Math. Soc., 34(1979), 149-162.

C. Hooley, On a new technique and its applications to the theory of numbers, Proc. London Math. Soc., 38(1979), 115-151 (see p. 125-128).

For further problems and results on sequences of integers see the excellent book of Hallerstam and Roth, "Sequences", 1966, Oxford.

A FORM INVARIANT MULTIVARIABLE POLYNOMIAL REPRESENTATION OF GRAPHS

E. V. KRISHNAMURTHY
Indian Institute of Science
Bangalore 560 012

ABSTRACT

A generalized adjacency matrix (called "representation matrix") is defined for a graph : the elements of this matrix are the edge labels. Treating these labels as independent variables, if the determinant is evaluated, the resulting multivariable polynomial parametrizes a graph. This serves as a basis to detect isomorphism, automorphism, subgraph isomorphism, and other graph properties.

Essentially, this paper exhibits the analogy that exists between forms among expressions and isomorphism among graphs. Consequently, many of the graph properties can be inferred from properties, such as symmetry, variable-separable-factorizability, and similarity of forms of the parametrizing polynomial.

It is shown how this formalism can be used for coding a graph. The decoding or reconstruction of a graph from its invariant polynomial code is also described. These will have many practical applications.

1. INTRODUCTION

Several authors [1] - [6] have considered the problem of characterizing a graph from certain similarity invariants constructed from the adjacency matrix of a graph, e.g. the characteristic polynomial, or its elementary divisors or eigenvalues or immanents [4]. These approaches ultimately did not yield any useful result, since non-isomorphic graphs too can have the same characteristic polynomial (and elementary divisors), etc. In fact, in a detailed study Meyer [5] has shown that such procedures fall short of characterizing even a simpler class of graphs called transition graphs

---

KEY WORDS AND PHRASES : Representation polynomial, isomorphism, automorphism, line graph, complexity, variable-separable-factorization of polynomial.

(directed graphs in which every node has out-degree 0 or 1) up to isomorphism.

In this paper we construct a more general invariant that does parametrize a graph up to isomorphism and that in particular cases reduces to the characteristic polynomial. This invariant is nothing but the determinant of the representation matrix of a graph, which is a more generalized adjacency matrix whose entries are the labels of the edges rather than just 1's and 0's. The use of a generalized adjacency matrix is not new. In fact it is well-known among network theorists and is called the "transmission matrix" [7][8]. Accordingly for our purpose it is only necessary to reinvoke some of the important properties and theorems [7][8] about the transmission matrix.

2. DEFINITIONS AND PRELIMINARIES

We consider finite, non-directed, connected, loopless graphs without multiple edges.

A non-directed graph G is given by a pair $G = (V,E)$ where V is the vertex set and E the edge set (a set of unordered pairs of elements of V). Two graphs $G_1 = (V_1,E_1)$ and $G_2 = (V_2,E_2)$ are said to be isomorphic if there is a bijection $\varphi : V_1 \to V_2$ such that $(v_1,v_2) \in E_1$ iff $(\varphi(v_1), \varphi(v_2)) \in E_2$ — in other words, there is a one-to-one edge matching from set E_1 to set E_2, preserving adjacency.

An automorphism of a graph G is an isomorphism of G with itself. In fact, the graph isomorphism problem is equivalent to the automorphism partitioning problem; given a graph $G(V_1,E_1)$, to partition V_1 in such a way that two vertices (v_1,v_2) belong to the same cell iff there is at least one automorphism of G mapping v_1 into v_2 (namely, the permutation of the vertex set which preserves adjacency).

In order to detect isomorphism (or automorphism) one therefore has to construct a graph invariant that is preserved under isomorphism. For this purpose we associate a matrix $R(G) = \{r_{ij}\}$ (called a representation matrix) with a graph $G(V,E)$ as follows :

Let $|V| = n$ (number of vertices in G) and $|E| = m$ (number of edges in G); number the vertices $v_1,v_2,\ldots,v_n$ in an arbitrary manner using numerals $1,2,\ldots,i,$ $j,\ldots,n$; and define $r_{ij} \in \underline{\underline{R}}$ (a commutative ring with identity 1) $(1 \leq i,j \leq n)$ as follows.

<u>Case (i) : $i \neq j$</u>. If $(v_i, v_j) \in E$, $r_{ij} = r_{ji}$ = label of the i-j edge; otherwise, set $r_{ij} = r_{ji} = 0$.

<u>Case (ii) : $i = j$</u>. Set $r_{ii} = 1$ (The unit element in the ring).

This means $R = \{r_{ij}\}$ is a more general adjacency matrix; the adjacency matrix is a particular case of the representation matrix, where all the non-zero non-diagonal terms are set to unity and the diagonals to zero. It is also noted that for non-directed graphs R is symmetric, containing at most $n(n-1)/2$ independent entries.

We now introduce the following definition and theorem.

<u>Definition 1</u>. Let $|R|$ denote the determinant of the representation matrix, defined in the usual way as for a matrix over a commutative ring with identity; let this multi-variable polynomial in $\frac{n(n-1)}{2}$ variables be denoted by $\varphi(G)$, called the representation polynomial of G.

<u>Theorem 1</u>. The graph G has p components iff $\varphi(G)$ is factorable into p unique variable separable factors or we can express

$$\varphi(G) = \prod_{i=1}^{p} \varphi_i \, (r_{i1}, r_{i2}, \ldots, r_{ik_p})$$

where r_{ik_p}'s form partitions of the $\frac{n(n-1)}{2}$ variables r_{ij}.

The proof of this theorem follows from the fact that $R(G)$ can be brought into block diagonal form with p diagonal blocks.

3. THE REPRESENTATION POLYNOMIAL AND ITS USES

From Theorem 1 it is clear that for every graph G, for any arbitrary lebelling of vertices and edges, the form of $\varphi(G)$ is an invariant. This forms the basis for detecting automorphism, isomorphism, subgraph isomorphism and other graph properties.

(i) <u>Automorphism</u>

When a graph has automorphisms (other than the identity), then $\varphi(G)$ should remain identical or unchanged (or $\varphi(G)$ is symmetric) for any specified interchange of two or more variables.

On the other hand, if the graph G has no automorphism other than the trivial identity, $\varphi(G)$ will not be identical for interchange of two variables.

(ii) Isomorphism

When two graphs G_1 and G_2 are isomorphic, the forms of $\varphi(G_1)$ and $\varphi(G_2)$ should be similar since there exists a match between the edge labels.

(iii) Subgraph isomorphism

Let $H(V_1,E_1)$ be the subgraph to be matched with some subgraph of $G(V_2,E_2)$; let $|V_1| = n_1$, $|E_1| = m_1$, $|V_2| = n_2$, $|E_2| = m_2$ and let $n_1 \le n_2$, $m_1 \le m_2$ and $\Delta = (m_2 - m_1)$.

Then for detecting subgraph isomorphism, we compute $\varphi(G)$ for every different set of Δ edges, by setting the values of the corresponding variables to zero; the resulting form should be similar with the form of $\varphi(H)$, if there exists a matching.

We will now illustrate these with examples.

a. Consider

G_1:

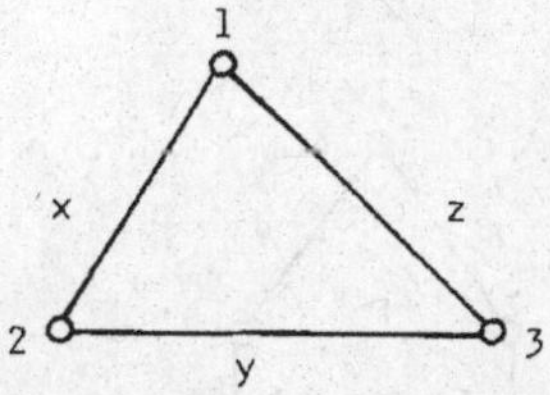

$$R(G_1) = \begin{bmatrix} 1 & x & z \\ x & 1 & y \\ z & y & 1 \end{bmatrix},$$

$$\varphi(G_1) = 1 - (x^2+y^2+z^2) + 2xyz.$$

b. Consider

G_2:

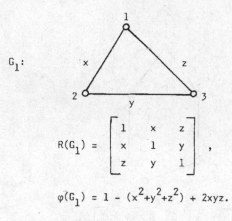

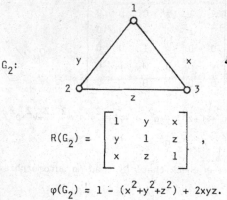

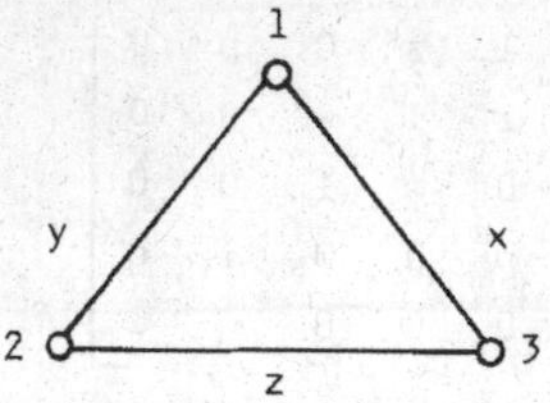

$$R(G_2) = \begin{bmatrix} 1 & y & x \\ y & 1 & z \\ x & z & 1 \end{bmatrix},$$

$$\varphi(G_2) = 1 - (x^2+y^2+z^2) + 2xyz.$$

c. Consider

G_3:

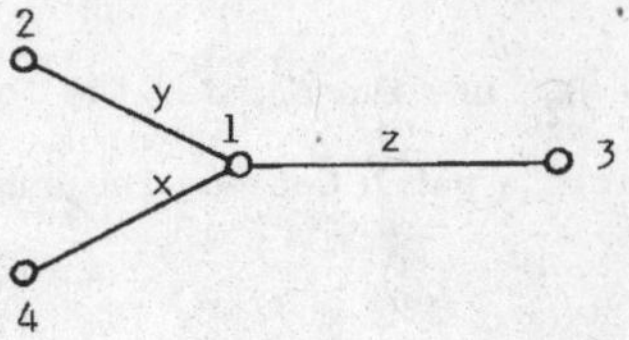

$$R(G_3) = \begin{bmatrix} 1 & y & z & x \\ y & 1 & 0 & 0 \\ z & 0 & 1 & 0 \\ x & 0 & 0 & 1 \end{bmatrix},$$

$$\varphi(G_3) = 1 - (x^2+y^2+z^2).$$

Remark. It is proved in the next section, that trees will contain only degree two terms in each variable.

(iv) Automorphism

Consider

G_4:

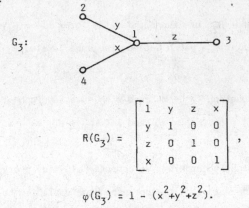

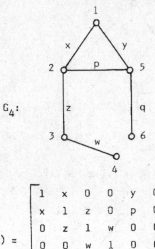

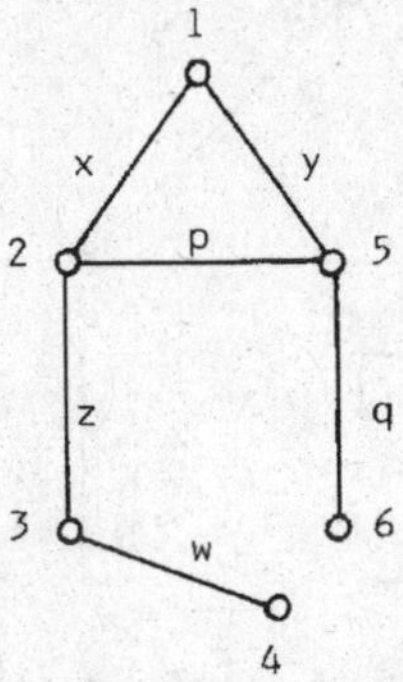

$$R(G_4) = \begin{bmatrix} 1 & x & 0 & 0 & y & 0 \\ x & 1 & z & 0 & p & 0 \\ 0 & z & 1 & w & 0 & 0 \\ 0 & 0 & w & 1 & 0 & 0 \\ y & p & 0 & 0 & 1 & q \\ 0 & 0 & 0 & 0 & q & 1 \end{bmatrix}$$

$$\varphi(G_4) = 1 - (x^2+y^2+p^2+q^2+w^2+z^2) + q^2(w^2+z^2+x^2) + w^2(x^2+y^2+p^2) + 2xpy(1-w)^2 + y^2z^2 - x^2w^2q^2 \qquad \ldots (1)$$

By a little manipulation one can check that G_4 has no automorphism other than the identity.

Setting $z = 0$, the resulting graph G_4' has

$$\varphi(G_4') = (1-w^2)[1-(x^2+y^2+p^2+q^2+2xpy-x^2q^2)]$$

Consider now the case when $w = 0$ which corresponds to

G_5:

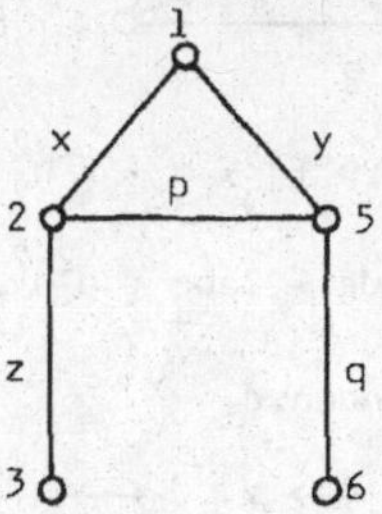

Setting $w = 0$ in (1), we get

$$\varphi(G_5) = 1 - (x^2+y^2+z^2+p^2+q^2) + q^2(z^2+x^2) + 2xpy + y^2z^2 \quad \ldots \quad (2)$$

It is clearly seen that (2) is symmetric under the transformation ($q \leftrightarrow z$ and $y \leftrightarrow x$) indicating the automorphism.

Now set $z = q = 0$, obtaining

G_6:

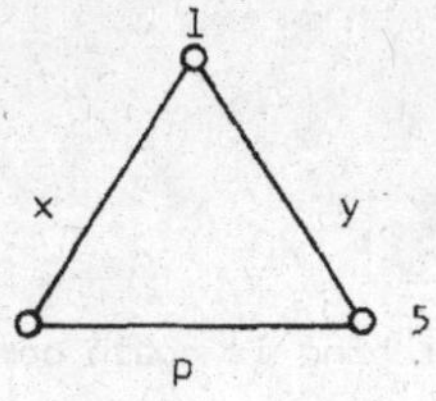

We find

$$\varphi(G_6) = 1 - (x^2+y^2+p^2) + 2xpy \quad \ldots \quad (3)$$

which is clearly seen to be isomorphic to G_3 and has several other automorphisms, e.g., ($x \leftrightarrow y$, $p \leftrightarrow p$).

(v) Isomorphism

Consider

G_7:

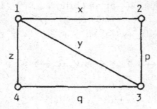

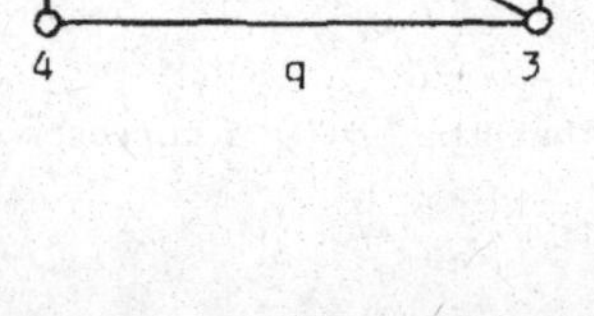

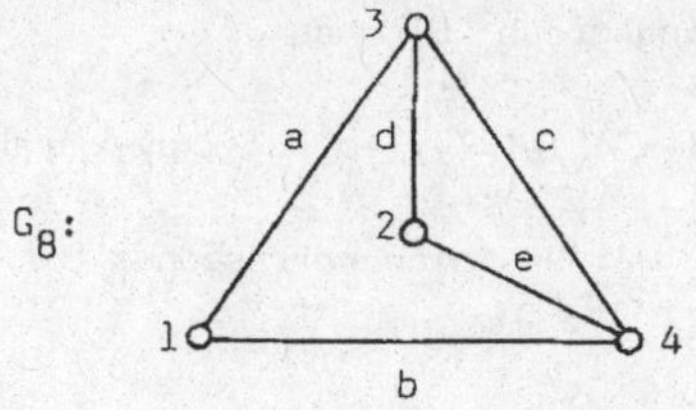

$$\varphi(G_7) = 1 - (q^2+p^2+x^2+z^2+y^2) + 2xyp + 2yzq + x^2q^2 + p^2z^2 - 2xzqp \qquad \ldots (4)$$

$$\varphi(G_8) = 1 - (c^2+d^2+e^2+a^2+b^2) + 2cde + 2abc + a^2e^2 + b^2d^2 - 2abed \qquad \ldots (5)$$

From (4) and (5) we obtain the correspondence

$$(a \longleftrightarrow p,\ b \longleftrightarrow x,\ c \longleftrightarrow y,\ d \longleftrightarrow q,\ e \longleftrightarrow z).$$

One can also arrive at the following automorphisms using symmetric function concept.

G_7 : (x) (y) (z) (p) (q) (identity)
(xp) (qz) (y)
(xz) (pq) (y)
(xq) (pz) (y)

G_8 : (a) (b) (c) (d) (e) (identity)
(ab) (de) (c)
(be) (ad) (c)
(bd) (ae) (c)

(vi) Subgraph matching

The procedure for subgraph matching is again combinatorial, but one can work with it, just as in algebra.

As an example consider the subgraph G_9 of G_8 obtained by setting $b = 0$ in (5). Thus G_9 is parametrized by

$$\varphi(G_9) = 1 - (a^2+c^2+d^2+e^2) + 2cde + a^2e^2 \qquad \ldots (6)$$

Compare the form of (6) with the form of $\varphi(G_{10})$ where G_{10} is the subgraph obtained from G_5 by deleting edge q or setting $q = 0$ in (2). Now

$$\varphi(G_{10}) = 1 - (x^2+y^2+z^2+p^2) + 2xpy + y^2z^2$$

indicating the correspondence $(x \longleftrightarrow d,\ y \longleftrightarrow e,\ z \longleftrightarrow a,\ p \longleftrightarrow c)$.

(vii) Spanning trees and chords

It is interesting to note that the $\varphi(G)$'s corresponding to trees will have only even degree or square terms. The proof of this statement rests on a theorem due to Mason [7] and Ash [8]. This theorem is applicable to directed graphs in which there are feed-back loops between nodes as given by the representation matrix. In fact the matrix R can also be interpreted as though it represents a directed graph with feed-back loops between the nodes rather than a non-directed graph.

For our requirements, Mason's theorem just gives the value of the determinant of R in terms of the products of the labels of edges in the loops, when $r_{ii} = 1$.

Theorem 2. Determinant R = 1 + (-1) (sum of appropriately-signed products of single directed loops) + $(-1)^2$ (sum of appropriately-signed products of labels of non-touching feed-back loops taken two at a time) + $(-1)^3$ (sum of appropriately signed products of labels of non-touching feed-back loops taken three at a time) +...+ $(-1)^m$(sum of appropriately signed products of non-touching feed-back loops taken m at a time).

Here the number m is the maximum number of non-touching feed-back loops. By "appropriately" signed product we mean the following convention :

If there is an even number of edges the sign is positive; otherwise, negative.

For instance consider the directed feed-back loop graph

G_{11}:

$$R(G_{11}) = \begin{bmatrix} 1 & x & 0 & 0 & w \\ x & 1 & y & z & 0 \\ 0 & y & 1 & 0 & 0 \\ 0 & z & 0 & 1 & 0 \\ w & 0 & 0 & 0 & 1 \end{bmatrix}$$

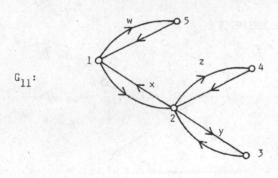

$$\varphi(G_{11}) = 1 - (x^2+y^2+z^2+w^2) + w^2(z^2+y^2)$$

We are now in a position to use Theorem 2 for computing the determinant of R for tree graphs.

Theorem 3. If G is a tree then all the variables in $\varphi(G)$ have degree two.

The proof is obvious from Theorem 2, since for every tree there corresponds a unique feed-back loop graph; in this graph every loop contains an even number of edges with identical labels, thus accounting for degree two of each such label. Also, as there is no cycle, no odd degree term can arise. (Also whenever there is a cycle the corresponding term will have coefficient 2).

Theorem 3 is useful for generating the spanning tree and a set of chords. The algorithm is as follows :

Set to zero those variables that would make odd-degree terms in $\varphi(G)$ vanish; these correspond to chords; the rest of the edges form the spanning tree.

As an example consider G_7 : The chords are either (a) p and q or (b) x and y or (c) y and q or (d) z and p. Note that by setting $x = p = 0$ or $z = q = 0$ G_7 will still be cyclic.

(viii) Blocks, cut-points and bridges

Consider G_{12} :

G_{12}:

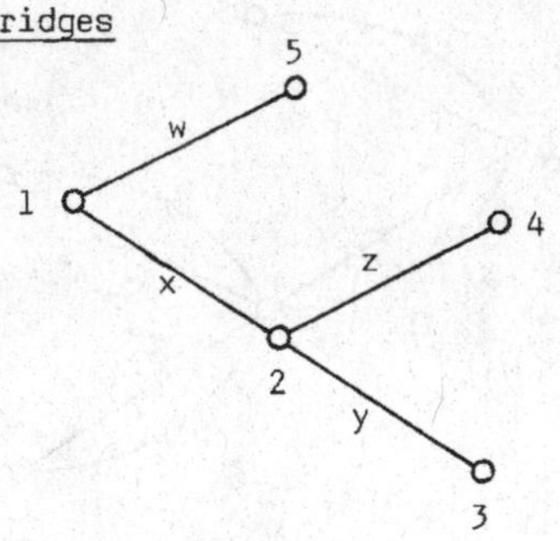

$$\varphi(G_{12}) = 1 - (x^2+y^2+z^2+w^2) + w^2(y^2+z^2)$$

If we set $x = 0$, G_{12} breaks up into two components and the resulting graph G_{14} has

$$\varphi(G_{14}) = (1-w^2)(1-y^2-z^2).$$

Thus Theorem 1 can be used for detecting blocks, cut-points and bridges.

(ix) Asymptotic properties

It was mentioned that every connected graph or block cannot be variable-separately factored over the commutative ring with identity. It is well known that over a ring or field almost all polynomials are irreducible or non-factorable and almost all polynomials are non-symmetric functions in many variables. Accordingly, it follows that (a) almost all graphs are connected, (b) almost all graphs are blocks, and (c) almost all graphs are identity graphs having no automorphism, other than the trivial automorphism. For a different approach to the problem see Erdös and Spencer [9]; Harary and Palmer [10].

(x) Trees, Boolean rings and partial order

We also saw that the representation polynomial $\varphi(G)$ for a tree graph G can contain only even degree edge labels, and each term composed of these edge labels must have a coefficient of 0 or ± 1. Therefore, if the representation matrix is assumed to be over a Boolean ring, the resulting $\varphi(G)$ for a tree would reduce to a multi-variable polynomial in the Boolean algebra. Accordingly, the detection of symmetry types of such functions (which are analogous to switching functions) and hence the automorphisms of G can be worked out using the standard algorithms of switching theory [11-15].

The fact that the Boolean ring forms a lattice exhibits the ordering property that naturally exists for trees; however, $\varphi(G)$ for a general graph cannot be identified with the lattice (due to the fact that the coefficients are 0, ± 1, ± 2 and both odd and even degree terms occur). Therefore there is no natural way to impose a hierarchy or ordering on the cycles.

In the next section we shall show how to compute the symmetry property of a tree graph.

(xi) Symmetry edge of a tree

If u and v are vertices in a graph G and there exists an automorphism σ such that $\sigma(u) = v$, then u and v are called similar vertices. This is an equivalence relation which partitions the vertices of G.

A symmetry edge is defined [16] as an edge which joins two similar vertices.

It is known that [16] every tree has at most one symmetry edge. We shall use our formalism to define the symmetry edge and a method of finding it.

Using Theorem 1, it is seen that the symmetry edge of a tree is that edge whose removal from the tree causes $\varphi(G)$ to be factored into two identical forms; in other words, if the variable corresponding to a symmetry edge is set to zero and the other variables are set equal to x (say), then the resulting $\varphi(G)$ will be a square function.

Example. Consider

G_{13}:

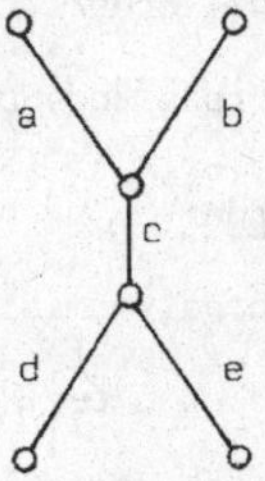

$$\varphi(G_{13}) = 1 - (a^2+b^2+c^2+d^2+e^2) + (a^2+b^2)(d^2+e^2)$$

Setting $c = 0$ and $a = b = d = e = x$, we find

$$\varphi(G_{13}) = (1 - (a^2+b^2))\ (1-d^2+e^2) = (1-2x^2)^2$$

4. REALIZABILITY OF A POLYNOMIAL AS A GRAPH

We now discuss how well the representation polynomial serves as a code for a graph. There are two aspects of a code, viz. (a) uniqueness and (b) reconstruction or decoding of the graph given its unique code.

We have already shown that the form of $\varphi(G)$ is unique up to isomorphism. We now ask the following questions regarding the reconstruction or decoding of G given $\varphi(G)$.

(1) Given any general polynomial in m variables, does there exist a graph whose representation polynomial coincides with the given polynomial?

(2) Given the representation polynomial of a general graph over a general commutative ring with identity, is it possible to reconstruct G or its isomorph? In other words, can we reconstruct the representation matrix using the available information in $\varphi(G)$?

The answer to question (1) is easy, viz., not every polynomial corresponds to a graph, as supported by counter-examples; e.g., consider $\varphi(G) = x$.

We now consider the second question, viz., the reconstruction of the graph G from $\varphi(G)$. The reconstruction algorithm involves very simple logical steps, and no matrix operations; it can be, if required, executed in parallel.

The algorithm for reconstructing G from $\varphi(G)$ is as follows :

First observe that $\varphi(G)$ contains terms with one, two, three, ... edges.

Step 1. Note down the edge labels from all the single element square terms; note down all the cycles from the terms containing 3 or more single degree edge labels, and form the cycle list B using single degree terms.

Step 2. Note down the pairs of edge labels occurring as squares in two-element terms. By definition, the edges in each term are non-adjacent or non-touching; therefore we can now form a topological non-adjacency matrix D for all such edges as follows : This matrix is of order $(m \times m)$; its rows $(1 \leq i \leq m)$ and columns $(1 \leq j \leq m)$ denote the m edge labels.

Set d_{ii} = *(undefined), $1 \leq i \leq m$

and $d_{ij} = d_{ji} = 1$ if edge (i,j) occurs as a degree two pair in any term

$d_{ij} = d_{ji} = 0$, otherwise.

Note that the entry $d_{ij} = 1$ whenever the distance between edges i and j is greater than or equal to 2. Thus the complement of D (denoted by $\overline{D}$) will describe those edges which are adjacent in G.

Step 3. Complement D to $\overline{D}$[†] thus :

$\overline{d}_{ii}$ = *(undefined), $1 \leq i \leq m$

$\overline{d}_{ij} = \overline{d}_{ji} = (1-d_{ij})$ for $1 \leq i, j \leq m$.

Step 4. The matrix D can now be used in conjunction with the cycle-list B to reconstruct G uniquely thus :

---

† The graph corresponding to $\overline{D}$ is identical to the line graph of G [17].

Form all the cycles listed in B. Then read each row $i(1 \leq i \leq m)$ of $\overline{D}$. For each j, if the jth entry is in the cycle, do nothing. Otherwise, we need to decide where j is to be linked to i. This is done as follows :

If i has a neighbor k and j is a neighbor of k, then link j to i at the end where i is linked to k. If, however, k is not a neighbor of j, then link j to the end where i is not linked to k.

These steps are continued until the $\overline{D}$ matrix is exhausted. This will reconstruct G uniquely.

Example. Consider G_4 :

$$\varphi(G_4) = 1 - (x^2+y^2+p^2+q^2+w^2+z^2) + q^2(w^2+z^2+x^2) + w^2(x^2+y^2+p^2) + 2xpy(1-w^2) + y^2z^2 - x^2w^2q^2.$$

Cycle-list : (xpy)

$D(G_4) =$

	x	y	p	q	w	z
x	*	0	0	1	1	0
y	0	*	0	0	1	1
p	0	0	*	0	1	0
q	1	0	0	*	1	1
w	1	1	1	1	*	0
z	0	1	0	1	0	*

$\overline{D}(G_4) =$

	x	y	p	q	w	z
x	*	1	1	0	0	1
y	1	*	1	1	0	0
p	1	1	*	1	0	1
q	0	1	1	*	0	0
w	0	0	0	0	*	1
z	1	0	1	0	1	*

Thus by using Step 4 we reconstruct G_4 :

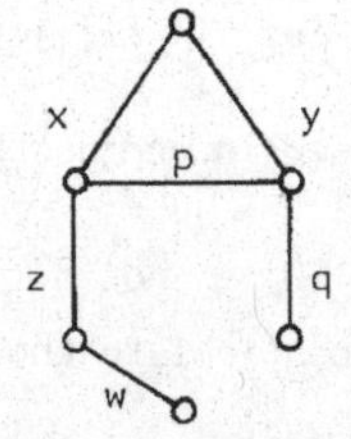

Remark. Since $\overline{D}$ corresponds to the line graph $L(G)$ of G, in general, it suffices to compare the two element terms of $\varphi(G_1)$ and $\varphi(G_2)$ to establish isomorphism of two general graphs G_1 and G_2 (except when $G_1 = K_{1,3}$ and $G_2 = K_3$; in such a case the two-element terms are absent in both $\varphi(G_1)$ and $\varphi(G_2)$ and $L(G_1)$ isomorphic to $L(G_2)$ does not imply G_1 is isomorphic to G_2).

5. APPLICATIONS

a. The polynomial $\varphi(G)$ can be used for establishing Isomorphism.

b. Also it can be used to compute the Vertex Chromatic number [18].

c. Also the algebraic approach for graphs is useful for symbolic manipulation by the computer.

REFERENCES

1. R.C. Read and D.G. Corneil, The graph isomorphism disease, J. Graph Theory, 1(1977), 339-363.
2. A. Mowshowitz, The adjacency matrix and the group of a graph, in New Directions in the Theory of Graphs, edited by F. Harary, Academic Press, 1973.
3. R.A. Bari and F. Harary, Graphs and Combinatorics, Lecture Notes in Mathematics 406, Springer Verlag, N.Y., 1974.
4. J. Turner, Generalized matrix functions and the graph isomorphism problem, Siam J. App. Math., 16(1968), 520-526.
5. J.F. Meyer, Algebraic isomorphism invariants for graphs of automata, in Graph Theory and Computing, edited by R.C. Read, Academic Press, N.Y., 1972.
6. B. Weisfeiler, On Construction and Identification of Graphs, Lecture Notes in Mathematics 558, Springer Verlag, N.Y., 1976.
7. W.K. Kim and R.T. Chien, Topological Analysis and Synthesis of Communication Networks, Columbia University Press, N.Y., 1962.
8. R.B. Ash, Topology and the solution of linear systems, J. Franklin Institute, 268(1959), 453-463.
9. P. Erdös and J. Spencer, Probabilistic Methods in Graph Theory, Academic Press, N.Y., 1973.
10. F. Harary and E. Palmer, Graphical Enumeration, Academic Press, N.Y., 1973.
11. S. Seshu and F.E. Hohn, Symmetric polynomials in Boolean Algebra, Proc. Symp. Theory Switching, Part II, Harvard University Press, Cambridge, Mass., 1959, 225-234.
12. D. Slepian, On the number of symmetry types of Boolean functions of n variables, Canad. J. Math., 5(1953), 185-193.
13. S.H. Caldwell, The recognition and identification of symmetric switching functions, Trans. A.I.E.E., 73(1954), 593-599.

14. R. Gould, The application of graph theory to the synthesis of contact networks, Proc. Symp. Theory Switching, Part I, Harvard University Press, Cambridge,Mass., 1959, 244-292.

15. W. Semon, Matrix methods in the theory of switching, Proc. Symp. Theory Switching, Part II, Harvard University Press, Cambridge, Mass., 1959, 13-50.

16. E.G. Whitehead, Enumerative Combinatorics, Courant Institute Lecture Notes 1971-1972, New York University Press, N.Y., 1973.

17. F. Harary, Graph Theory, Addison-Wesley, Reading, Mass., 1971.

18. M. Dhurandhar, Finding vertex chromatic numbers (To be published).

SOME COMBINATORIAL APPLICATIONS OF THE NEW LINEAR PROGRAMMING ALGORITHM

L. LOVASZ
Mathematical Institute
A. Jozsef University
Szeged

A. SCHRIJVER
Amsterdam

INTRODUCTION

One of the greatest successes in combinatorial optimization in recent years has been the algorithm of Khachiyan (1979), which solves the linear programming problem in polynomial time. The existence of such an algorithm has been a major unsolved problem in the theory of computational complexity, and has been an obstable in the way of the classification of various combinatorial problems with respect to the P-NP scheme, for example in scheduling theory. There are a large number of problems which have been reduced to linear programs; the polynomial solvability of the linear programming problem immediately implies the polynomial solvability of these problems.

It turns out, however, that there is a really wide class of combinatorial problems which cannot be reduced to linear programming and yet the <u>method</u> of the new linear programming algorithm can be applied to solve them, or more precisely, to reduce them to simpler combinatorial problems. It has to be noted that the basic idea of the new method, as Khachiyan remarks in his paper, is due to Shor (1970), who applied it in non-linear optimization. Since the applications of the method given in this paper are somewhere between linear and non-linear programming (or some of them are, in fact, non-linear), it is more apt to call this method the Shor-Khachiyan method or - rhyming with the classical simplex method - the Ellipsoid Method.

The method will be outlined in Section 1. Here we also describe the general approach to combinatorial optimization which makes the application of the method possible.

The main result formulated in Section 2 is that every submodular set-function can be minimized in polynomial time. This result, combined with the rather trivial Greedy

Algorithm and some more applications of the Ellipsoid Method, implies the polynomial solvability of a number of combinatorial optimization problems such as the matching, matroid intersection, optimum branching, optimum covering of directed cuts, etc.

In the third section we very briefly mention some other applications and conclude with some remarks on the prospects of this method.

1. THE ELLIPSOID METHOD AND POLYHEDRAL COMBINATORICS

Let $P \subseteq R^n$ be a polyhedron, determined by the inequalities

$$a_1^T x \leq b_1, \ldots, a_m^T x \leq b_m,$$

and let $c^T x$ be a linear objective function that we want to maximize over P. For the sake of simplicity assume that P is bounded and full-dimensional; say P is contained in the ball $S(0,R)$ of radius R about the origin, and it contains somewhere a ball of radius r.

We define a sequence of points $x_0, x_1, \ldots$ and a sequence of ellipsoids $E_0, E_1, \ldots$ such that x_i is the centre of E_i. Let $x_0 = 0$ and $E_0 = S(0,R)$. Assume that x_k and E_k are defined. Then let us check whether or not $x_k \in P$.

Case 1. If $x_k \in P$, then consider the half-ellipsoid which is the intersection of E_k with the halfspace $c^T x \geq c^T x_k$, and include it in an ellipsoid E_{k+1} with least possible volume. Let x_{k+1} be the centre of E_{k+1}.

Case 2. If $x_k \notin P$ then let $a_i^T x \leq b_i$ be a constraint which is violated by x_k. Include the intersection of E_k with the halfspace $a_i^T x \leq b_i$ in an ellipsoid E_{k+1} with least possible volume. Let x_{k+1} be the centre of E_{k+1}.

The following can be proved : if we look at those values of k for which $x_k \in P$ (call these briefly feasible values), then

$$\max\{c^T x_k : 1 \leq k \leq p,\ k \text{ feasible}\} \to \max\{c^T x : x \in P\},\ (p \to \infty), \qquad \ldots \quad (1)$$

and the convergence is exponentially fast. In a sense this fact may be considered as the polynomial solution of the optimization problem. If we want to get the precise solution, we have to know and use something of the arithmetical nature of the vertices of P (this is quite natural : if a vertex might have, say, irrational coordinates then it would be impossible to describe the answer). In combinatorial applications, for example, we quite often will know that the vertices of P are lattice points. The exact optimum then can be obtained simply by rounding.

The idea of the proof of the fast convergence in (1) is quite simple, although precise details are tedious. We look at the piece of P where the value of the objective function is not smaller than the maximum value at feasible points x_k found so far, and prove by induction on k that this piece is included in the current ellipsoid. It is easy to see that the volume of E_k drops exponentially fast with k, and hence the volume of the piece of P where the objective function is not smaller than the current record also must drop exponentially fast. Hence the difference between the current record and the true optimum of the objective function must also decrease exponentially fast.

Details of this argument can be found in [14].

Now we come to the combinatorial part of the paper. A general setting in which combinatorial optimization problems can be treated is the following. Given a finite subset $S \subseteq R^n$ and a linear objective function c^Tx, find

$$\max\{c^Tx : x \in S\}.$$

Since a linear function always assumes its optimum at a vertex of a polytope, we have

$$\max\{c^Tx : x \in S\} = \max\{c^Tx : x \in \text{conv}(S)\}.$$

Since conv(S) is a polytope, the left hand side is equivalent to a linear program. So if we can write up this program explicitly and then solve it by the Simplex or Ellipsoid Method, we are done.

This program, initiated among others by Edmonds, Fulkerson and Hoffman, has motivated the considerable amount of research that has concerned the facial structure of combinatorially defined polytopes. But even for classes of polytopes for which the facets are well known, this approach could not lead to satisfactory (polynomial) algorithms. The reason for this is that (with a very few exceptions) the number of facets of conv(S), i.e. the minimum number of inequalities needed to describe conv(S), is exponentially large in the size of the original combinatorial structure. So even to write up the linear program which we have to solve in order to find the desired optimum, takes exponential time. (This is not surprising if we think of the fact that S in itself is usually exponentially large in the size of the original combinatorial structure; if not, we can determine the optimum by evaluating the objective function at every element of S.)

It may be the most important feature of the Ellipsoid Method that it overcomes this difficulty in many cases. To see how, note that the inequalities defining P enter the algorithm only at one point : when we have to decide if $x_k \in P$ and if not, we need an inequality $a_i^T x \leqq b_i$ which is violated by x_k. One way to do this is to substitute x_k into each of the inequalities $a_i^T x \leqq b_i$; but since in combinatorial optimization problems these inequalities are usually very structured, we may well have a subroutine which checks $x_k \in P$, and picks a violated inequality if $x_k \notin P$, without explicitly writing up all inequalities defining P. So the Ellipsoid Method reduces the original combinatorial problem to another one, which (as we shall illustrate in the next section) is quite different from the original and often easier to solve.

One way to formulate this relation between combinatorial problems is in terms of anti-blocking polyhedra (see Fulkerson [12]). Let A be non-negative m × m matrix and b a non-negative n-vector. Let

$$P = \{x \in R_+^m : Ax \leqq b\}. \qquad \ldots \quad (2)$$

Then the <u>anti-blocker</u> of the polytope P is the polytope

$$P^* = \{y \in R_+^m : x^T y \leqq 1 \text{ for every } x \in P\}.$$

<u>Theorem 1.1</u>. Let K be a class of polytopes of type (2) such that there exists an algorithm to maximize an arbitrary linear objective function over polytopes in K in time polynomial in nm and the number of decimal digits in the coefficients of the objective function. Then there exists such an algorithm for the class of anti-blockers of polytopes in K.

2. MINIMIZING SUBMODULAR FUNCTIONS AND APPLICATIONS

Let f be a function defined on the subsets of a set S. The function f will be called <u>submodular</u> if

$$f(X \cap Y) + f(X \cup Y) \leq f(X) + f(Y)$$

holds for every pair of subsets of S. The main result which we state in this section is the following.

<u>Theorem 2.1</u>. Let f be a submodular function on the subsets of a finite set S. Assume that we have a subroutine to compute $f(X)$ in time less than T. Then the subset of S minimizing f can be determined in time polynomial in T and $|S|$.

As a first application we show how to find the value of a maximum flow between a,b of a network G. For $X \subseteq V(G)$, denote by $\delta(X)$ the capacity of the cut determined by X. It is a simple well-known fact that $\delta(X)$ is a submodular function. By the Max-flow-min-cut Theorem, the value of a maximum flow is

$$\min\{\delta(X) : a \in X \quad V(G) - b\}.$$

This minimum can be determined in polynomial time by minimizing the submodular function $\delta(X \cup \{a\})$ over the subsets of $V(G)$-a-b. It is, of course, also important to find an optimum flow, and this does not follow directly from the Ellipsoid Method. But if we have a polynomial algorithm to find the value of an optimum flow for every network, then there are several quite easy procedures to find a maximum flow. We leave this to the reader.

A more essential application is the following. Let G be a digraph. A <u>directed cut</u> is the set of edges connecting $X \subset V(G)$ to $V(G) - X$, provided $X \neq \phi$ or $V(G)$, and there is no edge connecting $V(G) - X$ to X. A theorem of Lucchesi and Younger asserts that the maximum number of edge-disjoint directed cuts is equal to the minimum number of edges covering all directed cuts. More recently Lucchesi [19], Karzanov [15] and Frank [10] gave polynomial algorithms to find this number. To show that such an algorithm can be based on the Ellipsoid Method, let us assign a variable x_j to every edge j and consider the polytope

$$0 \leq x_j \leq 1, \qquad \text{(for every edge } j)$$

$$\sum_{j \in D} x_j \geq 1 \qquad \text{(for every directed cut } D).$$

It easily follows from the Lucchesi-Younger Theorem that the vertices of this polytope are 0-1 vectors corresponding to those sets of edges which cover all directed cuts. So to determine the minimum number of edges covering all directed cuts we have to minimize the linear objective function Σx_j over the polytope P. In order to apply the Ellipsoid Method, we only need a polynomial subroutine which checks $x \in P$ and if the answer is in the negative, it finds a constraint which is violated. The first set of constraints is easily checked one by one, so we may assume that $x \geq 0$ and that we are only interested in checking whether or not all the directed cut-constraints are satisfied. This is clearly equivalent to the following problem :

Given a digraph and a weight on every edge, find a directed cut with minimum weight.

This can be reduced to the problem of minimizing a submodular function as follows. For every edge e of G, add a new edge which connects the endpoints of e in the reverse order, and let its weight be N, where $N > \Sigma x_j$. Define $\delta(X)$ as the sum of weights of edges connecting X to $V(G) - X$ in this new graph. Then $\delta(X)$ is submodular, and moreover, a non-empty set minimizing it necessarily determines a directed cut. So it suffices to minimize the submodular set-function $\delta(X)$ over the non-empty subsets of $V(G)$, which as we have seen, can be done in polynomial time.

By similar methods we can find algorithms which go with the minimax theorems in Edmonds-Giles [6] and Frank [9]. (These results generalize several minimax results, like polymatroid intersection (Edmonds [3,5], Lawler [17]), optimum branching (Edmonds [2]), packing of rooted cuts (Fulkerson [12]), max-flow-min-cut (Ford-Fulkerson [8]), packing of directed cuts (Lucchesi and Younger [20]), etc.)

Let us conclude with the discussion of the matching problem. This needs an extension of the submodular function minimization problem. The following result generalizes a recent algorithm of Padberg and Rao [21].

Theorem 2.2. Let f be a submodular set-function defined on the subsets of a finite set S. Assume that we have a subroutine to compute $f(X)$ in time at most T. Let H be a collection of subsets of S with the property that $X \in H, Y \notin H, X \cap Y \notin H$ implies that $X \cup Y \in H$. Also assume that we have a subroutine to check $X \in H$ in time at most T'. Then there is an algorithm to find

$$\min\{f(X) : X \in H\}$$

in time polynomial in $|S|$, T and T'.

An example of a collection H of subsets with the given property is the collection of all subsets with odd cardinality.

Let G be a graph with an even number of vertices and let us assign a non-negative weight w_j to each edge. We want to find a perfect matching with maximum total weight. By Edmonds [1] the convex hull of perfect matchings of G is given by the inequalities

$$x_j \geqq 0 \qquad \text{(for every edge } j)$$

$$\sum_{j \ni v} x_j = 1 \qquad \text{(for every vertex } v)$$

$$\sum_{j \in C} x_j \geqq 1 \qquad \text{(for every odd cut } C),$$

where an odd cut means the set of edges connecting X to $V(G) - X$ for some $X \subseteq V(G)$, $|X|$ odd. To be able to apply the Ellipsoid Method we have to be able to check if a vector x satisfies these inequalities. The first two kinds of constraints are easily checked one by one. To deal with the third kind, we need a subroutine to find an odd cut C for which $\Sigma_{j \in C}\, x_j$ is minimum. But this means to minimize a submodular function over the odd-size subsets of $V(G)$. By Theorem 2.2, this can be done in polynomial time.

3. CONCLUDING REMARKS

1. In the combinatorial applications discussed above we were only concerned with the speed of our algorithms up to polynomiality. As a matter of fact the running times, as far as we could estimate, are rather poor. For those applications where polynomial algorithms have been known before these are much faster than ours. The main point has been to solve all these problems using one technique, the Ellipsoid Method.

It is natural that such a general approach cannot compete with special-purpose algorithms. For those problems to which the Ellipsoid method seems to have yielded the first polynomial solutions, this fact should be a challenge to find better algorithms making better use of the specialities of the problem.

2. The Ellipsoid Method applies to convex bodies other than polyhedra, and this fact can also be utilized in combinatorics. An application of this kind is an algorithm to compute the independence number of a perfect graph in polynomial time [14]. More generally, one can compute a rather sharp upper bound for the independence number of an arbitrary graph (see [18]).

3. It may be interesting to point out that the majority of those combinatorial optimization problems which are known to be polynomially solvable, can be solved by a combination of the Ellipsoid Method and the Greedy Algorithm.

4. The Ellipsoid Method shows that the investigation into the facial structure of polytopes may be useful in designing algorithms in a more specific way than just "gaining insight". This is particularly significant if we consider that the description of facets of combinatorial polyhedra means a "good characterization" of the maximum value of linear objective functions. So this is a quite general situation where knowing a "good characterization" helps in designing a good algorithm. Thus the Ellipsoid Method may also contribute to the problem whether $NP \cap coNP = P$.

REFERENCES

1. J. Edmonds, Maximum matching and a polyhedron with 0,1-vertices, J. Res. Nat. Bur. Stan. B 69(1965), 125-130.
2. J. Edmonds, Optimum branchings, J. Res Nat. Bur. Stan. B 71(1967), 233-240.
3. J. Edmonds, Submodular functions, matroids, and certain polyhedra, in : Combinatorial Structures and their Appl. (Proc. Intern. Conf. Calgary, 1969; R. Guy, H. Hanani, N. Sauer, J. Schönheim, eds.), Gordon and Breach, N.Y. 1970, 69-87.
4. J. Edmonds, Edge-disjoint branchings, in : Combinatorial Algorithms (Courant Comp. Sci. Symp. Monterey, 1972; R. Rustin, ed.) Academic Press, N.Y. 1973, 91-96.
5. J. Edmonds, Matroid intersection, Annals of Disctrete Math., 4(1979), 39-49.
6. J. Edmonds and R. Giles, A min-max relation for submodular functions on graphs, Annals of Discrete Math., 1(1977), 185-204.
7. J. Edmonds and E.L. Johnson, Matching, Euler tours, and the Chinese postman, Math. Prog., 5 (1973), 88-124.

8. L.R. Ford and D.R. Fulkerson, Flows in Networks, Princeton Univ. Press, Princeton, N.J. 1962.

9. A. Frank, Kernel systems of directed graphs, Acta Sci. Math. Szeged 41(1979), 63-76.

10. A. Frank, How to make a digraph strongly connected? Combinatorica (submitted).

11. D.R. Fulkerson, Packing rooted directed cuts in weighted directed graphs, Math. Prog., 6(1974), 1-13.

12. D.R. Fulkerson, Anti-blocking polyhedra, J. Combinatorial Theory,B 12(1972), 50-71.

13. P. Gács and L. Lovász, Khachiyan's algorithm for linear programming, Math. Prog. Studies (submitted).

14. M. Grötschel, L. Lovász and A. Schrijver, The ellipsoid method and its consequences in combinatorial optimization, Combinatorica (submitted).

15. A.V. Karzanov, On the minimal number of arcs of a digraph meeting all its directed cut sets. (To appear).

16. L.G. Khachiyan, A polynomial algorithm in linear programming, Dokl. Akad. Nauk SSSR 244(1979), 1093-1096.

17. E.L. Lawler, Optimal matroid intersections, in : Combinatorial Structures and their Applications (Proc. Intern. Conf. Calgary, 1969; R. Guy, H. Hanani, N. Sauer, J. Schönheim, eds.), Gordon and Breach, N.Y. 1970, 233-235.

18. L. Lovász, On the Shannon capacity of a graph, IEEE Trans. on Inf. Theory, 25(1979), 1-7.

19. C.L. Lucchesi, A minimax equality for directed graphs, Thesis, University of Waterloo, 1976.

20. C.L. Lucchesi and D.H. Younger, A minimax relation for directed graphs, J. London Math. Soc., 17(1978), 369-374.

21. M.W. Padberg and M.R. Rao, Minimum cut-sets and b-matchings, (to appear).

22. N.Z. Shor, Convergence rate of the gradient desent method with dilatation of the space, Kibernetika 2(1970), 80-85.

IN SEARCH OF A COMPLETE INVARIANT FOR GRAPHS

K. BALASUBRAMANIAN
Department of Statistics
Loyola College
Madras 600 034

K. R. PARTHASARATHY
Department of Mathematics
Indian Institute of Technology
Madras 600 036

ABSTRACT

Let A be the adjacency matrix of an ordinary (simple) graph G and $A' = xI + \lambda A + (J-A-I)$ where I is the $n \times n$ identity matrix and J is the $n \times n$ matrix of 1's. Then we call $P(x,\lambda) = \mathrm{Per}(A')$ the permanent polynomial of G. A frame (2-matching) of a graph G is a spanning subgraph F of G whose components are single points, single lines, paths or cycles. If F has w_i paths P_i, $i = 1,\ldots,n$ and y_j cycles C_j we let $w(F) = \prod_{i=1}^{n} p_i^{w_i} \prod_{j=3}^{n} c_j^{y_j}$ the weight of F and call $F(p,c) = \sum_{F \in \underline{\underline{F}}} w(F)$, where $\underline{\underline{F}}$ is the family of all frames of G, the frame polynomial of G. We conjecture that either of these is a complete invariant for graphs, show their interrelation and present some evidence why the conjectures are plausible.

1. INTRODUCTION

We consider finite ordinary (simple) graphs (that is, without loops or multiple edges) and generally follow Harary [3] for notation and terminology.

A function $\theta : G \to K$ from the set of all graphs to a set K (usually a ring or a field) is a graph parameter or graph invariant if for isomorphic graphs G and H we have $\theta(G) = \theta(H)$. An invariant θ is said to be complete if $\theta(G) = \theta(H)$ implies G is isomorphic to H. So far only one complete graph invariant seems to have been discovered, namely a canonical code of a graph (see e.g., Shah et. al [7] or Read [5]). For any (labelled) graph G we take the upper triangular portion UTAM of the adjacency matrix A of G and arrange the strings of zeros and ones in the rows of UTAM into a single string and call the resulting (binary) number the code of G. This will give different codes even for isomorphic graphs. To get a canonical code we generate the code number of all the $n!$ graphs isomorphic to G and choose the maximum of this set of integers. It is easy to see that this canonical code is a complete invariant for the graph but it is as good as specifying the graph itself : the binary number is just the adjacency matrix written in a convenient form.

A more serious candidate for a complete invariant was the characteristic polynomial of a graph but the history of its failure is now too well-known. Here we propose two related polynomial invariants as possible candidates for a complete graph invariant and study some of their properties and interrelations.

2. THE PERMANENT POLYNOMIAL

One reason why the characteristic polynomial failed to be a complete invariant may be because the determinant is not the appropriate matrix function relevant for combinatorial studies as by its very definition (with positive and negative terms) many terms get cancelled out, thereby losing significant combinatorial information. This observation leads us to consider the permanent as possibly the more appropriate matrix function for combinatorial studies. For any square matrix M of order n the permanent is defined by

$$\mathrm{per}(M) = \sum_{\sigma\varepsilon S_n} m_{1\sigma_1} m_{2\sigma_2} \dots m_{n\sigma_n} \qquad \dots \quad (1)$$

where $\sigma = \begin{pmatrix} 1 & 2 & \dots & n \\ \sigma_1 & \sigma_2 & \dots & \sigma_n \end{pmatrix}$ is a typical permutation of the symmetric group S_n acting on the set $\{1,2,\dots,n\}$. For definition and various properties see e.g., Ryser [6] and Minc [4]. Equivalently,

$$\mathrm{per}(M) = \sum_{\sigma\varepsilon S_n} \mathrm{Pr}(P_\sigma M) \qquad \dots \quad (2)$$

where P_σ is the permutation matrix corresponding to σ and Pr is the product of the elements of $P_\sigma M$.

To get a polynomial, we convert the adjacency matrix A of the graph G to $A(\lambda) = (\lambda^{a_{ij}})$ and consider per $A(\lambda)$.

It is easy to verify that per $A(\lambda)$ is an invariant; for, if $G_1 \simeq G_2$, then their adjacency matrices A_1 and A_2 are related by $A_2 = PA_1P^T$ where P is a permutation matrix (and P^T denotes the transpose of P) and $A_2(\lambda) = PA_1(\lambda)P^T$ and per $PA_1(\lambda)Q$ = per $A_1(\lambda)$ for any two permutation matrices P and Q. Looking at the converse we see that $\mathrm{per}(PA_1(\lambda)P^T) = \mathrm{per}(PA_1(\lambda)) = \mathrm{per}\, A_1(\lambda)$, so that if PA_1 is the adjacency matrix of a graph nonisomorphic to the one represented by A_1, then per $A(\lambda)$ is not a complete invariant. We first observe that such a P cannot have odd order (the order of a permutation matrix P is the least positive integer m such that $P^m = I$).

Lemma 1. If A and PA are adjacency matrices of graphs G_1 and G_2 where P is a permutation matrix of odd order then $G_1 \simeq G_2$.

Proof. Let $P^{2m-1} = I$. Since A and PA are symmetric matrices, $PA = A^TP^T = AP^T$, so that $A = PAP$. By repeated application $A = P^rAP^r$ for any positive integer r. In particular, $A = P^mAP^m$ and $P^m(P^mAP^m)(P^m)^T = P^{2m}A = PA$ shows A and PA represent isomorphic graphs.

That there may exist an even order permutation matrix P such that PA is an adjacency matrix is borne out by the following examples, which incidentally provide a counterexample to show that per $A(\lambda)$ is not a complete invariant.

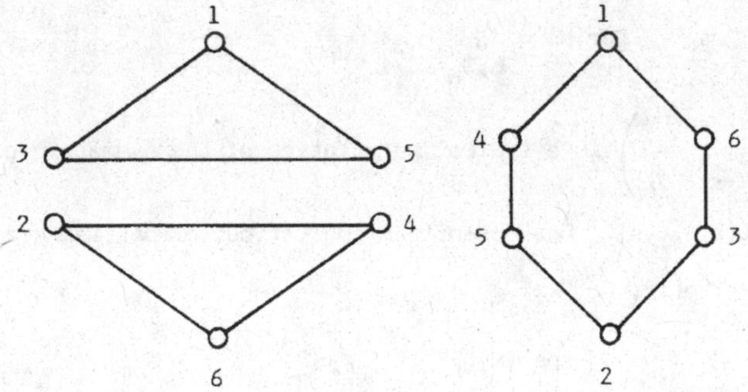

Figure 1

If A_1 and A_2 are adjacency matrices of G_1 and G_2 of Figure 1, it can be seen that $A_2 = PA_1$ where P is the permutation matrix corresponding to $\sigma = (12)(34)(56)$. We observe, in passing, that even when A and PA are adjacency matrices of nonisomorphic graphs they will have the same degree sequence.

To get over the difficulty created by the above counter example we modify the polynomial function $A(\lambda)$ and introduce the permanent polynomial. For a graph G of order n with adjacency matrix A we define the matrix $A(x,\lambda)$ by

$$\begin{aligned} A(x,\lambda) &= xI + \lambda A + (J-A-I) \\ &= (x-1)I + (\lambda-1)A + J \qquad \ldots \quad (3) \end{aligned}$$

where I is the $n \times n$ identity matrix and J is the $n \times n$ matrix of 1's.

We denote the permanent of $A(x,\lambda)$ by $P(x,\lambda)$ and call this the permanent polynomial of the graph G. To simplify writing we set $A' = A(x,\lambda)$.

The First Invariant Conjecture. The permanent polynomial $P(G;x,\lambda) = \text{per}(A') = \text{per}(xI+\lambda A+J-A-I)$ is a complete invariant for graphs.

Lemma 2. The permanent polynomial is a graph invariant.

Proof. If $G \cong H$ and have adjacency matrices A and B, then there exists a permutation matrix P such that $B = PAP^T$. But then, $B(x,\lambda) = P(A(x,\lambda))P^T$ and permanent is unaltered by pre or post multiplication by permutation matrices. Thus $P(G;x,\lambda) = P(H;x,\lambda)$.

We now observe that the counter example to the earlier conjecture arose from the fact that there are non-trivial permutation matrices P such that A and $PA = B$ are adjacency matrices of graphs and that this implied $B(\lambda) = PA(\lambda)$. But such a contingency cannot arise in the present case. For, $B(x,\lambda) = PA(x,\lambda)$ implies that

$$(x-1)I + (\lambda-1)B + J = P\{(x-1)I + (\lambda-1)A + J\} = (x-1)P + (\lambda-1)PA + J, \qquad \ldots \quad (4)$$

so that, equating coefficients of x and λ we get $P = I$ and $B = PA = A$. This does not, however, mean that this conjecture may not be falsified by a pair of non-isomorphic graphs with same degree sequence.

3. THE COEFFICIENT MATRIX C

To analyse the properties of the permanent polynomial we express it explicitly as a polynomial in x and λ with coefficients $c_{r,s}$

$$P(G;x,\lambda) = \sum_{r=0}^{n} \sum_{s=0}^{r} c_{r,s}\, x^{n-r} \lambda^{s}. \qquad \ldots \quad (5)$$

We can arrange the $c_{r,s}$ in an $(n+1) \times (n+1)$ matrix and the coefficient matrix C determines the permanent polynomial and conversely C is a lower triangular matrix. To derive an important property of these coefficients we homogenize (5) by introducing another variable μ and the resulting homogeneous permanent polynomial is

$$\begin{aligned} P(G;x,\lambda,\mu) &= \sum_{r=0}^{n} \sum_{s=0}^{r} c_{r,s}\, x^{n-r} \lambda^{s} \mu^{r-s} \\ &= \mathrm{per}(xI+\lambda A+\mu(J-A-I)) \qquad \ldots \quad (6) \\ &= \mathrm{per}(A''), \quad \text{say.} \end{aligned}$$

We prove

<u>Lemma 3</u>. If G^c is the complement of the graph G, then

$$P(G^c;x,\lambda,\mu) = P(G;x,\mu,\lambda) \qquad \ldots \quad (7)$$

<u>Proof</u>. If A is the adjacency matrix of G, then the adjacency matrix of G^c is $J-A-I$. Thus,

$$P(G^c;x,\lambda,\mu) = \mathrm{per}\{xI+\lambda(J-A-I)+\mu A\} = P(G;x,\mu,\lambda). \qquad \ldots \quad (8)$$

<u>Corollary 1</u>. $c_{r,s}(G^c) = c_{r,r-s}(G)$. This is simply the coefficient version of Lemma 3.

This fact that the permanent polynomial of a graph fixes that of its complement, appears to be quite significant. This property is not shared by the characteristic polynomial.

The following simple properties of the $c_{r,s}$'s are easily verified.

<u>Lemma 4</u>. For an (n,m) graph G,

$c_{0,0} = 1$, $c_{1,j} = 0$ for all j.

$c_{2,0} = \binom{n}{2} - m = m(G^c)$, $c_{2,1} = 0$, $c_{2,2} = m$.

$c_{3,0}$ = twice the number of triangles in G^c.

$c_{3,3}$ = twice the number of triangles in G. $\qquad \ldots \quad (9)$

$c_{4,4}$ = number of (not necessarily induced) subgraphs of G isomorphic to $2P_2$ plus twice the number of 4-cycles of G.

$C_{5,5}$ = twice the number of subgraphs isomorphic to (C_3+P_2) plus twice the number of 5-cycles of G, etc.

Using these simple properties we have verified that the conjecture is true for all graphs with at most 6 points (using the catalogue of graphs given in Harary [3]).

There are (n+1) linear equations satisfied by the $c_{r,s}$'s. These arise from the fact that for all m-point graphs the row sum of the C matrices are the same, viz., R_i, the row sum for the i-th row (coefficient of x^{n-i} in P(G,x,1) is simply the number of permutations of S_n with exactly (n-i) symbols fixed). This is the content of

Lemma 5. The coefficients $c_{r,s}$ are related by the (n+1) equations

$$\sum_{s=0}^{r} c_{r,s} = \frac{n!}{(n-r)!} \sum_{i=0}^{r} \frac{(-1)^{r-i}}{(r-i)!} \qquad \ldots (10)$$

Proof. $P(G;x,\lambda,\lambda) = \sum_r \sum_s c_{r,s}\, x^{n-r} \lambda^r = \mathrm{Per}\{xI + \lambda(J-I)\}$

$$= \mathrm{per}\{\lambda J + (x-\lambda)I\} = \sum_{r=0}^{n} \binom{n}{r} r!\, \lambda^r (x-\lambda)^{n-r}.$$

Thus,

$$\sum_s c_{r,s} = \text{coefficient of } \lambda^r x^{n-r} \text{ in the above expression}$$

$$= \sum_{i=0}^{r} \binom{n}{i} i!\, \binom{n-i}{n-r} (-1)^{r-i}$$

$$= \frac{n!}{(n-r)!} \sum_{i=0}^{r} \frac{(-1)^{r-i}}{(r-i)!},$$

a constant independent of the graph.

4. THE AUXILIARY MATRIX S

To get an explicit formula for the $c_{r,s}$ in terms of the matrix A we introduce the following definitions.

Definitions : For any matrix M let

(i) $S_r(M)$ be the sum of the permanents of all $r \times r$ submatrices of M.

(ii) $S_{(r)}(M)$ be the sum of the permanents of all $r \times r$ principal submatrices of M.

(iii) $S_{r,s}(M)$ be the sum of the permanents of all $s \times s$ submatrices of all $r \times r$ principal submatrices of M. We define $S_{r,o} = \binom{n}{r}$ for $r = 0,1,\ldots,n$.

Here an $r \times r$ principal submatrix of M is obtained by the deletion of $(n-r)$ rows and the same $(n-r)$ columns of M. If the deleted columns are $I = \{i_1, i_2, \ldots, i_{n-r}\}$, the resulting $r \times r$ principal submatrix of M is denoted by $A(I,I)$ or $A[N-I, N-I]$ where N is the index set of all the rows of M.

Lemma 6. When A is the adjacency matrix of a graph we have the following formula for $S_{r,s}(A)$:

$$S_{r,s}(A) = \frac{1}{(n-r)!\ s!(r-s)!} \left[\frac{\partial^{n-r+s} P}{\partial x^{n-r}\ \partial \lambda^{s}}\right] \quad \ldots \quad (12)$$

where $P = P(G;x,\lambda,\mu)$ and [ ] implies that the function within these brackets is evaluated at $(1,1,1)$.

Proof. This essentially follows from the following rule for differentiation of the permanent of an $n \times n$ matrix M whose elements m_{ij} are functions of a variable θ :

$$\frac{d}{d\theta}(\text{per } M) = \sum_{i=1}^{n} \sum_{j=1}^{n} M_{ij} \frac{d}{d\theta}(m_{ij})$$

where $M_{ij} = \text{per } M(i,j)$ is the permanental co-factor of m_{ij}.

Applying this to $A'' = A(x,\lambda,\mu)$ we get first, $\frac{\partial^{n-r} P}{\partial x^{n-r}} = (n-r)!\ S_{(r)}(A'')$ where $S_{(r)}(A'')$ is the sum of the permanents of all $r \times r$ principal submatrices of A'', the factor $(n-r)!$ occurs because of the possible permutations of the $(n-r)$ diagonal positions where the x's occur. Differentiating this s times with respect to λ we get

$$\frac{\partial^{n-r+s} P}{\partial x^{n-r} \partial \lambda^{s}} = (n-r)!\ s!\ S_{r,r-s}(A'')$$

in the notation introduced earlier. Now while evaluating the L.H.S. at $(1,1,1)$ each of the matrices of R.H.S. becomes an $(r-s) \times (r-s)$ matrix of 1's whose permanent is $(r-s)!$ There are as many such $(r-s) \times (r-s)$ matrices as there are different s independent positions in $S_{(r)}(A'')$; this is exactly $S_{r,s}(A)$. Hence the lemma. Using this we get the following formula for the P polynomial for various arguments :

Lemma 7. $$P(G;x,\lambda,1) = \sum_{r=0}^{n} \sum_{s=0}^{r} (r-s)!\ S_{r,s}(A)(x-1)^{n-r}(\lambda-1)^{s} \quad \ldots \quad (13)$$

Proof. Expanding $P(G;x,\lambda,1)$ by Taylor's theorem in the neighbourhood of $x = 1$, $\lambda = 1$ we get

$$P(G;x,\lambda,1) = \sum_{j=0}^{n} \frac{1}{j!} \left((x-1)\frac{\partial}{\partial x} + (\lambda-1)\frac{\partial}{\partial j}\right)^j P_{0|1}$$

where $P_0 = P(G;x,\lambda,1)$ and $|1$ denotes the substitution

$x = \lambda = 1$ in the derivatives

$$= \sum_{j=0}^{n} \frac{1}{j!} \sum_{k=0}^{j} \binom{j}{k}(x-1)^k(\lambda-1)^{j-k}\left[\frac{\partial^j P}{\partial x^k\, \partial\lambda^{j-k}}\right]$$

in earlier notation,

$$= \sum_{r=0}^{n} \sum_{s=0}^{n-r} \frac{1}{r!}\frac{1}{s!}(x-1)^r(\lambda-1)^s \frac{\partial^{r+s}P}{\partial x^r\, \partial\lambda^s}$$

by a suitable change of variables,

$$= \sum_{r=0}^{n} \sum_{s=0}^{n-r} (n-r-s)!\, S_{n-r,s}(A)\,(x-1)^r(\lambda-1)^s \quad \text{using (12)}$$

$$= \sum_{r=0}^{n} \sum_{s=0}^{r} (r-s)!\, S_{r,s}(A)\,(x-1)^{n-r}(\lambda-1)^s.$$

Corollary 2. $P(G;x,\lambda,\mu) = \sum_{r=0}^{n} \sum_{s=0}^{r} (r-s)!\, S_{r,s}(A)(x-\mu)^{n-r}(\lambda-\mu)^s\mu^{r-s}$ ... (14)

This is an obvious derivation from (13), since μ is just a homogenizing variable.

Corollary 3. $P(G;x+\mu,\lambda+\mu,\mu) = \sum_{r=0}^{n} \sum_{s=0}^{r} (r-s)!\, S_{r,s}(A)\, x^{n-r}\,\lambda^s\,\mu^{r-s}$ ... (15)

This follows by an obvious substitution in (14).

Formulae (14) and (15) together with definition (6) give the interrelations between the c_{ij}'s and $S_{r,s}$'s.

Thus, from (14) we have

$$\sum_{i=0}^{n} \sum_{j=0}^{i} c_{uj}\, x^{n-i}\,\lambda^j\,\mu^{i-j} = \sum_{r=0}^{n} \sum_{s=0}^{r} (r-s)!\, S_{r,s}(x-\mu)^{n-r}(\lambda-\mu)^s\,\mu^{r-s}$$

$$= \sum_{r=0}^{n} \sum_{s=0}^{r} (r-s)!\, S_{r,s}\left(\sum_{i=0}^{n-r}\binom{n-r}{i} x^i(-\mu)^{n-r-i}\right)\left(\sum_{j=0}^{s}\binom{s}{j}\lambda^j(-\mu)^{s-j}\,\mu^{r-s}\right)$$

Therefore

$$c_{n-i,j} = \sum_{r=0}^{n-i} \sum_{s=j}^{r} (r-s)!\binom{n-r}{i}\binom{s}{j}(-1)^{n-i-j-(r-s)} S_{r,s}$$

or,

$$c_{i,j} = (-1)^{i-j} \sum_{r=0}^{i} \sum_{s=j}^{r} (r-s)!\binom{n-r}{n-i}\binom{s}{j}(-1)^{r-s}\, S_{r,s}. \quad \text{... (16)}$$

In the other direction, from (15) we have

$$\sum_{r=0}^{n}\sum_{s=0}^{r}(r-s)!\,S_{r,s}\,x^{n-r}\lambda^{s}\mu^{r-s}=\sum_{i=0}^{n}\sum_{j=0}^{i}c_{ij}(x+\mu)^{n-i}(\lambda+\mu)^{j}\mu^{i-j}$$

$$=\sum_{i=0}^{n}\sum_{j=0}^{i}c_{ij}\,\mu^{i-j}\left(\sum_{r=0}^{i}\binom{n-i}{n-r}x^{n-r}\mu^{r-i}\right)\left(\sum_{s=0}^{j}\binom{j}{s}\lambda^{j}\mu^{j-s}\right)$$

giving

$$(r-s)!\,S_{r,s}=\sum_{i=0}^{s}\sum_{j=s}^{n}\binom{n-i}{n-r}\binom{j}{s}c_{ij} \qquad \ldots \quad (17)$$

Equations (16) and (17) being inversions of each other we develop (17) first.

For two matrices M_1 and M_2 of same order, if we define the product M_1*M_2 by

$$(M_1*M_2)_{ij}=(M_1)_{ij}\times(M_2)_{ij}, \qquad \ldots \quad (18)$$

we see that L.H.S. of (17) represents the matrix product $\Omega*S$ where

$$S=(S_{r,s}),\ r,s=0,1,\ldots,n \qquad \ldots \quad (19)$$

is an $(n+1)\times(n+1)$ matrix of permanent sums $S_{r,s}$ and

$$\Omega=(\omega_{r,s})\ \text{ where }\ \omega_{r,s}=(r-s)! \qquad \ldots \quad (20)$$

is a matching $(n+1)\times(n+1)$ matrix.

If we define by B and L the $(n+1)\times(n+1)$ matrices whose r^{th} row and s^{th} columns are, respectively

$$B_r=\left(\binom{n}{n-r},\binom{n-1}{n-r},\ldots,\binom{n-r}{n-r},0,0,\ldots0\right)$$

and

$$L^s=\left(0,0,\ldots,0,\binom{s}{s},\binom{s+1}{s},\binom{s+2}{s},\ldots,\binom{n}{s}\right)^T \qquad \ldots \quad (21)$$

then R.H.S. of (17) is seen to be B_rCL^s, so that we get

$$\Omega*S=BCL. \qquad \ldots \quad (22)$$

Now for any matrix M let us denote by M^-, M' and M^+ the matrices with $(i,j)^{th}$ elements

$$(M^-)_{i,j}=M_{i,n-j},\ (M')_{i,j}=M_{n-i,j},\ (M^+)_{i,j}=M_{n-i,n-j}. \qquad \ldots \quad (23)$$

We see that $L=(B^T)^+$, so that (22) becomes $\Omega*S=BC(B^T)^+$.

Thus we have proved the following

Theorem 1. The coefficient matrix C and the auxiliary matrix S are related by

$$(\Omega*S) = BC(B^T)^+ \qquad \ldots (24)$$

$$C = B^{-1}(\Omega*S)((B^T)^+)^{-1} \qquad \ldots (25)$$

These are simply equations (16) and (17) written in compact matrix form.

5. SOME SPECIAL FORMULAE

In this section we show how the use of the differential operator

$$\nabla = (\frac{\partial}{\partial x} + \frac{\partial}{\partial \lambda} + \frac{\partial}{\partial \mu}) \qquad \ldots (26)$$

on $P(G;x,\lambda,\mu) = \text{per}(A'') = \text{per}(xI+\lambda A+\mu(J-I-A))$ enables us to derive the formulae for $P(G)$ in some special cases.

By ∇^r we denote the operator obtained by expanding $(\frac{\partial}{\partial x} + \frac{\partial}{\partial \lambda} + \frac{\partial}{\partial \mu})^r$ by the Binomial theorem and interpreting the operator as usual.

Lemma 8. If $A'' = xI + \lambda A + \mu(J-I-A)$

$$S_{n-r}(A'') = \frac{1}{r!}\nabla^r P \quad \text{for } r = 0,1,\ldots,n \qquad \ldots (27)$$

where $S_0(A'')$ is defined as $\frac{1}{n!}\nabla^n P$.

Proof. We have

$$\frac{1}{r!}\nabla^r P = \frac{1}{r!}\sum_{i}\sum_{j}\sum_{k}{}_{i+j+k=r} \frac{r!}{i!j!k!} \frac{\partial^r P}{\partial x^r\,\partial\lambda^j\,\partial\mu^k}$$

and the result follows by interpretations of the derivatives as used in proving Lemma 6.

Our immediate use of the ∇ operator is in the following context. Let G be a graph with adjacency matrix A and $G_1 = K_1 \cup G$ and $G_2 = K_1 + G$ in Harary's notation ([3], p.21), then the adjacency matrices of G_1 and G_2 are, respectively

$$A_1 = \left[\begin{array}{c|c} 0 & \underset{\sim}{0} \\ \hline \underset{\sim}{0} & A \end{array}\right] \text{ and } A_2 = \left[\begin{array}{c|c} 0 & \underset{\sim}{1} \\ \hline \underset{\sim}{1} & A \end{array}\right].$$

This gives $A_1'' = \left[\begin{array}{c|c} x & \underset{\sim}{\mu} \\ \hline \underset{\sim}{\mu} & A'' \end{array}\right]$ and $A_2'' = \left[\begin{array}{c|c} x & \underset{\sim}{\lambda} \\ \hline \underset{\sim}{\lambda} & A'' \end{array}\right]$, so that their permanent polynomials $P(G_1;x,\lambda,\mu) = P_1$ and $P(G_2;x,\lambda,\mu) = P_2$ are given by $P_1 = xP + \mu^2 S_{n-1}(A'') = xP + \mu^2\nabla P = (x + \mu^2\nabla)P$, using (27) with $r = 1$ and $P_2 = xP + \lambda^2\nabla P = (x + \lambda^2\nabla)P$, where $P = P(G;x,\lambda,\mu)$. If we call G_1 as $\delta_1 G$ and G_2 as $\delta_2 G$ and these operations δ_1 and δ_2 are repeatedly performed, then we have

<u>Lemma 9</u>. $P(\delta_1^r G) = (x+\mu^2\nabla)^r P$ and $P(\delta_2^r G) = (x+\lambda^2\nabla)^r P$ ... (28)

where the power 'r' indicates repeated application of the relevant operator (and not binomial expansion and substitution).

<u>Corollary 4</u>. Since $P(K_1) = x$ we have

$$P(K_n) = \delta_2^{n-1} K_1 = (x + \lambda^2\nabla)^{n-1} x, \quad \text{... (29)}$$

$$P(\overline{K}_n) = \delta_1^{n-1} K_1 = (x + \mu^2\nabla)^{n-1} x, \quad \text{... (30)}$$

and

$$P(K_{1,n}) = P(\delta_2 \overline{K}_n) = (x + \lambda^2\nabla)(x + \mu^2\nabla)^{n-1} x. \quad \text{... (31)}$$

6. A MULTIPLICATIVE POLYNOMIAL

While the permanent polynomial $P(G;x,\lambda)$ and its homogenized version $P(G;x,\lambda,\mu)$ are conceptually the most natural ones they lack the very desirable property of being multiplicative over components of G. To get over this deficiency we define another polynomial having the $S_{r,s}$ as the direct coefficients and by a property of the $S_{r,s}$'s the polynomial gets the multiplicative property.

<u>Definition</u>. Let $Q(G;x,\lambda,\mu) = \omega\mathrm{per}(A'') = \omega P(G;x+\mu,\lambda+\mu,\mu)$ ... (32)

where ω is the linear operator on the vector space of all polynomials in μ defined by

$$\omega(\mu^r) = \frac{1}{r!}\mu^r \quad \text{... (33)}$$

<u>Lemma 10</u>. $Q(G;x,\lambda,\mu) = \sum_{r=0}^{n}\sum_{s=0}^{r} S_{r,s}(A)\, x^{n-r}\lambda^s\mu^{r-s}$ ... (34)

<u>Proof</u>. Follows from (15) and the definition of ω.

<u>Remark</u>. Q is completely specified by the matrix S introduced earlier.

We are now in a position to prove the multiplicative property of Q through

Lemma 11. If a matrix A is the direct sum $A_1 \oplus A_2$ of two matrices A_1 and A_2 then

$$S_{r,s}(A) = S_{r,s}(A_1 \oplus A_2) = \sum_k \sum_\ell S_{k,\ell}(A_1)\, S_{r-k,s-\ell}(A_2) \qquad \ldots \quad (35)$$

Proof. This is easily seen by observing that an $r \times r$ principal submatrix of $A_1 \oplus A_2$ can be made up of k rows of A_1 and $(r-k)$ rows of A_2 and the corresponding columns and that an $s \times s$ submatrix of such a submatrix can have ℓ rows in A_1 and $(s-\ell)$ rows in A_2 and that the permanent of the direct sum of two matrices is the product of their permanents.

Theorem 2. If a graph G has components G_1 and G_2 then

$$Q(G) = Q(G_1)\, Q(G_2) \qquad \ldots \quad (36)$$

Proof. If $n_1 = |V(G_1)|$, $n_2 = |V(G_2)|$ and $n = n_1 + n_2$ then we have from (34)

$$Q(G_1) = \sum_{r=0}^{n_1} \sum_{s=0}^{r} S_{r,s}(A_1) x^{n_1-r} \lambda^s \mu^{r-s},$$

$$Q(G_2) = \sum_{r=0}^{n_2} \sum_{s=0}^{r} S_{r,s}(A_2) x^{n_2-r} \lambda^s \mu^{r-s},$$

and

$$Q(G_1 \cup G_2) = \sum_{r=0}^{n} \sum_{s=0}^{r} S_{r,s}(A_1 \oplus A_2)\, x^{n-r} \lambda^s \mu^{r-s}$$

$$= Q(G_1)\, Q(G_2) \text{ using (35).}$$

Corollary 5. If G has k components G_i with Q-polynomials Q_i, then

$$Q(G) = \prod_{i=1}^{k} Q_i \qquad \ldots \quad (37)$$

Corollary 6. $Q(\overline{K}_n) = (x+\mu)^n \qquad \ldots \quad (38)$

Proof. $Q(\overline{K}_1) = Q(K_1) = \omega\,\mathrm{per}(xI + \mu J) = \omega(x+\mu) = x + \mu.$

$Q(\overline{K}_n) = (Q(\overline{K}_1))^n = (x+\mu)^n.$

One way to view the first invariant conjecture in terms of Q-polynomial is as follows : Let $\underline{\underline{G}}_n$ be the lattice of all labelled n-point graphs partially ordered by $G \le H$ iff G is a spanning subgraph of H. This is simply the Boolean lattice of all subsets of an $\binom{n}{2}$-set, the rank of an (n,m) graph being simply n. The action of S_n on the graphs splits up $\underline{\underline{G}}_n$ into equivalence classes, G being equivalent to H iff

G is isomorphic to H. If we let $\overline{\underline{\underline{G}}}_n$ denote the set of all equivalence classes of $\underline{\underline{G}}_n$, $\overline{\underline{\underline{G}}}_n$ is a poset partially ordered by $G \leq H$ iff G is isomorphic to a spanning subgraph of H. Let $\underline{\underline{P}}_n$ denote the set of all homogeneous polynomials of degree n in the three variables x, λ, μ with coefficients in nonnegative integers (Z^+). Then $\underline{\underline{P}}_n$ is a poset, partially ordered by $f \leq g$ in $\underline{\underline{P}}_n$ iff every coefficient in f is not greater than the corresponding coefficient in g. Now Q can be thought of as a function from $\overline{\underline{\underline{G}}}_n$ into $\underline{\underline{P}}_n$. The conjecture claims that this is a (1-1) function. The same can, in fact, be said about P, but unlike Q,P does not have the order preserving property proved in

<u>Theorem 3</u>. If H is a spanning subgraph of G, then

$$Q(H) \leq Q(G) \qquad \ldots \quad (39)$$

and equality holds here if and only if H is the same as G.

<u>Proof</u>. If A and B are any two (0,1)-matrices we can say $A \leq B$ iff $A_{ij} \leq B_{ij}$ for every $i, j = 1,\ldots,n$. It is clear that H is a spanning subgraph of G iff $A(H) \leq A(G)$. The latter implies $S_{r,s}(A(H)) \leq S_{r,s}(A(G))$ for all pairs r,s $(r \leq s)$. Then from (16) the result follows.

It is also clear that equality $G \simeq H$ holds if and only if $S_{r,s}(A(H)) = S_{r,s}(A(G))$ for all pairs r,s; that is $Q(G) = Q(H)$.

7. s-CHAINS, THE T-MATRIX AND THE T-POLYNOMIAL

We now adopt a different procedure to arrive at a polynomial equivalent to Q. For any (0,1)-natrix M we define an s-chain as a set of s 1's in independent positions, that is no two of which are in the same row or column of M. The rank $\rho(c)$ of an s-chain c is defined as the number of rows (equivalently columns) in the minimal principal submatrix containing the elements of c. If I is the set of row-suffixes and J is the set of column-suffixes of the elements of c, it is easy to see that $\rho(c) = |I \cup J|$ and that $|I| = |J| = s$. Thus we have

<u>Lemma 12</u>. For any s-chain c,

$$s \leq \rho(c) \leq 2s. \qquad \ldots \quad (40)$$

Proof. $s = |I| \leq |I \cup J| \leq |I| + |J| = 2s$.

Let $t_{r,s}$ denote the number of distinct s-chains of rank r in A. Then we define the T-polynomial and T-matrix of A by

$$\text{(i)} \quad T(A;x,\lambda,\mu) = \sum_{r=0}^{n} \sum_{s=o}^{r} t_{rs,s}\, x^{n-r} \lambda^{s} \mu^{r-s} \qquad \ldots \quad (41)$$

$$\text{(ii)} \quad T = (t_{r,s}) \; r,s = 0,1,\ldots,n. \qquad \ldots \quad (42)$$

If G is a graph and A is its adjacency matrix we call $T(A;x,\lambda,\mu)$ the T-polynomial of G and write it as $T(G;x,\lambda,\mu)$.

We link the Q-polynomial and the T-polynomial of a graph (equivalently the S and T matrices) through

Theorem 4. The S and T matrices of a graph G are related by

$$S = BT \qquad \ldots \quad (43)$$

where B is the matrix

$$B_{r,s} = \begin{cases} \binom{n-s}{r-s} & \text{for } 0 \leq s \leq r \\ 0 & \text{for } s > r \end{cases} \qquad \ldots \quad (44)$$

Proof. An s-chain c of rank u can contribute a 1 only to those $S_{r,s}$'s for which $r \geq u$. We get such a contribution 1 to every choice of a principal $r \times r$ submatrix N of A which contains the minimal principal $u \times u$ submatrix M of A containing the elements of c. Since the remaining (r-u) rows (columns) of such a matrix N can be chosen from the remaining (n-u) rows (columns) of A in $\binom{n-u}{r-u}$ distinct ways, c contributes $\binom{n-u}{r-u}$ to each such $S_{r,s}$. Thus if C_s is the set of s-chains of A,

$$S_{r,s}(A) = \sum_{c\in C_s} \binom{n-u}{r-u}, \qquad \ldots \quad (45)$$

on bringing the notation $t_{u,s}$ of the number of s-chains of rank u, we get

$$S_{r,s}(A) = \sum_{u=s}^{r} \binom{n-u}{r-u} t_{u,s}. \qquad \ldots \quad (46)$$

Using $r \geq s$, and making s run from 0 to n we complete the proof of the theorem.

Corollary 7. As the matrix B is easily seen to be invertible T is obtainable from S as $B^{-1}S$, then the Q-polynomial and T-polynomial are 'exchangeable'.

Lemma 13. The number of non-zero entries in the T matrix is at most $N = [\frac{n(n+4)}{4}]$ (where [x] is the greatest integer not exceeding x).

Proof. Since $s \leq r \leq \min(2s,n)$, the possible combinations of (r,s) values for which $t_{r,s}$ is defined are

(i) (s,s), (s+1,s),...,(2s,s) for $s \leq [n/2]$

and

(ii) (s,s), (s+1,s),...,(n,s) for $s > [n/2]$

Therefore

$$N = \sum_{s\leq[\frac{n}{2}]} (s+1) + \sum_{s>[\frac{n}{2}]} (n-s+1)$$

$$= \frac{(t+1)(t+2)}{2} - 1 + \frac{(n-t)(n-t+1)}{2} \quad \text{where } t = [\tfrac{n}{2}]$$

$$= \frac{n^2 - n(2t-1) + 2t^2 + 2t}{2}$$

$$= \begin{cases} \frac{n(n+4)}{4} & \text{when } t = \frac{n}{2} \\ \frac{n(n+4)-1}{4} & \text{when } t = \frac{n-1}{2}. \end{cases}$$

Corollary 8. The T-matrix has roughly (asymptotically) half the number of significant terms as the adjacentry matrix. Thus if the T-polynomial (T-matrix) is a complete invariant, it is likely to be quite a useful tool.

8. FRAMES AND FRAME POLYNOMIALS

Let us define a spanning subgraph of a graph G, whose components are singleton points or paths or cycles as a frame of G (A frame has been referred to as a 2-matching in the literature, see e.g. Berge [1],p.150). If a frame F has w_i paths P_i (i = 1,2,...,n) (here $P_1 = K_1$, $P_2 = K_2$) and y_j cycles C_j we denote by w(F) the product $p_1^{w_1} p_2^{w_2} \ldots p_n^{w_n} c_3^{y_3} c_4^{y_4} \ldots c_n^{y_n}$ and call it the weight of F. Clearly w(F) completely specifies F upto isomorphism. Let $\underline{\underline{F}}$ be the collection of all frames of G and $\underline{\underline{F}}^*$ be the collection of all equivalence classes of non-isomorphic frames of G. If F* is the equivalence class to which F belongs, all frames in F* have the same weight w(F). Let $|F^*| = c_F$. Then we define

$$F(G;\underset{\sim}{p},\underset{\sim}{c}) = \sum_{F^*\varepsilon\underline{\underline{F}}^*} c_F w(F) \qquad \ldots (47)$$

and call it the Frame polynomial of G.

We are now in a position to state our second major conjecture,

<u>The Second Invariant Conjecture</u> (Or,the <u>Frame Conjecture</u>). The multivariable frame polynomial $F(G;\underset{\sim}{p},\underset{\sim}{c})$ is a complete invariant for graphs.

Since the frame polynomial (47) completely specifies the frame-structure of G, the non-trivial part of the above conjecture is clearly in the nature of a <u>reconstruction conjecture</u>.

To relate this to our first conjecture we define a modified frame polynomial as follows : Let ℓ be the total number of lines in a frame F and k be the number of components of F. In terms of the weight vector we have

$$\ell = \sum_{2}^{n} (i-1)w_i + \sum_{3}^{n} jy_j$$

and

$$k = \sum_{1}^{n} w_i + \sum_{3}^{n} y_j.$$

We define a 4-variable reduced <u>frame polynomial by</u>

$$R(G;x,y,z,t) = \sum_{F^*\varepsilon \underline{F}^*} c_F x^{w_1} y^{w_2} z^{\ell} t^k \qquad \ldots \quad (48)$$

Clearly lot of information has been lost while reducing F to R. While R is derivable from F the converse process is not feasible (even when the frames consist only of cycles or only of paths, except possibly in some special cases).

We now show that the T-polynomial can be obtained rrom the reduced frame polynomial (the R-polynomial for short).

<u>Theorem 5</u>. The F-polynomial may be obtained from the R-polynomial by means of the substitution

$$x^{w_1} y^{w_2} z^{\ell} t^k \to \sum_{i=0}^{w_2} \binom{w_2}{i} 2^{k-w_1-i} x^{w_1} \lambda^{\ell+i} \mu^{n-w_1-\ell-i} \qquad \ldots \quad (49)$$

<u>Proof</u>. $x^{w_1} y^{w_2} z^{\ell} t^k$ represents a typical frame F of the graph G. To see its contribution to different terms $t_{r,s}$ we observe the following :

(i) Each path P_i, $i \geq 3$ corresponds to two i-chains with rank i.

(ii) Each cycle c_j, $j \geq 3$ corresponds to two j-chains with rank j.

(iii) Each of the w_2 P_2's can be taken as either 2 ones (two 1-chains of rank 2) or one 2-chain of rank 2.

(iv) When the contributions of the different components of F are combined, the dimensions of the chains (s) and their rank (r) get added up and the numbers of occurrences get multiplied.

The total rank is always the number of points other than the single points, that is $(n-w_1)$. If i P_2's are taken as 2-chains and the others as 1-chains, the total length s of the chain is $\ell+i$. This explains the term $x^{w_1}\lambda^{\ell+i}\mu^{n-w_1-\ell-i}$. The i P_2's can be chosen in $\binom{w_2}{2}$ ways, all components except the w_1 points and the i chosen P_2's contribute 2 each, hence the coefficienct $\binom{w_2}{2}2^{k-w_1-i}$. This proves the theorem.

Remark. The status of the converse of this theorem is not very clear.

9. CONCLUSION

We have introduced five different graph polynomials the P,Q,T,R and F polynomials and conjectured that any of these is a complete graph invariant. We have established the equivalence of the first three in Theorems 1 and 4. We have also established the implications $F \to R \to T$ in the sense that R is derivable from F and T from R. The converse implications are not clear. Thus, as things stand, the permanent conjecture (the first invariant conjecture) may be false even if the frame conjecture is true. But if the frame conjecture is false then, of course both the conjectures fall.

Defining a frame to be maximal if it is not a subframe of any other frame of the given graph, one may formulate the maximal frame conjecture as the assertion that the maximal frames of a graph determine the graph uniquely. This was however disproved by a computer-generated counter-example.

The validity of the first graph-invariant conjecture has been verified for graphs with at most 7 points by the generation of the C-matrices of these graphs with the aid of an IBM 370 computer. A PL/1 programme for the same may be obtained from the second author.

On the theoretical side it is easily seen that the F-polynomial is an instance of the general graph-polynomial introduced by Farrell in [2]. The consequences of this and the relation of these polynomials to the special polynomials like the dichromatic polynomials etc., are perhaps worth studying.

Note. The results reported here mostly from the last chapter of the Doctoral dissertation of the first author, under the guidance of the second author, submitted to the I.S.I., Calcutta.

REFERENCES

1. C. Berge, Graphs and Hypergraphs, Second Edition, North Holland, Amsterdam, 1976.
2. E.J. Farrell, On a general class of graph polynomials, J. Comb. Theory, B, 26(1979), 111-122.
3. F. Harary, Graph Theory, Addison-Wesley, 1969.
4. H. Minc, Permanents, Addison-Wesley, 1978.
5. R.C. Read, Teaching graph theory to a computer, in Recent progress in combinatorics (W.T.Tutte Ed.), Academic Press, 1969, 161-174.
6. H.J. Ryser, Combinatorical Mathematics, Carus Monographs No.14, 1963.
7. Y. Shah, G.I. Davida and M.K. McCarthy, Optimum features and graph isomorphism, IEEE Transactions Sys. Man and Cyb, SMC 4(1974), 313-319.

AFFINE TRIPLE SYSTEMS

DIJEN K. RAY-CHAUDHURI
Department of Mathematics
The Ohio State University
Columbus, Ohio 43210

1. INTRODUCTION

An incidence structure π is a triple $\pi = (P,L,I)$ where P and L are disjoint sets and $I \subseteq P \times L$. Elements of P and L are respectively called points and lines. If $(p,\ell) \in I$, we say that the point p and the line ℓ are mutually incident or that the point p is a point of the line ℓ or that the line ℓ contains the point p. The incidence structure π is called finite if and only if (iff) $P \cup L$ is finite. It is called linear if any two distinct points are contained in a unique line. Let $v > k > \lambda > 0$ be integers. A (v,k,λ) 2-design is an incidence structure with v point in which each line contains k points and any two distinct points are simultaneously contained in exactly λ lines. It is called symmetric if $|L| = |P|$. A subset F of P of a linear incidence structure π is called a flat if all points of any line containing two distinct points of F belong to F. Clearly P itself is a flat. For a subset P' of P, $\langle P'\rangle$ is the intersection of all flats containing P' and is calle the flat generated by P'. The rank of a flat F is the smallest cardinality of a set generating F and the geometric dimension of F is one less than its rank. Rank and geometric dimension of π are respectively the rank and geometric dimension of P. Flats of geometric dimensions 0,1 and 2 are respectively called points, lines and planes. A linear incidence structure π is called a projective plane if any two distinct lines of π contain a unique common point and π contains a set of 4 points no three of which are collinear. A finite projective plane of order n can also be define as a symmetric $(n^2+n+1,\ n+1,\ 1)$ 2-design. A linear incidence structure π is called an affine plane if (1) every line of π contains at least 2 points, (2) geometric dimer sion of π is 2 and (3) there is an equivalence relation called parallelism on the line of π such that given a line ℓ and a point p not on ℓ, there exists a unique line ℓ' containing p and parallel to ℓ. A finite affine plane of order n, can also be defined as an $(n^2,n,1)$ 2-design. A projective space is a linear incidence structure π such that the geometric dimension of π is at least two and every triple of noncollinear points of π generates a projective plane. An affine space is a linear incidence structure π such that (1) the geometric dimension of π is at least two, (2) every triple of noncollinear points of π generates an affine plane and (3) two distinct planes contained in a 3-flat of π (flat of geometric dimension 3) which have a common point have a common line.

Let K be a finite field of order q, V be a vector space of dimension $d \geq 3$ over K and W_i $(1 \leq i \leq d)$ be the set of linear subspaces of V of dimension i. Let A_i be the set of i-dimensional subspaces of V and their translates, $0 \leq i \leq d$. Let $PG(d-1,q)$ and $AG(d,q)$ respectively denote the incidence structures $(W_1,W_2,\subseteq)$ and $(A_0,A_1,\subseteq)$. Here incidence is by set inclusion. A fundamental theorem of projective geometry states that if π is a finite projective space of geometric dimension ≥ 3, then π is isomorphic to some $PG(d,q)$. Similarly if π is a finite affine space with geometric dim. ≥ 3, then for some d and q, π is isomorphic to $AG(d,q)$. Buekenhout [5] in 1969 proved a theorem showing that the axiom regarding the 3-flats in the definition of a finite affine space is redundant when the lines contain at least four points. More precisely, he showed that a finite linear incidence structure in which every line contains at least four points and every triple of noncollinear points generates an affine plane is an affine space. The conclusion of Buekenhout's theorem does not hold when the line sizes do not exceed three. A $(v,3,1)$ 2-design is called a steiner triple system. A steiner triple system in which every triple of noncollinear points generates an affine plane is called an affine triple system (ATS). Clearly an $AG(d,3)$ is an ATS. Affine triple systems which are not isomorphic to any $AG(d,3)$ are called Hall triple system (HTS). Hall [6] has proved several interesting theorems about affine triple systems. For instance he showed that a steiner triple system is an affine triple system iff for every point p of the system there exists an involution of the points which is an automorphism of S and whose only fixed point is p. Hall [7] constructed the first example of a Hall triple system with 81 points. In his paper [7], he gave a construction simpler than his original construction. Let K be the group represented by

$K = \langle a,b,c,t : t^2 = a^3 = b^3 = c^3 = 1, tat = a^{-1}, tbt = b^{-1}, tct = c^{-1} \rangle.$

Let $C = C_K(t)$ denote the centralizer of t in K. Then $[K : C] = 81$ and K can be represented faithfully as a permutation group on the 81 cosets of C in K. The element g of K is represented by the permutation $g^{\psi} : CH \mapsto Chg$.

Each of the 81 conjugates of the involution t in K is represented by an involution fixing one coset of C and distinct conjugates fix distinct cosets. The 81 points of the Hall's triple system (HTS) are the 81 cosets of C. A triple $\{Ch_i, CH_j, Ch_k\}$ is a line iff (Ch_j, CH_k) is a transposition of the involution that fixes Ch_i.

Affine triple systems arise in many other contexts. Affine triple systems can be viewed as perfect matroid designs. The following is one of many possible equivalent definitions of a finite matroid. A finite matroid is a pair $(P, \underline{\underline{F}})$ where $\underline{\underline{F}}$ is a set of subsets of P such that (1) $\underline{\underline{F}}$ is closed under intersection and contains P, (2) for all $F \in \underline{\underline{F}}$, and $p \in P-F$, there exists a unique set F' which contains p and covers F. For $F', F \in \underline{\underline{F}}$, F' is said to cover F if F' contains F properly and for no other element F'' of $\underline{\underline{F}}$, $F \subsetneqq F'' \subsetneqq F'$ holds. Sets belonging to $\underline{\underline{F}}$ are called the flats of the matroid. The rank of a flat can be defined as in the case of a linear incidence structure. A matroid is called a perfect matroid design if all flats of rank i have the same cardinality α_i, for every natural number i.

Let S be an affine triple system with point set P. If we take the empty set, the points, the lines, the affine planes of order 3 and P as flats, then we get a perfect matroid design based on S. The only known perfect matroid designs arise from projective spaces, affine spaces, t-designs and affine triple systems. Presently there is great interest in finding other perfect matroid designs.

Affine triple systems from an algebraic point of view can be regarded as Commutative Moufang exponent three loops (c-M exp-3 loop). A Moufang loop is an ordered pair $(G,\cdot)$ where G is a set and $\cdot$ is a binary operation on G satisfying (1) for all elements a and b of G, there exist elements x and y of G such that $a \cdot x = b$ and $y \cdot a = b$, (2) there exists an element e (called the identity) such that for all

$a \in G$, $e\cdot a = a\cdot e = a$, (3) for all $x,y,z, \in G$, $(x\cdot y)\cdot(z\cdot x) = (x\cdot(y\cdot z))\cdot x$ (the Moufang identity). If the operation $\cdot$ is commutative, the Moufang identity reduces to $(x\cdot y)\cdot(x\cdot z) = x\cdot(x\cdot(y\cdot z))$. Commutative Moufang loops differ from Abelian groups only in not satisfying the axiom of associativity. Associativity is replaced by the Moufang identity. A Moufang loop is said to be exponent three if for all elements $a \in G$, $a^3 = e$. Affine triple systems are indeed equivalent to c-M exp-3 loops. In a linear incidence structure a line can be viewed as the subset of its incident points. Let $(G,\underline{\underline{L}})$ be a steiner triple system where G is the set of points and $\underline{\underline{L}}$ is the set of lines, each line being a subset of exactly three points. Let e be a distinguished point. We define a binary commutative operation $\cdot$ on G (with respect to e) as follows.

$e\cdot e = e$, $e\cdot x = x$ for all $x \in G$. $x\cdot y = e$, $x\cdot x = y$ where $\{e,x,y\}$ is a line. If $\{e,x,y\}$ is not a line, $x\cdot y = z$ where $\{x',y',z\}$, $\{e,x,x'\}$ and $\{e,y,y'\}$ are lines.

The following theorem due to Bruck [3] as well as Young [11] establishes the basic connection between affine triple systems and c-M exp-3 loops.

Theorem 1. Let $S = (G,\underline{\underline{L}})$ be an affine triple system. Let $e \in G$. Then $(G,\cdot)$ is a finite c-M exp-3 loop. Conversely, if $(G,\cdot)$ is a finite c-M exp-3 loop, then $(G,\underline{\underline{L}})$ is an affine triple system where $\underline{\underline{L}} = \{\{x,y,x^{-1}y^{-1}\} : x \neq y \in G\}$.

The following theorem due to Bruck [3] as well as Hall [6] shows that the c-M exp-3 loop of an affine triple system is a group only when the affine triple system is isomorphic to an $AG(n,3)$, $n \geq 1$. More precisely,

Theorem 2. Let $S = (G,\underline{\underline{L}})$ be an affine triple system and $e \in G$. Then S is isomorphic to an $AG(n,3)$, $n \geq 2$ iff $(G,\cdot)$ is an elementary abelian 3-group of order 3^n.

An affine triple system is a Hall triple system iff the associated c-M exp-3 loop is not an abelian 3-group. The first example of a c-M exp-3 loop which is not a group was constructed by Zassenhaus (see Bol [2], p.423). In his construction, Zassenhaus takes $G = \mathbb{Z}_3 \times \mathbb{Z}_3 \times \mathbb{Z}_3 \times \mathbb{Z}_3$. For two elements $x = (x_1,x_2,x_3,x_4)$ and $y = (y_1,y_2,y_3,y_4)$ of G, the product $x\cdot y$ is defined by $x\cdot y = (x_1+y_1,\ x_2+y_2,\ x_3+y_3,\ x_4+y_4+(x_3-y_3)(x_1y_2-x_2y_1))$. Bruck [3] in his book made a systematic study of Moufang loops and other more general binary systems. Here we give a brief review of some of the interesting results on Moufang loops.

Throughout the remainder of this section $(G,\cdot)$ will denote a c-M loop with identity e and the loop product $x\cdot y$ will be shortened to xy. For $x,y,z \in G$ the <u>associator</u> (x,y,z) of x,y and z is the unique element of G such that $(x,y,z)[x(yz)] = (xy)z$. The <u>center of</u> G is the subloop $Z(G) = \{z \in G :$ for all $x,y \in G, (z,x,y) = (x,z,y) = (x,y,z) = e\}$. Of course, $Z(G)$ is an abelian group. A subloop H is <u>normal in</u> G if for all $x,y \in G$, $x(yH) = (xy)H$. When H is normal in G the quotient loop G/H is well-defined. With H_1,H_2 and H_3 normal subloops of G, the <u>associator subloop</u> (H_1,H_2,H_3) <u>of</u> H_1,H_2 <u>and</u> H_3 is the subloop generated by $\{(h_1,h_2,h_3) : h_i \in H_i\}$. The <u>lower central series of</u> G is defined recursively by $G_0 = G$ and $G_{i+1} = (G_i,G,G)$ and G is said to be lower centrally nilpotent of class n, written ($\underline{\text{nil}}(G) = n$) if n is the least integer for which $G_n = \{e\}$. $G_1 = (G,G,G)$ is called the <u>associator subloop of</u> G and is often denoted by G'. The upper central series for G is defined as $Z_0(G), Z_1(G),\ldots.$ where $Z_0(G) = \{e\}, Z_1(G) = Z(G), Z_i(G)$ is the inverse image of $Z(G/Z_{i-1}(G))$ under the canonical mapping $\varphi : G \to G/Z_{i-1}(G)$. G is said to be upper centrally nilpotent if for some integer r, $Z_r(G) = G$ and we write $\overline{\text{nil}}(G) = \text{Min}\{r : Z_r(G) = G\}$. Bruck showed that a c-M loop G is upper centrally nilpotent iff it is lower centrally nilpotent and in that case $\overline{\text{nil}}(G) = \underline{\text{nil}}(G) = \text{nil}(G)$(say).

<u>Theorem 3.</u>(Bruck-Skaby) If G is a c-M loop generated by n elements then G is centrally nilpotent and $\text{nil}(G) \leq n-1$.

<u>Corollary 4.</u> If G is a finitely generated c-M exp-3 loop then $|G| = 3^m$ for some m

<u>Theorem 5.</u>(Moufang) If G is a moufang loop then G is diassociative, that is, the subloop generated by any two elements of G is a group; and more generally, for all $x,y,z \in G$ such that $(x,y,z) = e$ the subloop $\langle x,y,z\rangle$ is a group.

The following properties of associators follow from the general result of Bruck that if G is a c-M loop then $(x,y,z) \in Z(\langle x,y,z\rangle)$.

Property A-1 : $(x,y,z) = (z,x,y)$ and $(x,z,y) = (x,y,z)^{-1}$.

Property A-2 : $(x^{-1},y,z) = (x,y,z)^{-1}$.

Notice that when G has exponent 3, $(x,z,y) = (x,y,z)^2$ and $(x^2,y,z) = (x,y,z)$. The following rule is called the expansion law for c-M loops :

$$(wx,y,z) = [(w,y,z)(x,y,z)][((w,y,z),w,x)((x,y,z),x,w)].$$

When $nil(G) = 2$ and hence $(G',G,G) = \{e\}$ we have

$$(wx,y,z) = (w,y,z)(x,y,z)$$

which we shall refer to as simplified expansion law.

With G a finitely generated c-M loop, $B \subseteq G$ is a <u>basis of</u> G if $G = \langle B\rangle$ but $\langle\overline{B}\rangle \neq G$ for all proper subsets $\overline{B} \subseteq B$. The main results of Beneteau [1] are these :

<u>Theorem 6</u>.(Beneteau) If G is a finitely generated c-M loop then every basis of G has the same cardinality which is defined to be the <u>dimension of</u> G and is denoted by $d(G)$. Furthermore, G/G' is an abelian group with $d(G/G') = d(G)$. When G has exponent 3, G/G' is an elementary abelian 3-group.

<u>Theorem 7</u>.(Beneteau) If G is a c-M loop with $nil(G) \geq 2$ and $|G| = 3^m$ then $|Z(G)| \leq 3^{m-3}$.

That $|Z(G)| \neq 1$ (and hence $|Z(G)| \geq 3$ when G has exponent 3) follows from the upper central nilpotence of G. Exterior algebra based on F, the galois field of order 3, proved to be a useful source of exp-3 Moufang loops. For every finite subset S of N, including ϕ, let X_S be an indeterminate symbol. Here N is the set of natural numbers. We abbreviate $X_{\{i\}}$ by X_i. For each nonnegative integer i, let $\Lambda^i = \{ \sum_{n=1}^{\ell} k_n X_{S_n}, \ell \in N, k_n \in F, |S_n| = i\}$. In other words Λ^i is the set of finite linear combinations with coefficients in F of indeterminates whose corresponding set has cardinality i. For each nonnegative j, let $\bigoplus_{i=j}^{\infty} \Lambda^i = \{ \sum_{n=1}^{\ell} k_n X_{S_n} : \ell \in N, k_n \in F, |S_n| \geq j, 1 \leq n \leq \ell\}$ and let $\underline{\underline{E}} = \bigoplus_{i=0}^{\infty} \Lambda^i$. $\underline{\underline{E}}$ endowed with the standard addition is a vector space over F. The additive identity element of $\underline{\underline{E}}$ is written as $\overline{0}$. Next define a multiplication on $\underline{\underline{E}}$ by extending the following rules to linear combinations via the distributive laws : $X_S \cdot X_T = \overline{0}$ if $S \cap T \neq \phi$ and $X_S \cdot X_T = (-1)^{p(S,T)} X_{S \cup T}$ if $S \cap T = \phi$ where $p(S,T) = |\{(s,t) : s \in S, t \in T, s > t\}|$. $\underline{\underline{E}}$ with the addition and multiplication defined above is an associative algebra with multiplicative identity element X_ϕ. Let $\underline{B} = \{\{a,x : a \in \Lambda', x \in \underline{E}\}$ and define $(a,x)*(\ell,y) = (a+\ell, x+y+(x-y)a\ell)$ where juxtaposition indicates multiplication in $\underline{E}$. Bruck proved that $(\underline{B},*)$ is a c-M exp-3 loop. For $i \in N$, let $\beta_i = (x_i, x_\phi)$ and for $d \in N$, let B_d be the subloop of $\underline{B}$ generated by $\{\beta_i : 1 \leq i \leq d\}$. Bruck proved that for all $d \geq 3$, $nil(B_d) = 1 + [d/2]$ and $|B_d| = 3^{2^{d-1}}$. For each nonnegative integer d the subset K_d of $\underline{B}$ is defined by $K_d = \{(\sum_{i=1}^{d} \ell_i X_i, \sum_{S \subseteq I_d} K_S X_S) : \ell_i \in F$ for $1 \leq i \leq d$, $K_S \in F$, $S \subseteq I_d\}$. Then K_d is a c-M exp-3 loop. Furthermore, $|K_d| = 3^{d+2^d}$, $B_d \subseteq K_d$ and $nil(K_d) = 1 + [d/2]$. Let $O(\underline{E})$ (the odd part of $\underline{E}$) be defined by $O(\underline{E}) = \bigoplus_{i=0}^{\infty} \Lambda^{2i+1} = \{\sum_{n=1}^{\ell} K_n X_{S_n} ; \ell \in N, K_n \in F, |S_n|$ is odd for $1 \leq n \leq \ell\}$. Let $M = (O(\underline{E}) \times O(\underline{E}), \cdot)$ where the operation $\cdot$ is defined by $(a,x)\cdot(b,y) = (a+b+aby-abx, x+y+axy-bxy)$. Malbos [8] proved that $(M,\cdot)$ is a c-M exp-3 loop. Let M_d be the subloop of M generated by $\{(x_i, x_{2i}) : 1 \leq i \leq d\}$. Malbos proved the interesting result that M_d is a c-M exp-3 loop with $d(M_d) = d$ and $nil(M_d) = d-1$. Results of Malbos establish that $nil(F_d) = d-1$, for $d \geq 2$ where F_d is the free c-M loop on d generators. J.D.H. Smith [9] independently proved the same result with techniques of nilpotent group theory and associator calculus.

2. COMMUTATIVE MOUFANG EXPONENT THREE LOOPS OF NILPOTENCE CLASS TWO

c-M exp-3 loops of nilpotence class 1 are the elementary abelian 3-groups. They correspond to AG(3,n) and they are completely understood. Naturally, as a next step one likes to describe completely nilpotence class 2 c-M exp-3 loops. Theorem 8 proved by Ray-Chaudhuri and Roth to a great extent accomplishes this. In this section we describe briefly the results proved by Ray-Chaudhuri and Roth on c-M exp-3 loops of nilpotence class 2.

Let $[\gamma_d, \gamma_{d-1}, \ldots, \gamma_1]$ be an ordered basis of G, a c-M exp-3 loop of nilpotence class 2. An arbitrary element of G has a unique representation in the form $g[\gamma_d^{\ell_d}(\gamma_{d-1}^{\ell_{d-1}} \ldots (\gamma_2^{\ell_2} \gamma_1^{\ell_1}) \ldots)]$. The bracketing is from left to right. For instance for $d = 4$ we will get $g[\gamma_4^{\ell_4}(\gamma_3^{\ell_3}(\gamma_2^{\ell_2} \gamma_1^{\ell_1}))]$. Here $g \in G_1 = (G,G,G)$ and $\ell_i \in \{0,1,2$ $1 \leq i \leq d$. For the sake of convenience of writing, we suppress the bracketing assuming that it is from left to right. The product of two elements

$$(g, \gamma_d^{\ell_d}, \ldots, \gamma_1^{\ell_1}) \text{ and } (g', \gamma_d^{\ell'_d}, \ldots, \gamma_1^{\ell'_1})$$

is given by

$$(gg'\gamma, \gamma_d^{\ell_d+\ell'_d}, \ldots, \gamma_1^{\ell_1+\ell'_1}).$$

Here γ is the twist factor which makes G different from a group in general. γ is defined in terms of the basic associators $(\gamma_t, \gamma_i, \gamma_j)$.

$$\gamma = \pi(\gamma_t, \gamma_i, \gamma_j)^{(\ell_t-\ell'_t)(\ell_i\ell'_j-\ell'_i\ell_j)}$$

where the product is over the $\binom{n}{3}$ terms, corresponding to $1 \leq j < i < t \leq d$. It should be observed since $(G_1, G, G) = e$, $(a,b,c) = e$ if at least one of the elements a, b and c itself is an associator. Thus the product $gg'\gamma$ does not depend on any kind of bracketing. To construct c-M exp-3 loops of nilpotence class 2, we formally reverse the process described above and show that it works. The detailed and exact description of the proof itself is very complicated.

For the purpose of our free construction of c-M exp-3 loops of nilpotence class 2, we start with a finite elementary abelian 3-group G_1 written multiplicatively with identity e and take generators $\gamma_1, \gamma_3, \ldots, \gamma_d$. For $i = 1, \ldots, d$, let $\gamma_i^3 = e$. $C_i = \{\gamma_i, \gamma_2^2, e\}$ is a cyclic group of order 3. Let us set $G = G_1 \times C_d \times C_{d-1} \times \ldots \times C_1$. We write $\overline{\gamma}_i$ for $(e, e, \ldots, \gamma_i, \ldots e)$. Eventually G_1 will be the associator subloop of our c-M exp-3 loop of nilpotence class 2.

To define the binary operation on G, we need to assign values to the $\binom{n}{3}$ triples $[\gamma_t, \gamma_i, \gamma_j]$ from G_1, i.e., we have a mapping V : the set of $\binom{n}{3}$ triples $[\gamma_t, \gamma_i, \gamma_j] \to G_1$. We write $V([\gamma_t, \gamma_i, \gamma_j])$ as $(\overline{\gamma}_t, \overline{\gamma}_i, \overline{\gamma}_j)$. Eventually $(\overline{\gamma}_t, \overline{\gamma}_i, \overline{\gamma}_j)$ will be the appropriate associator. We will also assume that $\langle(\overline{\gamma}_t, \overline{\gamma}_i, \overline{\gamma}_j)\rangle = G_1 \neq e$. Now we are ready to define products in G. The product of $(g, \gamma_d^{\ell_d}, \ldots, \gamma_1^{\ell_1})$ and $(g', \gamma_d^{\ell'_d}, \ldots, \gamma_1^{\ell'_1})$ is $(gg'\gamma, \gamma_d^{\ell_d+\ell'_d}, \ldots, \gamma_1^{\ell_1+\ell'_1})$, $\gamma = \Pi(\overline{\gamma}_t, \overline{\gamma}_i, \overline{\gamma}_j)^{(\ell_t-\ell'_t)(\ell_i\ell'_j - \ell_j\ell'_i)}$ where the product is over the $\binom{n}{3}$ terms corresponding to $1 \leq j < i < t \leq d$. We denote this product by $\cdot$.

<u>Theorem 8.</u> (Ray-Chaudhuri and Roth) $(G, \cdot)$ described above is a finite c-M exp-3 loop and every finite c-M exp-3 loop can be constructed in this manner.

If G_1 has order 3^m, G has order 3^{m+d}. Corresponding to different value functions V we get different c-M exp-3 loops. For all ordered pairs $(m,z) \neq (5,1)$ with $m \geq 4$ and $1 \leq z \leq m-3$, there exists a c-M exp-3 loop G of nilpotence class 2 of size 3^m and center of size 3^z. We are able to prove that for $m = 4$ and 5, there is exactly one nonassociative c-M exp-3 loop of size 3^m and each of these loops has nilpotence class 2. Roth [10] in his Ph.D. dissertation was able to establish that there are exactly three nonassociative c-M exp-3 loops of size 3^6 and each of these loops has nilpotence class 2. Roth [10] also gave a general construction for c-M exp-3 loops of nilpotence class 2. For $d \geq 4$, the free c-M exp-3 loop of nilpotence class 3 is denoted by $F_{3,d}$. Roth has proved that $|F_{3,d}| = 3^{d+\binom{d}{3}+4\binom{d+1}{5}}$. Roth also gave an explicit construction of $F_{3,4}$.

In conclusion, I mention a few directions for future research :

1. Prove analogues of the theorem of Ray-Chaudhuri and Roth for higher nilpotenc classes.

2. Find new perfect matroid designs (PMD) with lines of size 3. For instance, the case of planes of size 13 will be the next interesting case. There are 2 Steiner triples systems (STS) of size 13; one of them is more regular and has two orbits for blocks under Z_{13}. We should demand that the planes are the regular STS of order 13. Try to get a coordinatization and algebraic system corresponding to an STS in which ever plane is a regular STS of order 13.

3. Can a Steiner System i.e., a (v,k,1)- t-design be erected into a PMD of rank greater than t+1 ?

4. Is there a theorem about factorization of c-M exp-3 loops into irreducible c-M exp-3 loops ?

5. Intervals of length 4 in perfect matroid designs lead to 2-designs. Study the local designs of Hall triple systems.

6. Start a classification theory for rank 4 PMD's.

REFERENCES

1. L. Beneteau, Topics about Moufang Loops and Hall Triple Systems, to appear in "Simon Stevin".
2. G. Bol, Gewebe und Gruppen, Mathematisch Annalen, 114(1937), 414-431.
3. R.H. Bruck, A Survey of Binary Systems, Ergebnisse der Mathematik und ihrer Grenzgebiete, Band 20, Springer-Verlag, Berlin-Göttingen-Heidelberg, 1958.
4. R.H. Bruck, Some theorems on Moufang loops, Mathematische Zeitscrift, 73(1960), 59-78.
5. F. Buekenhout, Une caractérisation des espaces affins basée sur la notion de droite, Mathematische Zeitschrift, 111(1969), 367-371.
6. M. Hall, Automorphisms of Steiner triple systems, IBM Journal of Research and Development, 1960, 460-472.
7. M. Hall, Incidence axioms for affine geometry, Journal of Algebra, 21(1972), 535-547.
8. J-P. Malbos, Sur la classe de nilpotence des Boucles commutatives de Moufang et des espaces mediaux, C. R. Acad. Sc. Paris, Série A, 287(1978), 691-693.
9. J.D.H. Smith, On the nilpotence class of commutative Moufang loops, Mathematical Proceedings of the Cambridge Philosophical Society, 84(1978), 387-404.
10. R. Roth, Hall triple systems and commutative Moufang exponent three loops, Ph.D. dissertation 1979, The Ohio State University.
11. P. Young, Affine triple systems and matroid designs, Mathematische Zeitschrift, 132(1973), 343-366.

TABLES OF TWO-GRAPHS

F. C. BUSSEMAKER
R. A. MATHON
J. J. SEIDEL
University of Technology
Eindhoven

1. INTRODUCTION

The present paper contains the most important tables from the report [*].

The first objective is to determine all nonisomorphic two-graphs on $n \leq 9$ vertices, and all nonisomorphic two-graphs on $n \leq 10$ vertices whose full automorphism group does not fix any graph in its switching class ($\gamma \neq 0$). The numbers of these two-graphs are as follows :

n	4	5	6	7	8	9	10
$\gamma = \beta = 0$	3	7	14	54	224	2038	32728
$\gamma \neq 0,\ \beta = 0$	0	0	2	0	17	0	392
$\gamma \neq 0,\ \beta \neq 0$	0	0	0	0	2	0	0
total	3	7	16	54	243	2038	33120

Here γ and β denote the first and second cohomology invariant in the sense of Cameron [6].

The second objective is to collect information about the known regular two-graphs on $n \leq 50$ vertices.

Their numbers N(n) are as follows :

n	6	10	14	16	18	26	28	30	36	38	42	46	50
ρ_1	$\sqrt{5}$	3	$\sqrt{13}$	3	$\sqrt{17}$	5	3	$\sqrt{29}$	5	$\sqrt{37}$	$\sqrt{41}$	$\sqrt{45}$	7
$-\rho_2$	$\sqrt{5}$	3	$\sqrt{13}$	5	$\sqrt{17}$	5	9	$\sqrt{29}$	7	$\sqrt{37}$	$\sqrt{41}$	$\sqrt{45}$	7
N(n)	$\bar{1}$	$\bar{1}$	$\bar{1}$	$\bar{1}$	$\bar{1}$	$\bar{4}$	$\bar{1}$	6	91	11	18	80	18

Here ρ_1 and ρ_2 denote the eigenvalues of the regular two-graphs. A bar denotes that N(n) is complete.

Certain strongly regular graphs are associated with a regular two-graph. As an illustration of cospectrality, and of triviality of automorphism groups, we quote from [*] the numbers of nonisomorphic strongly regular graphs associated with the 91 regular two-graphs of order 36 :

(n,ρ_0,ρ_1,ρ_2)	from S(2,3,15)	from $L_3(6)$	total	trivial Aut
(36,5,5,-7)	16111	337	16448	15417
(36,-7,5,-7)	57	48	105	28
(35,-2,5,-7)	1817	38	1853	1576

Two-graphs are introduced in Chapter 2. Following Cameron [6], the relations between two-graphs, switching classes and Euler graphs are explained in terms of vector spaces over $\mathbb{F}_2$, which leads to the enumeration formulae 2.3 and 2.4, and to the Mallows-Sloane equicardinality Theorem 2.2. We refer to [*] for tables of two-graphs of the orders 4,...,9. In Chapter 3 the automorphism group of a two-graph is compared and confronted with the automorphism groups of the graphs in the corresponding switching class. Thus the significance of the nonvanishing of the first invariant γ is explained. The two-graphs with $n = 6,8,10$ and $\gamma \neq 0$ are exposed in Table A.

The regular two-graphs on $n \leq 50$ vertices are distinguished into the regular tw-graphs with $\rho_1 \neq -\rho_2$, the conference two-graphs with integral, and those with irrational eigenvalues, which are discussed in Chapters 4,5,6, and tabulated in Tables B,C,E, respectively. The emphasis in Chapter 4 is on the regular two-graphs with $n = 36$; the results of [4] are extended in [*]; Table B is a subset. The data of Table D are used for the construction of the conference two-graphs with $n = 50$, the main subject in Chapter 5 and Table G. The known results [15] about the case $n = 26$ are also included. The methods of construction for the conference matrices with $n = 30,38,42,46$ are described in Chapter 6. They are based on partitioning of the adjacency matrix in various ways, and on filling the blocks by circulants on multi-circulants. The actual results are collected in Table E.

The theory of two-graphs is surveyed in [17] and [19]. Earlier tables are contained in [11],[15], [1], [22], [7], of which the tables of [*] constitute a self-contained extension.

2. TWO-GRAPHS AND SWITCHING CLASSES

A two-graph (Ω,Δ) is a pair consisting of a finite set Ω and a set Δ of 3-subsets of Ω, such that each 4-subset of Ω contains an even number of 3-subsets from Δ. The elements of Ω are the vertices, those of Δ are the triples of the two-graph (Ω,Δ). For any $\omega \in \Omega$ the set Δ is determined by its triples containing ω. Indeed, $\{\omega_1,\omega_2,\omega_3\} \in \Delta$ whenever an odd number of the remaining 3-subsets from $\{\omega,\omega_1,\omega_2,\omega_3$ belongs to Δ. The complement of the two-graph (Ω,Δ) is the two-graph $(\Omega,\overline{\Delta})$, where $\overline{\Delta}$ is the component of Δ in $\binom{\Omega}{3}$.

Any graph (Ω,E) gives rise to a two-graph as follows. Let Δ be the set of the 3-subsets of Ω which carry an odd number of edges from the edge set E. Then (Ω,Δ) is a two-graph. Indeed, for any graph on 4 vertices there is an even number of subgraphs on 3 vertices having an odd number of edges. Conversely, with any given two-graph (Ω,Δ) there is associated a class of graphs on Ω, each of which gives rise to (Ω,Δ) in the way described above. Indeed, for any disjoint partition

$$\Omega = \Omega_0 \cup \Omega_1 \cup \Omega_2, \quad \Omega_0 = \{\omega\},$$

such a graph is defined by the edges

$$\{\omega,\omega_1\}, \ \{\omega_1,\omega_1'\}, \ \{\omega_2,\omega_2'\}, \ \{\omega_1,\omega_2\},$$

for all $\omega_1,\omega_1' \in \Omega_1$ and $\omega_2,\omega_2' \in \Omega_2$ satisfying

$$\{\omega,\omega_1,\omega_1'\} \in \Delta, \ \{\omega,\omega_2,\omega_2'\} \in \Delta, \ \{\omega,\omega_1,\omega_2\} \notin \Delta .$$

The class of all such graphs is called the switching class associated with the two-graph (Ω,Δ). The graph in the switching class defined by $\Omega_0 = \{\omega\}$, $\Omega_1 = \phi$, $\Omega_2 = \Omega\setminus\{\omega\}$ is called the descendant with respect to ω; this graph has ω as an isolated vertex. Associated with the void two-graph (Ω,ϕ) is the switching class consisting of all bipartite graphs on Ω. Clearly, the superposition mod 2 of any graphs (Ω,E) and (Ω,F) in the switching class of any two-graph (Ω,Δ) is a complete bipartite graph on Ω.

Essentially, in the above the set Ω is labeled. The labeled graphs on Ω are represented by the elements of the vector space V over $\mathbb{F}_2$ with standard basis $\binom{\Omega}{2}$. The complete bipartite graphs on Ω constitute a subspace B of dimension $|\Omega| - 1$.

The switching classes on Ω are the elements of the quotient space V/B. The Euler graphs on Ω (the graphs having even valency at each vertex) constitute another subspace Z, whose intersection with B has dimension 0 for odd $|\Omega|$, and dimension $|\Omega| - 2$ for even $|\Omega|$. The standard inner product of any pair of graphs equals the number of their common edges mod 2. Under this inner product the subspaces B and Z are maximally orthogonal. Hence, as a consequence of a theorem from linear algebra (cf. [10], p. 91), the dual vector space Z^* is isomorphic to V/B.

The labeled two-graphs are represented as elements of the vector space over $\mathbb{F}_2$ with standard basis $\binom{\Omega}{3}$. The defining property expresses that these elements have vanishing coboundary, cf. [21], [6], [19]. Hence the two-graphs constitute the subspace T of the cocycles. The correspondence between switching classes and two-graphs, mentioned earlier, is now expressed by the isomorphism of V/B and T, a consequence of the triviality of the cohomology. Summarizing we have the following theorem.

Theorem 2.1. $Z^* \simeq V/B \simeq T$, where V,B,Z,T are the $\mathbb{F}_2$-vector spaces of the labeled graphs, complete bipartite graphs, Euler graphs, and two-graphs on Ω, respectively.

By letting the symmetric group of Ω act on these vector spaces, we obtain the theorem of Mallows and Sloane [12], in the proof setting of Cameron [6].

Theorem 2.2. The isomorphism classes of the Euler graphs, the two-graphs, and the switching classes on n vertices are equal in number.

The following theorems yield explicit formulae, which are due to Robinson [16] and to Goethals [8], cf. Cameron [6]. Any permutation σ of Ω extends to a linear transformation σ of V. The elements of V which are fixed under σ constitute the kernel of the linear transformation $\sigma + 1$. Since $\sigma + 1$ leaves B invariant, it acts on V/B. Let $2^{v(\sigma)}$, $2^{b(\sigma)}$, $2^{t(\sigma)}$, and $2^{c(\sigma)}$ denote the number of fixed elements under σ in its action on V, B, V/B, and any coset C of B, respectively. Each coset of B fixed by σ contains a fixed vector [12] and hence exactly $2^{b(\sigma)}$ fixed vectors. It follows that

$$b(\sigma) = c(\sigma),\ t(\sigma) = v(\sigma) - b(\sigma).$$

Applying Burnside's lemma we obtain :

<u>Theorem 2.3</u>. The number of nonisomorphic two-graphs on n vertices equals

$$\frac{1}{n!} \sum_{\sigma \in \Sigma_n} 2^{v(\sigma)-b(\sigma)} .$$

<u>Theorem 2.4</u>. The number of nonisomorphic graphs in the switching class of a two-graph with automorphism group G equals

$$\frac{1}{|G|} \sum_{\sigma \in G} 2^{b(\sigma)} .$$

The interpretation of $v(\sigma)$ and $b(\sigma)$ in terms of the action of the permutation σ on the complete graph on Ω is the following, cf. [6]. $v(\sigma)$ is the number of edge-cycles, and $b(\sigma)$ is the number of vertex-cycles if every cycle has even length and one less otherwise.

As a consequence of the isomorphism of V/B and T we can provide any two graph with a spectrum of eigenvalues and multiplicities. For any labeled graph (Ω,E), let A denote its $(-1,1)$ adjacency matrix, defined by the entries $a_{ii} = 0$; $a_{ij} = -1$ for $\{i,j\} \in E$; $a_{ij} = 1$ for $\{i,j\} \notin E$. The switching class of (Ω,E) is represented by the adjacency matrices $D A D$, where D runs through the diagonal matrices of order n with diagonal entries ± 1. Clearly, A and $D A D$ have the same spectrum. The <u>spectrum of a two-graph</u> is defined to be the spectrum of the $(-1,1)$ adjacency matrix of any graph in its switching class.

This leads to the following geometric interpretation of two-graphs and their switching classes of graphs. Let the $(-1,1)$-adjacency matrix A have smallest eigenvalue $-\rho$, with multiplicity $n - d$. Write $\cos \varphi = 1/\rho$, $0 \leq \varphi \leq \pi/2$, and

$$\rho I + A = HH^t,$$

where H is an $n \times d$ matrix of rank d. The rows of H denote n vectors in $\mathbb{R}^d$ of equal length $\rho^{1/2}$ and at angles φ and $\pi - \varphi$, according as the corresponding vertices of the graph are nonadjacent and adjacent, respectively. The n lines spanned by these vectors are equiangular, that is, the angle between each pair of lines equals φ. Thus, a two-graph is represented by a set of <u>equiangular lines</u> in Euclidean space, and any graph in the corresponding switching class is represented by a set of unit vectors, one along each line. Conversely, any dependent set of equiangular lines defines a unique two-graph, cf. [17], Theorem 5.4. These notions are to be distinguished from the set of all unit vectors along equiangular lines, which corresponds to a double covering of the complete graph K_n. In a double covering of K_n each vertex is replaced by a pair of vertices, and two pairs are joined either direct or skew, cf. [6], [19]. The correspondence is established by letting vectors at acute (oblique) angles correspond to vertices which are joined direct (skew).

In [*] the isomorphism classes of the two-graphs on $n \leq 9$ vertices are exposed (those on $n \leq 7$ already occur in [11]). From these tables we read the following numbers of nonisomorphic two-graphs of order n, selfcomplementary two-graphs, two-graphs with integral spectrum, two-graphs having a transitive, and a trivial automorphism group, the numbers of cospectral pairs, triples, and quadruples of two-graphs :

n	4	5	6	7	8	9
two-graphs	3	7	16	54	243	2038
self-compl.	1	1	4	0	19	10
spec $\varepsilon\ \mathbb{Z}$	2	2	4	8	9	16
tra Aut	2	3	5	4	6	9
tri Aut	0	0	0	0	8	264
cosp. pairs	0	0	0	0	6	160
cosp. triples	0	0	0	0	1	22
cosp. quadr.	0	0	0	0	0	2

3. AUTOMORPHISM GROUPS OF TWO-GRAPHS

An automorphism of the two-graph (Ω,Δ) is a permutation of Ω which preserves Δ, that is, which preserves the corresponding switching class C of graphs. Clearly, the full automorphism group Aut(c) of any graph c in the switching class C is a subgroup of the full automorphism group Aut(C) of the switching class C, so

$$\mathrm{Aut}(c) < \mathrm{Aut}(C), \quad \text{for all} \quad c \in C.$$

In terms of the $\mathbb{F}_2$ vector space V of all graphs this reads as follows. Any permutation σ of Ω induces a linear transformation σ of V which leaves invariant the space B of all complete bipartite graphs. An automorphism σ of the switching class C is characterized by the property

$$(\sigma + 1)c \in B, \text{ for all } c \in C,$$

and an automorphsim σ of the graph c by

$$(\sigma + 1)c = 0.$$

We apply Lagrange's theorem to $Aut(c) < Aut(C)$ in two different situations.

First, for any $\omega \in \Omega$, let c_ω denote the descendant with respect to ω, that is, the graph in the switching class C which has ω as an isolated vertex. The orbit of c_ω under $Aut(C)$ is determined by the orbit of ω under $Aut(C)$.

<u>Theorem 3.1</u>. $|Aut(C)| = |Aut(c_\omega)| \cdot |\text{orbit of } \omega \in \Omega \text{ under } Aut(C)|$.

Second, for any graph c in the switching class $C = c + B$, and any $\sigma \in Aut(C)$, we have $\sigma(c) = c + b_\sigma$ for some $b_\sigma \in B$. The orbit of c under $Aut(C)$ is determined by the orbit of $\sigma \in B$ under $Aut(C)$, that is, by the orbit on B under $Aut(C)$ as an affine transformation of B.

<u>Theorem 3.2</u>. $|Aut(C)| = |Aut(c)| \cdot |\text{orbit of } \sigma \text{ on } B \text{ under } Aut(C)|$. As a consequence we have, cf. [17] Theorem 4.7 :

<u>Theorem 3.3</u>. Let $c_1,\ldots,c_s$ denote the nonisomorphic graphs in the switching class C. Then

$$\sum_{i=1}^{s} \frac{|Aut(C)|}{|Aut(c_i)|} = 2^{n-1}.$$

We now approach the key problem of the present chapter. Mallows and Sloane ([12], cf. [6] Theorem 3.4) proved that any automorphism of a switching class is an automorphism of some graph in the switching class, that is,

$$\forall_{\sigma\varepsilon Aut(C)}\ \exists_{c\varepsilon C}((\sigma + 1)c = 0).$$

However, the following property (*) is <u>not</u> <u>generally</u> true for any switching class C and any $G < Aut(C)$:

(*) $$\exists_{c\varepsilon C}\ \forall_{\sigma\varepsilon G}((\sigma + 1)c = 0).$$

If this property holds for $G = Aut(C)$, then $Aut(C) = Aut(c)$ for some $c \in C$. However, there exist switching C whose $Aut(C)$ is larger than $Aut(c)$ for each $c \in C$. In fact this makes switching classes and two-graphs interesting, and this is why they have been invented. They provide combinatorial structures whose Aut may go beyond the Aut of their graphs; they even may have 2-transitive automorphism groups. It is the aim of the present chapter to determine all two-graphs on $n \leq 10$ vertices whose full automorphism group does not satisfy the property (*) mentioned above.

Cameron [6] phrased the validity of property (*) for a switching class C and $G < \mathrm{Aut}(C)$ in terms of the one-dimensional cohomology group $H^1(G,B)$. Indeed, for any $c + b \in C$ the derivation $d : G \to B$, defined by

$$d(\sigma) = (\sigma + 1)(c + b), \quad \sigma \in G,$$

reduces to an inner derivation

$$d(\sigma) = (\sigma + 1)b, \quad \sigma \in G,$$

whenever c satisfies $(\sigma + 1)c = 0$ for all $\sigma \in G$. This implies that the switching class C defines an element $\gamma \in H^1(G,C)$, called the <u>first invariant</u> of C and G, and $\gamma = 0$ if and only if property (*) holds.

Cameron [6] also introduces the <u>second invariant</u> β of C and G, an element of the two-dimensional cohomology group $H^2(G,\mathbb{Z}_2)$. He shows that $\beta = 0$ if and only if G can be realized as a group of automorphisms of the double covering associated with C. We shall not enter further details, since for $n \leq 10$ there exists only one pair of complementary two-graphs with $\beta \neq 0$. The switching class of one of these contains the octagon graph. For details we refer to [19].

In our search for all two-graphs on $n \leq 10$ vertices with $\gamma \neq 0$, and those with $\beta \neq 0$, the following necessary conditions, thaken from [6], will prove useful.

<u>Theorem 3.4.</u> Let C be a switching class of graphs on n vertices, and let $G < \mathrm{Aut}(C)$. Suppose $\gamma \neq 0$, that is, suppose there is no graph $c \in C$ with $G < \mathrm{Aut}(c)$. Then

(a) $n \equiv 0 \pmod 2$,

(b) $|G| \equiv 0 \pmod 4$,

(c) all orbits of G on Ω have even size.

<u>Theorem 3.5.</u> Let C be a switching class of graphs on n vertices, let $G < \mathrm{Aut}(C)$, and let $\beta \neq 0$. Then

(a) $\gamma \neq 0$,

(b) $n \equiv 0 \pmod 8$,

(c) the largest and the smallest eigenvalue of C have even multiplicity.

In Table A the two-graphs with $\gamma \neq 0$, $n \leq 10$ are isplayed. Each two-graph is given by the $\frac{1}{2}(n-1)(n-2)$ relevant entries of the adjacency matrix of a graph in its switching class which contains an isolated vertex. These entries are listed as the coordinates of a vector which is the concatenation of the rows of the right upper part of the adjacency matrix. The self-complementary two-graphs are indicated by the letter s, and from the remaining two-graphs only one of (Ω,Δ) and $(\Omega,\bar{\Delta})$ is listed. Those with $n = 4,6,8$ were selected from [*] on the basis of $|\mathrm{Aut}(c)| < |\mathrm{Aut}(C)|$ for $\gamma \neq 0$, and Theorem 3.5(c) for $\beta \neq 0$.

The two-graphs with $\gamma \neq 0$, $n = 10$ were generated by a backtracking search. The underlying idea for an efficient search is the experimental fact that the switching classes contain only few graphs having a minimum number of edges, and that these graphs have relatively large automorphism groups. In fact, most often such a graph c has $|\mathrm{Aut}(c)| = |\mathrm{Aut}(C)|$ and hence is rejected. Together with the criteria of Theorem 3.4, this reduces the number of candidates for the switching classes with $\gamma \neq 0$ by a factor of about 50. There are 471 valency sequences (all vertex-degrees ≤ 4), yielding 24423 graphs. From these, only 271 satisfy the necessary conditions. The latter are put into canonical form and are tested for isomorphism. This leads to 206 non-isomorphic two-graphs with $\gamma \neq 0$, up to taking complements. Since 20 of them are self-complementary, we finally arrive at the total of 392 two-graphs with $\gamma \neq 0$, $n = 10$.

Summarizing we find the following numbers of nonisomorphic two-graphs on $n \leq 10$ vertices having the indicated properties for the invariants γ and β.

n	4	5	6	7	8	9	10
$\gamma = \beta = 0$	3	7	14	54	224	2038	32728
$\gamma \neq 0,\ \beta = 0$	0	0	2	0	17	0	392
$\gamma \neq 0,\ \beta \neq 0$	0	0	0	0	2	0	0
total	3	7	16	54	243	2038	33120

For $n = 6$ there are 2 two-graphs with $\gamma \neq 0$, both self-complementary. One is represented by the switching class of the pentagon graph with an additional isolated vertex and has the alternating group on 5 symbols as its full automorphism group. The othe is represented by the path of length 4 with an additional isolated vertex and has the Klein group as its full automorphism group.

For $n = 8$ there are 2 complementary two-graphs with $\beta \neq 0$, listed under identification number # 6. One is represented by the switching class of the octagon graph and has the holomorph of Z_8 as its full automorphism group, cf. [19]. The 17 two-graphs with $\gamma \neq 0$, $\beta = 0$ fall into 8 complementary pairs and one self-complementary two-graph. The two-graphs # 1, # 2, # 10 have the same |Aut|, number of graphs, and numbers [x,y] = (x graphs with |Aut| = y), and so do the two-graphs # 3, # 8, # 9.

For $n = 10$, $\gamma \neq 0$, the 206 two-graphs listed in Table A are ordered following their eigenvalues; this shows that there are many cospectral pairs, and even 2 cospectral triples, namely # 27, # 28, # 29 and # 151, # 152, # 153. Those with integral spectrum are # # 10,11,191,193. The two-graph # 191, represented by the Petersen graph, has the largest automorphism group, namely $Sp(4,2) \simeq \Sigma_6$ of order 720. As above for $n = 8$ the 206 two-graphs of Table A may be grouped into sets of two-graphs having the same |Aut|, number of graphs, and numbers [x,y] = (x graphs with |Aut| = y). For this information we refer to [*].

4. REGULAR TWO-GRAPHS WITH $\rho_1 + \rho_2 \neq 0$

We give two equivalent definitions for regular two-graphs.

A two-graph (Ω,Δ) is <u>regular</u> whenever each pair of elements of Ω is contained in the same number of triples of Δ.

A two-graph is <u>regular</u> whenever it has only two eigenvalues, that is, whenever its switching class consists of graphs whose (-1,1) adjacency matrix A satisfies

$$(A - \rho_1 I)(A - \rho_2 I) = 0, \quad \rho_1 > \rho_2 \quad \text{say.}$$

The equivalence of these definitions is seen as follows, cf. [17]. For any adjacent vertices x and y of a graph, let p(x,y) denote the number of vertices which are adjacent to x and nonadjacent to y. For any nonadjacent vertices u and v, let q(u,v) denote the number of vertices which are adjacent to u and nonadjacent to v. For any graph in the switching class of a regular two-graph, either definition amounts to the independence of p(x,y) + p(y,x) and q(u,v) + q(v,u) of the choice of {x,y} and {u,v}, respectively, and

$$n = 1 - \rho_1\rho_2,$$

$$p(x,y) + p(y,x) = -\tfrac{1}{2}(\rho_1-1)(\rho_2-1)$$

$$q(u,v) + q(v,u) = -\tfrac{1}{2}(\rho_1+1)(\rho_2+1).$$

Also a <u>strongly regular graph</u> admits two equivalent definitions, cf. [20]. It is a regular graph for which $p(x,y) = p(y,x)$ and $q(u,v) = q(v,u)$ are independent of the choice of $\{x,y\}$ and $\{u,v\}$. Equivalently, its (-1,1) adjacency matrix C of size n satisfies

$$(C - \sigma_1 I)(C - \sigma_2 I) = (n - 1 + \sigma_1\sigma_2)J, \quad C_j = \sigma_0 j,$$

with $\sigma_1 > \sigma_2$, say. The numbers $(n,\sigma_0,\sigma_1,\sigma_2)$ are the <u>parameters</u> of the strongly regular graph. They satisfy

$$(\sigma_0 - \sigma_1)(\sigma_0 - \sigma_2) = n(n - 1 + \sigma_1\sigma_2).$$

The switching class of a regular two-graph may contain strongly regular graphs. Their (-1,1) adjacency matrix A should satisfy

$$(A - \rho_1 I)(A - \rho_2 I) = 0, \text{ and } A_j = \rho_1 j \text{ or } A_j = \rho_2 j.$$

The descendent of any vertex of a regular two-graph on n vertices yields a strongly regular graph on $n - 1$ vertices, after removal of the isolated vertex. Its (-1,1) adjacency matrix B satisfies

$$(B - \rho_1 I)(B - \rho_2 I) = -J, \quad B_j = (\rho_1 + \rho_2)j.$$

The eigenvalues of regular two-graphs are subject to restrictions. From $trA = 0$ and $trA^2 = n - 1$ it follows that ρ_1 and ρ_2 are odd integers if $\rho_1 + \rho_2 \neq 0$, cf. [17]. In the present chapter we investigate nontrivial regular two-graphs with $n \leq 50$ and eigenvalues satisfying $\rho_1 + \rho_2 \neq 0$. Up to taking complements there are 3 possibilities, namely, $n = 16$, $\rho_1 = 3$, $\rho_2 = -5$; $n = 28$, $\rho_1 = 3$, $\rho_2 = -9$; $n = 36$, $\rho_1 = 5$, $\rho_2 = -7$.

For $n = 16$ there is only one regular two-graph, namely the complement of the symplectic two-graph $\Sigma(4,2)$, cf. [17]. Its switching class contains strongly regular graphs with valencies 6 and 10. Valency 6 is realized by the lattice graph $L_2(4)$ and by the exceptional Shrikhande graph. Valency 10 is uniquely realized by the Clebsch graph. All descendants yield the triangular graph $T(6)$, after removal of the isolated vertex.

For $n = 28$ there is only one regular two-graph, namely the complement of the orthogonal two-graph $\Omega^{-}(6,2)$, whose automorphism group is isomorphic to $Sp(6,2)$. Its switching class may contain strongly regular graphs with valencies 12 and 18. Valency 12 is realized by the triangular graph $T(8)$ and by the 3 exceptional Chang graphs. Valency 18 is impossible, cf. [20]. All descendants yield the Schläfli graph on 27 vertices, after removal of the isolated vertex.

For $n = 36$ a total of 91 nonisomorphic regular two-graphs is known, cf. [4]. We conjecture that no further such two-graphs exist. The possible parameter sets for strongly regular graphs in the switching class are (36,5,5,-7) and (36,-7,5,-7), with valencies 15 and 21, respectively. From [*] it follows that the first set is realized by 16448 graphs, and that the second set is realized by 105 graphs. By deleting an isolated vertex we obtain 1853 strongly regular graphs with parameters (35,-2,5,-7). We first discuss the generation and the analysis of the known regular two-graphs on 36 vertices in more detail.

A Steiner system $S(2,k,v)$ is a collection of k-subsets (called blocks) of a -set V such that every 2-subset is contained in exactly one k-subset. From $S(2,k,v)$ he Steiner graph $St_k(v)$ is obtained as follows. The vertices are the $b = v(v-1)/k(k-1)$ locks, and two vertices are adjacent whenever the corresponding blocks have one element n common. In particular, $St_k(2k^2-k)$ extended by an isolated vertex belongs to the witching class of a regular two-graph with parameters

$$n = 4k^2,\ \rho_1 = 2k - 1,\ \rho_2 = -2k - 1.$$

transversal design $TD(k,v)$ is a collection of k-vectors with coordinates in a v-set such that for each pair of coordinates every pair of elements occurs exactly once. $TD(k,v)$ is equivalent to a set of $k - 2$ mutually orthogonal Latin squares of size . From $TD(k,v)$ the Latin square graph $L_k(v)$ is obtained as follows. The vertices re the v^2 vectors, and two vertices are adjacent whenever the corresponding vectors gree in one coordinate. In particular, $L_k(2k)$ belongs to the switching class of a egular two-graph with parameters

$$n = 4k^2, \rho_1 = 2k - 1, \rho_2 = -2k - 1.$$

The known regular two-graphs on 36 vertices are constructed either from Steiner triple systems $S(2,3,15)$ or from Latin squares $TD(3,6)$ of order 6. There are 80 nonisomorphic $S(2,3,15)$ and 12 nonisomorphic $TD(3,6)$. These yield $80 + 11 = 91$ nonisomorphic regular two-graphs with $n = 36$, $\rho_1 = 5$, $\rho_2 = -7$, of <u>Steiner type</u> and <u>Latin square type</u>, respectively. Indeed, in [4] it was shown that only two of the Latin square graphs are switching equivalent.

In order to count the strongly regular graphs with parameters (35,-2,5,-7) we use the following consequence of Theorem 3.1.

<u>Theorem 4.1.</u> Let the action of $Aut(\Omega,\Delta)$ on Ω have the orbits $\Omega_1,\ldots,\Omega_s$. The descendants c_ω and $c_{\omega'}$ are isomorphic iff the corresponding vertices ω and ω' are in the same orbit, and

$$|Aut(\Omega,\Delta)| = |Aut(c_\omega)| \cdot |\Omega_i|, \quad \text{if} \quad \omega \in \Omega_i.$$

In order to count the strongly regular graphs with parameters (36,5,5,-7) and (36,-7, 5,-7) we use the following consequence of Theorem 3.2. Let $\rho \in \{\rho_1,\rho_2\}$ be one of the eigenvalues of the regular two-graph (Ω,Δ). Let c, with adjacency matrix A, be any graph in the switching class of (Ω,Δ). We wish to switch c into a regular graph with $\rho_0 = \rho$. This amounts to a search for egenvectors d, with entires ± 1, to the eigenvalue ρ of A. Indeed,

$$(A - \rho I)d = 0 \quad \text{iff} \quad DADj = cj, \; D = diag(d),$$

and DAD is the adjacency matrix of a regular graph, to be denoted by c_d. Let $\underline{\underline{D}}_\rho$ denote the set of all such eigenvectors d.

<u>Theorem 4.2.</u> Let the action of $Aut(\Omega,\Delta)$ on $\underline{\underline{D}}_\rho$ have the orbits $\underline{\underline{D}}_1,\ldots,\underline{\underline{D}}_t$. The strongly regular graphs c_d and $c_{d'}$ are isomorphic iff the corresponding eigenvectors d and d' are in the same orbit, and

$$|Aut(\Omega,\Delta)| = |Aut(c_d)| \cdot |\underline{\underline{D}}_i|, \quad \text{if} \quad d \in \underline{\underline{D}}_i.$$

For a given matrix A and eigenvector ρ, we construct the set $\underline{\underline{D}}$ by using the backtracking procedure Eigenvector of Paulus [15].

In [*] all information about the regular two-graphs with $n = 16,28,36$, and their strongly regular graphs is collected. The present Table B only contains the transitive strongly regular graphs. We use octal representation, that is, we convert binary vectors into octal numbers (for instance, the vector -++|+-+|+-- into the octal number 423).

In Table B, the numbering of the two-graphs with $n = 36$ is the same as in [4], # 1 through # 80 being of Steiner type, and # 81 through # 92 being of Latin square type (# 87 does not occur since the two-graphs # 87 and # 89 are isomorphic; we choose # 89 since the corresponding Latin square graph has an automorphism group which is 3 times larger than that of # 87). We mention in passing that the Steiner graphs and the Steiner triple systems agree in the order of their automorphism groups, except for # 1 where a factor 2 occurs.

We make some observations concerning [4] and [*]. The automorphism groups of the two-graphs ## 1,89,90,92 act transitively on the vertex set Ω. The groups of the two-graphs ## 8 through 80 and 81 through 92 are the same as the groups of the corresponding Steiner and Latin square graphs, hence these two-graphs all have $\gamma = 0$. For ## 1 through 7 these groups differ by a factor 36,8,2,2,2,2,2, respectively. This makes it conceivable that the two-graphs ## 1 through 7 have $\gamma \neq 0$.

All Steiner graphs can be switched regular with valency 15, but only 23 of them can be switched regular with valency 21 (namely ## 1 through 22 and # 61). All Latin square graphs can be switched both 15-regular and 21-regular. A total of 36 (out of 91) two-graphs has trivial automorphism group. For the strongly regular graphs (36,5,5,-7), (36,-7,5,-7), (35,-2,5,-7) the totals having trivial automorphism group are 15417 (out of 16448), 28 (out of 105), 1576 (out of 1853), respectively.

In Table B all transitive strongly regular graphs are listed which are associated with the regular two-graphs under consideration. The 2-transitive regular two-graph $\Omega^+(6,2)$, # 1 in Table B, contains three rank 3 graphs in its switching class, cf. [4], namely (36,5,5,-7) with group $O^-(6,2)$,(36,-7,5,-7) with group $G(2,2)$, and (35,-2,5,-7) with group $L(4,2)$. The two-graphs ## 89,90,92 of Latin square type yield transitive strongly regular graphs with parameters (36,5,5,-7) and (36,-7,5,-7); these are not rank 3 graphs.

5. CONFERENCE TWO-GRAPHS WITH INTEGRAL EIGENVALUES

A conference two-graph is a reular two-graph on n vertices having eigenvalues $\rho_1 = \sqrt{n-1}$, $\rho_2 = -\sqrt{n-1}$. With each conference two-graph there is associated a switching class of symmetric conference matrices. A conference matrix A of order n is an $n \times n$ marrix A, having zero's on the diagonal and entries ± 1 elsewhere, and satisfying $AA^t = (n-1)I$. Conference matrices have been introduced by Belevitch [2] in connection with the construction of networks for conference telephony. For the existence of a conference two-graph on n vertices it is necessary that $n \equiv 2 \pmod 4$ and that $n-1$ is a sum of square of two integers, cf. [11].

In the present chapter we shall discuss conference two-graphs with parameters

$$n = (2k-1)^2 + 1,\ \rho_1 = -\rho_2 = 2k-1,\quad k \geq 2.$$

Such two-graphs can be constructed from Steiner systems and from transversal designs. It is easily checked that both $St_k(2k^2-k+1)$ and $L_k(2k-1) \cup \{\omega\}$ where ω is an isolated vertex, belong to the switching calss of a conference two-graph of the order $n = (2k-1)^2 + 1$, cf. [17] and Chapter 4. Furthermore, the complement of any conference two-graph again is a conference two-graph.

We recall two further constructions for conference two-graphs, the first of which (to be called the PN construction) is due to Goethals and Delsarte and to Turyan (unpublished, cf. [17]). Let $V = V(2,q)$ denote the vector space of dimension 2 over $GF(q)$, where q is an odd prime power. Partition the one-dimensional subspaces into any two disjoint sets P and N of equal size $\frac{1}{2}(q+1)$. Then a self-complementary conference two-graph $(V \cup \{\infty\},\Delta)$ on q^2+1 vertices is defined by

$$\forall_{x,y \in V}((\{\infty,x,y\} \in \Delta) :\Longleftrightarrow (\langle x-y \rangle \in P)).$$

Let $D : V \times V \to GF(q)$ be a nondegenerate bilinear form on V, and let $q \equiv 1 \pmod 4$. The Paley two-graph (Ω,Δ) with parameters

$$n = q + 1,\quad \rho_1 = -\rho_2 = \sqrt{q}$$

is defined as follows. Ω is the set of the projective points of $PG(1,q)$, and Δ is the set of all triples $\{x,y,z\}$ from Ω for which $D(x,y)D(y,z)D(z,x)$ is a non-square in $GF(q)$. The Paley two-graph is self-complementary and admits $PSL(2,q)$ as a 2-transitive automorphism group.

There are 3 possibilities for conference two-graphs with integral eigenvalues on $n \leq 50$ vertices, namely, $n = 10$, $\rho_1 = -\rho_2 = 3$; $n = 26$, $\rho_1 = -\rho_2 = 5$; $n = 50$, $\rho_1 = -\rho_2 = 7$.

For $n = 10$ there is only one conference two-graph, namely the orthogonal two-graph $\Omega^+(4,2)$, whose automorphism group is $Sp(4,2)$. Its switching class contains the Petersen graph and its complement $St_2(5) = T(5)$. Deleting any isolated vertex we obtain $L_2(3)$.

For $n = 26$ there are exactly 4 conference two-graphs, cf. Weisfeiler [22]. Two of these are constructed from Latin squares of order 5, and two from Steiner triple systems of order 13. They have been investigated in [15], [1], [22], [7]. They are displayed in Table C. All four are self-complementary. They admit 15600,72,39,6 automorphisms, respectively (the first case corresponds to the Paley two-graph). The 4 switching classes contain 10 strongly regular graphs with parameters (26,-5,5,-5), one of which has a trivial automorphism group. Deleting an isolated vertex we obtain 15 strongly regular graphs (25,0,5,-5), two of which have a trivial group.

For $n = 50$ we have constructed a total of 18 conference two-graphs. They arise from the 7 nonequivalent Graeco-Latin squares of order 7 (Norton [14]) and from the 4 known Steiner systems $S(2,4,25)$ (Brouwer [3]). These are displayed in Table D. Notice that one of Norton's squares has been corrected. Following Brouwer there are no further $S(2,4,25)$ with $|Aut| \geq 5$. Table C contains the pairwise complementary two-graphs ## 3,4; 6,7; 8,9; 10,11, and the self-complementary # 5, # 1 (the Paley two-graph), # 2 (the other PN two-graph) arising from the Graeco-Latin squares, and the pairwise complementary ## 12,13; 15,16; 17,18, and the self-complementary # 14 arising from $S(2,4,25)$. There are 89 strongly regular graphs (49,0,7,-7), derived from the 18 conference two-graphs by deleting an isolated vertex. 36 of these have a trivial automorphism group. Due to difficulties with computer time, we have only partial results concerning the strongly regular graphs (50,-7,7,-7) in the 18 switching classes. Each switching class contains at least one strongly regular graph with these parameters.

6. CONFERENCE TWO-GRAPHS WITH IRRATIONAL EIGENVALUES

For conference two-graphs of order n, where $n-1$ is not an integral square, the following possibilities with $n \leq 50$ exist :

$$n = 6,14,30,38,42,46.$$

Notice, that the Paley two-graph exists for all these orders, except for $n = 46$. However, many further conference two-graphs of these orders have been constructed. Table E lists the known ones and their descendants $(n-1,\ 0,\ \sqrt{n-1},\ -\sqrt{n-1})$. For irrational eigenvalues there are no strongly regular graphs on n vertices in the switching class.

For $n = 6$ there is only one conference two-graph, namely $\Omega^-(4,2)$ with automorphism group $Sp(4,2) \simeq A_5$. For $n = 14$ and $n = 18$ only the Paley two-graph exists [18]. For $n = 30$ there are 6 nonisomorphic conference two-graphs known, and 41 descendants. This follows from the incomplete backtracking search in [1]; we have found no further two-graphs. For $n = 38$ we found 11 conference two-graphs and 82 descendants, and for $n = 42$ we found 18 conference two-graphs and 120 descendants. For $n = 38$ there are nonisomorphic pairs of complemented two-graphs; for $n = 42$ there is a non-Paley transitive two-graph. For $n = 46$, there have been constructed 80 conference two-graphs, yielding 1856 descendants (1344 of which have trivial automorphism group). It is interesting to observe that all known conference two-graphs of order 46 have small automorphism groups, $|\mathrm{Aut}| \leq 10$, and that none of them is self-complementary.

Our construction method for $n = 30,38,42$ is based on the assumption that an essential part of the adjacency matrix A of the conference two-graph can be written as an $n_1 \times n_1$ block matrix $[A_{ij}]$, where each block A_{ij} is a regular $n_2 \times n_2$ matrix with constant row and column sums r_{ij}. We consider such block-regular partitions for the matrices A_0, A_1, A_2 defined by

$$A = A_0 = \begin{bmatrix} 0 & j^t \\ j & A_1 \end{bmatrix} = \begin{bmatrix} 0 & 1 & j^t & j^t \\ 1 & 0 & j^t & -j^t \\ j & j & & \\ j & -j & & A_2 \end{bmatrix},$$

and denote them by type 0,1,2, respectively. Since $A^2 = (n-1)I$, the block-valencies r_{ij} satisfy certain conditions. For example, for a type 0 matrix we have

$$4(t_{ii}+r_{ii}) = (n_2-1)(4s_i-n+2) + 4s_i,\ 1 \le i \le n_1,$$

$$4(t_{ik}+r_{ik}) = n_2(2s_i + 2s_k - n + 2),\quad 1 \le i < k \le n_1,$$

where $0 \le r_{ii} \le n_2 - 1$, $0 \le r_{ik} \le n_2$ and

$$s_i = \sum_{j=1}^{n_1} r_{ij},\quad t_{ik} = \sum_{j=1}^{n_1} r_{ij}r_{jk}.$$

Similar relations hold for matrices of type 1 and 2. The search for conference two-graphs with $n = 30,38,42$ proceeds in two stages. In the first stage we generate block-valency matrices which is non-trivial cases are of a form exemplified by

a	b	c	c	c	c	d	e	f
	a	c	c	c	c	e	d	f
		g	h	i	h	j	j	k
			g	h	i	j	j	k
				g	h	j	j	k
					g	j	j	k
						ℓ	m	n
							ℓ	n
								0

to be denoted by (a,b)(c,c,c,c)(d,e)(f)(g,h,i,h)(j,j)(k)(ℓ,m)(n)(o). (the square super-blocks are circulants, and the nonsquare superblocks are constants). In the second stage we attempt to fill each block A_{ij} by a circulant matrix of the appropriate order and valency, by use of a systematic backtracking procedure.

For $n = 46$ we follow the constructions of [13]. We consider graphs in the switching class whose A is of type 1, where A_1 is partitioned into 5×5 regular blocks of size 9. Define

$$D = \begin{bmatrix} 0 & - & - \\ - & 0 & - \\ - & - & 0 \end{bmatrix},\ E = \begin{bmatrix} + & - & + \\ + & - & + \\ + & - & + \end{bmatrix},\quad F = \begin{bmatrix} - & + & + \\ + & - & + \\ + & + & - \end{bmatrix},\quad P = \begin{bmatrix} 0 & + & 0 \\ 0 & 0 & + \\ + & 0 & 0 \end{bmatrix},$$

$$B_0 = \begin{bmatrix} D & FP^2 & FP \\ FP & D & FP^2 \\ FP^2 & FP & D \end{bmatrix}, \quad B_1 = \begin{bmatrix} F & PF & FP^2 \\ FP^2 & F & FP \\ FP & FP^2 & F \end{bmatrix}, \quad B_2 = \begin{bmatrix} EP & EP & EP \\ EP^2 & EP^2 & EP^2 \\ E & E & E \end{bmatrix}.$$

We construct a basic set of 8 conference two-graphs by adding an isolated vertex to the graphs which are represented by the adjacency matrices

$$A_I = \begin{bmatrix} B_0 & B_1 & -B_2 & -B_2^T & B_1^T \\ B_1^T & B_0 & B_1 & -B_2 & -B_2^T \\ -B_2^T & B_1^T & B_0 & B_1 & -B_2 \\ -B_2 & -B_2^T & B_1^T & B_0 & B_1 \\ B_1 & -B_2 & -B_2^T & B_1^T & B_0 \end{bmatrix}, \quad A_{II} = \begin{bmatrix} B_0 & -B_1 & B_1 & B_2^T & -B_1^T \\ -B_1^T & B_0 & -B_1 & B_2 & B_2^T \\ B_2^T & -B_1^T & B_0 & -B_1 & B_2 \\ B_2 & B_2^T & -B_1^T & B_0 & -B_1 \\ -B_1 & B_2 & B_2^T & -B_1^T & B_0 \end{bmatrix},$$

$$A_{III} = \begin{bmatrix} B_0 & B_1 & -B_2 & B_2^T & -B_1^T \\ B_1^T & B_0 & B_1 & -B_2 & -B_2^T \\ -B_2^T & B_1^T & B_0 & -B_1 & B_2 \\ B_2 & -B_2^T & -B_1^T & B_0 & B_1 \\ -B_1 & -B_2 & B_2^T & B_1^T & B_0 \end{bmatrix}, \quad A_{IV} = \begin{bmatrix} B_0 & -B_1 & B_2 & -B_2^T & B_1^T \\ -B_1^T & B_0 & -B_1 & B_2 & B_2^T \\ B_2^T & -B_1^T & B_0 & B_1 & -B_2 \\ -B_2 & B_2^T & B_1^T & B_0 & -B_1 \\ B_1 & B_2 & -B_2^T & -B_1^T & B_0 \end{bmatrix},$$

and their complements. From these 8 basic conference two-graphs we obtain many more new two-graphs by permuting the diagonal blocks ([13], Theorem 4.2). Let Q denote the permutation matrix

$$Q = \begin{bmatrix} P & 0 & 0 \\ 0 & P^2 & 0 \\ 0 & 0 & I \end{bmatrix}.$$

Let $b^t = (b_1,\dots,b_5)$ denote the so called pointer vector, with integral components $b_i = \pm k$ iff

$$A_{ii} = \pm (Q^k)^t B_0 Q^k, \quad 1 \le i \le 5, \quad 1 \le k \le 3.$$

Hence each pointer vector b represents a permutation of the diagonal blocks of A

In [13], by an exhaustive search the admissable permutations are determined as follows

(1,-1,3,-2,3)	(1,-3,-1,-2,-2)	(1,-3,2,-3,-1)
(1,2,2,-1,3)	(2,-1,3,1,2)	(3,1,-1,3,-2)
(2,2,2,2,2)	(2,2,-1,3,1)	(3,1,2,2,-1)
(3,3,3,3,3)	(2,-3,-1,1,-3)	(3,-2,3,1,-1).

The first column of block permutations is applied to A_I, $-A_I$, A_{II}, $-A_{II}$, yielding 8 two-graphs with $|\mathrm{Aut}| = 10$ and 8 two-graphs with $|\mathrm{Aut}| = 2$. If all 12 block permutations are applied to $A_{III}, -A_{III}, A_{IV}, -A_{IV}$, then another 48 conference two-graphs are obtained, all with $|\mathrm{Aut}| = 2$. Many more two-graphs can be obtained by permuting off-diagonal blocks of the basic set as well. For example, putting

$$A_{12} = A_{21}^t = B_1Q, \quad A_{52} = A_{25}^t = -B_2Q^2 \quad \text{in } A_I, -A_{II}, A_{III}, -A_{IV},$$

$$A_{12} = A_{21}^t = -B_1Q, \quad A_{52} = A_{25}^t = B_2Q^2 \quad \text{in } -A_I, A_{II}, -A_{III}, A_{IV},$$

and applying the block permutations (3,3,3,3,3) to all matrices and (2,-3,-1,1,-3), (1,-3,2,-3,-1) to A_{III}, $-A_{III}$, A_{IV}, $-A_{IV}$, we obtain 16 new two-graphs with $|\mathrm{Aut}| = 3$. Table E only contains three regular two-graphs with $n = 46$, with $|\mathrm{Aut}| = 10,3,2$, respectively. For the other known two-graphs we refer to [*].

REFERENCES

1. V.L. Arlasarov, A.A. Lehman, M.S. Rosenfeld, The construction and analysis by a computer of the graphs on 25,26 and 29 vertices (in Russian), Instit. of Control Theory, Moscow, 1975.
2. V. Belevitch, Conference networks and Hadamard matrices, Ann. Soc. Sci. Bruxelles, Sér I 82(1968), 13-32.
3. A.E. Brouwer, private communication.
4. F.C. Bussemaker, J.J. Seidel, Symmetric Hadamard matrices of order 36, Ann. N.Y. Acad. Sci. 175(1970), 66-79; Report Techn. Univ. Eindhoven 70-WSK-02, 68,(1970).
5. P.J. Cameron, Automorphisms and cohomology of switching classes, J. Combin. Theory B 22(1977), 297-298.
6. P.J. Cameron, Cohomological aspects of two-graphs, Math. Zeitschr. 157(1977), 101-119.
7. D.G. Corneil, R.A. Mathon, Algorithmic techniques for the generation and analysis of strongly regular graphs and other combinatorial configurations, Ann. Discr. Math. 2(1978), 1-32.
8. J.M. Goethals, unpublished.
9. J.M. Goethals, J.J. Seidel, Orthogonal matrices with zero diagonal, Canad. J. Math., 19(1967), 1001-1010.
10. S. Lang, Algebra, Addison-Wesley, (1965).

11. J.H. van Lint, J.J. Seidel, Equilateral point sets in elliptic geometry, Proc. Kon. Nederl. Akad. Wet., Ser. A, 69(1966), (- Indag. Math. 28), 335-348.

12. C.L. Mallows, N.J.A. Sloane, Two-graphs, switching classes, and Euler graphs are equal in number, SIAM J. Appl. Math., 28(1975), 876-880.

13. R.A. Mathon, Symmetric conference matrices of order $pq^2 + 1$, Canad. J. Math., 30(1978), 321-331.

14. H.W. Norton, The 7 × 7 squares, Annals Eugenics 9(1939), 269-307.

15. A.J.L. Paulus, Conference matrices and graphs of order 26, Report Techn. Univ. Eindhoven 73-WSK-06 (1973), 89.

16. A.W Robinson, Enumeration of Euler graphs, in : Proof techniques in graph theory (ed. F. Harary), 47-53, Acad. Press (1969).

17. J.J. Seidel, A survey of two-graphs, Proc. Intern. Colloqu. Theorie Combinatorie (Roma 1973), Tomo I, Acad. Naz. Lincei, (1976), 481-511.

18. J.J. Seidel, Graphs and two-graphs, 5-th Southeastern Confer. on Combin., Graphs, Computing, Utilitas Math. Publ. Inc., Winnipeg (1974), 125-143.

19. J.J.Seidel, D.E. Taylor, Two-graphs, a second survey, Proc. Intern. Colloqu. Algebraic methods in graph theory, Szeged 1978, to be published.

20. J.J. Seidel, Strongly regular graphs, Proc. 7-th British Combin. Confer., Cambridge 1979.

21. D.E. Taylor, Regular 2-graphs, Proc. London Math. Soc., 35(1977), 257-274.

22. B. Weisfeiler, On construction and identification of graphs, Lecture Notes, 558, Springer, 1976.

* F.C. Bussemaker, R.A. Mathon, J.J. Seidel, Tables of two-graphs, Report Techn. Univ., Eindhoven 79-Wsk-05, (1979), 99.

```
TABLE A.   TWO-GRAPHS WITH NONZERO GAMMA OF ORDER 6,8,10
EACH TWO-GRAPH HAS TWO LINES:
LINE A:  TWO-GRAPH IDENTIFICATION NUMBER, UPPER HALF OF THE (-,+)-ADJACENCY MATRIX OF A DESCENDANT (FIRST
         ROW DELETED), EIGENVALUES, S = SELF-COMPLEMENTARY;
LINE B:  ORDER OF AUT, AUTOMORPHISM PARTITION, GENERATORS OF AUT.

TWO-GRAPHS WITH NONZERO GAMMA OF ORDER 6:

 1       ---+ -+- ++ -       -3.000 -2.236 -1.000  1.000  2.236  3.000    S
    4        122441          (23)(45)   (16)(45)
 2       --++ +-+ +- -       -2.236 -2.236 -2.236  2.236  2.236  2.236    S
   60        111111          (34)(56)   (23)(45)   (12)(34)

TWO-GRAPHS WITH NONZERO GAMMA OF ORDER 8:

 1       ----++ --+-+ +-+- -+- -+ +      -3.828 -2.464 -1.000 -1.000 -1.000  1.828  3.000  4.464
   16          12215555                  (78)   (56)   (23)(57)(68)   (14)(57)(68)
 2       ----++ --+-+ +-+- -+- -+ -      -3.828 -2.236 -1.000 -1.000 -1.000  1.828  2.236  5.000
   16          12331332                  (67)   (34)   (28)(36)(47)   (15)(36)(47)
 3       ----++ --+-+ +-+- ++- -+ -      -3.494 -2.464 -2.236 -1.000 -0.110  2.236  2.604  4.464
    4          12341342                  (28)(36)(47)   (15)(36)(47)
 4       ----++ -++-- ++++ --+ +- -      -3.000 -3.000 -3.000 -1.000  1.000  3.000  3.000  3.000    S
   16          11333311                  (34)(56)(78)   (36)(45)   (12)(35)(46)   (17)(28)(45)
 5       ----++ -++-- ++++ --+ +- +      -3.000 -3.000 -2.236 -1.828  1.000  2.236  3.000  3.828
    8          12331332                  (34)(67)   (28)(36)(47)   (15)(36)(47)   (15)(28)
 6       ----++ -++-- ++++ +-+ +- +      -3.000 -3.000 -1.828 -1.828  1.000  1.000  3.828  3.828
   32          11111111                  (23)(67)   (26)(37)(48)   (12)(35)(46)(78)
 7       ----++ -++-- ++++ +-- -- +      -3.000 -3.000 -1.000 -1.000 -1.000  1.000  3.000  5.000
   16          12241422                  (78)   (23)   (27)(38)(46)   (15)(46)
 8       ----++ -++-- ++++ +-- -+ +      -3.000 -2.604 -2.464 -1.000  0.110  1.000  3.494  4.464
    4          12215665                  (23)(67)   (14)(58)(67)
 9       ----++ -++-+ +++- +-- ++ +      -3.000 -2.604 -2.236 -1.828  0.110  2.236  3.494  3.828
    4          12215665                  (23)(67)   (14)(23)(58)
10       ----++ +++-+ ++-+ ++- +- -      -3.000 -2.464 -2.236 -1.000 -1.000  2.236  3.000  4.464
   16          12241422                  (78)   (23)   (27)(38)(46)   (15)(46)   (15)(27)(38)

TWO-GRAPHS WITH NONZERO GAMMA OF ORDER 10:

 1       -------- -----++ ---+-+ -++++ ++++ +++ -+ +      -5.340 -3.000 -1.622 -1.606 -1.000  1.000  1.000  1.962  3.000  5.606
   16            1133117887                               (56)   (34)(89)   (12)   (15)(26)(70)(89)
 2       -------- -----++ ---+-+ -++++ ++++ +++ ++ +      -5.340 -2.236 -2.123 -1.622 -1.000  1.000  1.000  1.962  2.236  6.123
   16            1133117887                               (56)   (34)(89)   (12)   (15)(26)(70)(89)
 3       -------- -----++ ---+-+ -++++ ++++ ++- ++ +      -5.230 -3.000 -2.236 -1.606 -0.492  1.000  1.000  2.236  2.722  5.606
   16            1133117887                               (56)   (34)(89)   (12)   (15)(26)(70)(89)
 4       -------- -----++ --++-- -++++ ++++ +-- -- +      -5.000 -4.123 -1.000 -1.000 -1.000  1.000  1.000  1.000  4.123  5.000    S
   64            1133117777                               (90)   (78)   (56)   (34)(79)(80)   (12)   (15)(26)(79)(80)
 5       -------- -----++ ----++ +++-+ ++-+ ++- +- +      -5.000 -3.828 -1.828 -1.000 -1.000  1.000  1.000  1.828  3.828  5.000    S
   64            1111555599                               (78)   (56)   (57)(68)(90)   (34)   (12)   (13)(24)(90)
 6       -------- -----++ --++-- -++++ ++++ +-- -+ +      -5.000 -3.686 -2.464 -1.000  0.059  1.000  1.000  1.000  4.464  4.627
   16            1133117878                               (56)   (34)(79)(80)   (12)   (15)(26)(79)(80)
 7       -------+ -----+- ----++ +++-- ++-+ ++- ++ -      -5.000 -3.494 -2.604 -1.000 -0.110  0.110  1.000  2.604  3.494  5.000    S
    4            1221566599                               (23)(67)(90)   (14)(58)(90)
 8       -------+ -----+- ----++ +++-- ++-- +++ ++ +      -5.000 -3.340 -2.236 -1.000 -1.000  0.378  1.000  2.236  3.962  5.000
   16            1221555599                               (78)   (56)   (23)(50)   (14)(57)(68)(90)
```

```
 9     --------- ------++ --++-- -++++ ++++ +-+ +- +    -5.000 -3.000 -3.000 -1.828  1.000  1.000  1.000  1.000  3.828  5.000
   32             1133117777                            (78)(90)  (56)  (34)(79)(80)  (12)  (15)(26)(79)(80)
10     --------- ------++ -----++ +++-+ ++-+ ++- +- -   -5.000 -3.000 -3.000 -1.000 -1.000  1.000  1.000  3.000  3.000  5.000   S
   64             1111555599                            (78)  (56)  (57)(68)(90)  (34)  (12)  (13)(24)(90)
11     -------++ ----+-+ --+-+- -++-- ++++ -+- -+ -     -5.000 -3.000 -3.000 -1.000 -1.000  1.000  1.000  3.000  3.000  5.000   S
   16             1133117777                            (34)(78)(90)  (25)(79)(80)  (12)(56)(90)
12     --------- ------++ -----++ +++-- ++-+ ++- ++ +   -5.000 -3.000 -3.000 -1.000 -0.236  1.000  1.000  1.000  4.236  5.000
   16             1111566599                            (67)(90)  (34)  (12)  (13)(24)(58)(90)
13     -------++ ---++-+ ++-+-+ +-+-+ ++-- --+ -- +     -5.000 -3.000 -2.722 -1.000 -1.000  0.452  1.000  3.000  3.000  5.230
   16             1232263621                            (45)  (37)(68)  (29)  (10)(24)(59)(68)
14     --------- ------++ --++-- -++++ ++++ +-+ ++ +    -5.000 -3.000 -2.604 -1.606  0.110  1.000  1.000  1.000  3.494  5.606
   16             1133117878                            (56)  (34)(79)(80)  (12)  (15)(26)(79)(80)
15     --------+ ----++- --+-+- -++-+ ++++ +++ ++ +     -5.000 -3.000 -2.236 -1.962 -1.000  1.000  1.622  2.236  3.000  5.340
    4             1233217799                            (34)(78)  (16)(25)(78)(90)
16     --------- ------++ --++-- -++++ ++++ +++ ++ +    -5.000 -3.000 -2.123 -1.000 -1.000  1.000  1.000  1.000  3.000  6.123
   64             1133117777                            (90)  (78)  (56)  (34)(79)(80)  (12)  (15)(26)(79)(80)
17     --------+ ----++- --+-+- -++-+ ++++ +++ ++ -     -5.000 -2.654 -2.236 -1.828 -1.677  1.000  1.418  2.236  3.828  4.913
    4             1233217799                            (34)(78)  (16)(25)(78)(90)
18     ------+++ --++--+ +-+-++ -+-++ +--+ --+ ++ +     -5.000 -2.464 -1.000 -1.000 -1.000 -1.000 -1.000  3.000  4.464  5.000
   64             1221127271                            (45)  (36)  (23)(68)(79)  (14)(50)(79)  (10)
19     --------+ ----++- --+-+- ++++- ++++ -++ ++ +     -4.913 -3.547 -1.876 -1.418 -1.000  0.142  1.677  2.654  3.000  5.280
    4             1233167762                            (34)(78)  (15)(20)(69)(78)
20     --------+ ----++- --+-+- ++++- ++++ +++ ++ +     -4.913 -3.000 -2.236 -1.418 -1.000  0.172  1.677  2.236  2.654  5.828
    4             1233167762                            (34)(78)  (15)(20)(69)(78)
21     --------+ ----++- --+-+- +++-+ +++- +++ ++ +     -4.759 -3.547 -2.236 -1.876 -0.305  0.142  2.064  2.236  3.000  5.280
    4             1233517752                            (34)(78)  (16)(20)(55)(78)
22     -------++ ---++-- --++++ +-+-+ -++- +-- -- +     -4.722 -3.000 -3.000 -1.508 -1.000  1.000  1.000  3.000  3.230  5.000
    4             1221557799                            (56)(90)  (14)(23)(78)(90)
23     -------++ ----+-+ -++--- ++-++ ++-- +++ -+ +     -4.722 -3.000 -3.000 -1.508 -0.236 -0.236  1.000  3.230  4.236  4.236
    4             1131166399                            (12)(45)(67)(90)  (14)(25)(38)(67)
24     --------- ----+++ -++-++ ++-++ +-++ +-+ -+ +     -4.722 -3.000 -2.236 -1.606 -1.508  1.000  1.000  2.236  3.230  5.606
   16             1131167376                            (45)  (38)(79)  (12)  (14)(25)(60)(79)
25     --------- ----+++ -++-++ ++-++ +-++ +++ -+ +     -4.722 -3.000 -2.123 -1.508 -1.000  1.000  1.000  1.000  3.230  6.123
   16             1131167376                            (45)  (38)(79)  (12)  (14)(25)(60)(79)
26     -------++ ----+-+ -++--- ++-++ +--+ -+- +- +     -4.684 -3.828 -2.014 -1.828 -1.000  1.000  1.828  2.014  3.828  4.684   S
    4             1134466399                            (38)(67)(90)  (12)(45)(90)
27     -------++ ----+-+ -++--+ ++-+- +--- -++ +- +     -4.684 -3.494 -2.604 -2.014 -0.110  0.110  2.014  2.604  3.494  4.684
    4             1134466399                            (38)(45)(90)  (12)(67)(90)
28     --------+ ----++- -++-+- +++-- +-+- +-- ++ +     -4.684 -3.494 -2.604 -2.014 -0.110  0.110  2.014  2.604  3.494  4.684   S
    4             1214466882                            (45)(67)(89)  (13)(20)(89)
29     --------+ ----++- -++-+- +++-- +-+- +-- -+ +     -4.684 -3.494 -2.604 -2.014 -0.110  0.110  2.014  2.604  3.494  4.684   S
    4             1214466882                            (45)(67)(89)  (13)(20)(89)
30     --------+ ----++- -++-+- +++-- +--- ++- ++ +     -4.684 -3.340 -2.236 -2.014 -1.000  0.378  2.014  2.236  3.962  4.684
    4             1214466882                            (45)(89)  (13)(20)(67)(89)
31     -------++ ----+-+ -++--+ ++-+- +--+ -+- +- -     -4.684 -3.340 -2.236 -2.014 -1.000  0.378  2.014  2.236  3.962  4.684
    4             1134466399                            (38)(45)(67)(90)  (12)(90)
32     -------++ ----+-+ -++--- ++-++ +--+ -+- +- -     -4.684 -3.000 -3.000 -2.014 -1.000  1.000  2.014  3.000  3.000  4.684   S
    4             1134466399                            (38)(67)(90)  (12)(45)(90)
33     -------++ ----+-+ -++--- ++-++ +--- -++ +- +     -4.684 -3.000 -3.000 -2.014 -0.236  1.000  1.000  2.014  4.236  4.684
    4             1134466399                            (38)(90)  (12)(45)(67)(90)
34     --------+ ----++- -++--- ++++- +-+- +-- ++ +     -4.684 -3.000 -3.000 -2.014 -0.236  1.000  1.000  2.014  4.236  4.684
    4             1214466882                            (67)(89)  (13)(20)(45)(89)
35     --------- ----+++ -++-++ ++-++ +--+ ++- +- +     -4.627 -3.000 -3.000 -2.236 -0.059  1.000  1.000  2.236  3.686  5.000
   16             1131167367                            (45)  (38)(70)  (12)  (14)(25)(69)(70)
36     -------++ ---++-+ ++-+-- +-+-+ ++++ --+ ++ +     -4.627 -3.000 -3.000 -2.064 -0.059  0.305  1.000  3.000  3.686  4.759
    4             1234163624                            (37)(68)  (15)(29)(40)(68)
37     --------- ----+++ -++-++ ++-++ +--+ +++ +- +     -4.627 -3.000 -3.000 -1.606 -0.059  1.000  1.000  1.000  3.686  5.606
   16             1131167367                            (45)  (38)(70)  (12)  (14)(25)(69)(70)
38     ------+++ --++--+ +-+-+- -+-+- +--+ +++ ++ -     -4.627 -3.000 -1.828 -1.000 -1.000 -1.000 -0.059  3.686  3.828  5.000
   16             1233332199                            (45)  (34)(56)(90)  (18)(27)(90)
```

```
39    -------+++ --++--+ +-+-+- -+-+- +--+ +++ ++ +     -4.627 -2.464 -2.464 -1.000 -1.000 -1.000 -0.059  3.686  4.464  4.464
   16            1233332199                             (45)   (34)(56)(90)   (18)(27)(90)
40    -------++ ----+-+ --+-+- +++-+ +++- -+- -+ +      -4.513 -3.686 -2.640 -1.000 -1.000  0.059  1.277  3.000  3.877  4.627
    4            1133557799                             (34)(56)(78)(90)   (12)(90)
41    -------++ ----+-+ --+-+- +++-- ++++ -+- -+ -      -4.513 -3.494 -2.640 -1.000 -1.000 -0.110  1.277  2.604  3.877  5.000
    4            1133557799                             (34)(78)(90)   (12)(56)(90)
42    --------+ ----++- --+++- ++-++ ++-+ ++- -- -      -4.513 -3.230 -2.640 -2.236 -1.000  1.277  1.508  2.236  3.877  4.722
    4            1212557887                             (56)(89)   (13)(24)(70)(39)
43    --------+ ----++- --+++- ++-++ ++-+ ++- +- -      -4.513 -3.000 -2.640 -1.962 -1.000  1.000  1.277  1.622  3.877  5.340
    4            1212557887                             (56)(89)   (13)(24)(70)(89)
44    -------++ ----+-+ --+-+- +++-+ +++- -+- -+ -      -4.513 -2.722 -2.640 -2.236 -1.000  0.492  1.277  2.236  3.877  5.230
    4            1133557799                             (34)(56)(78)(90)   (12)(90)
45    -------++ ----+-+ --+-+- +++-- ++++ -+- -+ +      -4.513 -2.640 -2.604 -2.464 -1.000  0.110  1.277  3.494  3.877  4.464
    4            1133557799                             (34)(78)(90)   (12)(56)(90)
46    -------++ ----+-+ --+-+- +++-+ +++- ++- -+ +      -4.491 -3.883 -2.134 -1.355 -1.000 -0.030  1.355  3.000  4.046  4.491
    4            1133557799                             (34)(56)(78)(90)   (12)(90)
47    --------+ ----++- --+++- ++-++ ++-+ +++ -- -      -4.491 -3.526 -2.236 -2.031 -1.355  1.355  1.365  2.236  4.192  4.491
    4            1212557887                             (56)(89)   (13)(24)(70)(89)
48    -------++ ----+-+ --+-+- +++-- ++++ ++- -+ -      -4.491 -3.494 -2.464 -1.355 -1.000 -0.110  1.355  2.604  4.464  4.491
    4            1133557799                             (34)(78)(90)   (12)(56)(90)
49    --------+ ----++- --+++- ++-++ ++-+ +++ +- -      -4.491 -3.000 -2.654 -1.677 -1.355  1.000  1.355  1.418  4.491  4.913
    4            1212557887                             (56)(89)   (13)(24)(70)(89)
50    -------++ ----+-+ --+-+- +++-- ++++ ++- -+ +      -4.491 -3.000 -2.604 -1.828 -1.355  0.110  1.355  3.494  3.828  4.491
    4            1133557799                             (34)(78)(90)   (12)(56)(90)
51    -------++ ----+-+ --+-+- +++-+ +++- ++- -+ -      -4.491 -3.000 -2.236 -2.064 -1.355  0.305  1.355  2.236  4.491  4.759
    4            1133557799                             (34)(56)(78)(90)   (12)(90)
52    -------++ ----+-+ +++--+ ++--+ +-+- -+- +- +      -4.464 -4.338 -1.000 -1.000 -1.000 -1.000  2.048  2.464  3.000  5.290
   16            1134444399                             (67)   (45)   (38)(46)(57)(90)   (12)(90)
53    --------+ ----++- -++--- ++++- +--+ -++ ++ +      -4.464 -4.236 -2.464 -1.000 -0.236  0.236  1.000  2.464  4.236  4.464    S
    4            1233167672                             (34)(68)(79)   (15)(20)(68)(79)
54    -------++ ---++-- -+-+++ +++-+ ---- +-- -- +      -4.464 -4.236 -1.962 -1.000 -1.000  0.236  1.622  2.464  3.000  5.340
    4            1224417272                             (23)(45)(79)(80)   (16)(28)(30)(79)
55    --------- ----+++ -++-++ ++-++ ++-- +-- -- +      -4.464 -4.123 -2.464 -1.000 -1.000  1.000  1.000  2.464  4.123  4.464    S
   64            1131166366                             (90)   (67)   (45)   (38)(69)(70)   (12)   (14)(25)(69)(70)
56    -------++ ---++-+ -+-+-+ ++-+- --+- -+- -+ +      -4.464 -3.946 -1.925 -1.000 -1.000 -1.000  2.464  3.000  3.000  4.871
    8            1133533599                             (34)(67)   (36)(47)(58)(90)   (12)(90)
57    -------++ ---++-- --++++ +-+-+ +-+- +-- -- +      -4.464 -3.828 -3.000 -1.000 -1.000  1.000  1.828  2.464  3.000  5.000
    8            1221557777                             (56)(78)(90)   (23)(79)(80)   (14)(70)(89)
58    --------- ----+++ -++-++ ++-++ ++++ +++ -- +      -4.464 -3.828 -2.123 -1.000 -1.000  1.000  1.000  1.828  2.464  6.123
   64            1131166366                             (90)   (67)   (45)   (38)(69)(70)   (12)   (14)(25)(69)(70)
59    --------- ----+++ -++-++ ++-++ ++-- +-+ -- +      -4.464 -3.686 -3.000 -1.828  0.059  1.000  1.000  2.464  3.828  4.627
   16            1131167367                             (45)   (38)(69)(70)   (12)   (14)(25)(69)(70)
60    --------+ ----++- -++--- ++++- +-++ +-+ ++ +      -4.464 -3.686 -3.000 -1.828  0.059  1.000  1.000  2.464  3.828  4.627
    8            1233166662                             (67)(89)   (34)(68)(79)   (15)(20)(68)(79)
61    -------++ ---++-- --++++ +-+-+ +-+- +-- -+ +      -4.464 -3.606 -2.464 -2.236 -1.000  1.000  2.236  2.464  3.606  4.464    S
    4            1221557878                             (23)(79)(80)   (14)(56)(79)(80)
62    -------++ ---++-- -+-+-+ ++-++ --++ -+- -- +      -4.464 -3.606 -1.962 -1.000 -1.000 -1.000  1.622  2.464  3.606  5.340
    4            1134334399                             (38)(47)(56)(90)   (12)(35)(68)(90)
63    --------+ ----++- -++--- ++++- +-++ +++ ++ +      -4.464 -3.547 -2.604 -1.876  0.110  0.142  1.000  2.464  3.494  5.280
    4            1233167672                             (34)(68)(79)   (15)(20)(68)(79)
64    --------- ----+++ -++-++ ++-++ ++-+ +++ -- +      -4.464 -3.494 -3.000 -1.606 -0.110  1.000  1.000  2.464  2.604  5.606
   16            1131167367                             (45)   (38)(69)(70)   (12)   (14)(25)(69)(70)
65    -------++ ---++-+ ++---+ +---+ --+- ++- +- -      -4.464 -3.472 -2.236 -1.000 -1.000 -1.000  2.236  2.464  3.000  5.472
   16            1134434499                             (78)   (45)   (36)(47)(58)(90)   (12)(90)
66    -------++ ---++-+ -+-+-+ ++-+- --+- -+- -+ -      -4.464 -3.472 -2.236 -1.000 -1.000 -1.000  2.236  2.464  3.000  5.472
    8            1133533599                             (34)(67)   (36)(47)(58)(90)   (12)(90)
67    -------++ ---++-- -+-+-+ ++-++ --++ -+- -- -      -4.464 -3.340 -3.230 -1.000 -1.000  0.378  1.508  2.464  3.962  4.722
    4            1134334399                             (38)(47)(56)(90)   (12)(35)(68)(90)
68    -------++ ---++-- ++---- +---+ --++ ++- ++ +      -4.464 -3.340 -2.722 -1.000 -1.000  0.378  0.492  2.464  3.962  5.230
    4            1133535399                             (34)(57)(68)(90)   (12)(36)(48)(57)
```

```
69      -------- ----+++ -++-++ ++-++ ++-+ ++- -- +     -4.464 -3.000 -3.000 -3.000  1.000  1.000  1.000  2.464  3.000  5.000
   32                1131166366                        (67)(90)  (45)   (38)(69)(70)   (12)    (14)(25)(69)(70)
70      ------++ ---++-- --++++ +-+-+ +-+- -+- -+ -     -4.464 -3.000 -3.000 -2.464 -1.000  1.000  2.464  3.000  3.000  4.464   S
   16                1111557777                        (56)(78)(90)   (23)(79)(80)   (12)(34)(90)
71      -------+ ----++- -++--- ++++- ++++ +++ ++ +     -4.464 -3.000 -3.000 -1.000 -1.000  0.172  1.000  2.464  3.000  5.828
   16                1233166662                        (89)    (67)   (34)(68)(79)   (15)(20)(68)(79)
72      ------++ ---++-- -+-+-+ +-++- +-++ -++ -- +     -4.464 -3.000 -2.654 -2.236 -1.677  1.418  2.236  2.464  3.000  4.913
    4                1134463699                        (45)(90)    (12)(37)(68)(90)
73      ------++ ---++-- -+-+-+ +-++- ++++ -++ -- +     -4.464 -3.000 -2.464 -2.236 -1.828  1.000  2.236  2.464  3.828  4.464
    4                1134463699                        (45)(90)    (12)(37)(68)(90)
74      ------++ ---++-+ ++--+- +--+- --+- +-+ -+ +     -4.464 -2.583 -1.000 -1.000 -1.000 -1.000 -1.000  2.464  3.000  6.583
  144                1133333399                        (78)    (56)    (45)    (34)(57)(68)(90)    (12)(90)
75      ------++ ---++-+ ++--+- +--+- --+- +-+ -+ -     -4.464 -2.236 -1.000 -1.000 -1.000 -1.000 -1.000  2.236  2.464  7.000
  144                1133333399                        (78)    (56)    (45)    (34)(57)(68)(90)    (12)(90)
76      ------++ ----+-+ +++--- ++--- +-++ -++ +- -     -4.338 -4.236 -2.236 -1.000 -1.000  0.236  2.048  2.236  3.000  5.290
   16                1134444399                        (67)    (45)    (38)(90)    (12)(46)(57)(90)
77      -------+ ----++- +++--- ++-+- ++-- ++- -+ +     -4.338 -4.236 -2.236 -1.000 -1.000  0.236  2.048  2.236  3.000  5.290
    4                1214554882                        (56)(89)    (13)(20)(47)(89)
78      ------++ ----+-+ +++--- ++--+ +-+- -++ +- +     -4.338 -3.828 -2.604 -1.000 -1.000  0.110  1.828  2.048  3.494  5.290
    4                1134554399                        (38)(56)(90)    (12)(47)(90)
79      -------+ ----++- +++-+- ++-+- ++-- +-- ++ +     -4.338 -3.828 -2.604 -1.000 -1.000  0.110  1.828  2.048  3.494  5.290
   16                1214444882                        (67)    (45)    (46)(57)(89)    (13)(20)(89)
80      ------++ ----+-+ +++--+ ++--+ +-+- -+- +- -     -4.338 -3.606 -2.236 -1.000 -1.000 -1.000  2.048  2.236  3.606  5.290
   16                1134444399                        (67)    (45)    (38)(46)(57)(90)    (12)(90)
81      ------++ ----+-+ +++--- ++--+ +-+- -++ +- -     -4.338 -3.494 -3.000 -1.000 -1.000 -0.110  2.048  2.604  3.000  5.290
    4                1134554399                        (38)(56)(90)    (12)(47)(90)
82      -------+ ----++- +++-+- ++-+- ++-- +-- -+ +     -4.338 -3.494 -3.000 -1.000 -1.000 -0.110  2.048  2.604  3.000  5.290
   16                1214444882                        (67)    (45)    (46)(57)(89)    (13)(20)(89)
83      ------++ ----+-+ +++--- ++--- +-++ -++ +- +     -4.338 -3.340 -3.000 -1.000 -1.000  0.378  1.000  2.048  3.962  5.290
   16                1134444399                        (67)    (45)    (38)(90)    (12)(46)(57)(90)
84      -------+ ----++- +++--- ++-+- ++-- ++- ++ +     -4.338 -3.340 -3.000 -1.000 -1.000  0.378  1.000  2.048  3.962  5.290
    4                1214554882                        (56)(89)    (13)(20)(47)(89)
85      ------++ ---++-+ ++---+ +--+- -+-+ ++- +- -     -4.236 -3.946 -2.236 -1.925 -1.000  0.236  2.236  3.000  3.000  4.871
    4                1134364699                        (35)(47)(68)(90)    (12)(90)
86      ------++ ---++-+ -+-+-+ ++-+- +++- ++- +- +     -4.236 -3.946 -2.236 -1.925 -1.000  0.236  2.236  3.000  3.000  4.871
    8                1133566511                        (34)(67)    (12)(90)    (19)(20)(58)(67)    (10)(29)(58)(67)
87      -------+ ---+++- ++--+- +-+-+ +-+- ++- -+ +     -4.236 -3.828 -3.000 -1.828 -0.236  0.236  1.828  3.000  3.828  4.236   S
    4                1233566521                        (34)(58)(67)    (10)(29)(58)
88      ------++ ---++-- ++---- +--++ -+-+ +++ +- +     -4.236 -3.686 -3.340 -1.000  0.059  0.236  0.378  3.000  3.962  4.627
    4                1133363699                        (34)(57)(68)(90)    (12)(35)(47)(68)
89      -----+++ --++--+ +-+-+- -+-++ ++-+ --+ ++ +     -4.236 -3.000 -2.464 -1.828 -1.000 -0.236  0.236  3.828  4.236  4.464
    4                1224167674                        (23)(68)(79)    (15)(40)(79)
90      ------++ ---++-+ ++--+- +--+- -++- +-+ -+ +     -4.236 -3.000 -2.123 -1.000 -1.000 -1.000  0.236  3.000  3.000  6.123
   16                1133363699                        (45)    (34)(57)(68)(90)    (12)(90)
91      ------++ ---++-+ ++--+- +--+- -++- +-+ -+ -     -4.236 -2.583 -2.236 -1.000 -1.000 -1.000  0.236  2.236  3.000  6.583
   16                1133363699                        (45)    (34)(57)(68)(90)    (12)(90)
92      ------++ --+++-- +-++-+ +-+-+ +--- --- +- +     -4.192 -3.828 -3.000 -1.365 -1.000  1.000  1.828  2.031  3.526  5.000
    4                1234466132                        (45)(67)    (18)(20)(35)(67)
93      -------+ ---+++- ++-++- +-+++ +-++ -+- ++ +     -4.192 -3.547 -2.236 -1.876 -1.365  0.142  2.031  2.236  3.526  5.280
    4                1231563652                        (37)(68)    (14)(20)(59)(68)
94      -----+++ --++--+ +-+-+- -++-- +++- --+ +- -     -4.192 -3.000 -3.000 -2.464 -1.365  1.000  2.031  3.000  3.526  4.464
    4                1234413882                        (45)(89)    (16)(20)(37)(89)
95      -------+ ---+++- ++-++- +-+++ ++++ -+- ++ +     -4.192 -3.000 -3.000 -1.365 -1.000  0.172  1.000  2.031  3.526  5.828
    4                1231563652                        (37)(68)    (14)(20)(59)(68)
96      ------++ ---++-+ ++---+ +-++- +-+- ++- +- +     -4.123 -3.828 -3.000 -1.828 -1.000  1.000  1.828  3.000  3.828  4.123   S
   16                1133555511                        (56)(78)    (34)(57)(68)    (12)(90)    (19)(20)(57)(68)    (10)(29)(57)(68)
97      ------++ --+++-- +-++-+ -+++- ++++ --+ +- -     -4.123 -3.494 -3.000 -2.604 -0.110  0.110  2.604  3.000  3.494  4.123   S
    4                1134437799                        (45)(78)(90)    (12)(36)(90)
98      ------++ ---++-+ ++-++- ++-+- ++-+ --+ -+ +     -4.123 -3.340 -3.000 -2.236 -1.000  0.378  2.236  3.000  3.962  4.123
    8                1134437711                        (45)(78)    (12)(90)    (19)(20)(36)(78)    (10)(29)(36)(78)
```

```
 99      -------++ --++++-- +-++++ -++++ ++-- --+ +- -    -4.123 -3.230 -3.000 -1.000 -1.000 -1.000  1.508  3.000  4.123  4.722
     16            1133337799                             (78)(90)  (45)   (36)   (12)(34)(56)(90)
100      -------++ --+++-+ +-++-+ -+++- +++- --- ++ -     -4.123 -3.000 -3.000 -3.000 -1.000  1.000  3.000  3.000  3.000  4.123   S
     16            1133331199                             (34)(56)  (35)(46)(90)  (12)(78)(90)  (17)(28)(90)
101      -------++ --++++-- +-++-+ -+++- ++++ --- ++ +    -4.123 -3.000 -3.000 -3.000 -0.236  1.000  1.000  3.000  4.123  4.236
      8            1134431199                             (45)(90)  (12)(36)(78)(90)  (17)(28)(36)(90)
102      -------++ --+++-+ +-+++- -+++- ++-+ --- ++ -     -4.123 -3.000 -3.000 -1.000 -1.000 -1.000  1.000  3.000  4.123  5.000
     32            1133331199                             (45)  (34)(56)(90)  (12)(78)(90)  (17)(28)(90)
103      -------++ --+++-+ +-+++- -+++- ++-+ --- ++ +     -4.123 -3.000 -2.464 -1.828 -1.000 -1.000  1.000  3.828  4.123  4.464
     32            1133331199                             (45)  (34)(56)(90)  (12)(78)(90)  (17)(28)(90)
104      -------++ --+++-- +-++++ -++++ ++-- --+ +- +     -4.123 -3.000 -2.236 -1.962 -1.000 -1.000  1.622  2.236  4.123  5.340
     16            1133337799                             (78)(90)  (45)   (36)   (12)(34)(56)(90)
105      --------+ ---+++- ++--++ +-+-+ +++- +++ -- -     -4.046 -3.686 -3.000 -2.236  0.030  0.059  2.134  2.236  3.883  4.627
      4            1234432881                             (45)(89)  (10)(27)(36)(89)
106      --------+ ---+++- ++--++ +-+-+ +++- +++ +- -     -4.046 -3.547 -3.000 -1.876  0.030  0.142  1.000  2.134  3.883  5.280
      4            1234432881                             (45)(89)  (10)(27)(36)(89)
107      -------++ ---++-+ ++--+- +-+-+ +-+- ++- -+ +     -3.962 -3.877 -3.000 -1.277 -1.000 -0.378  2.640  3.000  3.340  4.513
      4            1133565699                             (34)(57)(68)(90)  (12)(90)
108      -------++ ---++-+ ++--+- +-+-- +-++ +++ -- -     -3.962 -3.606 -3.000 -2.464 -0.378  1.000  1.000  3.340  3.606  4.464
      4            1133555599                             (34)(58)(67)(90)  (12)(56)(78)(90)
109      -------++ ---++-+ ++--+- +-+-- +-++ +++ -- +     -3.962 -3.000 -3.000 -2.464 -0.378 -0.236  1.000  3.340  4.236  4.464
      4            1133555599                             (34)(58)(67)(90)  (12)(56)(78)(90)
110      -------++ ---++-+ ++--+- +-+-+ +-+- ++- -+ -     -3.962 -3.000 -3.000 -2.236 -1.000 -0.378  2.236  3.000  3.340  5.000
      4            1133565699                             (34)(57)(68)(90)  (12)(90)
111      -------++ ---++-+ ++--+- +-++- +-+- +-+ -+ +     -3.962 -3.000 -3.000 -1.606 -1.000 -0.378  1.000  3.000  3.340  5.606
      8            1133555599                             (56)(78)  (34)(57)(68)(90)  (12)(90)
112      -------++ ---++-+ ++--+- +-++- +-+- +-+ -+ -     -3.962 -3.000 -2.236 -2.123 -1.000 -0.378  1.000  2.236  3.340  6.123
      8            1133555599                             (56)(78)  (34)(57)(68)(90)  (12)(90)
113      -------++ --++++-- +-++-+ -+++- ++++ +-+ +- +    -3.946 -3.828 -2.604 -1.925 -1.000  0.110  1.828  3.000  3.494  4.871
      4            1134437799                             (45)(78)(90)  (12)(36)(90)
114      -------++ ---++-+ ++---+ +-++- +++- ++- +- +     -3.946 -3.828 -2.604 -1.925 -1.000  0.110  1.828  3.000  3.494  4.871
      8            1133565611                             (34)(57)(68)  (12)(90)  (19)(20)(57)(68)  (10)(29)(57)(68)
115      -------++ ---++-+ ++--+- ++++- +++- +-+ -+ +     -3.946 -3.686 -1.925 -1.000 -1.000 -1.000  0.059  3.000  4.627  4.871
     16            1133555599                             (78)  (56)  (34)(57)(68)(90)  (12)(90)
116      -------++ --+++-+ +-++-+ -+++- +++- +-+ +- -     -3.946 -3.606 -2.236 -1.925 -1.000 -1.000  2.236  3.000  3.606  4.871
      8            1133337799                             (34)(56)  (35)(46)(78)(90)  (12)(90)
117      -------++ --+++-+ +-++-+ -+++- +++- +-- ++ -     -3.946 -3.494 -3.000 -1.925 -1.000 -0.110  2.604  3.000  3.000  4.871
      8            1133337799                             (34)(56)  (35)(46)(90)  (12)(78)(90)
118      -------++ --++++-- +-++-+ -+++- ++++ +-+ +- -    -3.946 -3.494 -3.000 -1.925 -1.000 -0.110  2.604  3.000  3.000  4.871
      4            1134437799                             (45)(78)(90)  (12)(36)(90)
119      -------++ --+++-+ +-+++- -+++- ++-+ +-- ++ -     -3.946 -3.494 -1.925 -1.000 -1.000 -1.000 -0.110  2.604  4.871  5.000
     16            1133337799                             (45)  (34)(56)(90)  (12)(78)(90)
120      -------++ ---++-+ ++-++- ++-+- ++-+ +-+ -+ +     -3.946 -3.340 -3.000 -1.925 -1.000  0.378  1.000  3.000  3.962  4.871
      8            1134437711                             (45)(78)  (12)(90)  (19)(20)(36)(78)  (10)(29)(36)(78)
121      -------++ --++++-- +-++-+ -+++- ++++ +-- ++ +    -3.946 -3.340 -3.000 -1.925 -1.000  0.378  1.000  3.000  3.962  4.871
      4            1134437799                             (45)(90)  (12)(36)(78)(90)
122      -------++ --+++-- +-++++ -++++ ++-- +-+ +- -     -3.946 -3.230 -2.236 -1.925 -1.000 -1.000  1.508  2.236  4.722  4.871
     16            1133337799                             (78)(90)  (45)   (36)   (12)(34)(56)(90)
123      -------++ --+++-- +-++++ -++++ ++-- +-+ +- +     -3.946 -3.000 -1.962 -1.925 -1.000 -1.000  1.000  1.622  4.871  5.340
     16            1133337799                             (78)(90)  (45)   (36)   (12)(34)(56)(90)
124      -------++ --+++-+ +-+++- -+++- ++-+ +-+ +- -     -3.946 -2.722 -2.236 -1.925 -1.000 -1.000  0.492  2.236  4.871  5.230
     16            1133337799                             (45)  (34)(56)(78)(90)  (12)(90)
125      -------++ --+++-+ +-+++- -+++- ++-+ +-- ++ +     -3.946 -2.604 -2.464 -1.925 -1.000 -1.000  0.110  3.494  4.464  4.871
     16            1133337799                             (45)  (34)(56)(90)  (12)(78)(90)
126      -------++ ---++-+ ++--+- +-+-+ +++- ++- -+ +     -3.933 -3.883 -2.555 -2.134 -1.000 -0.030  2.555  3.000  3.933  4.046
      4            1133565699                             (34)(57)(68)(90)  (12)(90)
127      -------++ ---++-- ++-+-+ +-++- +-++ -++ -- -     -3.933 -3.526 -2.555 -2.236 -2.031  1.365  2.236  2.555  3.933  4.192
      4            1134463699                             (45)(90)  (12)(37)(68)(90)
128      -------++ ---++-+ ++-+-- +-+++ +-+- -+- -+ -     -3.933 -3.494 -2.555 -2.464 -1.000 -0.110  2.555  2.604  3.933  4.464
      4            1134463699                             (37)(68)(90)  (12)(45)(90)
```

```
129     -------++ ---++-- ++-+-+ +-++- +-++ -++ -- +    -3.933 -3.000 -2.654 -2.555 -1.677  1.000  1.418  2.555  3.933  4.913
     4            1134463699                            (45)(90)    (12)(37)(68)(90)
130     -------++ ---++-+ ++-+-- +-+++ +-+- -+- -+ +    -3.933 -3.000 -2.604 -2.555 -1.828  0.110  2.555  3.494  3.828  3.933
     4            1134463699                            (37)(68)(90)   (12)(45)(90)
131     -------++ ---++-+ ++-+-+ +-++- +-+- -+- -+ -    -3.933 -3.000 -2.555 -2.236 -2.064  0.305  2.236  2.555  3.933  4.759
     4            1134463699                            (37)(45)(68)(90)   (12)(90)
132     -------++ ---++-+ ++--+- +-++- ++-+ +-+ +- -    -3.883 -3.828 -2.236 -2.134 -1.000 -0.030  1.828  2.236  4.046  5.000
     4            1133565699                            (34)(57)(68)(90)   (12)(90)
133     -------++ ---++-+ ++--+- +-++- +++- +-+ -+ +    -3.883 -3.547 -2.134 -1.876 -1.000 -0.030  0.142  3.000  4.046  5.280
     4            1133565699                            (34)(57)(68)(90)   (12)(90)
134     -------++ ---++-+ ++--+- +-+-+ +++- ++- -+ -    -3.883 -3.000 -2.464 -2.236 -2.134 -0.030  2.236  3.000  4.046  4.464
     4            1133565699                            (34)(57)(68)(90)   (12)(90)
135     -------++ ---++-+ ++--+- +-++- +++- +-+ -+ -    -3.883 -3.000 -2.236 -2.134 -1.000 -0.030  0.172  2.236  4.046  5.828
     4            1133565699                            (34)(57)(68)(90)   (12)(90)
136     -------++ ---++-- ++-+-+ +-++- ++++ -++ -- -    -3.877 -3.606 -3.000 -2.236 -1.277  1.000  2.236  2.640  3.606  4.513
     4            1134463699                            (45)(90)   (12)(37)(68)(90)
137     -------++ ---++-+ ++-+-- +-+++ +++- -+- -+ -    -3.877 -3.494 -3.000 -1.828 -1.277 -0.110  2.604  2.640  3.828  4.513
     4            1134463699                            (37)(68)(90)   (12)(45)(90)
138     -------++ ---++-+ ++-+-- +-+++ +++- -+- -+ +    -3.877 -3.000 -3.000 -2.604 -1.277  0.110  2.640  3.000  3.494  4.513
     4            1134463699                            (37)(68)(90)   (12)(45)(90)
139     -------++ ---++-- ++-+-+ +-++- ++++ -++ -- +    -3.877 -3.000 -3.000 -2.464 -1.277  1.000  1.000  2.640  4.464  4.513
     4            1134463699                            (45)(90)   (12)(37)(68)(90)
140     -------++ ---++-+ ++-+-+ +-++- +++- -+- -+ -    -3.877 -3.000 -3.000 -2.236 -1.277 -0.236  2.236  2.640  4.236  4.513
     4            1134463699                            (37)(45)(68)(90)   (12)(90)
141     ------+++ --++--+ +-+-++ +-+-+ +--- --- +- -    -3.828 -3.828 -3.828 -1.000 -1.000  1.828  1.828  1.828  3.000  5.000
    48            1221122221                            (45)(67)(89)   (23)(68)(79)   (26)(38)(50)(79)   (14)(50)(79)
142     ------+++ --++--+ +-+-++ +-+-+ +--- -+- +- -    -3.828 -3.828 -3.000 -2.464 -1.000  1.828  1.828  3.000  3.000  4.464
    16            1221127271                            (23)(68)(79)   (26)(38)(50)(79)   (14)(50)(79)   (15)(40)(68)
143     ------+++ --++--+ +-+-+- +-+-+ ++-- --- +- -    -3.828 -3.828 -3.000 -2.464 -1.000  1.828  1.828  3.000  3.000  4.464
     8            1224162624                            (23)(68)(79)   (27)(39)(40)(68)   (15)(40)(68)
144     --------+ ---+++- ++-++- +++++ ++++ --- ++ +    -3.828 -3.828 -3.000 -1.000 -1.000  0.172  1.828  1.828  3.000  5.828
    16            1231553552                            (89)   (56)   (37)(58)(69)   (14)(20)(58)(69)
145     ------+++ --++--+ +-+-++ +++++ +--- --+ +- +    -3.828 -3.828 -2.464 -1.000 -1.000 -1.000  1.828  1.828  4.464  5.000
    16            1221127271                            (23)(68)(79)   (26)(38)(50)(79)   (14)(50)(79)   (15)(40)(68)(79)
146     --------+ ---+++- ++-++- ++-++ ++-+ --- ++ +    -3.828 -3.686 -3.000 -3.000  0.059  1.000  1.828  3.000  3.000  4.627
     8            1231553552                            (56)(89)   (37)(58)(69)   (14)(20)(58)(69)
147     -------++ ---++-+ ++--+- +++-+ +++- +-+ +- -    -3.828 -3.686 -2.464 -2.236 -1.000  0.059  1.828  2.236  4.464  4.627
     4            1133566599                            (34)(58)(67)(90)   (12)(90)
148     -------++ --+++-- +-++-+ +-++- +--- --+ +- +    -3.828 -3.606 -3.000 -2.464 -1.000  1.000  1.828  3.000  3.606  4.464
     4            1231537527                            (36)(58)   (14)(29)(58)(70)
149     ------+++ --++--+ +-+-+- +-+-+ ++-- -+- +- -    -3.828 -3.606 -3.000 -2.236 -1.828  1.828  2.236  3.000  3.606  3.828     S
     4            1224167674                            (23)(68)(79)   (15)(40)(68)
150     --------+ ---+++- ++-++- ++-++ ++++ --- ++ +    -3.828 -3.547 -3.494 -1.876 -0.110  0.142  1.828  2.604  3.000  5.280
     4            1231563562                            (37)(58)(69)   (14)(20)(58)(69)
151     -------++ --+++-+ +++++- ++++- ---+ ++- +- +    -3.828 -3.472 -3.000 -1.000 -1.000 -1.000  1.828  3.000  3.000  5.472
    32            1134434411                            (78)   (45)   (36)(47)(58)   (12)(90)   (19)(20)(47)(58)   (10)(29)(47)(58)
152     ------+++ --++--+ +-+-++ +-++- +--- --+ +- -    -3.828 -3.472 -3.000 -1.000 -1.000 -1.000  1.828  3.000  3.000  5.472
     4            1231527375                            (26)(38)(50)(79)   (14)(50)(79)
153     -------++ ---++-+ ++---+ ++++- +++- ++- +- +    -3.828 -3.472 -3.000 -1.000 -1.000 -1.000  1.828  3.000  3.000  5.472
    32            1133555511                            (78)   (56)   (34)(57)(68)   (12)(90)   (19)(20)(57)(68)   (10)(29)(57)(68)
154     ------+++ --++--+ ++++++ +++++ +--+ --+ ++ +    -3.828 -3.472 -1.000 -1.000 -1.000 -1.000 -1.000  1.828  5.000  5.472
    64            1221166661                            (89)   (67)   (45)   (23)(68)(79)   (14)(50)(68)(79)   (10)
155     ------+++ --++--+ +-+-++ +-+-+ ++-- -+- +- -    -3.828 -3.000 -3.000 -3.000 -1.828  1.828  3.000  3.000  3.000  3.828     S
    16            1221166661                            (45)(67)(89)   (23)(68)(79)   (14)(50)(79)   (10)(69)(78)
156     -------++ --+++-- +-++-+ +++++ -+-- +++ -- +    -3.828 -3.000 -3.000 -2.064 -1.000  0.305  1.000  1.828  4.759  5.000
     4            1231535828                            (36)(57)   (14)(29)(57)(80)
157     ------+++ --++--+ +-+-++ ++++- +--- --+ +- +    -3.828 -3.000 -3.000 -1.828 -1.000 -1.000  1.828  3.000  3.828  5.000
     8            1221527275                            (23)(68)(79)   (26)(38)(50)(79)   (14)(50)(79)
158     -------++ --+++-- +-++-+ +++++ -+-- +++ -+ +    -3.828 -3.000 -2.722 -2.464 -1.000  0.492  1.000  1.828  4.464  5.230
     4            1231535828                            (36)(57)   (14)(29)(57)(80)
```

```
159      -----+++ --++--+ +-+-++ ++++- +--- -++ +- +     -3.828 -3.000 -2.464 -2.464 -1.000 -1.000  1.828  3.000  4.464  4.464
     8            1221527275                              (23)(68)(79)   (26)(38)(50)(79)    (14)(50)(79)
160      -----+++ --++--+ ++++++ +++++ +--+ -++ ++ +     -3.828 -3.000 -2.464 -1.000 -1.000 -1.000  0.172  1.828  4.464  5.828
    16            1221167671                              (45)   (23)(68)(79)   (14)(50)(68)(79)    (10)
161      ------++ -++++-- ++++++ +++-+ ++-+ ++- +- +     -3.828 -3.000 -2.123 -1.828 -1.000 -1.000  1.000  1.828  3.828  6.123
    32            1111555599                              (78)   (56)   (57)(68)(90)   (12)(34)(90)   (13)(24)(90)
162      -----+++ --++--+ ++++++ +++++ ++++ +++ ++ +     -3.828 -3.000 -1.000 -1.000 -1.000 -1.000 -1.000  1.828  3.000  7.000
    64            1221166661                              (89)   (67)   (45)   (23)(68)(79)   (14)(50)(68)(79)   (10)
163      ------++ ++++++- +++++-+ +++-+ +++- ++- ++ +    -3.828 -2.604 -2.583 -1.000 -1.000 -1.000  0.110  1.828  3.494  6.583
    16            1134444399                              (67)   (45)   (46)(57)(90)   (12)(38)(90)
164      ------++ ---+-++ ++--+- ++++- ++++ +-- ++ +     -3.686 -3.606 -3.340 -1.000 -1.000  0.059  0.378  3.606  3.962  4.627
     4            1133555599                              (34)(57)(68)(90)   (12)(56)(78)(90)
165      ------++ --+++-- +-++++ +++++ ++-- --+ +- -     -3.686 -3.526 -3.000 -2.031 -1.000  0.059  1.365  3.000  4.192  4.627
     4            1133557799                              (78)(90)   (12)(34)(56)(90)
166      ------++ ---++-+ ++--+- ++++- +++- +-+ -+ -     -3.686 -3.472 -2.236 -1.000 -1.000 -1.000  0.059  2.236  4.627  5.472
    16            1133555599                              (78)   (56)   (34)(57)(68)(90)   (12)(90)
167      ------++ --++++-+ +-+++- ++++- ++-+ --- ++ -    -3.686 -3.000 -3.000 -2.464 -1.000  0.059  1.000  3.000  4.464  4.627
     8            1133551199                              (34)(56)(90)   (12)(78)(90)   (17)(28)(90)
168      ------++ --++++-+ +-+++- ++++- ++-+ --- ++ +    -3.686 -3.000 -3.000 -1.828 -1.828  0.059  1.000  3.828  3.828  4.627
     8            1133551199                              (34)(56)(90)   (12)(78)(90)   (17)(28)(90)
169      ------++ --+++-- +-++++ +++++ ++-- --+ +- +     -3.686 -3.000 -2.654 -2.236 -1.677  0.059  1.418  2.236  4.627  4.913
     4            1133557799                              (78)(90)   (12)(34)(56)(90)
170      -----+++ --++--+ +-+-+- +-++- ++-- -++ -+ +     -3.606 -3.606 -3.606 -1.000 -1.000 -1.000  3.000  3.606  3.606  3.606
    12            1222112122                              (234)(586)(790)   (15)(37)(49)(68)
171      -----+++ --++--+ +-+++- ++-+- +--+ --+ ++ +     -3.606 -3.606 -2.722 -1.000 -1.000 -1.000  0.492  3.606  3.606  5.230
     4            1224462621                              (23)(45)(68)(79)   (10)(27)(39)(68)
172      ------++ --++++-+ +-+++- +-++- ++-+ +-+ -+ +    -3.606 -3.472 -3.000 -1.000 -1.000 -1.000  1.000  3.000  3.606  5.472
     8            1134466311                              (45)(67)   (12)(90)   (19)(20)(38)(67)   (10)(29)(38)(67)
173      -----+++ --+---+ +++++- ++++- ---- +++ +- -     -3.606 -3.472 -3.000 -1.000 -1.000 -1.000  1.000  3.000  3.606  5.472
    16            1234421443                              (89)   (45)   (26)(30)(48)(59)   (17)(30)(48)(59)
174      ------++ --+++-+ +-+++- +-++- +--+ --+ +- +     -3.606 -3.000 -3.000 -3.000 -1.000  1.000  1.000  3.000  3.606  5.000
    24            1133333399                              (45)(67)   (34)(56)(78)(90)   (12)(90)
175      ------++ --+++-+ +-+++- +-++- +--+ --+ +- -     -3.606 -3.000 -3.000 -2.236 -1.606  1.000  1.000  2.236  3.606  5.606
    24            1133333399                              (45)(67)   (34)(56)(78)(90)   (12)(90)
176      -----+++ --++--+ +-+-++ ++++- ++-- --+ +- +     -3.606 -3.000 -2.464 -2.236 -1.828 -1.000  2.236  3.606  3.828  4.464
     4            1221567675                              (23)(68)(79)   (14)(50)(79)
177      -----+++ -+++--- ++++++ ++--- +--- --- ++ +     -3.606 -3.000 -1.000 -1.000 -1.000 -1.000 -1.000  1.000  3.606  7.000
   144            1221555555                              (90)   (89)   (67)   (56)   (23)(58)(69)(70)   (14)(58)(69)(70)
178      ------++ ++++++-+ +++++-+ +++-+ +++- ++- +- -   -3.606 -2.583 -2.236 -1.000 -1.000 -1.000 -1.000  2.236  3.606  6.583
   144            1133333399                              (78)   (67)   (45)   (34)   (36)(47)(58)(90)   (12)(90)
179      ------++ --+++-- +-++++ +++++ ++-- +-+ +- -     -3.547 -3.526 -2.236 -2.031 -1.876  0.142  1.365  2.236  4.192  5.280
     4            1133557799                              (78)(90)   (12)(34)(56)(90)
180      ------++ --++++-+ +-+++- ++++- ++-+ +-- ++ -    -3.547 -3.494 -2.464 -1.876 -1.000 -0.110  0.142  2.604  4.464  5.280
     4            1133557799                              (34)(56)(90)   (12)(78)(90)
181      ------++ --+++-- +-++++ +++++ ++-- +-+ +- +     -3.547 -3.000 -2.654 -1.876 -1.677  0.142  1.000  1.418  4.913  5.280
     4            1133557799                              (78)(90)   (12)(34)(56)(90)
182      ------++ --++++-+ +-+++- ++++- ++-+ +-- ++ +    -3.547 -3.000 -2.604 -1.876 -1.828  0.110  0.142  3.494  3.828  5.280
     4            1133557799                              (34)(56)(90)   (12)(78)(90)
183      ------++ --+++-+ +-+++- ++++- ++-+ +-+ +- -     -3.547 -3.000 -2.236 -2.064 -1.876  0.142  0.305  2.236  4.759  5.280
     4            1133557799                              (34)(56)(78)(90)   (12)(90)
184      ------++ -++++-- ++++++ -++-+ +++- +-+ +- +     -3.494 -3.000 -3.000 -1.828 -1.606 -0.110  1.000  2.604  3.828  5.606
     8            1111557799                              (56)(78)(90)   (12)(34)(90)   (13)(24)(90)
185      ------++ -+++++-+ +++++- +++-- ++-+ ++- ++ -    -3.494 -3.000 -2.604 -2.123 -1.000 -0.110  0.110  2.604  3.494  6.123
     4            1133566599                              (34)(67)(90)   (12)(58)(90)
186      ------++ ++++++- +++++-+ +++-+ +++- ++- ++ -    -3.494 -3.000 -2.583 -1.000 -1.000 -1.000 -0.110  2.604  3.000  6.583
    16            1134444399                              (67)   (45)   (46)(57)(90)   (12)(38)(90)
187      -----+++ -+++--- ++++++ ++--- +-++ -++ ++ +     -3.472 -3.000 -2.722 -1.000 -1.000 -1.000  0.492  1.000  5.230  5.472
    16            1221566566                              (90)   (67)   (23)(58)(69)(70)   (14)(58)(69)(70)
188      ----++++ ++++--+ ++---+ +-++- -++- +++ +- -     -3.472 -3.000 -2.464 -1.000 -1.000 -1.000 -1.000  3.000  4.464  5.472
    64            1222552551                              (89)   (56)   (34)   (23)(47)(58)(69)   (10)(58)(69)
```

```
189      ------++ +++++-- ++++-- +++-+ +++- +++ ++ +     -3.340 -3.000 -2.583 -1.000 -1.000 -1.000  0.378  1.000  3.962  6.583
   16             1133553399                             (78)   (56)(90)   (34)   (12)(37)(48)(90)
190      ------++ -++++-+ +++++- +++-- ++-- +++ ++ +     -3.340 -3.000 -2.236 -2.123 -1.000 -1.000  0.378  2.236  3.962  6.123
   16             1133555599                             (78)   (56)   (34)(90)   (12)(57)(68)(90)
191      ----++++ -++--++ ++++-- --+-+ +-+- --+ +- -     -3.000 -3.000 -3.000 -3.000 -3.000  3.000  3.000  3.000  3.000  3.000    S
  720             1111111111                             (56)(78)(90)   (34)(79)(80)   (35)(46)(85)   (23)(57)(68)   (12)(70)(89)
192      ------++ -++++-- ++++++ -++-+ +++- --+ +- -     -3.000 -3.000 -3.000 -3.000 -1.828  1.000  1.000  3.000  3.828  5.000
   16             1111555599                             (56)(78)(90)   (57)(68)   (12)(34)(90)   (13)(24)(90)
193      ----++++ -++--++ ++++-- --+-+ +-+- -+- -+ -     -3.000 -3.000 -3.000 -3.000 -1.000 -1.000  3.000  3.000  3.000  5.000
   48             1111555555                             (56)(78)(90)   (34)(79)(80)   (23)(57)(68)(90)   (12)(70)(89)
194      -----+++ -+++--- ++++++ -+--+ +++- --+ +- +     -3.000 -3.000 -3.000 -3.000 -1.000 -0.236  1.000  3.000  4.236  5.000
    4             1221567576                             (23)(58)(60)(79)   (14)(58)(60)(79)
195      ----++++ -++--++ ++++-- +---+ ++-+ +-- ++ +     -3.000 -3.000 -3.000 -3.000 -0.236 -0.236 -0.236  4.236  4.236  4.236
   60             1111111111                             (234)(579)(680)   (25)(39)(47)(68)   (12)(34)(56)(90)
196      ------++ -++++-- ++++++ -++-+ +++- +-+ +- -     -3.000 -3.000 -3.000 -2.604 -1.606  0.110  1.000  3.000  3.494  5.606
    8             1111557799                             (56)(78)(90)   (12)(34)(90)   (13)(24)(90)
197      ------++ -++++-- ++++++ -++-- ++++ --+ +- +     -3.000 -3.000 -3.000 -2.464 -2.236  1.000  1.000  2.236  4.464  5.000
   24             1111117799                             (78)(90)   (35)(46)   (12)(34)(56)(90)   (13)(24)(56)(90)
198      -----+++ -+++--- ++++++ -+--+ +++- -+- ++ -     -3.000 -3.000 -3.000 -2.464 -2.064  0.305  1.000  3.000  4.464  4.759
    4             1221567765                             (23)(50)(69)(78)   (14)(50)(69)(78)
199      ----++++ -++--++ ++++-- --+-+ +-+- -+- -+ +     -3.000 -3.000 -3.000 -2.464 -1.828 -1.000  3.000  3.000  3.828  4.464
   16             1111555599                             (56)(78)(90)   (23)(57)(68)(90)   (12)(34)(78)(90)
200      ------++ -++++-- ++++++ -++-- ++++ +-+ +- +     -3.000 -3.000 -3.000 -2.464 -1.606  1.000  1.000  1.000  4.464  5.606
   24             1111117799                             (78)(90)   (35)(46)   (12)(34)(56)(90)   (13)(24)(56)(90)
201      ------++ -++++-- ++++++ -++-+ +++- +-- ++ +     -3.000 -3.000 -3.000 -2.236 -1.606 -0.236  1.000  2.236  4.236  5.606
    8             1111557799                             (56)(90)   (12)(34)(78)(90)   (13)(24)(78)(90)
202      ------++ -++++-- ++++++ +++-+ ++-+ ++- +- -     -3.000 -3.000 -3.000 -2.123 -1.000 -1.000  1.000  3.000  3.000  6.123
   32             1111555599                             (78)   (56)   (57)(68)(90)   (12)(34)(90)   (13)(24)(90)
203      ------++ -++++-- ++++++ +++-- ++-+ ++- ++ +     -3.000 -3.000 -3.000 -2.123 -1.000 -0.236  1.000  1.000  4.236  6.123
    8             1111566599                             (67)(90)   (12)(34)(58)(90)   (13)(24)(58)(90)
204      -----+++ -+++--- ++++++ ++--- +--+ -++ ++ +     -3.000 -3.000 -3.000 -2.064 -1.000  0.172  0.305  1.000  4.759  5.828
    4             1221567567                             (23)(58)(69)(70)   (14)(58)(69)(70)
205      ----++++ -++--++ ++++-- --+-+ +-+- ++- -+ +     -3.000 -3.000 -3.000 -1.828 -1.828 -1.828  3.000  3.828  3.828  3.828
   48             1111117777                             (56)(78)(90)   (34)(79)(80)   (35)(46)(89)   (12)(70)(89)   (13)(24)(78)
206      -----+++ ++++--- +++-++ ++-++ ++-+ +-+ -+ +     -3.000 -3.000 -1.000 -1.828 -1.000 -1.000  0.172  3.000  3.828  5.828
   16             1214444882                             (67)   (45)   (46)(57)(89)   (13)(20)(69)
```

```
      TABLE B.   TRANSITIVE STRONGLY REGULAR GRAPHS ASSOCIATED WITH REGULAR TWO-GRAPHS OF ORDER 16,28,36.
   EACH GRAPH HAS TWO OR MORE LINES:
   LINE(S) A:  TWO-GRAPH IDENTIFICATION NUMBER, PARAMETERS OF THE STRONGLY REGULAR GRAPH, UPPER HALF OF THE
               (-1,1)-ADJACENCY MATRIX IN OCTAL REPRESENTATION;
   LINE(S) B:  ORDER OF AUT, GENERATORS OF AUT.

GRAPHS ASSOCIATED WITH TWO-GRAPHS OF ORDER 16:

 1      (16,3,3,-5)        770006160100700007711122211171122117 1217
  1152        1:(6,7)(9,10)(12,13)(15,16);  2:(5,6)(8,9)(11,12)(14,15);  3:(3,4)(11,14)(12,15)(13,16);  4:(2,3)(8,11)(9,12)(10,13);
              5:(2,5)(3,6)(4,7)(9,11)(10,14)(13,15);  6:(1,2)(5,8)(6,9)(7,10);
 1      (16,3,3,-5)        7700061600446022142125025103255510362564
   192        1:(3,4)(5,6)(8,9)(11,14)(12,13)(15,16);  2:(2,3)(4,5)(6,7)(9,11)(10,12)(14,15);  3:(1,2)(5,8)(6,9)(7,10)(11,12)
              (13,14);
 1      (16,-5,3,-5)       7774077071633252661335731665333635317777
  1920        1:(4,5)(7,8)(9,10)(12,13);  2:(3,4)(6,7)(10,11)(13,14);  3:(3,6)(4,7)(5,8)(15,16);  4:(2,3)(7,9)(8,10)(14,15);
              5:(1,2)(9,12)(10,13)(11,14);
 1      (15,-2,3,-5)       776017070644644521137707 15251375537
   720        1:(4,5)(8,9)(11,12)(13,14);  2:(3,4)(7,8)(10,11)(14,15);  3:(2,3)(6,7)(11,13)(12,14);  4:(2,6)(3,7)(4,8)(5,9);
              5:(1,2)(7,10)(8,11)(9,12);

GRAPHS ASSOCIATED WITH TWO-GRAPHS OF ORDER 28:

 1      (28,3,3,-9)        7777000007701740017204170071021070610211150204225010211377740017417007210706211150422502113776017070 64464452
                           113770715251375537
 40320        1:(6,7)(12,13)(17,18)(21,22)(24,25)(26,27);  2:(5,6)(11,12)(16,17)(20,21)(23,24)(27,28);  3:(4,5)(10,11)(15,16)
              (19,20)(24,26)(25,27);  4:(3,4)(9,10)(14,15)(20,23)(21,24)(22,25);  5:(2,3)(8,9)(15,19)(16,20)(17,21)(18,22);
              6:(2,8)(3,9)(4,10)(5,11)(6,12)(7,13);  7:(1,2)(9,14)(10,15)(11,16)(12,17)(13,18);
 1      (27,-6,3,-9)       777774001777017407707163071553251255532213347217547131553252332651716315265451634376075643562334256057776773
                           736737777
 51840        1:(5,6)(8,9)(10,11)(13,14)(18,19)(26,27);  2:(4,5)(7,8)(11,12)(14,15)(19,20)(25,26);  3:(4,7)(5,8)(6,9)(16,17)(21,22)
              (23,24);  4:(3,4)(8,10)(9,11)(15,16)(20,21)(24,25);  5:(2,3)(10,13)(11,14)(12,15)(21,23)(22,24);
              6:(1,2)(13,18)(14,19)(15,20)(16,21)(17,22);

GRAPHS ASSOCIATED WITH TWO-GRAPHS OF ORDER 36:

 1      (36,5,5,-7)        7777700000017600377000041703607400001703770011463140360031463036070360600363607400172530063142560031462 46017
                           14506060631031417101417030036303006317052525225252512555252512515132455125245253213152312313142 65122614
 51840        1:(17,18)(19,20)(21,22)(23,24)(25,26)(27,28)(29,30)(31,32)(33,34)(35,36);  2:(9,10)(11,12)(13,14)(15,16)(25,27)
              (26,28)(29,31)(30,32)(33,35)(34,36);  3:(5,6)(7,8)(9,11)(10,12)(13,16)(14,15)(17,19)(18,20)(21,24)(22,23)(25,28)
              (26,27);  4:(5,7)(6,8)(13,15)(14,16)(21,23)(22,24)(29,33)(30,34)(31,35)(32,36);  5:(3,4)(7,8)(9,13)(10,14)(11,16)
              (12,15)(17,21)(18,22)(19,24)(20,23)(29,32)(30,31);  6:(3,5)(4,6)(11,13)(12,14)(19,21)(20,22)(25,29)(26,30)(27,31)
              (28,32);  7:(2,3)(5,9)(6,11)(7,10)(8,12)(14,15)(21,25)(22,26)(23,27)(24,28)(31,33)(32,34);  8:(1,2)(9,17)(10,18)
              (11,19)(12,20)(13,21)(14,22)(15,23)(16,24)(26,27)(30,31)(34,35);
 89     (36,5,5,-7)        777770000017600377000071403007700606014141710060141463001403031765252540065252560032712256103550272105 53021
                           6436444270211231144321431205033122407145632035474027113073034205612254412160515552712715245226512 4515
   432        1:(7,8)(9,10)(11,12)(13,14)(15,16)(17,18)(19,20)(21,22)(23,24)(25,26)(27,28)(29,30)(31,32)(33,34)(35,36);
              2:(3,5)(4,6)(9,13)(10,14)(11,15)(12,16)(17,21)(18,22)(19,23)(20,24)(25,35)(26,36)(27,28)(29,31)(30,32)(33,34);
              3:(2,7)(3,9)(4,11)(5,13)(6,15)(10,14)(12,16)(17,23)(18,25)(19,21)(20,26)(22,35)(24,36)(30,33)(32,34);
              4:(1,2)(3,4)(5,6)(9,19)(10,20)(11,17)(12,18)(13,23)(14,24)(15,21)(16,22)(27,34)(28,33)(29,31)(30,32);
              5:(1,3)(2,4)(7,25)(8,26)(11,18)(12,17)(13,28)(14,27)(15,29)(16,30)(21,31)(22,32)(23,33)(24,34)(35,36);
 90     (36,5,5,-7)        77777000000176003770000714030077006060141417100601414630014030317352724012552725002525300522525600532 470612E
                           30344670443071130245116220073106065062252252525525227151156130251226710232110706271153022270211063 7736
   648        1:(4,5,6)(11,13,15)(12,14,16)(19,21,23)(20,22,24)(25,27,29)(26,28,30)(31,35,33)(32,36,34);  2:(2,3)(5,6)(7,10)(8,9)
```

```
        (11,12)(13,16)(14,15)(19,26)(20,25)(21,30)(22,29)(23,28)(24,27)(31,33)(32,34);  3:(2,7)(3,9)(4,11)(5,15)(6,13)(8,10)
        (12,14)(20,33)(21,23)(22,35)(24,31)(25,29)(26,32)(28,34)(30,36);  4:(1,2)(5,6)(9,17)(10,18)(11,19)(12,20)(13,23)
        (14,24)(15,21)(16,22)(25,26)(27,30)(28,29)(31,33)(32,34);  5:(1,4)(2,5)(3,6)(7,32)(8,31)(9,33)(10,34)(13,20)(14,19)
        (15,26)(16,25)(17,35)(18,36)(23,27)(24,28);
92   (36,5,5,-7)     7777700000017600377000071403007700606014141710060141463C0140303176525254006525256003651145214254305230722122
                     706542416422141344611126146115051443424532740275320653110534306103521134033203456516533621304522515163
 1296   1:(7,8)(9,10)(11,12)(13,14)(15,16)(17,18)(19,20)(21,22)(23,24)(25,26)(27,28)(29,30)(31,32)(33,34)(35,36);
        2:(3,6)(4,5)(9,15)(10,16)(11,13)(12,14)(17,23)(18,24)(19,21)(20,22)(25,36)(26,35)(27,34)(28,33)(29,30)(31,32);
        3:(2,3)(4,5)(7,9)(8,10)(11,13)(12,14)(17,18)(19,27)(20,28)(21,25)(22,26)(23,29)(24,30)(31,32)(33,35)(34,36);
        4:(2,7)(3,9)(4,13)(5,11)(6,15)(17,21)(18,25)(19,23)(20,35)(22,26)(24,36)(27,29)(28,33)(30,34)(31,32);
        5:(1,2)(3,4)(5,6)(9,19)(10,20)(11,17)(12,18)(13,23)(14,24)(15,21)(16,22)(27,34)(28,33)(29,31)(30,32);
1    (36,-7,5,-7)    777777700001777600377007741703607477001703771435642167243413561720074737415735211565722566265133357223725151
                     6365262663275716227733306745544766470667342356721563427355073547133155166736735552572537536565553677714
 12096  1:(5,8)(6,9)(7,10)(11,12)(13,14)(15,16)(17,18)(19,20)(21,22)(23,24)(25,26)(27,28)(29,30)(31,32)(33,34)(35,36);
        2:(3,4)(5,6)(8,9)(11,12)(15,22)(16,21)(17,20)(18,19)(23,30)(24,29)(25,28)(26,27);  3:(3,5)(4,6)(8,11)(9,12)(10,13)
        (16,17)(18,19)(20,21)(24,27)(25,28)(26,29)(34,35);  4:(2,3)(6,7)(9,10)(11,15)(12,16)(13,17)(14,18)(21,22)(27,31)
        (28,32)(29,33)(30,34);  5:(1,2)(5,6)(8,9)(15,25)(16,26)(17,24)(18,23)(15,30)(20,29)(21,27)(22,28)(31,32);
89   (36,-7,5,-7)    777777700001777600377007741703607474631463147525252521732113627521335145653625453147266154371376416715226455
                     3672713570732316271375125132C17775315467472747764163270376566273251456537625747134757542476654377 67567
  144   1:(2,6)(3,8)(4,9)(5,11)(7,22)(10,18)(12,21)(13,20)(14,15)(17,19)(23,30)(24,31)(25,35)(27,33)(28,29)(32,36);
        2:(2,14)(3,18)(4,20)(5,21)(6,15)(7,22)(8,10)(9,13)(11,12)(17,19)(24,25)(26,34)(28,32)(29,36)(31,35);
        3:(1,2)(3,9)(4,11)(5,8)(7,14)(10,13)(15,30)(16,25)(17,24)(18,29)(19,27)(20,28)(21,26)(22,23)(31,33)(34,36);
        4:(1,3)(2,4)(5,6)(7,8)(9,15)(10,16)(11,23)(12,24)(13,31)(14,32)(17,18)(19,26)(20,25)(21,33)(22,34)(27,28)(29,35)
        (30,36);
90   (36,-7,5,-7)    7777777000017776003770077417036074746314631475252525217321136275213351456536254531472661543713732173312264 55
                     367361476274626627137512513201775727163656236776246313517734326565145653762572633571716417761437767567
  216   1:(3,4,5)(8,11,9)(10,12,13)(16,19,17)(18,20,21)(24,27,25)(26,28,29)(31,35,33)(32,36,34);  2:(2,14)(3,18)(4,21)(5,20)
        (6,15)(7,22)(8,10)(9,12)(11,13)(16,19)(24,25)(26,34)(28,36)(29,32)(33,35);  3:(1,2,6)(7,23,15)(10,26,18)(12,28,20)
        (13,29,21)(14,30,22)(16,24,31)(17,25,33)(19,27,35);  4:(1,3)(2,4)(5,6)(7,8)(9,15)(10,16)(11,23)(12,24)(13,31)(14,32)
        (17,18)(19,26)(20,25)(21,33)(22,34)(27,28)(29,35)(30,36);
92   (36,-7,5,-7)    7777777000017776003770077417C3607474631463147525252521732113627521335145653625453147266154371376227235226455
                     3675605770656347071375125132C17773235627564367672545165177646271613456537625746365707466176655177675 67
  432   1:(3,4,5)(8,11,9)(10,12,13)(16,19,17)(18,20,21)(24,27,25)(26,28,29)(31,35,33)(32,36,34);  2:(2,6)(3,8)(4,9)(5,11)
        (7,22)(10,18)(12,21)(13,20)(14,15)(17,19)(23,30)(24,31)(25,35)(27,33)(28,29)(32,36);  3:(2,14)(3,18)(4,21)(5,20)
        (6,15)(7,22)(8,10)(9,12)(11,13)(17,19)(24,27)(26,34)(28,36)(29,32)(31,33);  4:(1,2)(3,8)(4,9)(5,11)(7,14)(12,13)
        (15,30)(16,24)(17,27)(18,26)(19,25)(20,29)(21,28)(22,23)(32,36)(33,35);  5:(1,3)(2,4)(5,6)(7,8)(9,15)(10,16)(11,23)
        (12,24)(13,31)(14,32)(17,18)(19,26)(20,25)(21,33)(22,34)(27,28)(29,35)(30,36);
1    (35,-2,5,-7)    7777770000037740077600377000C0377703607417030360036076CC377001700170377C371463146363031463463630314631474637
                     12525274252526351322465513274631473146331315132313371247612647230654763266266522257636231774
 40320  1:(12,13)(14,15)(16,17)(18,19)(20,21)(22,23)(24,25)(26,27)(28,25)(30,31)(32,33)(34,35);  2:(8,9)(10,11)(16,17)(18,19)
        (20,22)(21,23)(24,26)(25,27)(28,30)(29,31)(32,34)(33,35);  3:(8,10)(9,11)(12,14)(13,15)(16,18)(17,19)(24,25)(26,27)
        (28,30)(29,31)(32,35)(33,34);  4:(4,5)(6,7)(14,15)(18,19)(20,24)(21,25)(22,26)(23,27)(28,32)(29,33)(30,34)(31,35);
        5:(4,6)(5,7)(12,16)(13,17)(14,18)(15,19)(22,23)(26,27)(28,32)(25,33)(30,35)(31,34);  6:(4,8)(5,9)(6,10)(7,11)(14,16)
        (15,17)(22,24)(23,25)(30,32)(31,33);  7:(2,3)(6,7)(10,11)(18,19)(20,28)(21,29)(22,30)(23,31)(24,32)(25,33)(26,34)
        (27,35);  8:(2,4)(3,5)(8,12)(9,13)(10,14)(11,15)(21,22)(24,28)(25,30)(26,29)(27,31)(33,34);  9:(1,2)(5,6)(9,10)
        (12,20)(13,21)(14,22)(15,23)(16,24)(17,25)(18,26)(19,27)(34,35);
```

```
     TABLE C.    CONFERENCE TWO-GRAPHS OF ORDER 10,26,50
  EACH TWO-GRAPH HAS FOUR OR MORE LINES:
  LINE(S) A:  TWO-GRAPH IDENTIFICATION NUMBER(S), UPPER HALF OF THE (-1,1)-ADJACENCY MATRIX OF A DESCENDANT (FIRST ROW
              DELETED) IN OCTAL REPRESENTATION;
  LINE(S) B:  ORDER OF AUT, AUTOMORPHISM PARTITION (THE NONTRIVIAL ORBITS);
  LINE(S) C:  GENERATORS OF NONTRIVIAL AUT;
  LINE(S) D:  PARAMETERS, NUMBER OF STRONGLY REGULAR GRAPHS OF THE INDICATED PARAMETERS,  [X,Y] MEANS X GRAPHS WITH
              ORDER OF AUT = Y.

TWO-GRAPHS OF ORDER 10:

 1        7411403645 63
      720          (1 2 3 4 5 6 7 8 9 10)
       1: (5,6)(7,8)(9,10);  2: (3,4)(7,9)(8,10);  3: (3,5)(4,6)(8,9);  4: (2,3)(5,7)(6,8);  5: (1,2)(7,10)(8,9);
       (10,-3,3,-3)          1    [1,120]
       (9,0,3,-3)            1    [1,72]

TWO-GRAPHS OF ORDER 26:

 1        77770000760176014606074414146300606367256126535052461325443634561665442326435127507702472663146245 36
     15600         (1 2 3 4 5 6 7 8 9 10 11 12 13 14 15 16 17 18 19 20 21 22 23 24 25 26)
       1: (7,8)(9,10)(11,12)(13,14)(15,16)(17,18)(19,20)(21,22)(23,24)(25,26);  2: (3,4)(5,6)(7,9)(8,10)(11,13)(12,14)(15,16)
       (17,23)(18,24)(19,21)(20,22)(25,26);  3: (3,5,4,6)(7,11,9,13)(8,12,10,14)(15,26,16,25)(17,22,23,20)(18,21,24,19);
       4: (3,7)(4,9)(5,11)(6,13)(8,10)(12,14)(15,17)(16,23)(20,25)(22,26);  5: (2,3)(4,5)(9,17)(10,18)(11,15)(12,16)(13,19)
       (14,20)(23,25)(24,26);  6: (1,2)(5,6)(11,13)(12,14)(15,26)(16,25)(17,22)(18,21)(19,24)(20,23);
       (26,-5,5,-5)          1    [1,120]
       (25,0,5,-5)           1    [1,600]
 2        777700007601760170000770707070115151425252513131363113072232641615730713262615162352454237435432632 47
       72          (1 3 4 9 10 11 18 19 20 24 25 26) (6 7 8 12 13 14 15 16 17 21 22 23)
       1: (6,7,8)(9,10,11)(12,14,13)(15,16,17)(18,20,19)(21,22,23)(24,26,25);  2: (3,4)(7,8)(10,11)(12,13)(15,21)(16,23)(17,22)
       (18,25)(19,24)(20,26);  3: (1,3)(7,8)(9,26)(10,24)(11,25)(12,22)(13,23)(14,21)(16,17)(18,19);  4: (1,9)(3,19)(4,24)
       (6,14)(7,23)(8,16)(12,22)(13,17)(18,26)(20,25);  5: (1,18,11,19)(3,26,25,24)(4,10,20,9)(6,17,15,23)(7,13,22,16)
       (8,21,12,14);
       (26,-5,5,-5)          2    [1,6]    [1,4]
       (25,0,5,-5)           4    [2,72]   [2,6]
 3        777700007601760170000770707070115151425252513131371221712643207347426534545261346355214543362265 5433
       39          (1 5 6 8 9 12 13 17 19 20 21 22 23) (2 3 4 7 10 11 14 15 16 18 24 25 26)
       1: (2,3,4)(5,6,8)(9,19,23)(10,18,26)(11,15,25)(12,20,22)(13,17,21)(14,16,24);  2: (1,5,9)(2,11,10)(3,7,26)(4,14,25)
       (6,23,13)(8,12,22)(16,18,24)(19,20,21);  3: (1,12,23,22,17,9,19,5,8,20,21,6,13)(2,16,15,3,4,26,18,14,25,10,11,7,24);
       (26,-5,5,-5)          2    [1,39]   [1,3]
       (25,0,5,-5)           2    [2,3]
 4        777700007601760170000770707070115151425252513131371221722623107547432332625434346355212643561457 0535
       6           (1 17 21) (2 18 26) (3 4 13) (5 12 14) (6 8 15 16 22 23) (9 11 19 20 24 25)
       1: (3,4)(6,8)(9,11)(12,14)(15,23)(16,22)(17,21)(18,26)(19,25)(20,24);  2: (1,17)(2,26)(3,13)(5,14)(6,23)(8,16)(9,24)
       (11,19)(15,22)(20,25);
       (26,-5,5,-5)          5    [2,6]    [2,2]    [1,1]
       (25,0,5,-5)           8    [2,6]    [4,2]    [2,1]

TWO-GRAPHS OF ORDER 50:

 1        777777700000000777600017776000176074007400776071463000360741760601714031463700146303171474000317146303636521652545416452 4
          652561606564256155456442427234707261172443325341175023452722652756226206527730646037213644115670532601473054453642612302 77
          051641056741651053551367201057535204257521764330427523446466543054751144550653074225726056751127343163423315561315451554 22
          34415353272750651422153263
```

```
117600          (1 2 3 4 5 6 7 8 9 10 11 12 13 14 15 16 17 18 19 20 21 22 23 24 25 26 27 28 29 30 31 32 33 34 35 36 37 38 39
                40 41 42 43 44 45 46 47 48 49 50)
   1: (9,10)(11,12)(13,14)(15,16)(17,18)(19,20)(21,22)(23,24)(25,26)(27,28)(29,30)(31,32)(33,34)(35,36)(37,38)(39,40)
   (41,42)(43,44)(45,46)(47,48)(49,50);  2: (3,4,5)(6,7,8)(9,15,11)(10,16,12)(13,17,19)(14,18,20)(21,25,23)(22,26,24)
   (27,41,33)(28,42,34)(29,40,32)(30,39,31)(35,45,49)(36,46,50)(37,44,48)(38,43,47);  3: (3,6)(4,7)(5,8)(9,21)(10,22)
   (11,23)(12,24)(13,14)(15,25)(16,26)(17,18)(19,20)(27,36)(28,35)(29,48)(30,47)(31,43)(32,44)(33,50)(34,49)(37,40)(38,39)
   (41,46)(42,45);  4: (3,9,4,15,5,11)(6,21,7,25,8,23)(10,17,16,19,12,13)(14,22,18,26,20,24)(27,40,41,32,33,29)
   (28,49,42,35,34,45)(30,31,39)(36,37,46,44,50,48)(38,47,43);  5: (2,3)(5,6)(7,8)(11,13)(12,14)(15,27)(16,28)(17,29)
   (18,30)(19,35)(20,36)(21,34)(22,33)(23,31)(24,32)(25,37)(26,38)(41,43)(42,44)(47,49)(48,50);  6: (1,2)(4,5)(7,8)(9,11)
   (10,12)(17,19)(18,20)(21,23)(22,24)(27,46)(28,45)(29,44)(30,43)(31,47)(32,48)(33,50)(34,49)(35,42)(36,41)(37,40)(38,39);
   (50,-7,7,-7)          3    [1,336]    [1,100]    [1,28]
   (49,0,7,-7)           1    [1,2352]
2        777777770000000007776000177760001770700070007770707070007007707601616017600177000707007770700000777007070715555155151540555
         525525255033331331313301545332626414331366646105554633525403032744612554466316031352355450645307206553322530543366266606156
         334063314147264613035352325056606133505460633143503652150534723413366454655323067135265163356235613213436033630650474245 32
         16540734633565072536602240
3528            (2 3 4 5 6 7 8 9 10 11 12 13 14 15 16 17 18 19 20 21 22 23 24 25 26 27 28 29 30 31 32 33 34 35 36 37 38 39 40
                41 42 43 44 45 46 47 48 49 50)
   1: (9,10,11)(12,14,13)(15,17,16)(18,20,19)(21,23,22)(24,25,26)(27,29,28)(30,32,31)(33,34,35)(36,38,37)(39,40,41)
   (42,44,43)(45,47,46)(48,49,50);  2: (3,4,6,8,7,5)(9,15,23,24,20,13)(10,17,22,25,19,12)(11,16,21,26,18,14)
   (27,45,49,38,40,30)(28,46,48,36,39,31)(29,47,50,37,41,32)(33,43,34,42,35,44);  3: (3,9,4,16,6,22,8,24,7,18,5,12)
   (10,13,15,11,21,17,25,23,20,26,14,19)(27,28,46,47,50,49,38,36,39,41,32,30)(29,33,45,44,48,35,37,42,40,34,31,43);
   4: (2,3)(4,5)(7,8)(9,12,10,14,11,13)(15,32,17,31,16,30)(18,29,20,28,19,27)(21,35,23,33,22,34)(24,38,25,37,26,36)
   (39,43,40,42,41,44)(45,49,47,50,46,48);
   (50,-7,7,-7)          7    [1,63]    [1,42]    [1,6]    [2,4]    [2,3]
   (49,0,7,-7)           2    [1,3528]    [1,72]
3,4      777777770000000007776000177760001777000000007777700770077007700601760001760177000077770000770000770077770071616161616160364
         646407640735207526407646172417252524251325251325132513132513634417311312661132332616152534613223524554435456443166232705 11
         515625544163465115133263432256133035251343456462236170156345305461534346612351313070716265432512363441661513632252352545 64
         61706330623651307263462663
 126            (2 3 4 5 6 7 8) (9 10 11 12 13 14 15 16 17 18 19 20 21 22 23 24 25 26 27 28 29 30 31 32 33 34 35 36 37 38 39
                40 41 42 43 44 45 46 47 48 49 50)
   1: (9,10)(11,13)(12,14)(15,16)(17,18)(19,20)(21,22)(23,24)(25,26)(27,28)(29,30)(31,32)(33,38)(34,36)(35,37)(39,40)
   (41,42)(43,44)(45,50)(46,48)(47,49);  2: (9,11)(10,14)(12,13)(15,17)(16,19)(18,20)(21,26)(22,24)(23,25)(27,29)(28,31)
   (30,32)(33,34)(35,36)(37,38)(39,44)(40,42)(41,43)(45,46)(47,48)(49,50);  3: (3,6,8)(4,5,7)(9,15,22,10,16,21)
   (11,20,24,13,19,23)(12,17,25,14,18,26)(27,50,43,28,45,44)(29,48,41,30,46,42)(31,47,39,32,45,40)(33,35,36,38,37,34);
   4: (2,3,4)(5,6,8)(10,11,12)(15,33,39)(16,34,41)(17,35,40)(18,36,43)(19,37,42)(20,38,44)(21,27,45)(22,29,47)(23,28,46)
   (24,31,49)(25,30,48)(26,32,50);
   (50,-7,7,-7)          12    [1,9]    [4,3]    [7,1]
   (49,0,7,-7)           3    [1,126]    [1,18]    [1,3]
5        777777770000000007776000177760001760740074007760714630003607417606017140314637001463031714740003171463036370252734256425405
         273505650565112215412672541304701367545672624445343565613032765057250721750532523244211033324710411171267726433245165704 75
         026653530321453534161160705726122625534521622734426153546123157042631747220627702232726515451263463360463701727427264360 32
         604574235064621433146532 77
  48            (2 3 4 5 6 7 19 20 23 24 25 26 29 30 33 34 37 38 43 44 45 46 49 50) (9 10 11 12 13 14 15 16 17 18 21 22 27 28
                31 32 35 36 39 40 41 42 47 48)
   1: (9,10)(11,12)(13,14)(15,16)(17,18)(19,20)(21,22)(23,24)(25,26)(27,28)(29,30)(31,32)(33,34)(35,36)(37,38)(39,40)
   (41,42)(43,44)(45,46)(47,48)(49,50);  2: (2,3,6,7,5,4)(9,13,36,47,42,16)(10,14,35,48,41,15)(11,28,22,32,40,17)
   (12,27,21,31,39,18)(19,29,23,37,50,46)(20,30,24,38,49,45)(25,34,43)(26,33,44);  3: (2,19,5,50,6,23)(3,30,4,45,7,38)
   (9,15,42,48,36,14)(10,28,41,17,35,32)(11,22,40)(12,47,39,13,21,16)(18,27,31)(20,34,49,43,24,25)(26,37,33,29,44,46);
   (50,-7,7,-7)          12    [1,8]    [1,6]    [2,4]    [8,2]
   (49,0,7,-7)           4    [2,48]    [2,2]
6,7      777777770000000007776000177760001770700070007770707070007007707601516115600177001247206770460001371117073161465533511414475
         113256340525664651361212731326624705435367421550363236255230076024526327261055116744160463655136141150656430255662323403 35
         516127034073462323126575544303367640525452526254613454614361456422636346165362265303670256545333451613545445336035306331643
         313223532170666665374022 40
  24            (2 3 4 5 7 8 15 16 18 21 22 23 27 30 31 33 34 35 45 46 47 48 49 50) (9 10 11 12 13 14 17 19 20 24 25 26 28 29
                32 36 37 38 39 40 41 42 43 44)
   1: (2,3)(4,5)(7,8)(9,14)(10,13)(11,12)(15,31)(16,30)(17,32)(18,27)(19,29)(20,28)(21,35)(22,34)(23,33)(24,38)(25,36)
   (26,37)(39,43)(40,44)(41,42)(45,50)(46,49)(47,48);  2: (2,4,8)(3,5,7)(9,42,24)(10,44,25)(11,43,26)(12,39,37)(13,40,36)
```

```
         (14,41,38)(15,47,23)(16,46,22)(17,29,20)(18,45,35)(19,28,32)(21,27,50)(30,49,34)(31,48,33);   3: (2,15,21)(3,31,35)
         (4,45,16)(5,50,30)(7,22,47)(8,34,48)(9,44,26)(10,29,12)(11,13,19)(14,40,37)(17,41,43)(18,23,49)(20,36,24)(25,38,28)
         (27,33,46)(32,42,39);
         (50,-7,7,-7)        13    [6,4]    [1,3]    [2,2]    [4,1]
         (49,0,7,-7)          4    [2,24]   [2,1]
8,9         7777777700000000777600017776000177070007000777070707000070077707601616017600177000707007770700000777007070755511515151550552
            2545252552337607413125114666474556006165365706304750513551505116263036153116225443532131462153653024275134344146323367272l5
            6304726502534566122135671314423042741751426335411455632206333263163631423636220671557522532611362164473434544616112437116 6
            44532274751165472634606062
            6              (2 3 4 5 6 7) (9 17 18 31 34 41) (10 16 19 32 35 39) (11 15 20 30 33 40) (12 13 14 45 46 47) (21 25 29 37 42
                           48) (22 26 28 38 43 50) (23 24 27 36 44 49)
         1: (2,3,5)(4,6,7)(9,31,18)(10,32,19)(11,30,20)(12,14,13)(15,33,40)(16,35,39)(17,34,41)(21,37,42)(22,38,43)(23,36,44)
         (24,27,49)(25,29,48)(26,28,50)(45,46,47);   2: (2,4)(3,6)(5,7)(9,17)(10,16)(11,15)(12,47)(13,46)(14,45)(18,41)(19,39)
         (20,40)(21,29)(22,28)(23,27)(24,44)(25,42)(26,43)(30,33)(31,34)(32,35)(36,49)(37,48)(38,50);
         (50,-7,7,-7)        27    [4,3]    [3,2]    [20,1]
         (49,0,7,-7)         10    [2,6]    [8,1]
10,11       7777777700000000777600017776000177460006000777470317001700770350101743561614420166140677015033056304747506403625414155652 6
            1401740637474134723254111655703232255322213611652135532121366545232446546453312607161374600257270322725130322556423656541 3
            6605727143422110334556702553346551452246072554346634631065422652652505661502732554465115407723614665505045714534066663610 7
            64641651172155727240362565
            6              (2 3 15 20 27 32) (4 5 43) (6 7 18 30 46 48) (8 42 44) (9 10 17 29 40 41) (11 12 13 14 21 33) (16 19 26 28 31
                           35) (22 24 25 34 37 38) (23 36 45 47 49 50)
         1: (2,3)(6,7)(9,10)(11,12)(13,14)(15,27)(16,28)(17,29)(18,30)(19,31)(20,32)(21,33)(22,34)(23,36)(24,37)(25,38)(26,35)
         (40,41)(45,47)(46,48)(49,50);   2: (2,15,32,3,27,20)(4,43,5)(6,48,30,7,46,18)(8,44,42)(9,17,40,10,29,41)
         (11,33,14,12,21,13)(16,35,31,28,26,19)(22,25,37,34,38,24)(23,49,47,36,50,45);
         (50,-7,7,-7)        16    [4,3]    [12,1]
         (49,0,7,-7)         11    [2,6]    [2,2]    [7,1]
12,13       7777777700000000777600017776000177700000000777770077007700770060176001760177000077770000770000770077770071616161616160364
            6464640707352520707646461617252525242525131325251313252513136217451313306136132612362515613234324354434556445166243703215 1
            5516255305544651343415634235161331070713432654162351233262163544613321562561323251235646245701561633106623516513430655265 4
            34236216613625325647062663
            1008           (1 2 3 4 5 6 7 8) (9 10 11 12 13 14 15 16 17 18 19 20 21 22 23 24 25 26 27 28 29 30 31 32 33 34 35 36 37 38
                           39 40 41 42 43 44 45 46 47 48 49 50)
         1: (9,10)(11,13)(12,14)(15,16)(17,18)(19,20)(21,22)(23,24)(25,26)(27,28)(29,30)(31,32)(33,34)(35,36)(37,38)(39,44)
         (40,42)(41,43)(45,50)(46,48)(47,49);   2: (9,11)(10,14)(12,13)(15,17)(16,19)(18,20)(21,23)(22,25)(24,26)(27,32)(28,30)
         (29,31)(33,38)(34,36)(35,37)(39,40)(41,42)(43,44)(45,46)(47,48)(49,50);   3: (3,6,7)(4,5,8)(9,15,26,12,20,21)
         (10,17,25,13,19,23)(11,16,24,14,18,22)(27,38,49,29,36,46)(28,33,47,31,35,45)(30,34,48,32,37,50)(39,44,43,41,42,40);
         4: (2,3,4)(5,8,7)(10,12,11)(15,33,45)(16,35,46)(17,34,47)(18,37,48)(19,36,49)(20,38,50)(21,27,39)(22,29,40)(23,28,41)
         (24,31,42)(25,30,43)(26,32,44);   5: (1,2)(3,4)(5,7)(6,8)(15,21)(16,22)(17,23)(18,24)(19,25)(20,26)(27,38)(28,37)(29,36)
         (30,35)(31,34)(32,33)(39,44)(40,43)(41,42)(45,50)(46,49)(47,48);
         (50,-7,7,-7)         5    [1,18]   [1,12]   [1,6]    [1,4]    [1,2]
         (49,0,7,-7)          2    [1,126]  [1,24]
14          7777777700000000777600017776000177460006000777470257001740760314431612071434651207407421553103263463372041227104057063662 6
            2321661466051533171117030237345034524556651324163626725124314633213315246312417274164315054475264551326612411632266152323 7
            12136053456555533051160163202755133163116460325373064503340743651526315550032721627136666302425543525455325357405317014067
            15322132761527077162262461
            150            (1 3 4 8 11 15 16 20 22 26 27 28 30 33 35 36 38 39 40 44 46 47 48 49 50) (2 5 6 7 9 10 12 13 14 17 18 19 21
                           23 24 25 29 31 32 34 37 41 42 43 45)
         1: (5,7)(6,9)(8,11)(10,12)(13,14)(15,16)(17,21)(18,24)(19,23)(20,22)(27,28)(29,37)(30,33)(31,32)(35,38)(39,44)(40,47)
         (43,45)(46,48)(49,50);   2: (3,20)(4,15)(5,12)(6,24)(7,23)(8,26)(9,13)(10,19)(14,18)(17,25)(27,39)(28,38)(29,41)(30,50)
         (32,34)(33,46)(35,44)(36,40)(42,43)(48,49);   3: (1,3,4,26,36)(2,25,41,42,34)(5,17,31,10,9)(6,7,21,32,12)(8,50,48,16,38)
         (11,49,46,15,35)(13,18,37,23,43)(14,24,29,19,45)(20,47,44,30,27)(22,40,39,33,28);
         (50,-7,7,-7)         2    [1,150]  [1,6]
         (49,0,7,-7)          2    [2,6]
15,16       7777777700000000777600017776000177070007000777070707000070077707601616017600177000707007770700000777007070715555155151540555
            5255252550333313313133014647274470143516655162055645363451030475506064661553260615646351505465072065355445305465334660613 3
            5511315121272323160653615230366060747052605551465063621304674552165551345636461271526351366662535126146350556603604730456 1
            16321334333665074556302240
            63             (2 3 4 5 6 7 8) (9 10 11 18 19 20 21 22 23 30 31 32 36 37 38 42 43 44 45 46 47) (12 13 14 15 16 17 24 25 26
                           27 28 29 33 34 35 39 40 41 48 49 50)
```

1: (9,10,11)(12,14,13)(15,17,16)(18,20,19)(21,23,22)(24,25,26)(27,29,28)(30,32,31)(33,34,35)(36,38,37)(39,40,41)
(42,44,43)(45,47,46)(48,49,50); 2: (3,6,7)(4,8,5)(9,21,19)(10,23,18)(11,22,20)(12,16,26)(13,17,25)(14,15,24)(27,48,39)
(28,50,41)(29,49,40)(30,46,36)(31,47,37)(32,45,38)(42,44,43); 3: (2,3,4)(5,8,6)(12,27,15)(13,28,16)(14,29,17)(18,36,45)
(19,37,46)(20,38,47)(21,30,42)(22,31,43)(23,32,44)(24,33,39)(25,34,40)(26,35,41)(48,50,49);
(50,-7,7,-7) 6 [2,9] [4,3]
(49,0,7,-7) 4 [1,63] [1,9] [2,3]

17,18 777777770000000077760001777600017707000700077707060740074C7706142166010756170454407622164471430271714522520453514640726533
415047056534074205517154017232065166572451711152452323070336720653247123316354342463256634252243721353444512315514364436 26
215666544760323230447666545643010056664575760116414356311275112633277402066525237452355422630351313255113710434661060361623
63641433270165257323676036

21 (2 7 14 15 18 25 26) (3 4 6 9 10 16 17 19 23 27 28 30 32 34 35 37 38 39 42 44 46) (5 20 21 31 33 40 43) (8 11
24 36 41 48 50) (12 13 22 29 45 47 49)
1: (3,17,23)(4,16,9)(5,20,21)(6,10,19)(7,15,25)(8,24,11)(12,13,22)(14,26,18)(27,28,30)(29,49,47)(31,40,33)(32,34,42)
(35,46,38)(37,44,39)(41,48,50); 2: (2,7,25,26,15,18,14)(3,9,34,30,35,37,6)(4,27,44,17,42,46,10)(5,31,21,40,33,43,20)
(8,50,41,36,11,24,48)(12,45,29,47,13,49,22)(16,39,32,19,28,23,38);
(50,-7,7,-7) 8 [2,3] [6,1]
(49,0,7,-7) 6 [1,21] [4,3] [1,1]

TABLE D.   THE 7*7 GRAECO-LATIN SQUARES, AND THE KNOWN S(2,4,25)

THE 7*7 GRAECO-LATIN SQUARES

NR. 1
CORRESPONDS WITH TWO-GRAPH NR.
10,11 OF ORDER 50 IN TABLE C

A1	B2	C3	D4	E5	F6	G7
B6	C1	D7	E2	F3	G4	A5
C4	D5	E1	F7	G6	A2	B3
D3	E6	F5	G1	A4	B7	C2
E7	F4	G2	A3	B1	C5	D6
F2	G3	A6	B5	C7	D1	E4
G5	A7	B4	C6	D2	E3	F1

NR. 2
CORRESPONDS WITH TWO-GRAPH NR.
3,4 OF ORDER 50 IN TABLE C

A1	B2	C3	D4	E5	F6	G7
E2	A3	G1	B7	C6	D5	F4
D3	F1	A2	G5	B4	C7	E6
C4	E7	F5	A6	G3	B1	D2
B5	D6	E4	F3	A7	G2	C1
G6	C5	D7	E1	F2	A4	B3
F7	G4	B6	C2	D1	E3	A5

NR. 3
CORRESPONDS WITH TWO-GRAPH NR.
8,9 OF ORDER 50 IN TABLE C

A1	B2	C3	D4	E5	F6	G7
B5	C1	D6	E7	F4	G2	A3
C7	D3	E2	F1	G6	A5	B4
D2	E4	F5	G3	A7	B1	C6
E6	F7	G1	A2	B3	C4	D5
F3	G5	A4	B6	C2	D7	E1
G4	A6	B7	C5	D1	E3	F2

NR. 4
CORRESPONDS WITH TWO-GRAPH NR.
1 OF ORDER 50 IN TABLE C

A1	B2	C3	D4	E5	F6	G7
B3	C4	D5	E6	F7	G1	A2
C5	D6	E7	F1	G2	A3	B4
D7	E1	F2	G3	A4	B5	C6
E2	F3	G4	A5	B6	C7	D1
F4	G5	A6	B7	C1	D2	E3
G6	A7	B1	C2	D3	E4	F5

NR. 5
CORRESPONDS WITH TWO-GRAPH NR.
2 OF ORDER 50 IN TABLE C

A1	B2	C3	D4	E5	F6	G7
B4	C5	D6	E7	F1	G2	A3
C7	D1	E2	F3	G4	A5	B6
D3	E4	F5	G6	A7	B1	C2
E6	F7	G1	A2	B3	C4	D5
F2	G3	A4	B5	C6	D7	E1
G5	A6	B7	C1	D2	E3	F4

NR. 6
CORRESPONDS WITH TWO-GRAPH NR.
5 OF ORDER 50 IN TABLE C

A1	B2	C3	D4	E5	F6	G7
B4	C1	D6	A7	F3	G5	E2
C7	D5	E1	F2	G4	B3	A6
D2	A3	F4	G1	C6	E7	B5
E6	F7	G2	C5	B1	A4	D3
F5	G6	B7	E3	A2	D1	C4
G3	E4	A5	B6	D7	C2	F1

NR. 7
CORRESPONDS WITH TWO-GRAPH NR.
6,7 OF ORDER 50 IN TABLE C

A1	B2	C3	D4	E5	F6	G7
B3	F4	E1	G5	C6	A7	D2
C4	A5	G6	E7	F2	D3	B1
D5	C7	B4	F1	G3	E2	A6
E6	G1	D7	C2	A4	B5	F3
F7	E3	A2	B6	D1	G4	C5
G2	D6	F5	A3	B7	C1	E4

THE KNOWN S(2,4,25)

NR. 1
CORRESPONDS WITH TWO-
GRAPH NR. 14 IN TAB.C

9	10	12	16
6	18	19	25
1	7	17	25
1	8	14	19
1	2	6	11
1	4	16	21
10	11	13	17
7	16	19	20
2	8	16	18
2	9	15	20
2	3	7	12
2	5	17	22
11	12	14	18
8	17	20	21
3	9	17	19
3	6	10	21
3	4	8	13
1	3	18	23
12	13	15	19
9	18	21	22
4	10	18	20
4	7	11	22
4	5	9	14
2	4	19	24
6	13	14	20
10	19	22	23
5	11	19	21
5	8	12	23
1	5	10	15
3	5	20	25
7	14	15	21
11	20	23	24
1	12	20	22
1	9	13	24
6	8	15	22
12	21	24	25
2	13	21	23
2	10	14	25
6	7	9	23
13	16	22	25
3	14	22	24
3	11	15	16
7	8	10	24
14	16	17	23
4	15	23	25
4	6	12	17
8	9	11	25
15	17	18	24
5	6	16	24
5	7	13	18

NR. 2
CORRESPONDS WITH TWO-
GRAPH NR. 12,13 IN TAB.C

1	5	9	25
13	17	21	25
1	7	13	19
1	6	15	16
1	3	4	20
5	13	16	18
2	6	10	25
14	18	22	25
2	8	14	20
2	7	16	17
2	4	5	21
6	14	17	19
3	7	11	25
15	19	23	25
3	9	15	21
3	8	17	18
3	5	6	22
7	15	18	20
4	8	12	25
16	20	24	25
4	10	16	22
4	9	18	19
4	6	7	23
8	16	19	21
5	11	17	23
5	10	19	20
5	7	8	24
9	17	20	22
6	12	18	24
6	11	20	21
6	8	9	13
10	18	21	23
7	12	21	22
7	9	10	14
11	19	22	24
1	8	22	23
8	10	11	15
12	13	20	23
2	9	23	24
9	11	12	16
1	14	21	24
3	10	13	24
1	10	12	17
2	13	15	22
4	11	13	14
1	2	11	18
3	14	16	23
5	12	14	15
2	3	12	19
4	15	17	24

NR. 3
CORRESPONDS WITH TWO-
GRAPH NR. 15,16 IN TAB.C

8	16	24	25
1	9	17	25
2	8	11	21
2	3	5	9
2	10	18	25
3	8	12	22
3	4	6	10
5	10	16	19
10	11	13	17
3	11	19	25
4	8	13	23
4	5	7	11
6	11	16	20
11	12	14	18
3	13	18	24
1	18	19	21
4	12	20	25
5	8	14	17
1	5	6	12
7	12	16	21
12	13	15	19
4	14	19	24
2	19	20	22
5	13	21	25
6	8	15	18
2	6	7	13
1	13	16	22
9	13	14	20
5	15	20	24
3	20	21	23
6	14	22	25
7	8	9	19
1	3	7	14
2	14	16	23
10	14	15	21
6	9	21	24
4	17	21	22
7	15	23	25
1	8	10	20
1	2	4	15
3	15	16	17
9	11	15	22
7	10	22	24
5	18	22	23
4	9	16	18
9	10	12	23
1	11	23	24
6	17	19	23
2	12	17	24
7	17	18	20

NR. 4
CORRESPONDS WITH TWO-
GRAPH NR. 17,18 IN TAB.C

22	23	24	25
1	9	17	22
1	10	19	23
1	11	21	24
1	13	18	25
1	2	4	8
8	9	11	15
3	15	17	18
2	10	18	22
2	11	20	23
2	12	15	24
2	14	19	25
2	3	5	9
9	10	12	16
4	16	18	19
3	11	19	22
3	12	21	23
3	13	16	24
3	8	20	25
3	4	6	10
10	11	13	17
5	17	19	20
4	12	20	22
4	13	15	23
4	14	17	24
4	9	21	25
4	5	7	11
11	12	14	18
6	18	20	21
5	13	21	22
5	14	16	23
5	8	18	24
5	10	15	25
1	5	6	12
8	12	13	19
7	15	19	21
6	14	15	22
6	8	17	23
6	9	19	24
6	11	16	25
2	6	7	13
9	13	14	20
1	15	16	20
7	8	16	22
7	9	18	23
7	10	20	24
7	12	17	25
1	3	7	14
8	10	14	21
2	16	17	21

```
... TWO GRAPHS OF ORDER 6,14,18,30,38,42,46
EACH TWO-GRAPH HAS FOUR OR MORE LINES:
LINE(S) A:  TWO-GRAPH IDENTIFICATION NUMBER(S), UPPER HALF OF THE (-1,1)-ADJACENCY MATRIX OF A DESCENDANT (FIRST ROW
            DELETED) IN OCTAL REPRESENTATION;
LINE(S) B:  ORDER OF AUT, AUTOMORPHISM PARTITION (THE NONTRIVIAL ORBITS);
LINE(S) C:  GENERATORS OF NONTRIVIAL AUT;
LINE D:     PARAMETERS, NUMBER OF STRONGLY REGULAR GRAPHS OF THE INDICATED PARAMETERS.  [X,Y] MEANS X GRAPHS WITH
            ORDER OF AUT = Y.

TWO-GRAPHS OF ORDER 6:

 1       6114
        60          (1 2 3 4 5 6)
      1: (3,4)(5,6);  2: (2,3)(4,5);  3: (1,2)(5,6);
      (5,0,SQRT(5),-SQRT(5))        1     [1,12]

TWO-GRAPHS OF ORDER 14:

 1       77006160446225247165133326
      1092          (1 2 3 4 5 6 7 8 9 10 11 12 13 14)
      1: (3,4,6,8,7,5)(9,13,14,11,12,10);  2: (2,3)(4,5)(6,10)(7,9)(8,11)(13,14);  3: (1,2)(4,5)(6,7)(9,13)(10,14)(11,12);
      (13,0,SQRT(13),-SQRT(13))        1     [1,78]

TWO-GRAPHS OF ORDER 18:

 1       7760034170063142521523505134267125261732474560
      2448          (1 2 3 4 5 6 7 8 9 10 11 12 13 14 15 16 17 18)
      1: (3,4)(5,7)(6,8)(9,10)(11,12)(13,15)(14,16)(17,18);  2: (3,5,9,8,4,7,10,6)(11,18,16,15,12,17,14,13);  3: (2,3)(4,5)
      (7,11)(8,13)(9,12)(10,14)(15,17)(16,18);  4: (1,2)(5,6)(7,8)(9,10)(11,18)(12,17)(13,16)(14,15);
      (17,0,SQRT(17),-SQRT(17))        1     [1,136]

TWO-GRAPHS OF ORDER 30:

 1       77776000037401770030703417022311463113052165151502742711552424327124327157260705035270435524741636703604331436271525163271
         53444471145730
      12180         (1 2 3 4 5 6 7 8 9 10 11 12 13 14 15 16 17 18 19 20 21 22 23 24 25 26 27 28 29 30)
      1: (3,4,12,7,5,10,9)(6,11,15,13,14,16,8)(17,26,28,21,25,23,18)(19,24,30,29,20,27,22);
      2: (3,6,4,11,12,15,7,13,5,14,10,16,9,8)(17,22,26,19,28,24,21,30,25,29,23,20,18,27);  3: (2,3)(4,5)(6,7)(8,9)(10,17)
      (11,21)(12,20)(13,18)(14,19)(15,22)(16,23)(24,25)(26,28)(27,29);  4: (1,2)(4,9)(5,7)(6,8)(10,12)(11,16)(14,15)(17,24)
      (18,30)(19,26)(20,25)(21,27)(22,28)(23,29);
      (29,0,SQRT(29),-SQRT(29))        1     [1,406]
 2       77776000037401770036402017641607070411446075144175425325117073244470652627212372514550366267426145543046551351352334522533
         44274316740564
        18          (1 4 12 15 19 22) (2 3 6 8 13 14 20 21 29) (5 7 9 18 23 26 27 28 30) (10 11 16 17 24 25)
      1: (2,8,14)(3,21,6)(4,19,15)(5,23,27)(7,30,18)(9,26,28)(11,17,25)(13,29,20);  2: (1,4)(2,3)(9,30)(10,17)(11,16)(12,15)
      (13,14)(18,23)(19,22)(20,21)(24,25)(27,28);  3: (1,12,22)(2,3,29)(4,19,15)(5,9,30)(6,13,14)(7,27,28)(8,21,20)(10,16,24)
      (11,17,25)(18,23,26);
      (29,0,SQRT(29),-SQRT(29))        4     [2,3]    [2,2]
 3       77776000037401770036402017607407070453144074455175014723130541346464644764463152256531623231532312547510534346552621365126
         12473617312170
        18          (1 4 10 11 16 17 20 23 24) (2 3 6 8 9 25 26 27 30) (5 7 12 15 19 22) (13 14 18 21 28 29)
      1: (2,30)(3,27)(4,17)(5,15)(7,19)(8,25)(10,16)(12,22)(13,18)(14,28)(20,24)(21,29);  2: (1,4)(5,7)(6,8)(9,30)(11,20)
      (12,15)(13,21)(14,18)(15,22)(16,23)(26,27)(28,29);  3: (1,10,11,17,23,24)(2,6,3,9,25,26)(4,16,20)(5,7,19,12,22,15)
```

```
(8,27,30)(13,21,29,28,14,18);  4: (1,11,23)(2,3,25)(4,20,16)(5,19,22)(6,9,26)(7,12,15)(8,30,27)(10,17,24)(13,29,14)
(18,21,28);
(29,0,SQRT(29),-SQRT(29))      4    [2,3]    [2,2]
4      777760000374017700346030174506030714414064747065213073152621655312452447343354266417111741722345614323125616266322423353335
       44156512734324
       9          (1 3 24) (2 16 23) (4 5 6 11 13 15 18 19 22) (7 8 9 17 20 21 26 28 30) (10 12 14) (25 27 29)
1: (4,5,6)(7,8,9)(10,12,14)(11,13,15)(17,20,21)(18,19,22)(25,29,27)(26,30,28);  2: (1,3,24)(2,16,23)(4,18,11)(5,19,13)
(6,22,15)(7,30,20)(8,28,21)(9,26,17)(10,14,12)(25,27,29);
(29,0,SQRT(29),-SQRT(29))      6    [4,3]    [2,1]
5      77776000037401770034603017450603071441406474526525222713145325513305241754537024654155507472231523462314561526612644636336
       44515522705724
       6          (2 24) (3 23) (4 5 6 11 12 15) (7 8 9 17 20 21) (10 13 14 25 28 30) (18 19 22 26 27 29)
1: (4,5,6)(7,8,9)(10,13,14)(11,12,15)(17,20,21)(18,19,22)(25,30,28)(26,29,27);  2: (2,24)(3,23)(4,11)(5,12)(6,15)(7,20)
(8,21)(9,17)(10,25)(13,30)(14,28)(18,26)(19,29)(22,27);
(29,0,SQRT(29),-SQRT(29))      8    [2,6]    [2,3]    [4,1]
6      7777600003740177003640201760740707045300766045750151471315034521352433076245345523226660535255126256135422631653323503457 5
       0545073406 5670
       2          (2 4) (5 7) (6 8) (9 17) (11 27) (12 28) (13 26) (14 29) (15 25) (16 24) (18 19) (20 21)
1: (2,4)(5,7)(6,8)(9,17)(11,27)(12,28)(13,26)(14,29)(15,25)(16,24)(18,19)(20,21);
(29,0,SQRT(29),-SQRT(29))      18    [6,2]    [12,1]
```

TWO-GRAPHS OF ORDER 38:

```
1      777777000000776001776001707007007700746046074611630571052227215363011557123031146456642227354146352465053037213434512 4175
       272624721406563354435611451227523511221527232311603637051355276431543734661526426623445122667335 2240
  25308           (1 2 3 4 5 6 7 8 9 10 11 12 13 14 15 16 17 18 19 20 21 22 23 24 25 26 27 28 29 30 31 32 33 34 35 36 37 38)
  1: (3,4,6,12,9,5)(7,13,17,20,16,10)(8,14,18,19,15,11)(21,22,30,27,31,25)(23,34,36,29,32,24)(26,35,38,37,28,33);
  2: (3,7,15,5,10,19,9,16,18,12,20,14,6,17,8,4,13,11)(21,23,35,25,24,26,31,32,33,27,29,28,30,36,37,22,34,38);  3: (2,3)
  (4,5)(6,11)(7,10)(8,9)(12,21)(13,24)(14,25)(15,22)(16,23)(17,27)(18,29)(19,26)(20,28)(30,33)(31,32)(34,37)(35,36);
  4: (1,2)(4,5)(6,9)(7,11)(8,10)(13,15)(14,16)(17,19)(18,20)(21,38)(22,35)(23,34)(24,36)(25,37)(26,30)(27,33)(28,31)
  (29,32);
  (37,0,SQRT(37),-SQRT(37))      1    [1,666]
2      777777000000776001776001746006007746216016074303231231066045315332432323062565057453212065452057354266125614066437616612 32
       2307655035143154510772275053510427543334446750131072365706612343333432314362623615174610171436520657
  57              (1 3 4 5 8 9 10 11 14 15 16 19 23 26 29 31 34 35 38) (2 6 7 12 13 17 18 20 21 22 24 25 27 28 30 32 33 36 37)
  1: (2,28,37)(3,23,5)(4,29,38)(6,7,25)(8,16,10)(9,35,31)(11,34,15)(13,18,36)(14,26,19)(17,32,21)(20,33,27)(22,30,24);
  2: (1,3,10)(4,34,16)(5,8,9)(6,32,18)(7,30,20)(11,31,15)(12,36,25)(13,37,22)(14,38,23)(17,33,27)(19,35,29)(21,24,28);
  3: (1,11,19)(2,6,20)(3,9,14)(4,10,29)(5,16,31)(7,17,18)(8,34,26)(12,27,28)(13,33,22)(15,38,23)(24,30,25)(32,37,36);
  (37,0,SQRT(37),-SQRT(37))      2    [2,3]
3      777777000000776001776001764004007764174036074102471431073132211333204536162435152354470156471506123065067072702327177223 14
       54047652544655431033531125650656120675501655256440573233306655720343323307161554461704527401723706 55
  19              (1 3 4 5 7 9 10 13 17 19 22 23 24 25 26 28 31 34 38) (2 6 8 11 12 14 15 16 18 20 21 27 29 30 32 33 35 36 37)
  1: (1,3,9,4,17,31,5,25,19,26,24,10,13,22,28,38,34,7,23)(2,14,35,16,15,18,8,37,36,20,6,30,11,33,12,21,32,29,27);
  2: (1,10,4,28,5,7,26,3,13,17,38,25,23,24,9,22,31,34,19)(2,30,16,12,8,29,20,14,11,15,21,37,27,6,35,33,18,32,36);
  (37,0,SQRT(37),-SQRT(37))      2    [2,1]
4      777777000000776001776001707007007706446007070724430150176504300767403136134070175614307526463164531256232252643506751533 50
       4661431516562263154251533243264275124541626435461331343231336561022467552467020533556245075466447253
  18              (3 4 5 6 7 8 9 10 11 12 13 14 15 16 17 18 19 20) (21 22 23 24 25 26 27 28 29 30 31 32 33 34 35 36 37 38)
  1: (3,4,6,8,13,17,20,16,11,12,10,14,18,19,15,9,7,5)(21,35,37,29,32,27,30,24,22,33,36,26,23,34,38,28,31,25);
  (37,0,SQRT(37),-SQRT(37))      4    [2,18]    [2,1]
5      777777000000776001776001746006007746216016074321034331066215101733411256120673545306434217523445153664251516662643342634 70
       23071172710546465346506621665211323135305163056250533553674007153734332521330632600372624544754273 52
  12              (1 4 14 24 30 33) (2 3 7 9 10 15 17 18 19 20 28 38) (5 6 8 11 13 16 21 22 25 29 32 36) (12 23 34 37) (26 27
                  31 35)
  1: (2,10)(3,18)(4,33)(5,13)(6,29)(7,17)(8,21)(9,38)(11,36)(12,23)(14,30)(15,28)(16,25)(19,20)(22,32)(26,27)(31,35)
  (34,37);  2: (1,4,14)(2,3,20)(5,21,11)(6,16,8)(7,9,19)(10,38,28)(12,37,34)(13,22,29)(15,17,18)(24,33,30)(25,32,36)
  (26,35,31);
  (37,0,SQRT(37),-SQRT(37))      5    [2,3]    [1,2]    [2,1]
6      777777000000776001776001746006007746017017074207210215466006764612661435414551661072057056234515217721262562616623162535 60
```

```
 9          (2 4 7 10 18 25 26 31 34) (3 5 6 11 16 20 22 30 36) (8 12 17 24 28 29 33 37 38) (9 13 14 15 19 21 27 32 35)
 1: (2,4,31,26,18,7,34,25,10)(3,6,5,30,16,36,20,22,11)(8,12,33,29,28,37,24,38,17)(9,35,15,13,21,27,14,32,19);
 (37,0,SQRT(37),-SQRT(37))      6     [2,9]    [4,1]
7     77777700000077600177600170700700770700700707071541011515640476053160133463071304623313227632762071557150235064432713612646
      25274234736242522131266625155111124671463524265252265463715452730342747703424507420174766721362464 54
 9          (2 4 5 10 11 22 23 25 26) (6 7 8 9 21 24 27 28 29) (12 13 14 15 16 17 30 31 32) (18 19 20 33 34 35 36 37 38)
 1: (2,4,5)(6,7,8)(9,21,24)(10,22,25)(11,23,26)(12,31,17)(13,32,16)(14,30,15)(18,35,38)(19,33,37)(20,34,36)(27,29,28);
 2: (2,10,23,4,22,26,5,25,11)(6,28,21,7,27,24,8,29,9)(12,13,15,31,32,14,17,16,30)(18,19,34,35,33,36,38,37,20);
 (37,0,SQRT(37),-SQRT(37))      6     [2,9]    [4,1]
8     77777700000077600177600174600600774621601607430303433146403155321256072612073250711321652623514626503766017361263126261235
      17643642626454164143171325632110350371640765315502165267361463036172234255153347124643223346365606 15
 3          (2 24 36) (3 10 22) (4 14 27) (5 12 28) (6 18 30) (7 17 32) (8 25 29) (9 13 19) (11 35 37) (15 16 23) (20 34
            38) (21 31 33)
 1: (2,24,36)(3,22,10)(4,14,27)(5,28,12)(6,30,18)(7,17,32)(8,29,25)(9,13,19)(11,35,37)(15,16,23)(20,38,34)(21,31,33);
 (37,0,SQRT(37),-SQRT(37))      14    [2,3]    [12,1]
9     77777700000077600177600170700700770700700707071541011515640476053160133463071314423217237432724265557152024664525456612631
      35714145222447326131613225116631132272562324666442264563751452352353247721460507420174765306233703 16
 3          (2 4 5) (6 7 8) (9 21 24) (10 22 25) (11 23 26) (12 17 31) (13 16 32) (14 15 30) (18 35 38) (19 34 37) (20 33
            36) (27 28 29)
 1: (2,4,5)(6,7,8)(9,21,24)(10,22,25)(11,23,26)(12,31,17)(13,32,16)(14,30,15)(18,35,38)(19,34,37)(20,33,36)(27,29,28);
 (37,0,SQRT(37),-SQRT(37))      14    [2,3]    [12,1]
10,11 77777700000077600177600170700700770700700707071541011515640476053160133463071324423213633532760475457112215464526352612266
      11124353525645243131664625151531113463616125546542227446375505271434370732346410742017476306352664455
 3          (2 4 5) (6 7 8) (9 21 24) (10 22 25) (11 23 26) (12 17 31) (13 16 32) (14 15 30) (18 35 38) (19 34 36) (20 33
            37) (27 28 29)
 1: (2,4,5)(6,7,8)(9,21,24)(10,22,25)(11,23,26)(12,31,17)(13,32,16)(14,30,15)(18,35,38)(19,34,36)(20,33,37)(27,29,28);
 (37,0,SQRT(37),-SQRT(37))      14    [2,3]    [12,1]
```

TWO-GRAPHS OF ORDER 42:

```
1     77777760000003774001777000360740360176046314146060741463007474632035252037202532572032721062157224217665512454600273317252
      35342532206121570545123516272232437375132213444761523441076452153165036136317134116614516231271173044351336556131453504732
      3465154742626233317023563405 14
 34440      (1 2 3 4 5 6 7 8 9 10 11 12 13 14 15 16 17 18 19 20 21 22 23 24 25 26 27 28 29 30 31 32 33 34 35 36 37 38 39
            40 41 42)
 1: (3,4,8,7)(5,13,17,9)(6,15,19,11)(10,14,20,21)(12,16,22,18)(23,26,38,29)(24,37,27,25)(28,36,31,35)(30,34,40,41)
 (32,33,42,39);  2: (3,5,14,12,6,4,13,20,16,15,8,17,21,22,19,7,9,10,18,11)
 (23,42,31,27,34,26,39,35,25,40,38,32,28,24,41,29,33,36,37,30);  3: (2,3)(4,9)(5,7)(6,11)(8,10)(13,25)(14,29)(15,30)
 (16,32)(17,23)(18,27)(19,28)(20,31)(21,24)(22,26)(33,37)(34,35)(36,42)(38,40)(39,41);  4: (1,2)(4,7)(5,11)(6,9)(10,12)
 (13,19)(14,18)(15,17)(16,21)(20,22)(23,40)(24,33)(25,42)(26,34)(27,39)(28,36)(29,41)(30,38)(31,35)(32,37);
 (41,0,SQRT(41),-SQRT(41))      1     [1,820]
2     77777760000003774001777000374600300177461270036074141207535003701314645546160670515721141165225264661553243152440354474723
      11325622642563324474161437517811136511505660724352534247706254645527561450507233047311454534712620475622555631346642163611
      3617433061255426573162022667 44
 168        (1 2 3 4 5 6 7 8 9 10 11 12 13 14 15 16 17 18 19 20 21 22 23 24 25 26 27 28 29 30 31 32 33 34 35 36 37 38 39
            40 41 42)
 1: (2,23,39,28)(3,12,31,9)(4,32,19,21)(5,15,41,42)(6,38,13,20)(7,17,24,11)(8,29,34,35)(10,36,18,37)(14,33,25,16)
 (22,27,40,30);  2: (1,2,5)(3,4,6)(7,13,11)(8,30,27)(9,32,26)(10,31,25)(12,29,41)(14,24,28)(15,17,16)(18,40,36)(19,37,33)
 (20,38,34)(21,39,35)(22,23,42);
 (41,0,SQRT(41),-SQRT(41))      1     [1,4]
3     77777760000003774001777000374600300177461270036074151066101643540633263244644036270533205526433245255127071154335031436134
      54545522561474312621531723525316163515551440670613515251055444524577144662543250364747206260564232731523167413663470604257
      711266432416073461434751275244
 21         (1 2 3 4 5 6 12 17 18 19 24 25 27 28 29 30 31 34 35 38 41) (7 8 9 10 11 13 14 15 16 20 21 22 23 26 32 33 36
            37 39 40 42)
 1: (1,2,27)(3,6,28)(4,25,5)(7,32,16)(8,11,26)(9,15,42)(10,36,13)(12,19,38)(14,21,33)(17,31,18)(20,22,37)(23,40,39)
 (24,35,30)(29,34,41);  2: (1,3,41)(2,19,24)(4,34,12)(5,27,17)(6,31,38)(7,26,21)(8,23,36)(9,37,11)(10,42,32)(13,22,14)
 (15,40,33)(16,39,20)(18,29,30)(25,28,35);
```

```
        (41,0,SQRT(41),-SQRT(41))       2    [2,1]
4          77777760000003774001777000377000000177760370174076C2037000370174001777000376307143471412465223463125152267054516256630433 2
           1651650744535314517434263234513531150353521532233245514323626660323064531327244287125113171621663554254705163251547212171 6
           132460725325037055232673344530
  14                (1 2 3 4 5 6 7) (8 12 14 16 18 22 24 26 29 32 33 37 38 41) (9 11 13 17 15 20 23 27 28 30 34 36 40 42) (10 15
                    21 25 31 35 39)
        1: (8,12)(9,11)(13,17)(14,16)(18,22)(19,20)(23,27)(24,26)(28,30)(29,32)(33,37)(34,38)(38,41)(40,42);
        2: (1,2,4,7,6,3,5)(8,29,38,37,22,24,16)(9,30,42,36,19,27,13)(10,31,39,35,21,25,15)(11,28,40,34,20,23,17)
        (12,32,41,33,18,26,14);
        (41,0,SQRT(41),-SQRT(41))       4    [2,2]    [2,1]
5          777777600000037740017770003746003001774614700340761603070370017401614077434606714347015244730615512455512265626512346254 62
           4532652331434545133225465134446455726212467175741601407133524265325174361026615066624616562427151246717603111546213333306 6
           324707154646543247067013316174
  14                (1 2 3 4 5 6 7) (8 17 20 25 30 35 38) (9 10 15 16 18 19 26 27 31 32 36 37 35 40) (11 12 13 14 21 22 23 24 28
                    29 33 34 41 42)
        1: (9,10)(11,12)(13,14)(15,16)(18,19)(21,22)(23,24)(26,27)(28,29)(31,32)(33,34)(36,37)(35,40)(41,42);
        2: (1,2,5,6,3,7,4)(8,17,30,25,35,20,38)(9,15,32,27,36,19,39)(10,16,31,26,37,18,40)(11,23,21,28,41,34,14)
        (12,24,22,29,42,33,13);
        (41,0,SQRT(41),-SQRT(41))       4    [2,2]    [2,1]
6,7        77777760000003774001777000377000000177760370174076020370003701740017770003763071434714124652234631251522656145162563460 332
           1655214745117454516634471235053550550652711551323231517006666615423066213467115073112526171351263702654552165321546452307 6
           605445624651351237131632726114
  14                (1 2 3 4 5 6 7) (8 12 14 16 18 22 24 26 29 32 34 37 38 42) (9 11 13 17 15 20 23 27 28 30 33 35 39 41) (10 15
                    21 25 31 36 40)
        1: (8,12)(9,11)(13,17)(14,16)(18,22)(19,20)(23,27)(24,26)(28,30)(29,32)(33,35)(34,37)(38,42)(39,41);
        2: (1,2,4,5,3,7,6)(8,29,22,24,34,42,14,12,32,18,26,37,38,16)(9,30,19,27,35,41,17,11,28,20,23,33,39,13)
        (10,31,21,25,36,40,15);
        (41,0,SQRT(41),-SQRT(41))       4    [2,2]    [2,1]
8          7777776000000377400177700037700000017776037017407602034610364170005354626316311510725450712564413507501747446073341127142 6
           6134615706226155137322551270447641454730325465706251426323333230616342537016414164475251664066531622714463133132554524307 5
           07526343605554136253463164315 4
  14                (1 8 12 17 19 26 41) (2 23 28 31 33 35 36) (3 4 10 16 18 30 37) (5 6 7 24 25 38 39) (9 11 13 14 15 29 34) (20
                    21 22 27 32 40 42)
        1: (3,4)(5,6)(8,12)(9,11)(13,15)(16,18)(17,19)(20,22)(23,33)(24,39)(25,38)(26,41)(27,40)(28,35)(29,34)(30,37)(31,36)
        (32,42);  2: (1,8)(2,31)(3,10)(4,30)(5,7)(6,24)(9,34)(12,41)(13,14)(15,29)(17,26)(18,37)(20,42)(21,32)(22,27)(23,36)
        (28,33)(38,39);
        (41,0,SQRT(41),-SQRT(41))       6    [6,2]
9          7777776000000377400177700037700000017776037017407601034614344171044364646312007632567011454057542526313440674563603065632 2
           1553070365155603744323507330615232455234370511612544713416164565154524625334516253603225750057047431646646166427156226271 4
           226735111137142464537334036530
  8                 (2 21 30 36) (3 4 13 17 23 32 33 41) (5 12 24 39) (6 11 27 34) (7 10 14 22 26 31 35 38) (8 9 18 20) (15 19 29
                    37) (25 28 40 42)
        1: (3,4)(5,12)(6,11)(7,10)(8,9)(13,17)(14,22)(15,19)(23,41)(24,39)(25,40)(26,35)(27,34)(28,42)(29,37)(30,36)(31,38)
        (32,33);  2: (2,21)(3,32)(4,33)(5,39)(6,34)(7,14)(10,22)(11,27)(12,24)(13,41)(15,29)(17,23)(18,20)(19,37)(25,28)(26,38)
        (31,35)(40,42);  3: (2,30)(3,13)(4,41)(5,24)(7,31)(8,18)(9,20)(10,35)(11,34)(14,38)(17,32)(19,29)(21,36)(22,26)(23,33)
        (25,42);
        (41,0,SQRT(41),-SQRT(41))       10    [2,8]    [6,2]    [2,1]
10         7777776000000377400177700037460030017746127003607415044702167311025503701711205734103712036417642537436071144313052147454 7
           4261707215446453333634115134551524732265446022552722606365524714615523151245251633311423125724572513256127444237131606621 7
           4351361045764425164447177525 40
  8                 (2 17 26 33) (3 4 20 21 25 29 36 37) (5 10 15 16) (6 7 32 40) (8 9 30 38) (11 12 13 14 23 24 41 42) (18 19 27
                    34) (28 31 35 39)
        1: (3,4)(6,7)(8,9)(11,13)(12,14)(15,16)(18,19)(20,21)(23,24)(25,36)(26,33)(27,34)(28,35)(29,37)(30,38)(31,39)(32,40)
        (41,42);  2: (2,17)(3,25)(4,36)(5,10)(6,32)(7,40)(8,30)(9,38)(11,14)(12,13)(18,34)(19,27)(20,29)(21,37)(23,41)(24,42)
        (28,31)(35,39);  3: (2,26)(3,21)(4,37)(5,16)(6,40)(9,30)(10,15)(11,41)(12,24)(13,23)(14,42)(17,33)(18,27)(20,25)(29,36)
        (31,35);
        (41,0,SQRT(41),-SQRT(41))       10    [2,8]    [6,2]    [2,1]
11,12      7777776000000377400177700037700000017776037017407601034614344171044364646312105613566050474247552112674061535344562033632 2
           1553070365155603714323607330615132455234570511611546732417104565154505624374515253603425750057047431646646165427166226471 4
           1267351111371414656625740365 30
```

```
37) (23 28 40 42)
1: (3,4)(5,12)(6,10)(7,11)(13,17)(14,22)(15,19)(18,20)(23,41)(24,39)(25,40)(26,34)(27,35)(28,42)(29,37)(30,36)(31,38)
(32,33);  2: (2,21)(3,32)(4,33)(5,39)(6,35)(7,14)(8,9)(10,27)(11,22)(12,24)(13,41)(15,29)(17,23)(19,37)(25,28)(26,38)
(31,34)(40,42);  3: (2,30)(3,13)(4,41)(5,24)(7,31)(8,18)(9,20)(10,35)(11,34)(14,38)(17,32)(19,29)(21,36)(22,26)(23,33)
(25,42);
(41,0,SQRT(41),-SQRT(41))      10    [2,8]    [6,2]    [2,1]
13    7777776000000377400177700037070034017707203401607434523211140374360354256460605616547421514417452635045165133035063227256251730630441652331327213511652467205766243314752112371641417226715044551630706617214701536711642702166550622733074311451055541427332355650637555106137260
      7          (1 3 9 19 35 36 38) (2 10 28 29 30 33 41) (4 14 16 21 24 31 34) (5 7 12 18 20 25 26) (6 11 13 23 27 32 40) (8
                 15 17 22 37 39 42)
1: (1,3,38,36,9,35,19)(2,28,29,10,41,33,30)(4,16,31,14,24,21,34)(5,18,26,7,20,25,12)(6,11,23,32,13,27,40)
(8,39,22,37,42,17,15);  2: (1,35,36,3,19,9,38)(2,33,10,28,30,41,29)(4,21,14,16,34,24,31)(5,25,7,18,12,20,26)
(6,27,32,11,40,13,23)(8,17,37,39,15,42,22);
(41,0,SQRT(41),-SQRT(41))      6    [6,1]
14    7777776000000377400177700037460030017746127003607414146014144354001727244465515020737042527631112730353451340116033625625314505344542525557445075233242564527211726142314331535622145666550702270732265106572317020535473343151535226613151174072156322766005154271631366044717434
      7          (1 2 3 7 20 33 40) (4 8 19 26 28 34 36) (5 10 18 24 29 35 41) (6 11 22 23 30 31 32) (9 13 15 16 17 25 42) (12
                 14 21 27 37 38 39)
1: (1,2,33,20,7,3,40)(4,8,28,19,36,26,34)(5,41,10,18,35,24,29)(6,23,11,32,22,30,31)(9,17,25,42,15,16,13)
(12,21,37,39,27,38,14);  2: (1,3,20,2,40,7,33)(4,26,19,8,34,36,28)(5,24,18,41,29,35,10)(6,30,32,23,31,22,11)
(9,16,42,17,13,15,25)(12,38,39,21,14,27,37);
(41,0,SQRT(41),-SQRT(41))      6    [6,1]
15    7777776000000377400177700037460030017747007403607414606303140374342620276430107670765012447056447124551626155524626165125326243467262723444125221535555411157503552643567025212465625150331630217216730370627205437414324652332505553313116350631163255316256265044576566454724034
      6          (1 4 9 10 33 34) (2 22 25) (3 17 29) (5 11 41) (6 20 27) (7 8 15 18 30 31) (12 16 19 28 32 42) (13 14 21 26
                 37 38) (23 24 39) (35 36 40)
1: (1,4)(7,8)(9,10)(11,41)(12,42)(13,14)(15,31)(16,32)(17,29)(18,30)(19,28)(20,27)(21,26)(22,25)(23,24)(33,34)(35,36)
(37,38);  2: (1,9,34)(2,25,22)(3,17,29)(4,33,10)(5,41,11)(6,20,27)(7,31,30)(8,18,15)(12,28,16)(13,37,21)(14,26,38)
(19,42,32)(23,24,39)(35,40,36);
(41,0,SQRT(41),-SQRT(41))      10    [6,2]    [4,1]
16    7777776000000377400177700037460030017747007403607414606303140374252164274620522654317051154472426543461522365462626265225313225470722653066144652531125151457604375244704744671421721556152465351324336443055206357052625157025707444660563307313233512545431561313140366426735434
      6          (1 22 31) (2 3 7 8 34 36) (4 20 30) (5 18 28) (6 14 23) (9 10 16 19 25 29) (11 12 21 32 37 38) (13 15 17 24
                 26 27) (33 35 40) (39 41 42)
1: (2,3)(7,8)(9,10)(11,12)(13,24)(14,23)(15,26)(16,25)(17,27)(18,28)(19,29)(20,30)(21,32)(22,31)(33,35)(34,36)(37,38)
(41,42);  2: (1,22)(2,7)(3,36)(4,30)(5,28)(6,14)(8,34)(9,16)(10,29)(11,21)(12,38)(13,26)(15,17)(19,25)(24,27)(32,37)
(35,40)(39,42);
(41,0,SQRT(41),-SQRT(41))      10    [6,2]    [4,1]
17    777777600000037740017770003707003401770630340160763106321114417406531534605600724136514153336212167014266516552446563045575541542456072351510245057464312536371646072130073333227155205150555143644771062174507062716262462752530262654570067426313133754266544702626357065025624
      5          (2 4 7 13 14) (3 15 16 36 42) (5 9 21 30 39) (6 10 20 25 35) (8 18 29 32 38) (11 19 24 27 37) (12 28 31 33 41)
                 (17 22 23 34 40)
1: (2,4,14,13,7)(3,15,36,42,16)(5,9,39,30,21)(6,25,10,35,20)(8,38,32,29,18)(11,37,27,19,24)(12,33,41,28,31)
(17,23,40,22,34);
(41,0,SQRT(41),-SQRT(41))      10    [2,5]    [8,1]
18    7777776000000377400177700037460030017746127003607415044702167311015163221720172543542246216616246761457022657310133417007507030743440575515173446162521661266514062373345534127107543115501726532370132165075250434637053221633154232526334154451515473724632303304536535252334330
      4          (2 25 28 40) (3 10 11 26) (4 20 27 41) (5 12 14 24) (6 22 36 39) (7 29 37 38) (8 16 21 42) (9 15 19 33) (13
                 18 30 31) (17 23 34 35)
1: (2,25)(3,10)(4,27)(5,14)(6,22)(7,38)(8,42)(9,19)(11,26)(12,24)(13,30)(15,33)(16,21)(17,34)(18,31)(20,41)(23,35)
(28,40)(29,37)(36,39);  2: (2,28,25,40)(3,26,10,11)(4,20,27,41)(5,24,14,12)(6,36,22,35)(7,37,38,29)(8,21,42,16)
(9,15,19,33)(13,31,30,18)(17,23,34,35);
(41,0,SQRT(41),-SQRT(41))      12    [2,4]    [10,1]
```

TWO-GRAPHS OF ORDER 46:

```
1,2       77777774000000177700017774000770700160037706111702360741442263054631452115243252523223124625452722154433066544133026455671
          51211516632056465331213423055364646607364433245544272426257161455104731270663261646314320766342024771645410763607425676013
          23513350275111514273244706134545463075431570566020742277243572411544333474655474733120
          10          (2 5 12 21 26 28 41 42 43 44) (3 7 10 11 23 27 30 35 38 40) (4 9 13 17 18 22 25 29 32 45) (6 15 19 20 37) (8
                      14 16 24 31 33 34 36 39 46)
      1: (2,5,28,42,26,21,44,41,43,12)(3,7,38,35,27,40,10,30,23,11)(4,45,22,18,25,32,9,17,29,13)(6,19,37,15,20)
      (8,36,24,14,31,46,33,34,16,39);
      (45,0,SQRT(45),-SQRT(45))      6     [1,10]    [1,2]    [4,1]
9,10      77777774000000177700017774000770700160037706111702360741444443510721652226225447113223224645512722314463146511270256134665
          31051317311117114572143424172532645507554514622750471075125561253046522665252143666762034474322625227746010770712315150764
          45350607473111344474622704362434663265435544076322136316543520722531253076677413752240
          3           (2 32 38) (3 19 40) (4 21 30) (5 6 44) (7 9 46) (8 10 45) (17 29 41) (18 33 36) (20 34 43) (22 28 37) (23 35
                      39) (24 31 42)
      1: (2,32,38)(3,40,19)(4,21,30)(5,6,44)(7,46,9)(8,10,45)(17,41,29)(18,36,33)(20,34,43)(22,37,28)(23,35,39)(24,31,42);
      (45,0,SQRT(45),-SQRT(45))      22    [10,3]    [12,1]
25,26     77777774000000177700017774000770700160037707101700360761423443014630761063251350465146523007503605554666464130547413037204
          37771122450634611714552633611213624514342576230565151056333151461453452266232266614650572152005627474434563270552664264113
          35752222426422574253457243521435474231707663067170466062407665314720334355466506252227
          2           (2 19) (3 32) (4 46) (5 20) (6 34) (7 9) (8 37) (10 17) (12 43) (13 36) (15 26) (16 44) (18 39) (21 35) (22
                      28) (25 27) (29 42) (31 33) (38 41) (40 45)
      1: (2,19)(3,32)(4,46)(5,20)(6,34)(7,9)(8,37)(10,17)(12,43)(13,36)(15,26)(16,44)(18,39)(21,35)(22,28)(25,27)(29,42)
      (31,33)(38,41)(40,45);
      (45,0,SQRT(45),-SQRT(45))      26    [6,2]     [20,1]
```

DESIGNS, ADJACENCY MULTIGRAPHS AND EMBEDDINGS : A SURVEY

S. S. SHRIKHANDE
Flat No.3, Pukhraj, South Ambajali Road
Laxmi Nagar, Nagpur

N. M. SINGHI
School of Mathematics
Tata Institute of Fundamental Research
Colaba, Bombay 400 005

0. ABSTRACT

In this survey article we will discuss various known results on embedding of some interesting designs and characterizations of Adjacency multigraphs proved using 'Claws and Clique' techniques. We also describe some unsolved problems.

1. PRELIMINARY CONCEPTS AND DISCUSSION OF THE PROBLEMS

A <u>design</u> $D = (X,\beta)$ is a pair such that X is a finite set and β is a family of subsets of X. Elements of X are called <u>points</u> or <u>treatments</u> while those of β are called <u>blocks</u>. If for each $B \in \beta$, the cardinality of B, $|B| = 2$ then the design D is called a multigraph. For any $x,y \in X$, the <u>multiplicity</u> $m(x,y)$ of x and y in a design D is the number of blocks in D containing both x and y. To specify the design we will also use the notation $m_D(x,y)$ for $m(x,y)$. For $x = y$, $m(x,x)$ is the number of blocks containing x and will also be denoted by r_x. A design D is called <u>linear</u> if $m(x,y) \leq 1$ for all $x,y \in X$, $x \neq y$. A multigraph is completely defined if we are given the set of points X, the multiplicity function $m(x,y)$, $x \neq y$, $x,y \in X$. The treatments of a multigraph are called <u>vertices</u> while blocks are called <u>edges</u>. If $m(x,y) \neq 0$, we say that vertices x and y are joined by $m(x,y)$ edges, otherwise they are said to be not joined. A linear multigraph is simply called a <u>graph</u>.

Let $D = (X,\beta)$ be a design. The <u>dual</u> D^1 of D is the design for which β is the set of treatments and blocks are of the form $\langle x \rangle$, $x \in X$, defined by $\langle x \rangle = \{B \in \beta \mid x \in B \text{ in } D\}$. Note that the dual of D^1 is isomorphic to D.

Associated with each design $D = (X,\beta)$ are the following three multigraphs.

(a) The <u>adjacency-multigraph</u> A(D) defined as follows : X is the set vertices of A(D). Two distinct vertices $x,y \in X$, $x \neq y$ are joined by $m_D(x,y)$ edges.

(b) The <u>Intersection-multigraph</u> G(D) defined as follows : β is the set of vertices G(D). Two vertices B_1 and B_2, $B_1 \neq B_2$ are joined by $|B_1 \cap B_2|$ edges.

(c) The <u>Incidence graph</u> I(D) defined as follows : The vertex set of I(D) is the set $X \cup \beta$. Two vertices x and B are joined by an edge iff $x \in X$, $B \in \beta$ and $x \in B$ in D.

<u>Remark 1.1</u>. (i) $A(D) = G(D^1)$ and $A(D^1) = G(D)$.

(ii) D is linear <=> A(D) is linear <=> D^1 is linear <=> G(D) is linear.

(iii) D is a multigraph if and only if (abbreviated, iff) A(D) = D. If D is a multigraph then G(D) is also called line graph of D.

(iv) I(D) is a bipartite linear graph with respect to the partition (X,β) of vertices. Conversely any bipartite graph G with a given bipartition defines a unique design $D = (X,\beta)$ such that G = I(D) and (X,β) is the given bipartition.

Given a design $D = (X,\beta)$ one can define two matrices associated with it.

(1) <u>Incidence matrix of D</u> : Suppose $X = \{x_1,x_2,\ldots,x_v\}$ and $\beta = \{B_1,B_2,\ldots,B_b\}$, then the incidence matrix of D is $v \times b$ matrix $N = (n_{ij})$ defined by

$$n_{ij} = \begin{cases} 1 & \text{if } x_i \in B_j, \\ 0 & \text{otherwise.} \end{cases}$$

(2) <u>Adjacency matrix of D</u> : Adjacency matrix of D is a symmetric $v \times v$ matrix $A = (a_{ij})$ defined by

$$a_{ij} = \begin{cases} 0 & \text{if } i = j \\ m(x_i,x_j) & \text{if } i \neq j \end{cases}$$

<u>Remark 1.2</u>. (a) $NN^t = A + E$ where E is the diagonal matrix $<r_{x_1},r_{x_2},\ldots,r_{x_v}>$.

(b) Adjacency matrix of the graph A(D) is the same as adjacency matrix of the design D.

(c) Adjacency matrix of D^1 is the adjacency matrix of intersection multigraph G(D) of D.

(d) Adjacency matrix of incidence graph I(D) is given by

$$\begin{bmatrix} 0 & N \\ N^t & 0 \end{bmatrix}.$$

In this paper we will discuss various known results related to problems of essentially two different types :

Type (1) Embedding Problems.

Type (2) Characterizations of adjacency graphs of some nice classes of designs.

A typical example of embedding problem is the embedding of a quasi-residual design in a symmetric BIBD (see Section 2). In fact most of the results we discuss have arisen in the study of BIBDs and related designs. It may also be interesting to note that most of the time, a problem of type 1 give rise to, in a natural way, a problem of type 2. Typical examples of problems of type 2 are charactorizations of Pseudo-geometric graphs or Pseudo-net graphs discussed in Sections 2,3 and interest lies in getting good sufficient conditions only in terms of parameters for embedding or characterizations. We will not prove any results in this paper; however, we note that proofs of most of the results given here are essentially based on the now well known technique of 'Claws and Clique' started by Hoffman, Bruck and Bose [3,8]. In fact our aim in this paper is to collect all results proved by this technique, in the last 8 to 10 years. We essentially try to list those results proved using this technique which are not included in excellent articles [9,21,34,35] dealing with similar problems. However for completeness we have also included some of the older results.

In Section 2 we define Partial geometric designs and give a characterization of adjacency graphs of these designs and get various applications to embedding and characterization problems. Section 3 is devoted to the problem of embedding orthogonal arrays, or equivalently affine resolvable designs. In Section 4 we discuss problems corresponding to non-regular case (i.e., designs in which block size is not constant).

2. PARTIAL GEOMETRIC DESIGNS

A design $D = (X,\beta)$ is said to be a partial geometric design (r,k,t,c) (hereafter called $PGD(r,k,t,c)$) iff the following conditions are satisfied :

(i) $r_x = r$ for all $x \in X$.

(ii) $|B| = k$ for all $B \in \beta$.

(iii) Given $x \in X$ and $B \in \beta$ the number

$$p_3(x,B) = \sum_{y \in B} m(x,y) = \begin{cases} t & \text{if } x \notin B \\ r+k-1+c & \text{if } x \in B. \end{cases}$$

We note that if $x \in B$, $p_3(x,B) \geq r+k-1$, hence $c \geq 0$. See [6] for basic properties of PGD(r,k,t,c).

Given any multigraph $G = (V,E)$, let us define $P_i(v_1,v_2)$, $v_1, v_2 \in V$ to be number of distinct paths of length i from v_1 to v_2. In particular $P_1(v_1,v_2) = m(v_1,v_2)$. To specify the graph sometimes we will also write $P_i^G(v_1,v_2)$ for $P_i(v_1,v_2)$.

Remark 2.1 It can be easily seen that for any design $D = (X,\beta)$ with incidence graph I and $x \in X$, $B \in \beta$,

$$P_3^I(x,B) = \sum_{y \in B} m(x,y) = \sum_{x \in C} |B \cap C|.$$

This also shows that the dual of a partial geometric design (r,k,t,c) is a partial geometric design (k,r,t,c).

Examples of partial geometric designs :

(i) (r,k,t) partial geometries defined by Bose [3] are precisely (r,k,t,c) partial geometric designs with $c = 0$.

(ii) Let $D = (X,\beta)$ be a design such that $|X| = v$, $|\beta| = b$ and $|B| = k$ for each $B \in \beta$. Further let $r_x = r$ for all $x \in X$. Then for a given integer λ, D is a PGD(r,k,t,c) with $t = \lambda k$, $c = (\lambda-1)(k-1)$ iff D is a BIBD(v,b,r,k,λ). In particular D is a PGD$(k,k,\lambda k, (\lambda-1)(k-1))$ iff D is a SBIBD(v,k,λ). (For the usual definitions of BIBDs and basic properties see [14,32] and also [6]).

(iii) A PGD(r,k,t,c) with $c = 0$ and $t = r-1$ is called an (r,k)-net [8]. Existence of an (r,k)-net is equivalent to the existence of (r-2) mutually orthogonal latin squares of order k. We will discuss in Section 3 certain more general PGD's called (r,k,μ)-nets.

(iv) If D is a semi-regular group divisible design (SRGD) with usual parameters and $v = mn$ treatments partitioned in m groups of n each (see [14,32] for the definition of SRGD etc.), then D is a PGD(r,k,t,c) with $t = (\frac{k}{m} - 1)\lambda_1 + (k - \frac{k}{m})\lambda_2$ and $c = (\frac{k}{m} - 1)(\lambda_1 - 1) + (k - \frac{k}{m})(\lambda_2 - 1)$ (for details and some other examples see [6]).

We now define regular and strongly regular multigraphs. Given a multigraph $G = (X,E)$ and $x \in X$, the degree of x denoted deg x, and the loop degree of x denoted $\ell(x)$, are respectively defined by

$$\deg x = \sum_{y \in X-\{x\}} m(x,y),$$

$$\ell(x) = \frac{1}{2} \sum_{y \in X-\{x\}} m(x,y).(m(x,y)-1).$$

A multigraph $G = (X,E)$ is said to be regular of degree n_1 and loop degree ℓ if $\deg x = n_1$ and $\ell(x) = \ell$ for all $x \in X$. For a multigraph $G = (X,E)$ we will define u(G) by

$$u(G) = \max\{m(x,y) \mid x,y \in X,\ x \neq y\}.$$

A multigraph $G = (X,E)$ is said to be a strongly regular multigraph with parameters $(v,n_1,\ell;\alpha_i,0 \leq i \leq u(G))$ if

(i) G is regular with degree n_1 and loop degree ℓ.

(ii) For $x,y \in X, m(x,y) = i$, $x \neq y$, we have $P_2(x,y) = \alpha_i$.

Examples of strongly regular multigraphs :

(i) A linear graph G is strongly regular with parameters $(v,n_1,0;\alpha_o,\alpha_1)$ iff G is a strongly regular graph in the sense of Bose [3] with parameters $(v,n_1,\alpha_1,\alpha_o)$, ($\alpha_o$ and α_1 in this case, are also denoted by p^o_{11} and p^1_{11}).

(ii) If $D = (X,\beta)$ is a PGD(r,k,t,c) then the Adjacency-multigraph $G = A(D)$ is a strongly regular multigraph with parameters $(v,n_1,\ell;\alpha_m,\ 0 \leq m \leq u(G))$ where

(a) $v = \frac{k}{t}[(r-1)(k-1) + t-c]$ and $vr \equiv 0 \pmod{k}$

(b) $n_1 = r(k-1)$, $\ell = \frac{1}{2}rc$

(c) $\alpha_m = rt + m(k-r-t+c-1)$.

Any strongly regular multigraph G with parameters $(v,n_1,\ell;\alpha_m,0 \leq m \leq u(G))$ satisfying conditions (a),(b),(c) above is called a Pseudo-geometric multigraph (r,k,t,c).

Such a multigraph is called a geometric multigraph (r,k,t,c), if it is the adjacency multigraph of some PGD(r,k,t,c). Pseudo-geometric multigraphs have been studied in [6], where it is shown that all pseudo-geometric multigraphs (r,k,t,c) with large r, actually geometric multigraphs (r,k,t,c) (see Theorem 2.6 of this section).

(iii) Let $G = K_v^\lambda = (X,E), |X| = v$, be the λ-complete multigraph, i.e., every unordered pair (x,y) of vertices is joined by λ edges. Let L(G) be the line graph of G (i.e., intersection graph of G). Then L(G) is a strongly regular multigraph with parameters $(b,n_1,\ell;\alpha_0,\alpha_1,\alpha_2)$ where

$$b = |E| = \tfrac{1}{2}\lambda v(v-1),$$

$$n_1 = 2(\lambda-1) + 2\lambda(v-2),$$

$$\ell = \lambda - 1,$$

$$\alpha_m = 4\lambda+m(\lambda v-2\lambda-4), \quad 0 \le m \le 2.$$

In fact, $L(K_v^\lambda)$ is a geometric graph $(2,\lambda(v-1),2\lambda,\lambda-1)$. For $\lambda = 1$, $L(K_v^\lambda)$ is the well known triangular association scheme. Theorem 2.9 of this section shows that any strongly regular multigraph $(b,n_1,\ell;\alpha_0,\alpha_1,\alpha_2)$ defined by the above equations is in fact isomorphic to multigraph L(G) defined above unless $v = 8$. In case $v = 8$ exceptions do occur [10,11,20].

(iv) Let $G = K_{m,n}^\mu$ be the μ-complete regular multipartite multigraph on nm vertices, i.e., the vertex set of G is partitioned into m sets $X_1 \cup X_2 \cup \ldots \cup X_m$ of n elements each and two vertices x and y are joined iff $x \in X_i$ and $y \in X_j$, $i \neq j$; further if this is the case $m(x,y) = \mu$. G is a strongly regular multigraph with parameters $(v,n_1,\ell;\alpha_0,\alpha_\mu)$ where

$$v = mn, \; n_1 = \mu(m-1)n,$$
$$\ell = \tfrac{1}{2}\mu(\mu-1)(m-1)n \quad \text{and} \quad \alpha_i = n\mu^2(m-1) - n\mu i, \; 0 \le i \le \mu(G) \qquad \ldots \quad (2.2)$$

When $m = 2$, the line graph of $G, L(K_{2,n}^\mu)$ is also a strongly regular multigraph with parameters $(v',n',\ell';\alpha_i', 0 \le i \le 2)$ where $v' = \mu n^2$, $n_1' = 2(\mu-1) + 2\mu(n-1) = 2\mu n - 2$, $\ell' = 2(\mu-1)$, $\alpha_i' = 4\mu + i(\mu(n-1)-4)$. For $\mu = 1$, $L(K_{2,n}^1)$ is also called the L_2-graph. In Section 3 we will study certain more general strongly regular multigraphs. From these results it can be easily proved that any strongly regular multigraph G with parameters $(v',n_1',\ell';\alpha_i',0 \leq i \leq 2)$ defined by the above equations is actually isomorphic to graph $L(K^{\mu}_{2,n})$ unless $n = 4$. In case $n = 4$, exceptions do occur [38] (see remarks after Theorem 3.7). Before stating various uniqueness theorems we state a conjecture whose truthfulness will help very much in the study of embedding of (n,r,μ)-nets discussed in Section 3.

Conjecture 2.3. There exists a polynomial $p(x,y)$ such that if $n > p(m,\frac{\mu-1}{n})$ then any strongly regular multigraph G with parameters $(v,n_1,\ell;\alpha_i,0 \leq i \leq u(G))$ given by equation (2.2) is isomorphic to the graph $K^{\mu}_{m,n}$.

Note that we do not assume that $u(G) \leq \mu$. To prove that Conjecture 2.3 essentially we have to show that multiplicity of an edge in a multigraph G with parameters given by equation (2.2) is either o or μ for large n. Using the methods developed in [6,42,43] the following partial result can be proved regarding the Conjecture 2.3.

Theorem 2.4. There exists a polynomial $p(x,y,z)$ such that if G is any multigraph with parameters given by equation (2.2) and $u(G) = u + \mu$ then G is isomorphic to $K^{\mu}_{m,n}$, provided $n > p(m,\frac{\mu-1}{n},u)$.

We now consider partial geometric designs. Let $D = (X,\beta)$ be any design with incidence matrix N. It can be easily seen that the design D is a partial geometric design (r,k,t,c) iff N satisfies $NJ = rJ$, $JN = kJ$, $NN^tN = nN + tJ$ where $n = r+k+c-t-1$ and J's are matrices of all 1's of appropriate orders. These equations imply in particular that eigenvalues of NN^t are either o,n or rk. The following interesting theorem due to Bose, Bridges and M.S. Shrikhande [7] shows that the converse is also true.

Theorem 2.5. Let D be a design with incidence matrix N such that $NJ = rJ$ and $JN = kJ$, then D is a partial geometric design if and only if NN^t has at most one non zero eigenvalue distinct from the eigenvalue rk.

We now state the characterization theorem for the adjacency multigraphs of PGD's. As mentioned earlier the theorem was first proved in [6]. The present version is due to Neumaier [31] which improves the bounds given in [6] in some cases.

Theorem 2.6. Let G be a pseudo-geometric multigraph (r,k,t,c). Then G is a geometric multigraph (r,k,t,c) provided

$$k > \max\{t(c+1),(r-1)(2r-2r+1) + 4rc + t - c, \tfrac{1}{2}r(r-1)(rt+1) + \tfrac{1}{2}r^2c+t-c\}$$

(see also Theorem 4.15 of Section 4).

The following theorem is an immediate corollary of the above theorem.

Theorem 2.7. Any pseudo-geometric multigraph $(k,r,\lambda k,(\lambda-1)(k-1))$, $k \neq 1$, is the intersection multigraph of a unique $BIBD(v,b,r,k,\lambda)$, $v = \frac{r}{\lambda}(k-1) + 1$, $b = \frac{1}{k}vr$ provided

$$r > \max\{k(k-1)\lambda^2 - (k^2-2k), 2(k-1)(k^2\lambda-3k+2k\lambda+1) +\lambda, \tfrac{1}{2}(k^2-1)(k^2\lambda-k+2) + \lambda\}.$$

Taking $c = o$ in Theorem 2.6 we get the following well known theorem of Bose on Partial geometries [3].

Theorem 2.8. Let G be a pseudo-geometric graph (r,k,t). Then G is a geometric multigraph (r,k,t) provided

$$k > \tfrac{1}{2}[r(r-1) + t(r+1)(r^2-2r+2)].$$

Theorem 2.9. Let $v \neq 8$ and G be any strongly regular multigraph with parameters $(b,n_1,\ell;\alpha_o,\alpha_1,\alpha_2)$ where

$$b = \tfrac{1}{2}\lambda v(v-1), \; n_1 = 2(\lambda-1) + 2\lambda(v-2),$$
$$\ell = \lambda - 1, \; \alpha_m = 4\lambda+m(\lambda v-2\lambda-4), \; 0 \leq m \leq 2.$$

Then G is isomorphic to the multigraph $L(K_v^\lambda)$.

The case $\lambda = 1$ of the above theorem was proved by Connor [13] for $v > 8$, by Shrikhande [37] for $v \leq 6$ and by Hoffman for $v = 7$ [19]. The general theorem follows easily from the case $\lambda = 1$. In case $v = 8$, exceptions do occur. In fact it was demonstrated by Hoffman [20] and by Chang [10,11] that there are precisely 3 nonisomorphic strongly regular multigraphs with the above parameters with $\lambda = 1$ which are not isomorphic to $L(K_8^1)$, when $v = 8$. From this it follows that for each λ with $v = 8$ there are precisely three nonisomorphic strongly regular multigraphs which are not isomorphic to $L(K_8^\lambda)$ but have the same parameters. We also note that cases $v > 8$ of the above theorem can also be derived from Theorem 2.6.

We now discuss the application of Theorem 2.6 to the well known problem of embedding of quasi-residual designs.

Let $D = (X,\beta)$ be a SBIBD(v_1,k_1,λ_1). Let B be a block of D. Define two new designs $R_B = (X-B,\beta_1)$ and $D_B = (B,\beta_2)$ where

$$\beta_1 = \{C - (C \cap B) | C \in \beta, C \neq B\}, \quad \beta_2 = \{C \cap B | C \in \beta, C \neq B\}.$$

It can be easily seen that R_B and D_B (see [14,32] for details) are BIBD's. R_B is called the <u>residual design</u> of D and D_B the <u>derived design</u> of D. The parameters (v,b,r,k,λ) of R_B satisfy the following equations

$$v = \frac{k}{\lambda}(k+\lambda-1), \quad b = \frac{(k+\lambda)}{\lambda}(k+\lambda-1), \quad r = k+\lambda. \quad \ldots \quad (2.10)$$

Any BIBD(v,b,r,k,λ) satisfying equations (2.10) is called a quasi-residual design. It can be easily seen that a BIBD (v,b,r,k,λ) is a quasi-residual design iff $r = k+\lambda$.

An SBIBD (v,k,λ) with $\lambda = 1$ is called a <u>projective plane</u> of order $k-1$. The residual design of a projective plane is called an <u>affine plane</u>. A classical result of geometry states that every quasi-residual design with $\lambda = 1$ is an affine plane (see [14,32]). M. Hall and Connor [16,12] (see also [39]) have shown that every quasi-residual design with $\lambda = 2$ is also a residual design. Bhattacharya [2] had given an example of a quasi-residual design with $\lambda = 3$ which is not a residual design. Stanton [47] had shown that there exist infinitely many BIBD's which are quasi-residual designs but not residual designs. However, it was proved in [6,45,46] using characterization Theorem 2.6 that for given λ there can be only finitely many quasi-residual designs which are not residual designs. We state the theorem as given in [31] which improves the bound given in [6].

<u>Theorem 2.11</u>. A quasi-residual BIBD(v,b,r,k,λ) is embeddable in a unique SBIBD as a residual design, provided $k > 76$, if $\lambda = 3$; and $k > \frac{1}{2}(\lambda^2-1)(\lambda^3-\lambda^2-\lambda+2)$ if $\lambda \neq 3$.

The above theorem follows easily from Theorem 2.7 and the theorem given below which was first proved by Lawless [24] for the case $\lambda = 3$ and in [6] for $\lambda > 3$.

<u>Theorem 2.12</u>. If $D = (X,\beta)$ is a quasi-residual design (v,b,r,k,λ) then $|B_1 \cap B_2| \leq \lambda$ for all $B_1, B_2 \in \beta$, $B_1 \neq B_2$, if $k \geq 2\lambda^3 - 4\lambda^2 + 4\lambda - 1$.

3. (n,r,μ)-NETS

In this section we will study some problems related to embedding of (n,r,μ)-nets.

Let $D = (X,\beta)$ be a design. A subset $R \subseteq \beta$ is called a <u>parallel-class</u> of D iff R is a partition of X. The design D is said to be a <u>resolvable design</u> if there exists a partition $\{R_1,R_2,\ldots,R_r\}$ of β such that each R_i is a parallel class. The partition $\{R_1,R_2,\ldots,R_r\}$ is called a resolution of D.

A design $D = (X,\beta)$ is said to be an (n,r,μ)-net iff the following conditions are satisfied :

(i) There exists a resolution $\{R_1,R_2,\ldots,R_r\}$ of D such that $|R_i| = n$ for all i, $1 \leq i \leq r$.

(ii) $B \in R_i$ and $C \in R_j$, $i \neq j \rightarrow |B \cap C| = \mu$.

(Note that (i) and (ii) imply that $|B| = n\mu$ for all $B \in \beta$ and $|X| = n^2\mu$.)

(n,r,μ)-nets are also called affine-resolvable designs (see for example [41,42]). For various known results on constructions of (n,r,μ)-nets, some interesting properties and their relationships with orthogonal arrays etc., see [42,17,32]. Hanani [17] actually studies the dual and he calls them transversal designs. For some recent constructions using the ideas of auxilary matrices etc., see [15,23,36]. The following two results are well known (see [41,42]).

<u>Theorem 3.1</u>. Let $D = (X,\beta)$ be a design. Then the following are equivalent :

(i) D is an (n,r,μ)-net.

(ii) D is a resolvable $PGD(r,k,t,c)$ with $k = n\mu$, $t = (r-1)\mu$, $c = (r-1)(\mu-1)$.

(iii) The dual D^1 of D is a group divisible design with parameters $(v_1,b_1,r_1,k_1,\lambda_1,\lambda_2)$, where $v_1 = nr$, $b_1 = n^2\mu$, $k_1 = r$, $r_1 = n\mu$, $\lambda_1 = 0$, $\lambda_2 = \mu$, $m = r$, $n = n$.

<u>Theorem 3.2</u>. The existence of an (n,r,μ)-net D implies that

$$r \leq n^2\left(\frac{\mu-1}{n-1}\right) + n + 1.$$

Further if equality holds in the above inequality then $x = (\mu-1)/(n-1)$ is an integer and D is also a BIBD with parameters (v,b,r,k,λ), where

$v = n^2\mu$, $b = n(n^2x+n+1)$, $r = n^2x+n+1$, $k = n^2\mu$, $\mu = (n-1)x+1$ and $\lambda = nx+1$ ...(3.3)

An (n,r,μ)-net with $r = n^2(\frac{\mu-1}{n-1}) + n + 1$ is also called a complete-net. It is also known as Affine Resolvable BIBD (v,b,r,k,λ), where the parameters are given by equations (3.3) (see [42]).

(n,r,μ)-nets with $\mu = 1$ are precisely the well known (r,k)-nets of Bruck [8]. A complete (n,r,μ)-net with $\mu = 1$ is clearly an affine plane of order n. Our aim here is to list the results related to the following three problems :

Problem 3.4. Given a pseudo-geometric multigraph $G\ (r,k,t,c)$ with $k = n\mu$, $t = (r-1)\mu$, $c = (r-1)(\mu-1)$ when does there exist a partial-geometric design $D(r,k,t,c)$ such that $G = A(D)$?

Problem 3.5. When is a partial-geometric design (r,k,t,c) with $k = n\mu$, $t = (r-1)\mu$, $c = (r-1)(\mu-1)$ resolvable, i.e., when is it an (n,r,μ)-net?

Problem 3.6. Given an (n,r,μ)-net $D = (X,\beta)$ with $r < n^2(\frac{\mu-1}{n-1}) + n + 1 = q$ when can it be embedded in a complete-net, i.e., when does there exist a complete-net $E = (X,\beta^1)$ with parameters (n,q,μ) such that $\beta \subseteq \beta^1$?

It is clear that Problems 3.4, 3.5, and 3.6 are related. These problems have been studied, in great detail, for the case $\mu = 1$. The interest in this case arose because of their relationship to the problem of embedding a set of mutually orthogonal latin squares into a complete such set. The first general result in this direction was proved by Bruck [8].

Theorem 3.7. If $n > \frac{1}{2}(d^4-2d^3+2d^2+d-2)$ then any pseudo-geometric graph (d,n,t), with $t = n-1$, is the adjacency graph of a unique (d,n,μ)-net with $\mu = 1$.

The case $d = 2$ of the above theorem was first proved in [38] where it was shown that in fact the result is true for all $n \neq 4$. It is this result which gives uniqueness of $L(K^{\mu}_{2,n})$ for $n \neq 4$. The following is an immediate corollary of the Theorem [8].

Theorem 3.8. If $n > \frac{1}{4}(d^4-2d^3+2d^2+d-2)$ then any $(n,n+1-d,1)$-net with $\mu = 1$ can be embedded in a complete $(n,n+1,1)$-net, i.e., an affine plane of order n.

For $\mu \neq 1$ Problems (3.6) and (3 7) are far from being solved. Complete results are known only for $d \leq 3$.

Theorem 3.9. Any (n,r,μ)-net with $r = n^2(\frac{\mu-1}{n-1}) + n + 1 - d = q - d$ can be embedded in a unique complete (n,q,μ)-net if

(i) $d = 1$ or

(ii) $d = 2$ and $n \neq 4$ or

(iii) $d = 3$ and $n \geq 104$.

The cases $d = 1$ and $d = 2$ with $n = 2,3$ of above theorem are proved in [40,41], while $d = 2$ and $n \geq 5$ are proved in [43]. The case $d = 3$ is considered in [44]. In fact in [44],the following general theorem is proved using the techniques of [6] and then the above theorem for the case $d = 3$ is derived from it.

Theorem 3.10. If $n \geq 2(d-1)^2(2d^2-2d+1)$ then any pseudo-geometric multigraph (d,k,t,c) with $k = n\mu$, $t = (d-1)\mu$ and $c = (d-1)(\mu-1)$ is the adjacency multigraph of a unique partial geometric design (d,k,t,c).

Theorem 3.10 is actually not stated in [44]. However, it can be easily seen that it follows from results proved in [44]. The following result is proved in [43].

Theorem 3.11. If $D = (X,\beta)$ is an (n,r,μ)-net with $r = n^2(\frac{\mu-1}{n-1}) + n + 1 - d$ and if $x,y \in X$, $x \neq y$, then

$$n(\frac{\mu-1}{n-1}) + 1 - d \leq m(x,y) \leq (\frac{\mu-1}{n-1}) + 1$$

provided $n \geq 2d(d-1)$.

The following theorem is an immediate consequence of Theorems 3.10 and 3.11 (see [44]).

Theorem 3.12. If $D = (X,\beta)$ is an (n,r,μ)-net with $r = n^2(\frac{\mu-1}{n-1}) + n + 1 - d$ then there exists a unique BIBD $E = (X,\beta^1)$ with parameters (v,b,r,k,λ) given by equations (3.4) such that $\beta \subseteq \beta^1$, provided $n \geq 2(d-1)^2(2d^2-2d+1)$.

Theorem (3.12) gives a partial answer to Problem (3.6) for the cases $d \geq 4$, in the sense that any (n,r,μ)-net with $n \geq 2(d-1)^2(2d^2-2d+1)$ is embeddable in a unique BIBD(v,b,r,k,λ) with parameters given by (3.3). What is not known is whether this

BIBD is resolvable. It seems appropriate to make the following conjectures.

Conjecture 3.13. There exists a polynomial $p(x,y)$ such that for any given integers $n,r,\mu,\frac{\mu-1}{n-1}$ with $n > p(r,\frac{\mu-1}{n-1})$, any $PGD(r,k,t,c)$ with $k = n\mu$, $t = (n-1)$, and $c = (n-1)(\mu-1)$ is resolvable, i.e., it is an (n,r,μ)-net.

Conjecture 3.14. There exists a polynomial $q^*(x,y)$ such that for any given integers $n,d,\mu,\frac{\mu-1}{n-1}$ with $n > q^*(\frac{\mu-1}{n-1},d)$ any (n,r,μ)-net with $r = n^2(\frac{\mu-1}{n-1}) + n + 1 - d = q - d$ is embeddable in a unique complete (n,q,μ)-net.

Truthfulness of Conjecture 2.3 implies that of Conjecture 3.13. Also if Conjecture 3.13 is true then Conjecture 3.14 will be true; this follows from Theorem (3.10)-(3.12).

4. DESIGNS WITH NON-CONSTANT BLOCK-SIZE

An (r,λ) design is a design $D = (X,\beta)$ satisfying the following conditions :

(i) each treatment occurs in exactly r blocks,

(ii) each pair of distinct treatments occurs in exactly λ blocks.

BIBDs are precisely those (r,λ) designs in which all blocks are of the same size. The following theorem can be proved easily.

Theorem 4.1. An (r,λ) design on v treatments and b blocks is a $BIBD(v,b,r,k,\lambda)$ iff

$$b = \frac{r^2v}{\lambda(v-1) + r}$$

Our main interest here is essentially in studying $(r,1)$ designs which are obtained by removing certain treatments from a projective plane. In fact we will describe various known results on the following problem :

Problem 4.2. Given an $(n+1,1)$-design $D = (X,\beta)$ with $n^2+n+1-\alpha$ treatments, when does there exist a projective plane $E = (X^1,\beta^1)$ with parameters $(n^2+n+1,n+1,1)$ such that $X \subseteq X^1$ and $\beta = \{B \mid B \neq \phi,\ B = C \cap X$ for some $C \in \beta^1\}$, i.e., when can we embed D in a projective plane?

Let $G = (X,E)$ be a multigraph. We will say that G is a pseudo (r,λ)-multigraph of index v if there exist integers r,λ,v such that for each $B \in X$, there is an integer k_B satisfying the following :

(i) $\deg_G B = k_B(\lambda-1)$.

(ii) Total number of paths of length 2 starting from B is given by

$$P_2(B) = \sum_{C,D:C\neq D} m_G(B,C)\, m_G(C,D) = (r-1)k_B[\lambda(v-k_B) + r - 2].$$

(iii) If $B,C \in \beta$, $B \neq C$, $m(B,C) = i$ then

$$P_2(B,C) = \sum_{D\neq B,C} m(B,D)\, m(D,C) = \lambda k_B k_C + i(r-k_B-k_C-\lambda).$$

(iv) $\sum_{B\in X} k_B = rv.$

If $\lambda = 1$, G is a graph.

It can be easily seen that if $G = G(D)$ is the intersection multigraph of an (r,λ)-design $D = (X,\beta)$ then G is a pseudo (r,λ)-multigraph of index v where $v = |X|$ and $k_B = |B|$ for each $B \in \beta$. A pseudo (r,λ)-multigraph of index v is called (r,λ)-multigraph of index v if $G = G(D)$ for some (r,λ)-design D.

Theorem 4.3. Let G be a pseudo $(r,1)$-graph of index v, then G is an $(r,1)$-graph of index v if $r \geq 2(v-r)^2 + 4(v-r) + 2$ or $r \leq v$.

The following theorem which was first proved by Vanstone and McCarthy [25] can be deduced easily from the above theorem (see [26]).

Theorem 4.4. Let D be an $(n+1,1)$ design on $n^2+n+1-\alpha$ treatments. Then D can be embedded in a unique projective plane $(n^2+n+1,n+1,1)$ if $n \geq 2(\alpha-n+1)^2 + 4(\alpha-n+1) + 1$, or $n \leq \alpha - 1$.

An oval in a projective plane of even order n is a set of $(n+2)$ points no three of which are on the same line. The case $n \geq 2$ of the following theorem due to Bose and Shrikhande [5] follows from Theorem 4.4.

Theorem 4.5. Let $D = (X,\beta)$ be an $(n+1,1)$-design on n^2-1 treatments n even, such that each block of D is of size $n-1$ or $n+1$. Then D can be embedded in a unique projective plane $E = (X^1,\beta^1)$ such that the set of treatments E which are not in X form an oval in E.

Not much is known about Problem 4.2 when

$$n \leq 2(\alpha-n+1)^2 + 4(\alpha-n+1) + 1$$

in general. However, some results are known for some special designs. Mullin and Vanstone [28] defined a pseudo-parallel complement $PPC(n,\alpha)$ to be ab $(n+1,1)$-design, $D = (X,\beta)$ with $|X| = n^2 - \alpha n$ and $|\beta| \leq n^2 + n - \alpha$ in which there are at least $n - \alpha$ blocks of size n. A pseudo-intersecting complement $PIC(n,\alpha)$ is an $(n+1,1)$-design $D = (X,\beta)$ with $|X| = n^2 - \alpha n + \alpha - 1$ and $b \leq n^2 + n - \alpha$ in which there are at least $n - \alpha + 1$ blocks of size $n - 1$. The following theorem is due to Mullin and Vanstone [28].

<u>Theorem 4.6.</u> If $n > \frac{1}{2}(\alpha^4 - 2\alpha^3 + 2\alpha^2 + \alpha - 2)$ then any $PPC(n,\alpha)$ can be embedded in a projective plane of order n.

We now state a similar theorem for $PIC(n,\alpha)$. The theorem is due to Mullin and Vanstone [28] for $\alpha \geq 4$. The cases $\alpha = 2,3$ are contained in [29]. The case $\alpha = 2$ was first proved by P. de Witte.

<u>Theorem 4.7.</u> Any $PIC(n,\alpha)$ can be embedded in a projective plane of order n if

(i) $n > \frac{1}{2}(\alpha^4 - 2\alpha^3 + 2\alpha^2 + \alpha - 2)$ for $\alpha \geq 4$,

(ii) $n \geq 14$ for $\alpha = 3$ and

(iii) $n \geq 8$ for $\alpha = 2$.

Examples of $PIC(5,3)$ are given in [29] which cannot be embedded in a projective plane of order 5.

One should be able to generalise some of the above results for (r,λ) designs with $\lambda \geq 2$. We close the discussion of (r,λ) designs by stating a few conjectures and a theorem due to Mullin and Vanstone [27] for general λ.

<u>Conjecture 4.8.</u> There exists a polynomial $P(x,y)$ such that if $G = (X,E)$ is any pseudo (r,λ)-multigraph of index v, then G is an (r,λ)-multigraph of index v provided $r > P(k,\lambda)$, where $k = \text{minimum } \{k_B | B \in X\}$.

<u>Conjecture 4.9.</u> There exists a polynomial $P(x,y)$ such that any (r,λ)-design can be embedded in $SBIBD(v,r,\lambda)$ with $v = (r/\lambda)(r-1) + 1$ provided $r > P(k_1,\lambda)$ and $k_1 \geq 0$, where $k_1 = r - \min\{k_B | B \in \beta\}$.

The following interesting theorem due to Mullin and Vanstone [27] gives a partial answer to above question.

Theorem 4.10. Let D be any (r,λ) design of index v_1, $v_1 = \frac{r}{\lambda}(r-1) + 1 - a$, such that each block of D is of size at most r. Then D can be embedded in a SBIBD(v,r,λ) with $v = \frac{r}{\lambda}(r-1) + 1$ provided $r > a(a+1)\lambda - a$.

We now define some more general designs, which were first studied by Wolf [48].

Let $D = (X,\beta)$ be a design with incidence matrix N. Then D is said to be a weak partial geometric design (weak PGD) (Wolf calls them gradk onstante SPG) with parameters (v,n,r,λ) where r,n,λ are real numbers if

(i) $|X| = v$,

(ii) $NJ = rJ$ and $NN^tN = nN + \lambda JN, \lambda > 0$.

The following theorem can be easily proved.

Theorem 4.11. A weak PGD(v,n,r,λ) is an (r,λ)-design iff $n = r - \lambda$.

PGD(r,k,t,c) is clearly a weak PGD(v,n,r,λ) with $\lambda = \frac{t}{k}$, $n = r+k-t-c-1$, and $v = \frac{k}{t}[(r-1)(k-1) + t-c]$. The following interesting theorem due to Neumaier [31] shows that the converse is also true under some parameteric conditions.

Theorem 4.12. Let $D = (X,\beta)$ be a weak PGD(v,n,r,λ) and $\lambda < 1$. If either

(i) D is linear, or

(ii) $t = \frac{\lambda(\lambda v+n)}{r}$ is an integer with $\lambda(n+1-r) < (1-\lambda)(t+1)$

then D is a PGD(r,k,t,c) with $k = \frac{\lambda v+n}{r}$, $t = \frac{\lambda(\lambda v+n)}{r}$, $c = n+1+t-r-k$.

The following result is due to Wolf [48] (see also [31]).

Theorem 4.13. Let G be the adjacency multigraph of a weak PGD(v,n,r,λ). Then G is a strongly regular multigraph with parameters $(v,n_1,\ell;\alpha_i\text{'s})$ where

$$n_1 = \lambda v - r + n,$$
$$\ell = \lambda(\lambda-1)v-r(r-1) + (r+\lambda-1), \qquad \ldots \quad (4.14)$$
$$\alpha_i = \lambda(\lambda v+n) + i(n-2r).$$

Theorem 4.15. Let G be a strongly regular multigraph with parameters $(v,n_1,\ell;\alpha_i's)$ given by equations (4.14). Then G is the adjacency multigraph of a unique weak PGD(v,n,r,λ) provided

$$n > \max\{2(r-1)(\mu+1-r) + 8\ell, \frac{r(r-1)}{2}(\mu+1) + r\ell+r-1\}$$

where $\mu = \lambda(\lambda v+n)$.

Theorem 2.5 for PGDs is essentially obtained from Theorems 4.12 and 4.15 (see [31]).

We conclude this paper by describing two very general problems. The solution to these problems will help very much in studying various embedding and characterization problems discussed in this paper.

For each integer k, let us define two families of multigraphs as follows :

F_k = {G|G is a multigraph such that the minimum eigenvalue of the adjacency matrix of G is at least $-k$}.

I_k = {G|G is the adjacency multigraph of design $D = (X,\beta)$ in which $r_x = k$ for all $x \in X$}.

It can be easily seen that $I_k \subseteq F_k$. Also it is clear that families F_k and I_k are hereditary in the sense that if $G \in F_k$ (resp., $G \in I_k$), $G = (X,E)$, then each induced sub-multigraph H of G, i.e., multigraph $H = (X^1,E^1)$, $X^1 \subseteq X$ and $E^1 = \{B \in E | B \subseteq X^1\}$ is also in F_k(resp.,I_k). Hence the set of multigraphs which are not in F_k(resp.,I_k) has minimal elements.

Minimal multigraphs not in a given hereditary family F are said to be <u>forbidden multigraphs</u> for F.

<u>Problem 4.16</u>. Describe the family M_k of all forbidden multigraphs for the family F_k, k is a positive integer (in fact one could ask the question for any positive real number k).

<u>Problem 4.17</u>. Describe the family L_k of all forbidden multigraphs for the family I_k, k is a positive integer.

The answer to Problem 4.16 is completely known for $k = 2$. In fact using the result of Beineke [1] one can easily see that the family L_2 consists of following 12 graphs (see Harary [18]).

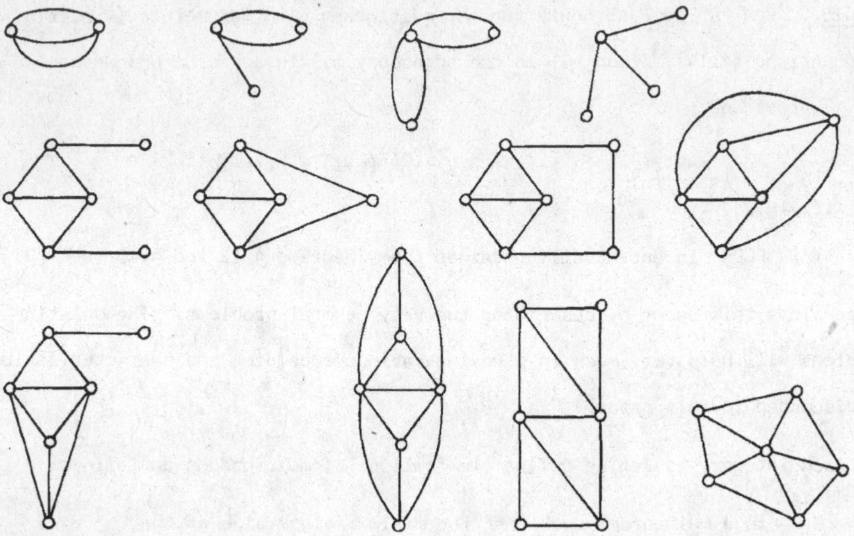

For $k \geq 3$ the families L_k are infinite.

For some interesting partial results on families L_k, $k \geq 3$, see [30,21]. The paper [9] gives an excellent account of various known interesting results on families F_2 and their relationship with I_2. For results on family M_2, see [33]. Not much is known about families M_k, $k \geq 3$ (see [21,22]).

REFERENCES

1. L.W. Beineke, Derived graphs and digraphs, Beitrage Zur Graphentheorie (H. Sachs et. al. (Eds.)), Teubner, Leipzig, 1968, 17-33.
2. K.N. Bhattacharya, A new balanced incomplete block design, Science and Culture, 1944, 508.
3. R.C. Bose, Strongly regular graphs, partial geometries and partially balanced designs, Pacific J. Math., 13(1963), 389-419.
4. R.C. Bose, Graphs and Designs, Finite Geometric Structures and their Applications, CIME advanced Summer Institute, 1972, Ed. A. Barlotti, Edizioni Cremonese, Roma 1973.
5. R.C. Bose, S.S. Shrikhande, Embedding the complement of an oval in a projective plane of even order, Discrete Math., 1973.
6. R.C. Bose, S.S. Shrikhande, N.M. Singhi, Edge regular multigraphs and partial geometric designs with an application to the embedding of quasi-residual designs, Academia Nationale Dei Lincei, Atti del convegni Lincei, 17(1976), 49-81.
7. R.C. Bose, W.G. Bridges, M.S. Shrikhande, Characterization of partial geometric designs, Discrete Math., 16(1976), 1-7.

8. R.H. Bruck, Finite nets II. Uniqueness and imbedding, Pacific J. Math., 13(1963), 421-457.

9. P.J. Cameron, J.M. Goethals, J.J. seidel, E.E. Shult, Line graphs, root systems and elliptic geometry, J. of Alg., 43(1976), 305-327.

10. Li-Chien Chang, The uniqueness and non-uniqueness of triangular association schemes, Science Record, Math., New Sr.3(1959), 604-613.

11. Li-Chien Chang, Association schemes of partially balanced designs with parameters $v = 28$, $n_1=12$, $n_2 = 15$, and $p_{11}^2 = 4$, Science Record, Math., New Sr.4, 12-18.

12. W.S. Connor, On the structure of balanced incomplete block designs, Ann. Math. Stat., 23(1952), 57-71.

13. W.S. Connor, The uniqueness of the triangular association scheme, Ann. Math. Stat., 29(1958), 262-266.

14. P. Dembowski, Finite Geometrics, Springer-Verlag, Berlin and New York, 1968.

15. D.A. Drake, D. Jungnickel, Klingenberg structures and partial designs II, regularity and uniformity, Pacific J. Math., 77(1978), 389-415.

16. M. Hall, Jr., W.S. Connor, An embedding theorem for balanced incomplete block designs, Canad. J. Math., 6(1953), 35-41.

17. H. Hanani, On transversal designs in Proceedings of advanced study Inst. on Combinatorics (Brukelen 1974), Part 1, Amsterdam, Mathematical Centre, 1974, 42-52.

18. F. Harary, Graph Theory, Addison-Wesley, Reading, 1972.

19. A.J. Hoffman, On the uniqueness of triangular association scheme, Ann. Math. Stat., 31(1960), 492-497.

20. A.J. Hoffman, On the exceptional case in a characterization of the arcs of complete graph, IBM J. Res. Develop., 4, 497-504.

21. A.J. Hoffman, Eigen values of graphs, studies in graph theory, Part II, ed. D.R. Fulkerson, M.A.A., 225-245.

22. A.J. Hoffman, $-1-\sqrt{2}$?, Combinatorial Structures and their Applications, Gordon and Breach, 1970, 173-176.

23. D. Jungnickel, On difference matrices, resolvable transversal designs and generalised Hadamard matrices, Math. Z., 167(1978), 49-60.

24. J.F. Lawless, Block intersections in quasi-residual designs, Aequationes Math., 5(1970), 40-46.

25. D. McCarthy, S.A. Vanstone, Embedding (r,1) designs in finite projective planes, Descrete Math., 19(1977), 67-76.

26. D. McCarthy, N.M. Singhi, S. Vanstone, A Graph - Theoretical Approach to Embedding (r,1)-Designs, Graph Theory and related Topics, Ed. J.A. Bondy and U.S.R. Murty, Academic Press, 1979.

27. R.C. Mullin, S.A. Vanstone, The approximation of SBIBD's by regular pairwise balanced designs : case $\lambda \geq 2$.

28. R.C. Mullin, S.A. Vanstone, A generalization of theorem of Totten, J. Aust. Math. Soc., 22(1976), 494-500.

29. R.C. Mullin, N.M. Singhi, S.A. Vanstone, Embedding the affine complement of three intersection lines in a finite projective plane, J. Aust. Math. Soc., 24(series A) (1977), 458-464.

30. R.N. Naik, S.B.Rao, S.S. Shrikhande, N.M. Singhi, Intersection graphs of k-uniform linear hypergraphs, submitted for publication.

31. Ar. Neumaier, Quasi-residual 2-designs, $1\frac{1}{2}$-designs and strongly regular multigraphs (submitted for publication)

32. D. Raghava Rao, Constructions and Combinatorial Problems in Design of Experiments, John Wiley, 1971.

33. S.B. Rao, N.M. Singhi, K.S. Vijayan, Forbidden subgraphs for generalised line graphs (submitted for publication).

34. D.K. Ray-Chaudhuri, Uniqueness of association schemes, Academic National Dei Lincei, Atti del conveni Lencei, 17(1976), 465-479.

35. D.K. Ray-Chaudhuri, Geometric incidence structures, Proc. 6th British Comb. Conf., 1977, 87-116.

36. S.S. Sane, A recursive construction of symmetric (s,r,μ)-nets (submitted for publication).

37. S.S. Shrikhande, On characterization of triangular association scheme, Ann. Math. Stat., 30(1959), 39-47.

38. S.S. Shrikhande, The uniqueness of the L_2 association scheme, Ann. Math. Stat., 30(1959), 781-798.

39. S.S. Shrikhande, Relations between incomplete block designs, contributions to probability and statistics, Essays in honour of Harold Hotelling, Stanford U. Press, 1960, 388-395.

40. S.S. Shrikande, and Bhagwandas, A note on embedding for Hadamard matrices in Contributions to Statistics and Probability, Essays in memory of Samarendra Nath Roy, University of North Carolina Press, 1967.

41. S.S. Shrikhande, and Bhagwandas, On embedding of orthogonal arrays of strength two, combinatorial mathematics and its applications, University of North Carolina Press, 1969, 256-273.

42. S.S. Shrikhande, Affine resolvable balanced incomplete block designs : A survey, Aequationes Math., 14(1976), 251-269.

43. S.S. Shrikhande, N.M. Singhi, A note on embedding of orthogonal arrays of strength two, J. Statist. Planning inf., 3(1979), 267-271.

44. S.S. Shrikhande, N.M. Singhi, Embedding of orthogonal arrays of strength two and deficiency greater than two, J. Statist. Planning inf. (to appear).

45. N.M. Singhi, S.S. Shrikhande, Embedding of quasi-residual designs, Geometriac Dedicata, 2(1974), 509-517.

46. N.M. Singhi, S.S. Shrikhande, Embedding of quasi-residual designs with $\lambda = 3$, Utilitas Mathematica, 4(1973), 35-53.

47. R.G. Stanton, Interconnections of related BIBD's, J. Comb. Theory, 3(1969), 387-391.

48. K.E. Wolf, Punktstrabile und Semipartial geometrische Inzidenzstrukturen, Mitt. Math. Sem, Giessen, 135(1978).

ON THE ADJUGATE OF A SYMMETRICAL BALANCED INCOMPLETE BLOCK DESIGN WITH $\lambda = 1$

G. A. PATWARDHAN

and

M. N. VARTAK
Department of Mathematics
Indian Institute of Technology, Bombay 400 076

1. INTRODUCTION AND PRELIMINARY RESULTS

The 2-adjugate matrix of the (0,1)-matrix N of order $v \times b$ is the (0,1,-1)-matrix of order $\binom{v}{2} \times \binom{b}{2}$, the elements of which are the determinants of all possible 2×2 minor matrices of N, arranged in lexicographic order. If, however, these minor determinants are reduced modulo 2, then the 2-adjugate matrix is again the (0,1)-matrix of order $\binom{v}{2} \times \binom{b}{2}$, denoted by N^*.

Let N be the incidence matrix of a given design $D \equiv D(v,b,r_i,k_j)$. The new (0-1)-matrix N^* obtained as above can be interpreted as the incidence matrix of a new design, called 'The 2-adjugate mod 2 design of the given design N'. The application of this technique to the class of unreduced Balanced Incomplete Block Designs is investigated in detail by the authors in [3].

In this paper, a general association scheme of '2-adjugate mod 2' designs is described. As a special case, the association scheme of 2-adjugate mod 2 design derived from a symmetrical B.I.B. design with $\lambda = 1$ is considered in detail. It is proved that its transformed design is a P.B.I.B. (partially Balanced Incomplete Block) design with at the most six associate classes and having an unsymmetrical association scheme.

Let the treatments of the design N^* be denoted by $1^*, 2^*, \ldots, \binom{v}{2}^*$ and those of N by $1,2,\ldots,v$. The problem of identifying any treatment j^* of N^* by means of a unique pair (i,u) of treatments of $N (1 \leq i < u \leq v)$ from which it arises when N^* is formed as the 2-adjugate mod 2 designs of N, is solved by means of the following Lemmas [3] :

Lemma 1.1. Let j be any integer such that $1 \leq j \leq \binom{v}{2}$. Then there exist two unique integers r and m such that $0 \leq r \leq v-2$, $0 < m < v-r$ and such that

$$(1.1) \qquad j = \frac{r(2v-r-1)}{2} + m .$$

Lemma 1.2. If the treatments of the design N^* are written in the lexicographical order, and if an integer j, $1 \leq j \leq \binom{v}{2}$, is uniquely written as in (1.1), then the treatment j^* of N^* is obtained by 2-adjugation of the treatments $r + 1$ and $r + m + 1$ of N and we may uniquely identify the treatment j^* of N^* by the pair $[(r+1), (r+m+1)]$ of treatments of N.

We may formally denote this correspondence by

$$j^* \sim (r+1, r+m+1).$$

2. ASSOCIATION SCHEME

Let j^* and u^* be any two treatments of N^* and let $j^* \sim (\ell,m)$ and $u^* \sim (p,q)$ where ℓ,m,p and q are treatments of N. We can, in general, envisage the following possibilities about the occurrence of the treatments $\ell,m.p$ and q in the blocks of N on which the association scheme of the derived design can be based :

(α) Either $\ell = p$(or q) or $m = p$(or q) and q(or p) is a distinct treatment not occurring along with ℓ or m in any block of N.

(β) Either $\ell = p$(or q) or $m = p$(or q) and q(or p) is a distinct treatment such that ℓ,m and q(or p) all occur together in one or more blocks of N.

(γ) All the treatments ℓ,m,p and q are distinct; ℓ,m occur together in one or more blocks of N and p,q occur together in one or more of the remaining blocks of N.

(δ) All the treatments, ℓ,m,p and q are distinct; ℓ,m and p(or q) occur together in one or more blocks of N and q(or p) does not occur together with ℓ,m and p(or q) in any block of N.

(ε) All the treatments, ℓ,m,p and q are distinct; p,q and ℓ(or m) occur together in one or more blocks of N and m(or ℓ) does not occur together with p,q and ℓ(or m) in any block of N.

(ϕ) All the treatments ℓ,m,p,q are distinct; ℓ and p(or q) occur together in one or more blocks of N, while m and q(or p) occur together in one or more blocks of N, which contain neither ℓ nor p(or q).

These are the various possibilities which are useful in defining association schemes of most of the 2-adjugate mod 2 designs. We shall, in fact, make the following general definition of association scheme of 2-adjugate mod 2 design N* derived from a given B.I.B. design N.

Definition 2.1. Let $j^* \sim (\ell,m)$ and $u^* \sim (p,q)$ as before. Then

(i) j^* and u^* are called first associates of each other if there are only three distinct treatments among ℓ,m,p,q and only one of the pairs (ℓ,m), (p,q) occurs together in one or more blocks of N, none of which contains the third distinct treatment.

(ii) j^* and u^* are called second associates of each other if there are only three distinct treatments among ℓ,m,p and q and all of them occur together in one or more blocks of N.

(iii) j^* and u^* are called third associates of each other if ℓ,m,p,q are all distinct and the pair (ℓ,m) occurs together in one or more blocks of N and the pair (p,q) occurs together in one or more of the remaining blocks of N.

(iv) u^* is called a fourth associate of j^* if ℓ,m,p,q are all distinct and if ℓ,m and p (or q) occur together in one or more blocks of N.

(v) u^* is called a fifth associate of j^* if ℓ,m,p,q are all distinct and if p,q and ℓ(or m) occur together in one or more blocks of N.

(vi) j^* and u^* are called sixth associates of each other if ℓ,m,p,q are all distinct and all of them occur together in one or more blocks of N.

(vii) j^* and u^* are called seventh associates of each other if ℓ,m,p,q are all distinct, ℓ and p(or q) occur together in one or more blocks of N, while m and q(or p) occur together in one or more of the remaining blocks of N.

Generally speaking, therefore, we should expect a 2-adjugate mod 2 design with at the most seven associate classes. In special cases, this number of associate classes can be reduced to even two. It may be remarked that, in the association scheme described above, the relation of association is not necessarily symmetrical (cf. Nair [1]). Thus, if (p,q) is a fourth (or fifth) associate of (ℓ,m), then (ℓ,m) need not be a fourth (or fifth) associate of (p,q).

We now show that, given a B.I.B. design N with $\lambda = 1$, in the transformed design N*, the possibility of two treatments being seventh associates of each other, does not exist. Let ℓ,m,p,q be distinct treatments of N such that ℓ and p (but not q) occur in a unique block B_1 of N and m and q (but not p) occur in another unique block B_2 of N. Further, a block B of N exists containing the treatments ℓ and m.

Case (i). B_1 is same as B but B_2 is not same as B.

In this case, ℓ,m and p occur together in one block while q occurs in another. Thus the treatment (p,q) of N* becomes a fourth associate of the treatment (ℓ,m) of N*.

Case (ii). B_2 is same as B but B_1 is not same as B.

In this case, ℓ,m and q occur together in one block while p occurs in another. Thus the treatment (p,q) becomes a fourth associate of the treatment (ℓ,m).

Case (iii). B_1,B_2 and B are identical.

In this case, ℓ,m,p and q occur together in the same block and then (p,q) and (ℓ,m) become sixth associates of each other.

Case (iv). B_1,B_2 and B are all distinct.

Since any given treatment occurs with every other treatment exactly once, we also have a unique block B' containing p and q together. This block contains neither ℓ nor m and hence is different from B. Since ℓ and m already occur in B, (p,q) becomes a third associate of (ℓ,m).

B_1 and B_2 cannot be identical since $\lambda = 1$.

This exhausts all the possible cases. Thus, if $\lambda = 1$, the possibility of two given treatments being seventh associates of each other is ruled out.

3. TRANSFORM OF SYMMETRICAL B.I.B. DESIGN HAVING $\lambda = 1$

For a symmetrical B.I.B. design with $\lambda = 1$, we shall have the other parameters expressed in terms of a single positive integral variable t as follows :

$$v = b = t^2 + t + 1,\ r = k = t + 1. \tag{3.1}$$

Whenever t is a prime or a power of a prime, say p^n, and the finite projective plane $PG(2,p^n)$ exists, we can construct the symmetrical B.I.B. design (3.1) by taking lines of $PG(2,p^n)$ as blocks. We shall, however, work out the 2-adjugate mod 2 design of (3.1) in terms of the general integral parameter. The following theorem is proved in this connection.

Theorem 3.1. The 2-adjugate mod 2 design N^* of the symmetrical B.I.B. design (3.1) is a symmetrical P.B.I.B. design with at most six associate classes and with parameters (3.2) given below, whenever they have meaningful values.

$$\begin{aligned}
& v^* = b^* = t(t+1)\ (t^2+t+2)/2,\ r^* = k^* = t(t+2) \\
& n_1^* = 2t^2,\ \lambda_1^* = t + 2 \\
& n_2^* = 2(t-1),\ \lambda_2^* = t, \\
& n_3^* = t^2(t-1)^2/2,\ \lambda_3^* = 2 \\
& n_4^* = t^2(t-1),\ \lambda_4^* = 2 \\
& n_5^* = t^2(t-1),\ \lambda_5^* = 2 \\
& n_6^* = (t-1)\ (t-2)/2,\ \lambda_6^* = 0.
\end{aligned} \tag{3.2}$$

$$(p_{ju}^{*1}) = \begin{bmatrix}
t^2-t+1 & t-1 & (t-1)^2 & t-1 & t-1 & 0 \\
t-1 & 0 & 0 & 0 & t-1 & 0 \\
(t-1)^2 & 0 & (t-1)^3(t-2)/2 & (t-1)^3 & (t-1)^2(t-2)/2 & 0 \\
(t-1) & 0 & (t-1)^3 & (t-1)^2 & (t-1)^2 & 0 \\
t-1 & t-1 & (t-1)^2(t-2)/2 & (t-1)^2 & (t-1)(t^2-2)/2 & (t-1)(t-2)/2 \\
0 & 0 & 0 & 0 & (t-1)(t-2)/2 & 0
\end{bmatrix}$$

$$(p^{*2}_{ju}) = \begin{bmatrix} t^2 & 0 & 0 & t^2 & 0 & 0 \\ 0 & t-1 & t^2(t-1)(t-2)/2 & 0 & 0 & t-2 \\ 0 & 0 & t^2(t-1)(t-2)/2 & 0 & t^2(t-1)/2 & 0 \\ t^2 & 0 & 0 & t^2(t-2) & 0 & 0 \\ 0 & 0 & t^2(t-1)/2 & 0 & t^2(t-1)/2 & 0 \\ 0 & t-2 & 0 & 0 & 0 & (t-2)(t-3)/2 \end{bmatrix}$$

$$(p^{*3}_{ju}) = \begin{bmatrix} 4 & 0 & 2(t-1)(t-2) & 2(t-2) & 4(t-1) & 0 \\ 0 & 0 & 2(t-2) & 2 & 0 & 0 \\ 2(t-1)(t-2) & 2(t-2) & \frac{(t-1)(t^3-5t^2+11t-8)}{2} & (t-1)(t-2)^2 & (t-1)(t-2)^2 & \frac{(t-2)(t-3)}{2} \\ 2(t-2) & 2 & (t-1)(t-2)^2 & 2t^2-5t+4 & 2(t-1)(t-2) & (t-2) \\ 4(t-1) & 0 & (t-1)(t-2)^2 & 2(t-1)(t-2) & 2(t-1)(t-2) & 0 \\ 0 & 0 & (t-2)(t-3)/2 & t-2 & 0 & 0 \end{bmatrix}$$

$$(p^{*4}_{ju}) = \begin{bmatrix} 2 & 0 & 2(t-1)^2 & 2(t-1) & 2(t-1) & 0 \\ 2 & 0 & 0 & 0 & 2(t-2) & 0 \\ (t-1)(t-2) & t-1 & (t-1)^2(t-2)^2/2 & (t-1)^2(t-2) & (t-1)(2t^2-5t+4)/2 & (t-1)(t-2)/2 \\ t^2-2 & t-1 & (t-1)^2(t-2) & (t-1)(t-2) & (t-1)(t-2) & 0 \\ 2(t-1) & 0 & (t-1)^2(t-2) & 2(t-1)^2 & (t-1)(t-2) & 0 \\ t-2 & 0 & 0 & 0 & (t-2)(t-3)/2 & 0 \end{bmatrix}$$

$$(p^{*5}_{ju}) = \begin{bmatrix} 2 & 2 & (t-1)(t-2) & t^2-2 & 2(t-1) & t-2 \\ 0 & 0 & (t-1) & t-1 & 0 & 0 \\ 2(t-1)^2 & 0 & (t-1)^2(t-2)^2/2 & (t-1)^2(t-2) & (t-1)^2(t-2) & (t-2)(t-3)/2 \\ 2(t-1) & 0 & (t-1)^2(t-2) & (t-1)(t-2) & 2(t-1)^2 & 0 \\ 2(t-1) & 2(t-2) & (t-1)(2t^2-5t+4)/2 & (t-1)(t-2) & (t-1)(t-2) & (t-2)(t-3)/2 \\ 0 & 0 & (t-1)(t-2)/2 & 0 & 0 & 0 \end{bmatrix}$$

$$(p^{*6}_{ju}) = \begin{bmatrix} 0 & 0 & 0 & 2t^2 & 0 & 0 \\ 0 & 4 & 0 & 0 & 0 & 2(t-3) \\ 0 & 0 & t^2(t-1)(t-3)/2 & 0 & t^2(t-2) & 0 \\ 2t^2 & 0 & 0 & t^2(t-3) & 0 & 0 \\ 0 & 0 & t^2(t-1) & 0 & 0 & 0 \\ 0 & 2(t-3) & 0 & 0 & 0 & (t-3)(t-4)/2 \end{bmatrix}$$

<u>Proof</u>. First of all, it can be verified (cf. Rao [2]) that although $n_4^* = n_5^* = t^2(t-1)$ and $\lambda_4^* = \lambda_5^* = 2$, the fourth and fifth associate classes cannot be combined, and remain distinct. For the same reason, the third and fourth or the third and fifth associate classes cannot be combined.

Since $v = b = t^2 + t + 1$ and $v^* = b^* = \binom{v}{2}$, it is obvious that v^* and b^* have the values as shown in (3.2). Consider now any treatment j^* of N^* and let $j^* \sim (\ell,m)$. The rows of N, corresponding to the treatments ℓ and m of N will have the following structure.

$$\begin{array}{llcccc} & \text{Partition} & \text{I} & \text{II} & \text{III} & \text{IV} \\ & \ell & 1 & 1\ldots.1 & 0\ldots.0 & 0\ldots.0 \\ (3.3) & & & & & \\ & m & 1 & 0\ldots.0 & 1\ldots.1 & 0\ldots.0 \\ & & \underbrace{}_{\lambda = 1} & \underbrace{}_{r - \lambda} & \underbrace{}_{r - \lambda} & \underbrace{}_{b - 2r + \lambda} \end{array}$$

The partitions shown in (3.3) contain numbers of columns as indicated, corresponding to the respective blocks of N. Of course, these columns would not always occur in the same order as shown above, but the actual order of occurrence of these columns of N for a particular pair (ℓ,m) is immaterial, because we are reducing all the 2×2 minors obtained by application of the 2-adjugation process to them modulo 2 and because their combinatorial structure would be the same for all orders of occurrences.

It is now obvious that the number of 1's in the row obtained by the 2-adjugation mod 2 process of these two rows will be

$$(3.4) \qquad (r-\lambda)^2 + 2\lambda(r-\lambda) = r^2 - \lambda^2,$$

because the 1's are obtained by combining the columns in partition I with each column in partitions II and III and combining each column of partition II with each column of partition III, and from no other combination of columns in (3.3). In view of (3.1), (3.4) reduces to $t(t+2)$. Since the number of 1's in the row in N^* corresponding to $j^* \sim (\ell,m)$ gives us the number of replications of j^* in the design N^*, it follows that $r^* = t(t+2)$. Further, since $v^* = b^*$, the arguments show that

$$(3.5) \qquad r^* = k^* = t(t+2)$$

as given in (3.2).

Consider now two treatments j^* and u^* of N^* and let them determine uniquely the pairs of treatments (ℓ,m) and (p,q) of N respectively, through Lemma 1.2. We shall assume here that $\ell = p$ or $m = p$ and q is a distinct treatment not belonging to any block containing P and m (and hence p). Since, however $\lambda = 1$, it is clear that there is a unique block in which ℓ and m occur together and then the condition imposed above requires that q should not occur in this unique block. In this case the treatments j^* and u^* of N^* are defined to be first associates of each other (cf. Section 2). Since there are (v-k) other treatments $\theta_3, \theta_4, \ldots, \theta_k$ of N, it is clear that j^* has $2(v-k)$ first associates determined by the pairs

$$(\ell, \theta_{k+1}), \ldots, (\ell, \theta_v), (m, \theta_{k+1}), \ldots, (m, \theta_v) \tag{3.6}$$

of N.

Now the rows in N corresponding to the three treatments, ℓ, m, and q which define the first associate treatments $j^* \sim (\ell,m)$ and $u^* \sim (\ell,q)$ of N^* will clearly have the following structure in N :

$$(3.7)\qquad \begin{array}{cccccc} \text{Partition} & \text{I} & \text{II} & \text{III} & \text{IV} & \text{V} \\ \ell & 1\,1\,0 & 1\ldots.1 & 0\ldots.0 & 0\ldots.0 & 0\ldots.0 \\ m & 1\,0\,1 & 0\ldots.0 & 1\ldots.1 & 0\ldots.0 & 0\ldots.0 \\ q & 0\,1\,1 & 0\ldots.0 & 0\ldots.0 & 1\ldots.1 & 0\ldots.0 \\ & \underbrace{\qquad}_{3\lambda = 3} & \underbrace{\qquad}_{r-2\lambda} & \underbrace{\qquad}_{r-2\lambda} & \underbrace{\qquad}_{r-2\lambda} & \underbrace{\qquad}_{b-3r+3\lambda} \end{array}$$

It is clear that the treatments j^* and u^* defined above are first associates of each other. Moreover, remarks similar to those made in connection with (3.3) apply to the partitioning of the columns of N as indicated in (3.7).

When we apply the 2-adjugation mod 2 process to the rows corresponding to (ℓ,m) and (ℓ,q), we shall get pairs of simultaneous 1's when we combine the columns of partition I with one another, and also when each column of partition II combines with the last column of partition I. No other combination of columns from various partitions in (3.7) can give us a pair of simultaneous 1's for the treatments j^* and u^* of N^*. This shows that any two first associates of the design N^* occur together in

$3\lambda^2 + \lambda(r-2\lambda) = r(r+\lambda)$ blocks of N^*. This, together with (3.1) gives

$$(3.8) \qquad n_1^* = 2(v-k) = 2t^2, \quad \lambda_1^* = r(r+\lambda) = t+2$$

is given in (3.2).

Consider once again two treatments j^* and u^* of N^* and let the pairs of treatments of N they uniquely determine through Lemma 1.2 be (ℓ,m) and (p,q) respectively. We now suppose that only three of $\ell,m.p$ and q are distinct; i.e., we have pairs like (ℓ,m), (ℓ,q) or (ℓ,m), (m,q) but we assume that all the three distinct treatments, ℓ,m,q occur in one block of N. They cannot occur together in more than one block, because $\lambda = 1$. Further, if $t \geq 2$, there is at least one block containing ℓ,m and one or more distinct treatments of N. In this case, we defined the treatments j^* and u^* of N^* to be second associates of each other (cf. Section 2). If $(\ell,m,\theta_3,\theta_4,\ldots,\theta_k)$ is the unique block in N containing the pair of treatments (ℓ,m), it is clear that the treatment $j^* \sim (\ell,m)$ of N^* will have $2(k-2) = 2(t-1)$ second associates :

$$(3.9) \qquad (\ell,\theta_3), (\ell,\theta_4),\ldots,(\ell,\theta_k), (m,\theta_3), (m,\theta_4),\ldots,(m,\theta_k).$$

The rows in N corresponding to the three treatments ℓ,m and q which define the second associate treatments $j^* \sim (\ell,m)$ and $u^* \sim (\ell,q)$ of N^* will clearly have the following structure in N :

(3.10)

Partition	I	II	III	IV	V
ℓ	1	1....1	0....0	0....0	0....0
m	1	0....0	0....0	0....0	0....0
q	1	0....0	0....0	1....1	0....0
1	1	r - 1	r - 1	r - 1	b-3r+2

Remarks similar to those made in connection with (3.3) apply to the partitioning of the columns of N as indicated in (3.10). When we apply the 2-adjugation mod 2 process to the rows corresponding to (ℓ,m)

and (ℓ,q) in (3.10), we shall get pairs of simultaneous 1's when we combine the column in partition I with each column in partition II and with no other combination of columns in (3.10). This shows that any two second associates of the design N^* occur together in $r - 1$ blocks of N^*. Hence we get

(3.11) $$n_2^* = 2(k-2) = 2(t-1), \quad \lambda_2^* = r - 1 = t,$$

as given in (3.2).

Consider, next, two treatments j^* and u^* of N^* and let the pairs of treatments of N uniquely determined by them through Lemma 1.2 be respectively (ℓ,m) and (p,q). We shall suppose that the treatments ℓ,m,p and q are all distinct and the pair (ℓ,m) occurs together in one block and the pair (p,q) occurs together in a different block of N. Since $\lambda = 1$, these blocks are unique. In this case, the treatments j^* and u^* of N^* are defined to be third associates of each other (cf. Section 2). To obtain the number of third associates of a given treatment $j^* \sim (\ell,m)$ of N^*, we proceed as follows :

Let $(\ell,m,\theta_3,\theta_4,\ldots,\theta_k)$ be the unique block of N containing the treatments ℓ and m together. Then there are $(v-k)$ other treatments, viz., $\theta_{k+1},\theta_{k+2},\ldots,\theta_v$ of N which belong to the other blocks of N. We can form $(v-k)(v-k-1)/2$ pairs out of these $(v-k)$ treatments, out of which $2(r-1).\frac{(k-1)(k-2)}{2}$ pairs do not become third associates of (ℓ,m), because the treatment ℓ occurs in $(r-1)$ other blocks of N. Every such blocks contains $(k-1)$ distinct treatments, none of which belongs to the block $(\ell,m,\theta_3,\theta_4,\ldots,\theta_k)$. These $(k-1)$ treatments form $(k-1)(k-2)/2$ pairs, none of which is a third associate of (ℓ,m). The total number of such pairs of treatments, belonging to the $(r-1)$ blocks containing ℓ becomes $(r-1).(k-1)\,(k-2)/2$. Similar argument can be made with the treatments belonging to the $(r-1)$ blocks containing m, and we get an equal number $(r-1)(k-1)(k-2)/2$ of treatment pairs none of which also will be a third associate of (ℓ,m). Deleting all such pairs from $(v-k)(v-k-1)/2$, the remainder

(3.12) $$\tfrac{1}{2}(v-k)\,(v-k-1) - (r-1)\,(k-1)\,(k-2)$$

gives us the total number of third associates of (ℓ,m).

Now let $j^* \sim (\ell,m)$ and $u^* \sim (p,q)$ be third associates of each other. Then the rows in N corresponding to the four treatments ℓ,m,p and q of N will have the following structure in N :

(3.13)

Partition	I	II	III	IV	V	VI	VII	VIII	IX	X	XI
ℓ	1	0	1	1	0	0	1...1	0...0	0...0	0...0	0...0
m	1	0	0	0	1	1	0...0	1...1	0...0	0...0	0...0
p	0	1	1	0	1	0	0...0	0...0	1...1	0...0	0...0
q	0	1	0	1	0	1	0...0	0...0	0...0	1...1	0...0
	$\lambda=1$	$\lambda=1$	$\lambda=1$	$\lambda=1$	$\lambda=1$	$\lambda=1$	$r-3\lambda$	$r-3\lambda$	$r-3\lambda$	$r-3\lambda$	$b-4r+6\lambda$

The same type of remarks as those made in connection with (3.3) apply to the partitioning of the columns of N indicated in (3.13). When we apply the 2-adjuation mod 2 process to the rows (ℓ,m) and (p,q) we shall get pairs of a simultaneous 1's when we combine each column of partition III with each column of partition VI and each column of partition IV with each column of partition V. There is no other combination of columns in (3.13) which would lead to a pair of simultaneous 1's in the process of 2-adjugation mod 2 applied to the rows (ℓ,m) and (p,q) in (3.13). It therefore follows that any two third associates of the design N* occur together in $\lambda_3^* = 2\lambda^2$ blocks of N*. In view of (3.1), we therefore have

$$n_3^* = t^2(t-1)^2/2, \quad \lambda_3^* = 2\lambda^2 = 2. \tag{3.14}$$

Once again consider two treatments j^* and u^* of N* and let them determine uniquely the pairs of treatments (ℓ,m) and (p,q) of N respectively through Lemma 1.2. We shall suppose here that all the four treatments ℓ,m,p and q of N are distinct and that ℓ,m and one of p and q occur together in a block of N and the remaining out of p and q is contained in some of the remaining blocks of N. Since $\lambda = 1$, it follows that there is a unique block of N which contains together ℓ,m and one of p and q. For the sake of definiteness, let ℓ,m and p occur together in a unique block of N and q occur in some of the remaining blocks of N. In this case, the treatment u^* is defined to be a fourth associate of the treatment j^* of N* (cf. Section 2).

Let $(\ell, m, \theta_3, \theta_4, \ldots, \theta_k)$ be the unique block of N in which the treatments ℓ, m of N occur together. Then there are $(v-k)$ other treatments, viz. $\theta_{k+1}, \theta_{k+2}, \ldots, \theta_v$ of N, which belong to the other blocks of N. It is now clear that the fourth associates of $j^* \sim (\ell, m)$ in N^* will be obtained from the pairs :

(3.15) $(\theta_3,\theta_{k+1}), (\theta_3,\theta_{k+2}),\ldots,(\theta_3,\theta_v),\ldots,(\theta_k,\theta_{k+1}),(\theta_k,\theta_{k+2}),\ldots,(\theta_k,\theta_v)$

of treatments of N. Obviously there are $(k-2)(v-k)$ pairs in (3.15).

Now let $u^* \sim (p,q)$ be a fourth associate of $j^* \sim (\ell, m)$ and for the sake of definiteness, let the treatments ℓ, m, p of N occur together in a unique block of N $(\lambda = 1)$ and q occur in some of the remaining blocks of N. Then the rows in N corresponding to these four treatments of N will have the following structure in N :

(3.16)

Partition	I	II	III	IV	V	VI	VII	VIII	IX
ℓ	1	1	0	0	1....1	0....0	0....0	0....0	0....0
m	1	0	1	0	0....0	1....1	0....0	0....0	0....0
p	1	0	0	1	0....0	0....0	1....1	0....0	0....0
q	0	1	1	1	0....0	0....0	0....0	1....1	0....0
	$\lambda=1$	$\lambda=1$	$\lambda=1$	$\lambda=1$	$r-2\lambda$	$r-2\lambda$	$r-2\lambda$	$r-3\lambda$	$b-4r+5\lambda$

Remarks similar to those made in connection with (3.3) apply to the partitioning of the columns of N as indicated in (3.16). Now, when we apply the 2-adjugation mod 2 process to the rows (ℓ, m) and (p,q), we shall get pairs of simultaneous 1's when we combine the column of partition I with the columns of partitions II and III. No other combination of columns in (3.16) can give us a pair of simultaneous 1's when we apply the 2-adjugation mod 2 process to the rows (ℓ, m) and (p,q) of (3.16). It thus follows that fourth associates of a given treatment of the design N^* occur with it in $\lambda_4^* = 2\lambda^2$ blocks of N^*. In view of (3.1), we therefore have

(3.17) $$n_4^* = (k-2).(v-k) = t^2(t-1), \ \lambda_4^* = 2\lambda^2 = 2.$$

Once again consider two treatments j^* and u^* of N^* and let them determine uniquely the pairs of treatments (ℓ, m) and (p,q) of N respectively through Lemma 1.2. We shall suppose here that all the four treatments, ℓ, m, p and q of N are distinct and that p,q and one of ℓ and m occur together in a block of N and the remaining treatment out of ℓ and m is contained in some of the remaining blocks of N. Since $\lambda = 1$, it is clear that there is a unique block of N which contains together p,q and one of ℓ and m. For the sake of definiteness, let p,q and ℓ occur together in a unique block of N and m occur in some of the remaining blocks of N. In this case, the treatment u^* of N^* is defined to be a fifth associate of the treatment j^* of N^* (cf. Section 2).

Let $(\ell,m,\theta_3,\theta_4\cdots\theta_k)$ be the unique block of N in which the treatments ℓ and m of N occur together. Then there are $r-1$ other blocks of N, given by

$$\begin{array}{c}(\ell,\theta_2^1,\ldots,\theta_k^1)\\(\ell,\theta_2^2,\ldots,\theta_k^2)\\\cdots\cdots\cdots\cdots\\(\ell,\theta_2^{r-1},\ldots,\theta_k^{r-1})\end{array}$$

which contain ℓ, and also $r-1$ other blocks of N, given by

$$\begin{array}{c}(m,\Psi_2^1,\ldots,\Psi_k^1)\\(m,\Psi_2^2,\ldots,\Psi_k^2)\\\cdots\cdots\\\cdots\cdots\\(m,\Psi_2^{r-1},\ldots,\Psi_k^{r-1})\end{array}$$

which contain m. It is now clear that the fifth associates of $j^* \sim (\ell,m)$ of N^* will be obtained from the pairs

$$\begin{array}{c}(\theta_2^1,\ \theta_3^1),\ldots,(\theta_{k-1}^1,\ \theta_k^1)\\\cdots\cdots\\\cdots\cdots\\(\theta_2^{r-1},\ \theta_3^{r-1}),\ldots,(\theta_{k-1}^{r-1},\ \theta_k^{r-1})\end{array} \tag{3.18}$$

which are $(r-1)\ (k-1)\ (k-2)/2$ in number, and also from the pairs

$$\begin{array}{c}(\Psi_2^1,\ \Psi_3^1),\ldots,(\Psi_{k-1}^1,\ \Psi_k^1)\\\cdots\cdots\\\cdots\cdots\\(\Psi_2^{r-1},\Psi_3^{r-1}),\ldots,(\Psi_{k-1}^{r-1},\Psi_k^{r-1})\end{array} \tag{3.19}$$

which are $(r-1)(k-1)(k-2)/2$ in number. Obviously there are $(r-1)(k-1)(k-2)$ pairs in (3.18) and (3.19) together, which will form all the possible fifth associates of $j^* \sim (\ell,m)$.

Now let $u^* \sim (p,q)$ be a fifth associate of $j^* \sim (\ell,m)$. For the sake of definiteness, let the treatments ℓ,p,q of N occur together in a unique block of N and m occur in some of the remaining blocks of N. Then the rows in N corresponding to these four treatments of N will have the following structures :

(3.20)

Partition	I	II	III	IV	V	VI	VII	VIII	IX
ℓ	1	0	1	0	1....1	0....0	0....0	0....0	0....0
m	0	1	1	1	0....0	1....1	0....0	0....0	0....0
p	1	1	0	0	0....0	0....0	1....1	0....0	0....0
q	1	0	0	1	0....0	0....0	0....0	1....1	0....0
	$\lambda=1$	$\lambda=1$	$\lambda=1$	$\lambda=1$	$r-2\lambda$	$r-3\lambda$	$r-2\lambda$	$r-2\lambda$	$b-4r+5\lambda$

Remarks similar to those made in connection with (3.3) apply to the partitioning of the columns of N as indicated in (3.20). Now when we apply the 2-adjugation mod 2 process to the rows (ℓ,m) and (p,q) we shall get pairs of simultaneous 1's when we combine the column of partition I with the columns of partitions II and IV. No other combination of columns in (3.20) can give us a pair of simultaneous 1's when we apply the 2-adjugation mod 2 process to the rows (ℓ,m) and (p,q) of (3.20). It thus follows that any fifth associate of a given treatment of the design N* occurs with it in $\lambda_5^* = 2\lambda^2$ blocks of N*. In view of (3.1) we therefore have

(3.21) $$n_5^* = (k-1)(k-2)(r-1) = t^2(t-1),\ \lambda_5^* = 2\lambda^2 = 2.$$

Consider, finally, two treatments j^* and u^* of N* and let the pairs of treatments of N uniquely determined by j^* and u^* through Lemma 1.2 be (ℓ,m) and (p,q) respectively. We shall suppose in this case that all the four treatments ℓ,m,p and q of N are distinct and that all of them occur together in a block of N. However, since $\lambda = 1$, it follows that there would be a unique block of N containing all the four treatments ℓ,m,p and q. In this case, the treatments j^* and u^* of N* are defined to be sixth associates of each other (cf. Section 2).

Let, as before, $(\ell,m,\theta_3,\theta_4,\ldots,\theta_k)$ be the unique block containing the pair (ℓ,m) of treatments of N. Now the sixth associates of the treatment $j^* \sim (\ell,m)$ of N^* would be obtained from the pairs

(3.22) $$(\theta_3,\theta_4),\ (\theta_3,\theta_5),\ldots,(\theta_3,\theta_k),\ldots,(\theta_{k-1},\theta_k)$$

of treatments of N, and clearly there are $(k-2)(k-3)/2$ pairs in (3.22).

Further let $j^* \sim (\ell,m)$ and $u^* \sim (p,q)$ be sixth associates of each other, then ℓ,m,p and q are all distinct treatments belonging to a unique block of N. The rows in N corresponding to these four treatments of N will have the following structures in N :

(3.23)

Partition	I	II	III	IV	V	VI
ℓ	1	1....1	0....0	0....0	0....0	0....0
m	1	0....0	1....1	0....0	0....0	0....0
p	1	0....0	0....0	1....1	0....0	0....0
q	1	0....0	0....0	0....0	1....1	0....0
	$\lambda = 1$	$r - \lambda$	$r - \lambda$	$r - \lambda$	$r - \lambda$	$b-4r+3\lambda$

The same type of remarks as those made in connection with (3.3) apply to the partitioning of the columns of N indicated in (3.23). It is easy now to verify that no combination of two columns in (3.23) would give a pair of simultaneous 1's for the 2-adjugation mod 2 of rows (ℓ,m) and (p,q). This shows that any two sixth associates of the design N^* cannot occur in any block of N. In view of (3.1), we have thus proved that

(3.24) $$n_6^* = (k-2)(k-3)/2 = (t-1)(t-2)/2,\ \lambda_6^* = 0.$$

It remains to calculate the matrices of the secondary parameters p^{*i}_{ju} for the design N^* where p^{*i}_{ju} is the number of treatments common between the j-th associates of the treatment θ_1^* and u-th associates of the treatment θ_2^* where θ_1^* and θ_2^* are i-th associates in the design N^* $(i,j,u = 1,2,\ldots,6)$. We shall get the matrix (p^{*1}_{ju}) by detailed arguments, while other matrices can be obtained by similar methods.

Let θ_1^* and θ_2^* be any two first associates in N^*. Then the pairs of treatments of N, uniquely determined by θ_1^* and θ_2^* through Lemma 1.2 would be respectively of the form (ℓ,m) and (ℓ,q), where q is a treatment of N not contained in the unique block of N containing ℓ and m. Let the blocks of N containing the pairs (ℓ,m) and (ℓ,q) be respectively

$$B_1 = (\ell,m,\theta_3,\theta_4,\ldots,\theta_k)$$

$$B_2 = (\ell,q,\theta_3,\theta_4,\ldots,\theta_k).$$

Let the other (r-2) blocks of N containing the treatment ℓ be

$$(\ell,\theta_2^2,\theta_3^2,\ldots,\theta_k^2)$$
$$\cdots\cdots$$
$$\cdots\cdots$$
$$(\ell,\theta_2^{r-1},\theta_3^{r-1},\ldots,\theta_k^{r-1})$$

and the other (r-1) blocks of N containing the treatment m be

$$(m, \psi_2^1, \psi_3^1,\ldots,\psi_k^1)$$
$$\cdots\cdots$$
$$\cdots\cdots$$
$$(m, \psi_2^{r-1}, \psi_3^{r-1},\ldots,\psi_k^{r-1})$$

Then the pairs of treatments of N which give rise to the various associate classes of $\theta_1^* \sim (\ell,m)$ and $\theta_2^* \sim (\ell,q)$ are as shown in the Tables 1 and 2 given at the end of this paper.

It is now obvious that the pairs of treatments of N which give rise to first associates in N^* of both the first associate treatments $\theta_1^* \sim (\ell,m)$ and $\theta_2^* \sim (\ell,q)$ of N^* are

$$(m,q),(\ell,\theta^2),\ldots,(\ell,\theta_k^{r-1})$$

and this shows that

(3.25) $$p_{11}^{*1} = 1 + (k-1)(r-2) = t^2 - t + 1.$$

Similarly it is clear that the pairs of treatments of N which give rise to first associates in N^* of θ_1^* and second associates in N^* of θ_2^* are

$$(\ell,\theta_3),(\ell,\theta_4),\ldots,(\ell,\theta_k)$$

and this shows that

(3.26) $$p^{*1}_{12} = k - 2 = t - 1.$$

Further, the pairs of treatments of N given below give rise to first associates in N* of θ^*_1 and third associates in N* of θ^*_2 :

$$(m,\theta_3^2),(m,\theta_4^2),\ldots,(m,\theta_k^2),\ldots,(m,\theta_k^{r-1}).$$

In view of (3.1), this gives

(3.27) $$p^{*1}_{13} = (k-2)\ (r-2) = (t-1)^2.$$

Similarly the pairs of treatments of N given below give rise to first associates in N* of θ^*_1 and fourth associates in N* of θ^*_2 :

$$(m,\theta_3^1),(m,\theta_4^1),\ldots,(m,\theta_k^1).$$

In view of (3.1), this shows that

(3.28) $$p^{*1}_{14} = (k-2) = t - 1.$$

Further, the pairs of treatments of N which give rise to first associates in N* of θ^*_1 and fifth associates in N* of θ^*_2 are

$$(m,\theta_2^2),(m,\theta_2^3),\ldots,(m,\theta_2^{r-1}).$$

In view of (3.1), this shows that

(3.29) $$p^{*1}_{15} = r - 2 = t - 1.$$

Further

$$p^{*1}_{11} + p^{*1}_{12} + p^{*1}_{13} + p^{*1}_{14} + p^{*1}_{15} = 2t^2 - 1 = n - 1.$$

Hence

(3.30) $$p^{*1}_{16} = 0$$

as is obvious otherwise also.

Now, the pairs of treatments of N which give rise to second associates in N* of θ^*_1 and first associates in N* of θ^*_2 are

$$(\ell,\theta_3),(\ell,\theta_4),\ldots,(\ell,\theta_k).$$

Hence we must have

(3.31) $$p^{*1}_{21} = k - 2 = t - 1.$$

Similar verifications can be made with the use of Tables 1 and 2. Then the following results are obtained :

(3.32) $$p^{*1}_{22} = p^{*1}_{23} = p^{*1}_{24} = 0.$$

Further, there are $k - 2 = t - 1$ pairs of treatments of N given by

$$(m,\theta_3),(m,\theta_4),\ldots,(m,\theta_k)$$

which give rise to second associates in N^* of θ^*_1 and fifth associates of θ^*_2 in N^*. Hence

(3.33) $$p^{*1}_{25} = t - 1.$$

Since this exhausts all the $2(t-1)$ second associates of θ^*_1 in N^*, it follows that

(3.34) $$p^{*1}_{25} = 0.$$

From Tables 1 and 2, values of the other paramenters p^{*1}_{ju} for $j,u = 2,3,4,5,6$ can be deduced and will be seen to be in agreement with those given in (3.2). In the same manner, the remaining association matrices can be determined.

This completes the proof of the theorem.

<u>Table 1</u>

Associate classes of $\theta^*_1 \sim (\ell,m)$

1st Associates	(ℓ,q), (ℓ,θ^1_3), $(\ell,\theta^1_4),\ldots,(\ell,\theta^1_k)$, (m,q), (m,θ^1_3), $(m,\theta^1_4),\ldots,(m,\theta^1_k)$, (ℓ,θ^2_2), $(\ell,\theta^2_3),\ldots,(\ell,\theta^2_k),\ldots,(\ell,\theta^{r-1}_k)$, (m,θ^2_2), $(m,\theta^2_3),\ldots,$ $(m,\theta^2_k),\ldots,(m,\theta^{r-1}_k)$
2nd Associates	(ℓ,θ_3), $(\ell,\theta_4),\ldots,(\ell,\theta_k)$, (m,θ_3), $(m,\theta_4),\ldots,(m,\theta_k)$.

Table 1 (contd.)

3rd Associates	(q,θ_3^2), (q,θ_4^2),...,(q,θ_k^{r-1}),...,$(\theta_k,\theta_{k-1}^{r-1})$, (θ_2^2,θ_3^2), (θ_2^2,θ_4^2),..., (θ_2^2,θ_k^2),...,$(\theta_2^2,\theta_k^{r-1})$,...,$(\theta_k^2,\theta_{k-1}^{r-1})$,...,$(\theta_2^{r-2},\theta_2^{r-1})$,..., $(\theta_2^{r-2},\theta_3^{r-1})$,...,$(\theta_2^{r-2},\theta_k^{r-1})$,...,$(\theta_k^{r-2},\theta_k^{r})$.
4th Associates	(θ_3,q), (θ_3,θ_3^1), (θ_3,θ_4^1),...,(θ_3,θ_k^1), (θ_3,θ_2^2), (θ_3,θ_3^2),..., (θ_3,θ_k^2),...,$(\theta_3,\theta_2^{r-1})$, $(\theta_3,\theta_3^{r-1})$,...,$(\theta_3,\theta_k^{r-1})$,...,$(\theta_k,\theta_2^{r-1})$, $(\theta_k,\theta_3^{r-1})$,...,$(\theta_k,\theta_k^{r-1})$.
5th Associates	(q,θ_3^1), (q,θ_4^1),...,(q,θ_k^1),...,$(\theta_{k-1}^1,\theta_k^1)$, (θ_2^2,θ_3^2), (θ_2^2,θ_4^2),..., (θ_2^2,θ_k^2),...,$(\theta_{k-1}^2,\theta_k^2)$, ... $(\theta_2^{r-1},\theta_3^{r-1})$, $(\theta_2^{r-1},\theta_4^{r-1})$,...,$(\theta_2^{r-1},\theta_k^{r-1})$,...,$(\theta_{k-1}^{r-1},\theta_k^{r-1})$, (q,θ_2^2), (q,θ_2^3),...,(q,θ_2^{r-1}),...,$(\theta_2^{r-2},\theta_2^{r-1})$, (θ_3^1,θ_3^2), (θ_3^1,θ_3^3),..., $(\theta_3^1,\theta_3^{r-1})$,...,$(\theta_3^{r-2},\theta_3^{r-1})$, (θ_k^1,θ_k^2), (θ_k^1,θ_k^3),...,$(\theta_k^1,\theta_k^{r-1})$,..., $(\theta_k^{r-2},\theta_k^{r-1})$.
6th Associates	(θ_3,θ_4), (θ_3,θ_5),...,(θ_3,θ_k),...,(θ_{k-1},θ_k)

Table 2

Associate classes of $\theta_2^* \sim (\ell,q)$

1st Associates	(ℓ,m), (ℓ,θ_3), (ℓ,θ_4),...,(ℓ,θ_k), (m,q), (q,θ_3), (q,θ_4),...,(q,θ_k), (ℓ,θ_2^2), (ℓ,θ_3^2),...,(ℓ,θ_k^2),...,(ℓ,θ_k^{r-1}), (q,θ_2^2), (q,θ_3^2),...,(q,θ_k^2),..., (q,θ_k^{r-1})
2nd Associates	(ℓ,θ_3^1), (ℓ,θ_4^1),...,(ℓ,θ_k^1), (q,θ_3^1), (q,θ_4^1),...,(q,θ_k^1).

Table 2 (contd.)

3rd Associates	(m,θ_3^2), (m,θ_4^2),...,(m,θ_k^2),...,$(\theta_k,\theta_{k-1}^{r-1})$, (θ_2^2,θ_3^2), (θ_2^2,θ_4^2),..., (θ_2^2,θ_k^2),...,$(\theta_2^2,\theta_k^{r-1})$, ... (θ_k^2,θ_3^2), (θ_k^2,θ_4^2),...,$(\theta_k^2,\theta_{k-1}^{r-1})$, ... $(\theta_2^{r-2},\theta_3^{r-1})$, $(\theta_2^{r-2},\theta_4^{r-1})$,...,$(\theta_2^{r-2},\theta_k^{r-1})$,..., $(\theta_k^{r-2},\theta_{k-1}^{r-1})$
4th Associates	(m,θ_3^1), (m,θ_4^1),...,(m,θ_k^1), (θ_3,θ_3^1), (θ_3,θ_4^1),...,(θ_3,θ_k^1) .... (θ_k,θ_3^1), (θ_k,θ_4^1),...,(θ_k,θ_k^1),...,(θ_3^1,θ_3^2), (θ_3^1,θ_3^2),..., (θ_3^1,θ_k^2),...,$(\theta_3^1,\theta_2^{r-1})$, $(\theta_3^1,\theta_3^{r-1})$,...,$(\theta_3^1,\theta_k^{r-1})$,...,$(\theta_k^1,\theta_2^{r-1})$, $(\theta_k^1,\theta_3^{r-1})$,...,$(\theta_3^1,\theta_k^{r-1})$.
5th Associates	(m,θ_3), (m,θ_4),...,(m,θ_k),...,(θ_{k-1},θ_k), (θ_2^2,θ_3^2), (θ_2^2,θ_4^2),..., (θ_2^2,θ_k^2),...,$(\theta_{k-1}^2,\theta_k^2)$,...,$(\theta_2^{r-1},\theta_3^{r-1})$, $(\theta_2^{r-1},\theta_4^{r-1})$,...,$(\theta_2^{r-1},\theta_k^{r-1})$, ...,$(\theta_{k-1}^{r-1},\theta_k^{r-1})$, ... (m,θ_2^2), (m,θ_2^3),...,(m,θ_2^{r-1}),...,$(\theta_2^{r-2},\theta_2^{r-1})$, (θ_3,θ_3^2), (θ_3,θ_3^3),..., $(\theta_3,\theta_3^{r-1})$,...,$(\theta_3^{r-2},\theta_2^{r-1})$, (θ_k,θ_k^2), (θ_k,θ_k^3),...,$(\theta_k,\theta_k^{r-1})$,..., $(\theta_k^{r-2},\theta_k^{r-1})$.
6th Associates	(θ_3^1,θ_4^1), (θ_3^1,θ_5^1),...,(θ_3^1,θ_k^1),...,$(\theta_{k-1}^1,\theta_k^1)$.

REFERENCES

1. C.R. Nair, A new class of designs, Jour. Amer. Stat. Assn., 59(1964), 817-833.
2. C.R. Rao, General methods of analysis for incomplete block designs, Jour. Amer. Stat. Assn., 42(1947), 541-561.
3. M.N. Vartak and G.A. Patwardhan, The 2-adjugate mod 2 class of designs, Jour. Combinatorial Theory (to appear).

CHARACTERIZATION OF POTENTIALLY CONNECTED INTEGER-PAIR SEQUENCES

NIRMALA ACHUTHAN
Indian Statistical Institute
203 B. T. Road, Calcutta 700 035

ABSTRACT

Given a pseudograph, multigraph or graph G, we can associate with it a sequence of unordered integer-pairs $S_G = (c_1, c_2, \ldots, c_q)$, where $q = |E(G)|$, constructed as follows : If the edges of G are labelled $1, 2, \ldots, q$, then for the sth edge (u,v) of G, $c_s = (a_s, b_s)$ where a_s, b_s are the degrees of u and v. An integer-pair sequence S is said to be graphic if there exists a graph G for with $S_G = S$. In this paper we characterize potentially connected integer-pair sequences.

1. INTRODUCTION AND DEFINITIONS

In this paper we consider finite, undirected graphs without loops or multiple edges.

Let G be a graph with vertex set $V(G)$ and edge set $E(G)$. An edge connecting vertices u_i and u_j will be denoted by (u_i, u_j). We associate with G, a sequence of unordered integer-pairs, $S(G) = (c_1, c_2, \ldots, c_q)$, constructed as follows, where $q = |E(G)|$. If the edges of G are numbered 1 to q, then, for the sth edge (u_i, u_j) of G, $c_s = (a_s, b_s)$ where a_s and b_s are the degrees of u_i and u_j. (In this paper (r,s) denotes the unordered pair of elements r and s.) This concept was introduced by Hakimi and Patrinos [1].

An integer-pair sequence is said to be potentially connected if there exists a connected graph realizing S.

In this paper we characterize potentially connected integer-pair sequences. This solves a problem in [1].

Let $p(G)$, $q(G)$ and $\nu(G)$ denote the numbers of vertices, edges and connected components of G respectively. Let $d_G(u)$ denote the degree of a vertex u in G. A pendant vertex of G is a vertex with degree one. A vertex of G is said to be a cyclic vertex if it lies on some cycle of G.

Definitions of the terms used can be found in Harary [2].

Let $S = ((a_1,b_1), (a_2,b_2),\ldots,(a_q,b_q))$ be an integer-pair sequence. We shall use the following notation.

S_1 is the sequence $(a_1,b_1,a_2,b_2,\ldots,a_q,b_q)$,

S_2 is the set of distinct members of S_1,

$S_3 = S_2 - \{1\} = \{d_1,d_2,\ldots,d_k\}$,

$k(r,s)$ is the number of times the unordered pair (r,s) appears in S,

$n(r)$ is the number of times r appears in S_1, and

$\ell(r) = \frac{n(r)}{r}$.

If S is pseudographic then Hakimi and Patrinos have shown that $\ell(r)$ denotes the number of vertices with degree r in any realization of S. In particular $n(1)$ denotes the number of pendant vertices in any realization of S.

Let $I = \{1,2,\ldots,k\}$. For $I_1,I_2 \subseteq I$ with $I_1 \cap I_2 = \phi$, we write

$$k(I_1) = \sum_{i,j\epsilon I_1} k(d_i,d_j),$$

$$k(I_1,I_2) = \sum_{i\epsilon I_1,j\epsilon I_2} k(d_i,d_j)$$

and

$$\ell(I_1) = \sum_{i\epsilon I_1} \ell(d_i).$$

Let H be the graph defined as follows : $V(H) = I$ and $(i,j) \in E(H)$ if and only if $k(d_i,d_j) \geq 1$. For $I_1 \subseteq I$, $H[I_1]$ denotes the subgraph of H induced on the vertices $i \in I_1$. If G is a realization of S, let $G(I_1)$ denote the subgraph of G induced on the vertices with degree d_i, $i \in I_1$. Clearly $k(I_1)$ denotes the number of edges of $G(I_1)$, $\ell(I_1)$ denotes the number of vertices of $G(I_1)$ and $k(I_1,I_2)$ denotes the number of edges between $G(I_1)$ and $G(I_2)$.

For a given integer-pair sequence S, we have defined the parameters $k(r,s)$, $\ell(r)$, $n(r)$, the set I and the graph H. These will be used with the suffix S, if necessary, to avoid ambiguity.

Let G be a graph realizing S. If u,v,x,y are distinct vertices of G such that (u,x), $(v,y) \in E(G)$ and (u,y), $(v,x) \notin E(G)$ and $d_G(u) = d_G(v)$, then we can replace the two edges (u,x) and (v,y) by the edges (u,y) and (v,x) resulting in a graph G', say. We denote (see Rao and Taneja [3]) such an interchange of edges by $I(u,x,v,y)$ and such a process of obtaining G' from G by

$$G \to I(u,x,v,y) \to G'.$$

Clearly, G' is a realization of S.

By the shrinking of a subgraph G^* of G we mean replacing G^* by a single vertex A, and joining A to every vertex u outside G^* for which there is a vertex v of G^* such that (u,v) is an edge of G.

Throughout this paper, S denotes the integer-pair sequence $((a_1,b_1),(a_2,b_2),\ldots,(a_q,b_q))$.

The following result is easy to prove.

<u>Result 1</u>. For $I_1 \subseteq I$, if $G(I_1)$ is connected, then, $H[I_1]$ is connected. Equivalently $\nu(H[I_1]) \leq \nu(G(I_1))$. Further equality is attained if for all u,v belonging to two distinct components of $G(I_1)$ we have $d_G(u) \neq d_G(v)$.

2. NECESSITY

<u>Theorem 1</u>. Let G be a connected graph with $p(G) \geq 3$ and let S be the integer-pair sequence of G. Then $(1,1) \notin S$ and the following conditions are satisfied.

(i) $q \geq \ell(I) + n(1) - 1$.

(ii) For $I_1 \subsetneq I$, if $k(I_1) - \ell(I_1) + k(I_1, I-I_1) = \alpha$, then $H[I-I_1]$ has at most $\alpha + 1$ connected components.

<u>Proof</u>. Clearly $(1,1) \notin S$. Since $\ell(I) + n(1)$ and q are respectively the numbers of vertices and edges in G and G is a connected graph, we have $q \geq \ell(I) + n(1) - 1$. This establishes (i).

For proving (ii), let $I_1 \subsetneq I$ and $k(I_1) - \ell(I_1) + k(I_1, I-I_1) = \alpha$. Let $\mu = \upsilon(G(I_1))$ and $\lambda = \upsilon(G(I-I_1))$. It is enough to prove that $\lambda \leq \alpha + 1$. Firstly we have

$$k(I_1) \geq \ell(I_1) - \mu. \qquad \ldots \quad (1)$$

Combining this with $k(I_1) - \ell(I_1) + k(I_1, 1-I_1) = \alpha$, we have

$$\mu + \alpha \geq k(I_1, I-I_1). \qquad \ldots \quad (2)$$

Since G is connected, we have, $k(I_1, I-I_1) \geq \lambda + \mu - 1$. This together with (2) gives $\lambda \leq \alpha + 1$ and this proves the theorem.

Remark 1. For $I_1 = \phi$, condition (ii) implies that the graph H is connected.

Remark 2. From (ii) it immediately follows that, for any $I_1 \subsetneqq I$, $k(I_1) \geq \ell(I_1) - k(I_1, I-I_1)$.

3. SUFFICIENCY

In the following we shall prove that the conditions (i) and (ii) given in the statement of Theorem 1 are sufficient for any graphic integer-pair sequence to have a connected realization.

Before going into the proof of sufficiency, we shall give the notations and definitions used in this paper.

Henceforth S will denote a graphic integer-pair sequence satisfying conditions (i) and (ii).

Define

$\underline{\underline{G}}_1(S) = \{G : G$ realizes S and has minimum number λ of components$\}$.

For $G \in \underline{\underline{G}}_1(S)$, let

$I^*(G) = \{i : i \in I$ and there is a cyclic vertex with degree d_i in $G\}$.

Let

$\underline{\underline{G}}_2(S) = \{G : G \in \underline{\underline{G}}_1(S)$ and $|I^*(G)|$ is maximum$\}$.

In what follows, G denotes a graph from the class $\underline{\underline{G}}_2(S)$, that is, G has the minimum number λ of components and $|I^*(G)|$ is maximum. We shall also write I^* for $I^*(G)$. Let $G_1, G_2, \ldots, G_\lambda$ be the connected components of G. Define, for $1 \leq j \leq \lambda$,

$I_j = \{i : i \in I^*$, there is a vertex of degree d_i in $G_j\}$.

If $\lambda = 1$, there is nothing to prove. So let $\lambda \geq 2$. Then from (i) we have $I^*(G) = \phi$. The following propositions are easy to prove.

Proposition 1. If $i \in I_j$, then there is a cyclic vertex with degree d_i in G_j, $1 \leq j \leq \lambda$.

Proposition 2. If $j \neq 1$, then $I_j \cap I_1 = \phi$.

Remark 3. From Proposition 2 it follows that I^* is partitioned into λ sets, $I_1, I_2, \ldots, I_\lambda$. For $1 \leq j \leq \lambda$, G_j contains all the vertices with degree d_i, for all $i \in I_j$. Possibly some of the I_j's are empty. Without loss of generality let us assume that $I_j \neq \phi$ for $1 \leq j \leq r$, and $I_j = \phi$ for $r+1 \leq j \leq \lambda$.

For $1 \leq j \leq r$, let λ_j and μ_j be the numbers of connected components of $G_j(I_j)$ and $G_j(I-I_j)$. Let $H_1^j, H_2^j, \ldots, H_{\mu_j}^j$ and $F_1^j, F_2^j, \ldots, F_{\mu_j}^j$ be the connected components of $G_j(I_j)$ and $G_j(I-I_j)$ respectively. From the definition of I_j, it can be seen that no vertex of $G_j(I-I_j)$ lies on a cycle and so F_t^j is a tree for $1 \leq j \leq r$ and $1 \leq t \leq \mu_j$. Further there is at most one edge between a H_s^j and a F_t^j.

Definition. We say that $G_j(I_j)$ satisfies condition C if and only if there are two vertices u and v belonging to two distinct components of $G_j(I_j)$ with $d_G(u) = d_G(v)$.

This is sometimes expressed as 'there are two distinct components of $G_j(I_j)$ with a common degree'.

Theorem 2. There is at least one j for which $G_j(I_j)$ satisfies condition C.

Proof. Suppose the theorem is not true, that is, no two components of $G_j(I_j)$, for $1 \leq j \leq r$, have a common degree. Then from Result 1 for $1 \leq j \leq r$, we have $\nu(H[I_j]) = \lambda_j$. Now adding this equation for $1 \leq j \leq r$, we have

$$\nu(H[I^*]) = \sum_{j=1}^{r} \lambda_j \qquad \ldots \quad (3)$$

Let us now construct a bipartite graph L_j from G_j, for $1 \leq j \leq r$, as follows : The vertices of L_j are $x_1^j, x_2^j, \ldots, x_{\lambda_j}^j$ and $y_1^j, y_2^j, \ldots, y_{\mu_j}^j$; the vertex x_s^j corresponds to H_s^j and the vertex y_t^j corresponds to F_t^j. We shall join x_s^j and y_t^j if and only if there is an edge between H_s^j and F_t^j. Clearly L_j is a tree and so it has $\lambda_j + \mu_j - 1$ edges.

Thus using the fact that there is at most one edge between a H_s^j and a F_t^j, we have,

$$k(I_j, I-I_j) = \lambda_j + \mu_j - 1 \qquad \ldots (4)$$

Adding (4) for $1 \leq j \leq r$ and rearranging the terms we have

$$\sum_{j=1}^{r} \mu_j = \sum_{j=1}^{r} k(I_j, I-I_j) - \sum_{j=1}^{r} \lambda_j + r = k(I^*, I-I^*) - \sum_{j=1}^{r} \lambda_j + r \qquad \ldots (5)$$

as $k(I_j, I-I_j) = k(I_j, I-I^*) + k(I_j, I^*-I_j)$ and $k(I_j, I^*-I_j) = 0$ for $1 \leq j \leq r$. Now

$$\nu(G(I-I^*)) = \sum_{j=1}^{r} \mu_j + \lambda - r = k(I^*, I-I^*) - \sum_{j=1}^{r} \lambda_j + \lambda \quad \text{(using 5)} \qquad \ldots (6)$$

Since $G(I-I^*)$ is a forest, we have

$$\begin{aligned} k(I-I^*) &= q(G(I-I^*)) \\ &= \ell(I-I^*) - \nu(G(I-I^*)) \\ &= \ell(I-I^*) - k(I^*, I-I^*) + \sum_{j=1}^{r} \lambda_j - \lambda \quad \text{(using 6)} \end{aligned} \qquad \ldots (7)$$

Now from condition (ii) of Theorem 1 we must have

$$\nu(H[I^*]) \leq \sum_{j=1}^{r} \lambda_j - \lambda + 1 \qquad \ldots (8)$$

But from (3) we have $\nu(H[I^*]) = \sum_{j=1}^{r} \lambda_j$. combining this with (8) we have $\lambda \leq 1$ which is a contradiction. This proves the theorem.

Without loss of generality let us assume that $G_1(I_1)$ satisfies condition C. We shall denote the components of $G_1(I_1)$ and $G_1(I-I_1)$ by $H_1, H_2, \ldots, H_{\lambda_1}$ and $F_1, F_2, \ldots, F_{\mu_1}$ respectively.

<u>Definition</u>. A vertex v of F_t with degree d_s is said to be of Type I if there is a vertex with degree d_s in G_j, for some $j \neq 1$.

Let G_1' denote the graph obtained from G_1 by shrinking H_γ into a single vertex x_γ, for $1 \leq \gamma \leq \lambda_1$. Clearly G_1' is a tree and there is a unique path between x_γ and x_δ for all γ and δ.

Let us write $J_1 = \{1, 2, \ldots, \lambda_1\}$. Let $R \subseteq J_1$ satisfying the following conditions:

(a) There are two elements α and β in R such that H_α and H_β have a common degree.

(b) If $\gamma \in R$ and $\delta \in J_1 - R$ then H_γ and H_δ do not have a common degree and there is a vertex of Type I on the path between x_γ and x_δ.

Clearly J_1 satisfies (a) and (b). Let $\Gamma_1 \subseteq J_1$ satisfying (a) and (b), and with minimum size.

Now in a similar fashion we can define Γ_j, for all j such that $G_j(I_j)$ satisfies condition C. Without loss of generality let us assume that

$$|\Gamma_1| = \min|\Gamma_j|.$$

and let us denote Γ_1 (which implicitly depends on G) by $\Gamma(G)$.

<u>In what follows</u> G <u>will denote a graph in</u> $\underline{\underline{G}}_2(S)$ <u>for which</u> $|\Gamma(G)|$ <u>is minimum, that is, if</u> $G' \in \underline{\underline{G}}_2(S)$ <u>then</u> $|\Gamma(G)| \leq |\Gamma(G')|$. <u>Sometimes</u> Γ <u>will be used in place of</u> $\Gamma(G)$.

Let us write without loss of generality $\Gamma(G) = \{1,2,\ldots,\theta\}$. Define $M(\Gamma)$ as the union of paths in L_1 (recall the definition of L_1 given in the proof of Theorem 2) between x_γ and x_δ for $1 \leq \gamma, \delta \leq \theta$. Clearly $M(\Gamma)$ is a tree. Let us now construct a subgraph $G^*(\Gamma)$ of G using the tree $M(\Gamma)$ as follows :

Replace a vertex x_γ of $M(\Gamma)$ by the corresponding H_γ and vertex y_t by the tree T_t described below :

Let $x_{\gamma_1}, x_{\gamma_2}, \ldots, x_{\gamma_p}$ be the vertices adjacent to y_t in $M(\Gamma)$. Let $e_1, e_2, \ldots, e_p$ be the edges between F_t and $H_{\gamma_1}, H_{\gamma_2}, \ldots, H_{\gamma_p}$ and let $a_1, a_2, \ldots, a_p$ be the end vertices of $e_1, e_2, \ldots, e_p$ in F_t. Now define T_t as the union of paths between a_i and a_j, for $1 \leq i, j \leq p$. If (x_γ, y_t) is an edge of $M(\Gamma)$, then replace it by the appropriate edge between H_γ and T_t. Let $G^*(\Gamma)$ be the resulting graph. This graph is sometimes referred to as G^*. Clearly G^* is connected.

<u>Definition</u>. A vertex v belonging to some T_t of G^* is said to satisfy condition D if and only if there exist pieces P_γ and P_δ containing H_γ and H_δ respectively such that H_γ and H_δ have a common degree.

Lemma 1. The graph $G^*(\Gamma)$ defined above satisfies the following conditions :

(1) There are two elements α and β in Γ such that H_α and H_β have a common degree.

(2) If v is a vertex belonging to some T_t of G^*, then there are at least two pieces at v (in G^*) with some H_γ's.

(3) If u is a vertex belonging to some H_γ in G^* with degree d_1 (≥ 2), then there is no vertex with degree d_1 in $G - G^*$.

(4) G^* does not have any vertices of Type I.

Proof. Condition (1) follows directly from the definition of Γ and (2) follows from the construction of G^*.

Let v be a vertex with degree d_1 belonging to some H_γ of G^*. Then from Proposition 2 it follows that there is no vertex of degree d_1 in any component G_j, for $j \geq 2$. Also since Γ satisfies (b), it follows that there is no vertex of degree d_1 in $G_1 - G^*$ Thus G^* satisfies 3.

We now prove that G^* satisfies (4). If possible let v be a vertex with degree d_s of Type I belonging to some T_t of G^*. Observe that pendant vertices of T_t have an edge to some H_γ. The following claim can easily be proved.

Claim. The vertex v does not satisfy condition D.

Thus it follows that any degree that appears in a H_γ belonging to a piece at v of G^* does not appear in any other piece at v. Without loss of generality let us assume that $\alpha = 1$, $\beta = 2$ and that H_1 and H_2 are contained in the piece P_1 at v.

Let us construct $\hat{\Gamma} \subseteq J_1$ as follows :

$$\hat{\Gamma} = \{\gamma : H_\gamma \text{ belongs to } P_1\}.$$

Clearly $\hat{\Gamma} \subseteq J_1$ and satisfies (a) and (b). Note that $|\hat{\Gamma}| < |\Gamma|$. This is a contradiction to the minimality of Γ which proves the lemma.

Definition. A vertex v with degree d_s belonging to some T_t of G^* is said to be of Type II if there is a vertex of degree d_s in $G_1 - G^*$.

We shall now modify G^* and get a subgraph $G^{**}(\Gamma)$ which will satisfy conditions (1) to (4) and (5), where (5) is given below :

(5) G^{**} does not have any vertex of Type II.

If in G^* there is no vertex of Type II then we take $G^{**} = G^*$. So let v be a vertex of Type II belonging to T_t with degree d_s. Let v' be a vertex of $G_1 - G^*$ with degree d_s.

Let Q be a path between v and v'. The following are easily proved.

(c) There is no vertex of Type I on Q.

(d) If u is a vertex on Q belong to H_{j_1}, then $j_1 \in \Gamma$.

Now let v_1 and v_1' be vertices adjacent to v and v' respectively such that v_1 belongs to G^* and v_1' is not on Q. Replace the edges (v,v_1) and (v',v_1') by the edges (v,v_1') and (v',v_1). Let

$$\hat{G}^* = \{G^* - (v,v_1)\} \cup Q \cup \{(v',v_1)\}.$$

This graph $\hat{G}^*$ satisfies the conditions (1) to (4). If it satisfies (5) then we take $\hat{G}^* = G^{**}$. Otherwise we shall repeat the above procedure. Proceeding this way, since G is finite, we finally get a graph G^{**} which satisfies conditions (1) to (5).

Now if $G^{**} = G_1$, then by (3) and (4) of Lemma 1, H is not connected. Thus it follows that G^{**} is a proper subgraph of G_1.

We shall now prove the sufficiency of conditions (i) and (ii).

<u>Theorem 3</u>. Let S be a graphic integer-pair sequence with $(1,1) \notin S$ satisfying the following conditions :

(i) $q \geq \ell(I) + n(1) - 1$

(ii) For $I_1 \subsetneq I$, if $k(I_1) - \ell(I_1) + k(I_1, I-I_1) = \alpha$,

then $H[I-I_1]$ has at most $\alpha + 1$ components. Then S is potentially connected.

<u>Proof</u>. We shall prove the theorem by induction on k, the size of I. For $k = 1$, the theorem is easy to prove. Let us assume that the theorem is true for any sequence S' with $|I_{S'}| < k$ and let S be a graphic integer-pair sequence satisfying (i) and

(ii) and with $|I_S| = k$.

Let us assume that the theorem is true for any sequence S' with $|I_{S'}| < k$ and let S be a graphic integer-pair sequence satisfying (i) and (ii) and with $|I_S| = k$.

Let G be a graph in $\underline{\underline{G}}_2(S)$ for which $|\Gamma(G)|$ is minimum, that is, if $G' \in \underline{\underline{G}}_2(S)$ then $|\Gamma(G)| \leq |\Gamma(G')|$. Sometimes $\Gamma(G)$ is denoted by Γ.

We shall first briefly describe our method of proof. We shall get a connected subgraph $\hat{G}$ with the following properties :

(P_1) Any degree $d_s (\geq 2)$ that appears in $\hat{G}$ does not appear in $G - \hat{G}$,

(P_2) There are at least two distinct degrees in $\hat{G}$ which are ≥ 2.

Now shrink $\hat{G}$ into a single vertex A and introduce some pendant vertices into the graph and join them to A so that the degree of A is different from all the degrees in $G - \hat{G}$. Let G' be the resulting graph and S' its integer-pair sequence. From (P_2) it follows that $|I_{S'}| \leq k - 1$. Next we shall prove that S' satisfies (i) and (ii) given in the statement of the theorem. Then by induction hypothesis, S' is potentially connected and from any connected realization of S' we get a connected realization of S.

Now let us proceed to prove the theorem. Let $G_1, G_2, \ldots, G_\lambda$ be the components of G and let $G^{**} \subsetneqq G_1$ satisfying conditions (1) to (5). Clearly G^{**} satisfies (P_1) and (P_2). Let $\hat{G}$ be the graph G^{**} together with the pendant vertices of G which are joined to vertices of G^{**}. It can easily be seen that $\hat{G}$ satisfies (P_1) and (P_2). Let $\hat{I} = \{i : \exists\, u \in V(G)$ with $d_G(u) = d_i \geq 2\}$.

Now shrink $\hat{G}$ to a single vertex A. Introduce $\beta (\geq 0)$ new pendant vertices into the graph and join them to A. This integer β is the smallest integer such that $k(\hat{I}, I-\hat{I}) + \beta \neq$ any degree in $G - G^{**}$. Let G' be the resulting graph and S' the integer-pair sequence of G'. Clearly G' does not have any multiple edges and $d_{G'}(A) = k(\hat{I}, I-\hat{I}) + \beta$ and $d_{G'}(A) \neq d_i$ for any $i \in I - \hat{I}$. Let us write $d_{G'}(A) = d_o$. For any other vertex u of G', $d_{G'}(u) = d_G(u)$.

If A is a pendant vertex of G', then it can easily be proved that S'

satiesfies (i) and (ii). So we shall assume that $d_{G'}(A) \neq 1$. Let us write $K = I_{S'}$. Then $K = (I - \hat{I}) \cup \{o\}$. Clearly

$$p(G') = \ell(I - \hat{I}) + n_{S'}(1) + 1,$$

$$q(G') = k(I - \hat{I}) + k(\hat{I}, I-\hat{I}) + n_{S'}(1).$$

Now it can be proved that S' satisfies (i) and (ii). To prove (ii) one can consider the two cases, that is, $0 \in I_1$ and $0 \notin I_1$ separately.

Now clearly $|K| < |I|$. By the induction hypothesis there is a connected graph G' realizing S'. Let A be the vertex of G' with degree d_o. Now let G'' be the graph obtained from G' by dropping β pendant vertices adjacent to A. Let us now replace A of G'' by $\hat{G}$. The vertices of G'' adjacent to A in G' are joined to the appropriate vertices of $\hat{G}$. This gives a connected realization of S and completes the proof.

Combining Theorems 1 and 3 we get the following

Theorem 4. Let $S = ((a_1,b_1),(a_2,b_2),\ldots,(a_q,b_q))$ be a graphic integer-pair sequence with $(1,1) \notin S$. Then S is potentially connected if and only if the following conditions are satisfied.

(i) $q \geq \ell(I_1) + n(1) - 1$.

(ii) For $I_1 \subsetneqq I$ if $k(I_1) - \ell(I_1) + k(I_1, I-I_1) = \alpha$, then $H[I-I_1]$ has at most $\alpha + 1$ connected components.

From the characterization of potentially connected integer-pair sequences the following theorem can easily be proved.

Theorem 5. Let $S = ((a_1,b_1),(a_2,b_2),\ldots,(a_q,b_q))$ be a graphic integer-pair sequence with $(1,1) \notin S$. Then there is a tree realizing S if and only if the following are satisfied.

(i) $q = \ell(I) + n(1) - 1$.

(ii) For $I_1 \subsetneqq I$, if $k(I_1) - \ell(I_1) + k(I_1, I-I_1) = \alpha$, then $H[I-I_1]$ has at most $\alpha + 1$ components.

ACKNOWLEDGEMENTS

I thank Drs. A.R. Rao and S.B. Rao for the useful discussions I had with them. I thank N.R. Achuthan for the discussions I had with him at various stages of preparation of this paper.

REFERENCES

1. S.L. Hakimi and A.N. Patrines, Relations between graphs and integer-pair sequences, Discrete Mathematics, Vol. 15, (1976), 347-358.
2. F. Harary, Graph Theory,(Addison-Wesley, Reading, Mass., 1969).
3. S.B. Rao and A. Taneja, Characterization of unipseudographic and unimultigraphic integer-pair sequences, Tech. Report No. 8/79, Indian Statistical Institute, Calcutta.

CONSTRUCTION AND COMBINATORIAL PROPERTIES OF ORTHOGONAL ARRAYS WITH VARIABLE NUMBER OF SYMBOLS IN ROWS

BASUDEB ADHIKARY
Department of Statistics
Calcutta University
35, Ballygunge Circular Road, Calcutta 700 019

PREMADHIS DAS
Kalyani University

1. INTRODUCTION

Orthogonal array with variable number of symbols in rows (OAVS) was introduced by Rao [11] as an array of r rows and N columns, where the i-th row of the array contains s_i different symbols, $i = 1,\ldots,r$ such that for any selection of d rows, every d-plet arising from the combination of elements from the selected rows occurs equally often in the selected $d \times N$ submatrix. This constant may however depend on the selected rows.

This OAVS is denoted by $(N,r,s_1 \times s_2 \times \ldots \times s_r,d)$. Let the set of symbols appearing in the i-th row be denoted by S_i, $i = 1,2,\ldots,r$. If $n(i_1', i_2' \ldots i_d')$ denotes the number of times the d-plet $(i_1', i_2', \ldots, i_d')$ occurs in the $d \times N$ submatrix formed from the i_1-th,...,i_d-th rows of the array, then in order that the array will be an OAVS, $n(i_1', i_2', \ldots, i_d')$ is to be constant for any combination of $i_1', i_2', \ldots, i_d'$, subject to the restriction $i_1' \in S_{i_1}$, $i_2' \in S_{i_2}, \ldots, i_d' \in S_{i_d}$. It may be noted that an $OA(N,r,s,d)$ is an $OAVS(N,r,s \times s \times \ldots \times s,d)$ (Rao [10]).

In this paper we shall study methods of construction of OAVS and study their combinatorial properties.

2. METHODS OF CONSTRUCTION

2.1 <u>Methods of Composition</u> : Chakraborti [6] has utilised the existence of t orthogonal arrays $A(N_i,r_i,s_i,d_i)$, $i = 1,2,\ldots,t$, to construct an array with $N = \prod_{i=1}^{t} N_i$ columns and $r = \sum_{i=1}^{t} r_i$ rows in which a set of r_i rows contains s_i different symbols. We can identify this resultant array as an OAVS $(N,r,s_1^{r_1} \times s_2^{r_2} \times \ldots \times s_t^{r_t},d)$ where $d = \min(d_1,d_2,\ldots,d_t)$ and utilise this result to get further generalisations.

Theorem 2.1.1. Existence of an OA $A_1(N_1,r_1,s_1,d_1)$ and an OA $A_2(N_2,r_2,s_2,d_2)$ implies the existence of an OAVS $A(N_1\ N_2,r_1 + r_2,s_1^{r_1} \times s_2^{r_2},d)$ where $d = \min(d_1,d_2)$.

Method of Construction. With each column of A_1 having r_1 entries adjoin a column of A_2 with r_2 entries to obtain a column of $r_1 + r_2$ entries. The N_1N_2 columns so formed will give a solution of OAVS $A(N_1N_2,r_1 + r_2,s_1^{r_1} \times s_2^{r_2},d)$ when $d = \min(d_1,d_2)$.

Theorem 2.1.2. Existence of an OAVS $A_1(N_1,r_1,s_1 \times s_2 \times \ldots \times s_{r_1},d_1)$ and OAVS $A_2(N_2,r_2,s'_1 \times s'_2 \times \ldots \times s'_{r_2},d_2)$ implies the esixtence of OAVS $A_3(N_1N_2,r_1 + r_2,s_1 \times \ldots \times s_{r_1} \times s'_1 \times \ldots \times s'_{r_2},d)$ when $d = \min(d_1,d_2)$.

Example. Consider OAVS $A_1(6,2,2 \times 3,2)$ given by

$$A_1 = \begin{bmatrix} 0 & 0 & 0 & 1 & 1 & 1 \\ 0 & 1 & 2 & 0 & 1 & 2 \end{bmatrix}$$

and OAVS $A_2(8,3,4 \times 2^2,2)$ given by

$$A_2 = \begin{bmatrix} 0 & 0 & 1 & 1 & 2 & 2 & 3 & 3 \\ 0 & 1 & 0 & 1 & 0 & 1 & 0 & 1 \\ 0 & 1 & 1 & 0 & 0 & 1 & 1 & 0 \end{bmatrix}.$$

By the above method we shall get a solution of

$$A_3(48,4,2^2 \times 3 \times 4,2).$$

2.2. Product of Arrays : Following Bush [5] we can show that the product of an OA $A_1(N_1,r_1,s_1,d_1)$ and an OA $A_2(N_2,r_1,s_2,d_2)$ is an OA $A_3(N_1N_2,r_1,s_1 \times s_2,d)$ when $d = \min(d_1,d_2)$. Hence we can state the following :

Theorem 2.2.1. Existence of an OAVS $A_1(N_1,r_1,s_1 \times s_2 \times \ldots \times s_{r_1},d_1)$ and an OAVS $A_2(N_2,r_1,s'_1 \times s'_2 \times \ldots \times s'_{r_1},d_2)$ implies the existence of an OAVS $A_3(N_1N_2,r_1,s_1s'_1 \times s_2s'_2 \times \ \ldots \ \times s_{r_1}s'_{r_1},d)$ when $d = \min(d_1,d_2)$.

The proof proceeds along the lines of Bush [5] by converting the two arrays into a double entry array.

Example. Consider OAVS $A_1(6,2,2 \times 3,2)$ given by

$$A_1 = \begin{bmatrix} 0 & 0 & 0 & 1 & 1 & 1 \\ 0 & 1 & 2 & 0 & 1 & 2 \end{bmatrix}$$

and OAVS $A_2(4,2^2,2)$ given by

$$A_2 = \begin{bmatrix} 0 & 0 & 1 & 1 \\ 0 & 1 & 0 & 1 \end{bmatrix}.$$

The product of these gives a solution of OAVS A_3 $(24,2,4 \times 6,2)$.

2.3. OAVS from orthogonal F-squares. For definition of F-square or frequency square design and orthogonality of F-squares, see Hedayat and Seiden [7].

Theorem 2.3.1. If there exists a set of t mutually orthogonal F-squares $F_i(n;\lambda_{i1} = \lambda_{i2} = \dots = \lambda_{is_i} = \lambda_i)$, $i = 1,2,\dots,t$, then an OAVS $(n^2, t+2, s_1 \times s_2 \times \dots \times s_t \times n^2, 2)$ exists.

Proof. Let a set of t mutually orthogonal F-squares $F_1(n;\lambda_{11} = \lambda_{12} = \dots = \lambda_{1s_1} = \lambda_1)$, $F_2(n;\lambda_{21} = \lambda_{22} = \dots = \lambda_{2s_2} = \lambda_2),\dots,F_t(n;\lambda_{t1} = \lambda_{t2} = \dots = \lambda_{ts_t} = \lambda_t)$ exist. Form a $t \times n^2$ array where the first row is formed by the juxtaposition of n rows of F_1; the second row is formed by the juxtaposition of n rows of F_2 and so on. To the set of t rows formed as above, one can add two more rows, one corresponding to the rows and one corresponding to the columns to get the required OAVS. Conversely let the OAVS $(n^2, t+2, s_1 \times s_2 \times \dots s_t \times n^2, 2)$ be given. Without any loss of generality assume that the two rows each of which contains n different symbols be the first two rows. Now use these two rows to represent rows and columns and transform each of the remaining rows to an F-square. These F-squares are obviously othogonal.

2.4. Method of Juxtaposition. Existence of an OA (N,r,s,d) implies the existence of an OAVS $(2N,r,2s \times s^{r-1},d)$.

This method is due to Addleman [1]. We shall briefly discuss his method for making some important observations. Consider any OA $A_1(N,r,s,d)$ in s symbols $0,1,2,\dots,s-1$. Get A_2 from A_1 by replacing the symbol i in the first row by $(s+i)$, $i = 0,1,\dots,s-1$. Consider the array $A = (A_1 : A_2)$. Clearly A is an OAVS $(2N,r,2s \times s^{r-1},d)$. Consider for example the OA $A_1(9,4,3,2)$ given by

$$A_1 = \begin{bmatrix} 0 & 0 & 0 & 1 & 1 & 1 & 2 & 2 & 2 \\ 0 & 1 & 2 & 0 & 1 & 2 & 0 & 1 & 2 \\ 0 & 1 & 2 & 1 & 2 & 0 & 2 & 0 & 1 \\ 0 & 2 & 1 & 1 & 0 & 2 & 2 & 1 & 0 \end{bmatrix} .$$

From A_1 the above construction gives

$$A_2 = \begin{bmatrix} 3 & 3 & 3 & 4 & 4 & 4 & 5 & 5 & 5 \\ 0 & 1 & 2 & 0 & 1 & 2 & 0 & 1 & 2 \\ 0 & 1 & 2 & 1 & 2 & 0 & 2 & 0 & 1 \\ 0 & 2 & 1 & 1 & 0 & 2 & 2 & 1 & 0 \end{bmatrix} .$$

Hence OAVS $A(18,4,6 \times 3^3,2)$ can be constructed, viz., $(A_1 : A_2)$.

In the above OAVS A, if we replace the symbol 5 in the first row by 4 and symbol 2 in the fourth row by 0 then it can be easily verified that the concept of proportional frequency advocated by Placket [8] and Addleman [2] is satisfied. Hence this will give a main effect plan (Resolution III design). But clearly this is not an OAVS. This incidentally establishes the fact that though existence of an OAVS$(N,r, s_1 \times s_2 \times \ldots \times s_r,d)$ implies the existence of a resolution d design, the converse is not true. The condition of orthogonal array being much more stringent, though existence of an OA (N,r,s,d) implies the existence of a resolution d design involving r factors each at s levels, the converse is not true.

<u>Corollary 2.4.1</u>. By the above method of Juxtaposition a new OAVS$(2N,r,s_1 \times s_2 \times \ldots \times s_{i-1} \times 2s_i \times s_{i+1} \times s_r,d)$ can also be constructed from a given OAVS$(N,r,s_1 \times s_2 \times \ldots \times s_i \times \ldots \times s_r,d)$.

2.5. <u>OAVS from incomplete block designs</u>. Bose, Srikhande and Bhattacharya (1953) have shown that the existence of a SRGD design with parameters $v = mn$, $b = n^2\lambda$, $r = n$, $k = m$, $\lambda_1 = 0$ and $\lambda_2 = \lambda$ is equivalent to the existence of an OA$(n^2\lambda,m,n,2)$: A similar relation can be found between an OAVS and balanced block design (BBD), see Adhikari [3].

<u>Theorem 2.5.1</u>. The existence of a BBD with parameters $v,b,k = m$, $v_1,v_2,\ldots,v_m$, $r_1,r_2,\ldots,r_m$, $\lambda_{jj} = 0$ for $j = 1,\ldots,m$ and λ_{ij}, $i \neq j = 1, 2,\ldots,m$ is equivalent

to the existence of an $OAVS(b,m,v_1 \times v_2 \times \ldots \times v_m,2)$.

Proof. Suppose a BBD $(b,v,k = m, v_1,\ldots,v_m, \lambda_{11} = \lambda_{22} = \ldots = \lambda_{mm} = 0$ and $\lambda_{ij}, i \neq j = 1,2,\ldots,m)$ exists. Take the blocks as columns so that the design gives an $m \times b$ array. Denote the treatments of the first group as $1,2,\ldots,v_1$, of the second group as $v_1 + 1,\ldots,v_1 + v_2$ and so on. Now as $\lambda_{jj} = 0$ for $j = 1,2,\ldots,m$ so no block contains two treatments from the same group and as the block size is equal to the number of groups, each block contains one and only one treatment from each group. Now if necessary permute the treatments in the blocks so that the treatments of the first group are in the first row, treatments in the second group are in the second row and so on.

Now if we choose any two rows i and j then every 2-plet is represented and is replicated the same number of times (λ_{ij}) in the $2 \times b$ sub-array formed by the i-th and j-th rows and b columns of resulting array, $i \neq j = 1,2,\ldots,m$. So the array is an $OAVS\ (b,m,v_1 \times \ldots \times v_m,2)$.

Conversely suppose an $OAVS\ (b,m,v_1 \times v_2 \times \ldots \times v_m,2)$ exists. Let S_i be the set of elements for the i-th row, $i = 1,\ldots,m$, and its elements be $v_{i-1}+1,v_{i-1}+2,\ldots,v_{i-1}+v_i$, $i = 1,2,\ldots,m$, where $v_o = 0$.

Then it is clear that the columns of the array are the blocks of BBD with the said parameters where the elements of S_i play the role of treatments of the i-th group.

Example 2.5. Consider the array

$$\begin{bmatrix} 1 & 1 & 2 & 2 & 3 & 3 & 4 & 4 & 1 & 1 & 2 & 2 & 3 & 3 & 4 & 4 \\ 5 & 6 & 5 & 6 & 5 & 6 & 5 & 6 & 7 & 8 & 7 & 8 & 7 & 8 & 7 & 8 \\ 9 & 10 & 10 & 9 & 9 & 10 & 10 & 9 & 9 & 10 & 10 & 9 & 9 & 10 & 10 & 9 \\ 11 & 12 & 11 & 12 & 12 & 11 & 12 & 11 & 11 & 12 & 11 & 12 & 12 & 11 & 12 & 11 \end{bmatrix}$$

It is an $OAVS\ (16,4,4^2 \times 2^2,2)$. The columns give also the blocks of a BBD with parameters $v = 12$, $v_1 = 4$, $v_2 = 4$, $v_3 = 2$, $v_4 = 2$, $r_1 = r_2 = 4$, $r_3 = r_4 = 8$. The λ_{ij}-parameters are given in the following table:

i \ j	1	2	3	4
1	0	1	2	2
2		0	2	2
3			0	4
4				0

Incidentally the above example is a solution of GD design with variable replications introduced by Adhikari [4] which forms a sub-class of BBD.

ACKNOWLEDGEMENT

The authors are thankful to the referee for some helpful comments.

REFERENCES

1. S. Addleman, Orthogonal main-effect plans for asymmetrical factorial experiments, Technometrics, 4,(1962), 21-46.
2. S. Addleman, Symmetrical and asymmetrical fractional plans, Technometrics, 4, (1962), 47-58.
3. B. Adhikary, On the properties and construction of balanced block designs with variable replications, Calcutta Stat. Assoc. Bull., 14 (1965), 36-64.
4. B. Adhikary, Group divisible designs with variable replications, Calcutta Stat. Assoc. Bull., 16,(1967), 73-92.
5. K.A. Bush, A generalization of theorem due to Macnish, Ann. Math. Stat., 23, (1952), 293-295.
6. I.M. Chakravorty, Practical replication in asymmetrical factorial designs and partially balanced arrays, Sankhyā, 17,(1956), 143-164.
7. A. Hedayat and E. Seiden, F-square and orthogonal F-square designs : A generalisation of latin square and orthogonal latin squares design, Ann. Math. Stat., 41, (1970), 2035-2044.
8. R.L. Placket, Some generalisations in the multifactorial design, Biometrika, 33, (1946), 328-334.
9. D. Raghavarao, Constructions and Combinatorial Problems in Design of Experiments, John Wiley and Sons, New York, 1971.
10. C.R. Rao, Hypercubes of strength "d" leading to confounded designs in factorial experiments, Bull. Cal. Math. Soc., 38 (1946), 67-78.
11. C.R. Rao, Some Combinatorial Problems of Arrays and Applications to Design of Experiments in A Survey of Combinatorial Theory (ed. J.N. Srivastava and others), North Holland Publishing Co., Amsterdam, 1973, pp 349-359.

CONSTRUCTION OF GROUP DIVISIBLE ROTATABLE DESIGNS

BASUDEB ADHIKARY
Department of Statistics
Calcutta University
35 Ballygunge Circular Road, Calcutta 700 019

RAJENDRANATH PANDA
Kalyani University

INTRODUCTION

Since Box and Hunter (1957) introduced the concept of rotatability as a desirable condition for fitting response surfaces, methods for constructing these designs have been discussed by many authors. A comprehensive bibliography is available in a review article by Meed and Pike (1975).

Herzberg (1966) first generalized the concept of rotability to what she called cyclindrical rotability. She later on (1967) extended this concept further to rotatability of type 1, type 2 and type 3.

Das and Dey (1967) independently generalized the concept of SORD and introduced GDSORD. However the conditions of GD rotatability are identical to the conditions of cyclindrical rotatability of type 3. Adhikary and Sinha (1967) pointed out certain limitations in the analysis of GDSORD as worked out by Das and Dey (1967) as well as in the methods of construction suggested by Das and Dey (1967) and Dey and Nigam (1968). However, Adhikary and Sinha (1976) restricted themselves to the case when the factor space was divided into two groups only.

We propose to develop the analysis and construction of GDSORD and GDTORD when the factor space is divided into s groups with unequal numbers of factors. These designs may be considered as particular cases of rotatable designs when the variance function remains constant for all orthogonal rotations of the type

$$P = \begin{bmatrix} M_1 & 0 & 0 & \cdots & 0 \\ 0 & M_2 & 0 & \cdots & 0 \\ \vdots & & & & \\ 0 & 0 & 0 & \cdots & M_s \end{bmatrix}$$

where $M_{\mu}^{v_\mu \times v_\mu}$ is an orthogonal matrix, $\mu = 1,2,\ldots,s$. However, in this paper we shall restrict ourselves mainly to the construction of GDTORD, indicating briefly the construction of GDSORD, while analysis of these designs will find its place somewhere else.

It may be noted that the number of design points required for group divisible rotatability is much less than that for rotatability. Also in some cases of GDSORD and GDTORD, which we have worked out, when rotatability is achieved by adding further points (and hence the experimentation may be performed sequentially) the number of design points required equals the existing minimum.

2. DEFINITIONS AND NOTATION

Let $N_1^{nk \times b'}$ be any matrix of order $nk \times b'$. Let $N_2^{v \times b}$ be another matrix of order $v \times b$ each column of which contains k 1's and $v-k$ 0's. Let us partition this matrix as

$$N_1 = \begin{bmatrix} N_{11} \\ N_{12} \\ \vdots \\ N_{1k} \end{bmatrix}$$

where each $N_{1i}^{n \times b'}$, $i = 1,2,\ldots,k$ is a matrix with n rows and b' columns. Consider the matrix N obtained by replacing the i-th 1 of any column of N_2 by $N_{1i}^{n \times b'}$, $i = 1,2,\ldots,k$ and the $v-k$ 0's of that column by the submatrices $0 : n \times b'$. The matrix $N^{nv \times bb'}$ thus obtained will be a matrix of nv rows and bb' columns. This matrix N has been defined by Adhikary (1972) as the restricted kronecker product of the two matrices N_1 and N_2 and is denoted by $N_1 \otimes N_2$.

In Adhikary and Panda (1977) we have generalised this definition and considered N_2 as a matrix in 1, -1 and 0 with exactly k elements either 1 or -1 and $(v-k)$ elements 0. We have defined the restricted kronecker product of the two matrices N_1 and N_2 as the matrix obtained by replacing the ith non-zero entry of any column of N_2 by $N_{1i}^{n \times b'}$ if the element is 1, by $-N_{1i}^{n \times b'}$ if it is -1 and by $0^{n \times b'}$ if the element is 0.

3. METHODS OF CONSTRUCTION OF GDSORD

The conditions of Group Divisible Second Order Rotatability may be stated as :

(i) $\Sigma x_{iu}^{\alpha_1} x_{ju}^{\alpha_2} x_{ku}^{\alpha_3} x_{\ell u}^{\alpha_4} = 0$ where $\Sigma \alpha_i \leq 4$ and any α_i odd.

(ii) $\Sigma x_{iu}^2 = \lambda_2^{(\mu)}$ for any factor belonging to the μ-th group.

(iii) $\Sigma x_{iu}^4 = 3\lambda_4^{(\mu)}$ for any factor belonging to the μ-th group.

(iv) $\Sigma x_{iu}^2 x_{ju}^2 = \lambda_4^{(\mu)}$ for any two factors belonging to the μ-th group.

(v) $\frac{\lambda_4^{(\mu)}.N}{[\lambda_2^{(\mu)}]^2} \quad \frac{v_\mu}{v_\mu+2}$, [non-singularity condition] where N is the number of design points.

3.1 Construction from GD design : Adhikary and Sinha (1976) have constructed GDSORD from GD design, having $m = 2$. We have considered the same method for constructing GDSORD having m (> 2) groups of factors. The details are not reported here.

3.2 Construction from Partially Balanced Incomplete Block Design by the Restricted Kronecker Product method : Consider 2^{p_1} points obtained from the $\frac{1}{2^{nk-p_1}}$ replicate of 2^{nk}-experiment so that no effect or interaction less than five factors is confounded and replace 0's by -1's. Let this design matrix be $N_{21}^{nk \times 2^{p_1}}$. For these 2^{p_1}-points,

(i) $\Sigma x_{iu}^{\alpha_1} x_{ju}^{\alpha_2} x_{k4}^{\alpha_3} x_{\ell u}^{\alpha_4} = 0$ if $\Sigma \alpha_i \leq 4$ and any α_i odd.

(ii) $\Sigma x_{iu}^2 = \text{constant} = 2^{p_1} = \Sigma x_{iu}^4 = \Sigma s_{iu}^2 x_{ju}^2$.

Let $N_{22}^{v \times b}$ be the incidence matrix of the PBIBD with the parameters $v,b,r,k,\lambda_1,\lambda_2; n_1,n_2$. Take the restricted kronecker product of N_{21} with N_{22} and call the resultant matrix $N_2^{nv \times 2^{p_1}b}$.

Here we have v groups of factors, each group consisting of n factors, corresponding to the v treatments of PBIBD. Take $N_1^{nv \times 1}$ where

$$N_1 = (\overbrace{a_1\ a_1\ \ldots\ a_1}^{n}\ \ldots\ \overbrace{a_v\ a_v\ \ldots\ a_v}^{n}).$$

Take the restricted kronecker product of N_1 with N_2 and call the resultant matrix $N*^{nv \times 2^{p_1}b}$.

If we consider the columns of N* as the design points, then for these design points, we have

(i) $\Sigma x_{iu}^{\alpha_1} x_{ju}^{\alpha_2} x_{ku}^{\alpha_3} x_{\ell u}^{\alpha_4} = 0$ if $\Sigma \alpha_i \leq 4$ and any α_i odd.

(ii) Σx_{iu}^2 = constant $= r.2^{p_1}.a_\mu^2$ if i belongs to the μ-th group, $\mu = 1,2,\ldots,v$.

(iii) $\Sigma x_{iu}^4 = r.2^{p_1}.a_\mu^4$ if i belongs to the μ-th group, $\mu = 1,2,\ldots,v$.

(iv) $\Sigma x_{iu}^2 x_{ju}^2 = r.2^{p_1} a_\mu^4$ if i and j both belong to the μ-th group;

$= \lambda_1.2^{p_1}.a_\mu^2 a_{\mu'}^2$, if i belongs to μ-th group, j belongs to μ'-th group and μ-th, μ'-th treatments of PBIBD are 1st associate to each other;

$= \lambda_2.2^{p_1}.a_\mu^2 a_{\mu'}^2$, if i belongs to μ-th group, j belongs to μ'-th group and μ-th and μ'-th treatments of PBIBD are 2nd associate to each other.

With this set of $2^{p_1}b$ points, we take an additional set of $2nv$ design points given by

$$N^{**\,nv \times 2nv} = \begin{bmatrix} N_{11} \otimes I_n & 0 & \cdots & 0 \\ 0 & N_{12} \otimes I_n & \cdots & 0 \\ \cdots & \cdots & \cdots & \cdots \\ 0 & 0 & \cdots & N_{1v} \otimes I_n \end{bmatrix}$$

where $N_{1\mu}^{1 \times 2} = (b_\mu, -b_\mu)$. For this set of $b.2^{p_1} + 2nv$ points by (N* N**), we have

(i) $\Sigma x_{iu}^{\alpha_1} x_{ju}^{\alpha_2} x_{ku}^{\alpha_3} x_{\ell u}^{\alpha_4} = 0$ if $\Sigma \alpha_i \leq 4$ and any α_i odd.

(ii) $\Sigma x_{iu}^2 = r.2^{p_1} a_\mu^2 + 2b_\mu^2$ if i belongs to μ-th group, $\mu = 1,2,\ldots,v$.

(iii) $\Sigma x_{iu}^4 = r.2^{p_1} a_\mu^4 + 2b_\mu^4$ if i belongs to μ-th group, $\mu = 1,2,\ldots,v$;

$\Sigma x_{iu}^2 x_{ju}^2$ remains unchanged.

If we take $r.2^{p_1} a_\mu^4 + 2b_\mu^4 = 3.r.2^{p_1} a_\mu^4$ then we get GDSORD in nv factors divided into v groups by adding some central points, if necessary.

Example. $N_{21}^{8 \times 64}$ is the design matrix of $\frac{1}{2^2}$ - replicate of 2^8-experiment with the level combinations ± 1 with the identity equation

$$I = 12345 = 45678 = 123678.$$

$N_{22}^{8 \times 6}$ is the incidence matrix of the PBIBD with the parameters $v = 8$, $b = 6$, $r = 3$, $k = 4$, $\lambda_1 = 3$, $\lambda_2 = 1$.

Here $n = 2$, $k = 4$. Then

$$N_2^{16 \times 384} = N_{21} \otimes N_{22},$$

$$N_1^{16 \times 1} \quad \text{where} \quad N_1' = (a_1\ a_1\ a_2\ a_2\ \cdots\ a_8\ a_8)$$

For these $384 + 32 = 416$ points, we have

(i) $\Sigma\, x_{iu}^{\alpha_1} x_{ju}^{\alpha_2} x_{ku}^{\alpha_3} x_{\ell u}^{\alpha_4} = 0$ if $\Sigma\, \alpha_i \leq 4$ and any α_i odd.

(ii) $\Sigma\, x_{iu}^2 = 192\, a_\mu^2 + 2b_\mu^2$ if i belongs to μ-th group, $\mu = 1,2,\ldots,8$.

(iii) $\Sigma\, x_{iu}^4 = 192\, a_\mu^4 + 2b_\mu^4$ if i belongs to μ-th group, $\mu = 1,2,\ldots,8$.

(iv) $\Sigma\, x_{iu}^2\, x_{ju}^2 = 192 a_\mu^4$ if i and j belong to the μ-th group, $\mu = 1,2,\ldots,8$.

$= 192 a_\mu^2\, a_{\mu'}^2$, if i belongs to μ-th group, j belongs to μ'-th group and μ-th, μ'-th treatments of PBIBD are 1st associate to each other, $\mu \neq \mu'$.

$= 64 a_\mu^2\, a_{\mu'}^2$, if i belongs to μ-th, j belongs to μ'-th group and μ-th, μ'-th treatments of PBIBD are 2nd associate to each other.

We choose b_μ's such that

$$b_\mu^4 = 192 a_\mu^4.$$

For these 416 design points, non-singularity condition is satisfied.

Thus we have a GDSORD on 16 factors in 8 groups in 416 design points.

3.3 Methods have also been developed for construction of GDSORD (consisting of equal or unequal number of factors in each group) from Balanced Block Designs (Adhikari(1965)).

4. METHODS OF CONSTRUCTION OF GDTORD

The conditions of Group Divisible third order rotatability may be stated as :

$A : \Sigma\, x_{iu}^{\alpha_1} x_{ju}^{\alpha_2} x_{ku}^{\alpha_3} x_{\ell u}^{\alpha_4} x_{mu}^{\alpha_5} x_{nu}^{\alpha_6} = 0$ where $\Sigma\, \alpha_i \leq 6$ and any α_i odd.

$B(i) : \Sigma\, x_{iu}^2 = \lambda_2^{(\mu)}$ for any factor belonging to the μ-th group.

$B(ii) : \Sigma\, x_{iu}^4 = 3\lambda_4^{(\mu)}$ for any factor belonging to the μ-th group.

$B_1 : \Sigma\, x_{iu}^6 = 15\, \lambda_6^{(\mu)}$ for any factor belonging to the μ-th group.

$C : \Sigma\, x_{iu}^2\, x_{ju}^2 = \lambda_4^{(\mu)}$ for any two factors belonging to the μ-th group

$= \theta_1^{(\mu\mu')}$ for two factors one from μ-th group and other from μ'-th group.

$C_1(i) : \Sigma\, x_{iu}^2\, x_{ju}^4 = 3\lambda_6^{(\mu)}$ for any two factors belonging to the μ-th group.

$C_1(ii) : \Sigma\, x_{iu}^2\, x_{ju}^2\, x_{ku}^2 = \lambda_6^{(\mu)}$ for any three factors belonging to the μ-th group

$= \theta_2^{(\mu\mu'\mu'')}$ for three factors one from μ-th group, one from μ'-th group and one from μ''-th group.

$D : \Sigma\, x_{iu}^4 = 3\Sigma\, x_{iu}^2\, x_{ju}^2$ for factor belonging to the μ-th group.

$D_1(i) : \Sigma\, x_{iu}^6 = 5\Sigma\, x_{iu}^2\, x_{ju}^4$ for factor belonging to the μ-th group.

$D_1(ii) : \Sigma\, x_{iu}^2\, x_{ju}^4 = 3\Sigma\, x_{iu}^2\, x_{ju}^2\, x_{ju}^2\, x_{ku}^2$ for factor belong to the μ-th group.

$E : \dfrac{\lambda_4^{(\mu)}.N}{\{\lambda_2^{(\mu)}\}^2} > \dfrac{v_\mu}{v_\mu+2}$ and $E_1 : \dfrac{\lambda_2^{(\mu)}\, \lambda_6^{(\mu)}}{\{\lambda_4^{(\mu)}\}^2} > \dfrac{v_\mu+2}{v_\mu+4}$.

4.1 <u>Construction from GD design</u> : Let $N_2^{v \times b}$ be the incidence matrix of a GD design with parameters $v,b,r,k,\lambda_1,\lambda_2,m,n$. Then $\overline{N}_2^{\,v \times b}$ is the incidence matrix of the complementary design in the block size $v-k$. We assume that μ-th set of factors corresponds to the μ-th group of treatments, $\mu = 1,2,\ldots,m$. So we have m groups of n factors each.

For simplicity, we may assume that $k \geq v-k$.

Let $N_{31}^{k \times 2^p}$ be the design matrix of $\dfrac{1}{2^{k-p}}$ replicate of 2^k-experiment with the level combinations ± 1 where the identity equation does not involve any effect or any interaction of order less than six. Take the restricted kronecker product of N_{31} with N_2 and call the resultant matrix $N_{2(1)}^{v \times 2^{p_b}} = N_{31} \otimes N_2$. Let $N_{32}^{v-k \times 2^{p_1}}$ be the design matrix of $\dfrac{1}{2^{v-k-p_1}}$ replicate of 2^{v-k}-experiment with the level combinations ± 1 where the identity equation does not involve any effect or any interaction of order less than six. Let

$$N_{2(2)}^{v \times 2^{p_1}b} = N_{32} \otimes \overline{N}_2.$$

Take

$$N_2^{*\, v \times 2^{p+1}b} = (N_{2(1)}, N_{2(2)} \text{ replicated } 2^{p-p_1} \text{ times}),$$

and $N_1^{v \times 1}$ where

$$N_1^1 = (\overbrace{a_1\ a_1 \ldots a_1}^{n} \ldots \overbrace{a_m\ a_m \ldots a_m}^{n}).$$

Take the restricted kronecker product of N_1 with N_2^* and call the resultant matrix $N^{*\, v \times 2^{p+1}b} = N_1 \otimes N_2^*$.

If we consider the columns of N^* as the design points, then for m groups of n factors, the group of factors corresponds to the group of treatments of GD design, conditions A, B(i), B(ii) and B_1 of GDTORD are satisfied.

They are as follows :

B(i) : $\Sigma x_{iu}^2 = b.2^p.a_\mu^2$ if i belongs to μ-th group, $\mu = 1,2,\ldots,m$.

B(ii) : $\Sigma x_{iu}^4 = b.2^p.a_\mu^4$ if i belongs to μ-th group, $\mu = 1,2,\ldots,m$.

B_1 : $\Sigma x_{iu}^6 = b.2^p.a_\mu^6$ if i belongs to μ-th group, $\mu = 1,2,\ldots,m$.

C : $\Sigma x_{iu}^2 x_{ju}^2 = (b-2r+2\lambda_1).2^p.a_\mu^4$ if i and j both belong to the μ-th group, $\mu = 1,2,\ldots,m$

$= (b-2r+2\lambda_2).2^p.a^2 a_{\mu'}^2$, if i belongs to μ-th and j belongs to μ'-th group, $\mu \neq \mu' = 1,2,\ldots,m$.

C_1(i): $\Sigma x_{iu}^2 x_{ju}^4 = (b-2r+2\lambda_1).2^p.a_\mu^6$ if i and j both belong to the μ-th group

$= (b-2r+2\lambda_2).2^p.a_\mu^2 a_{\mu'}^4$, if i belongs to μ-th and j belongs to μ'-th group, $\mu \neq \mu' = 1,2,\ldots,m$.

C_1(ii) : $\Sigma x_{iu}^2 x_{ju}^2 x_{ku}^2 = (b-3r+3\lambda_1).2^p.a_\mu^6$ if all the three factors i,j and k belong to the μ-th group

$= (b-3r+\lambda_1+2\lambda_2).2^p.a_\mu^4 a_{\mu'}^2$, if i and j both belong to the μ-th group and k belongs to the μ'-th group, $\mu \neq \mu' = 1,2,\ldots,m$

$= (b-3r+3\lambda_2).2^p.a_\mu^2\ a_{\mu'}^2 a_{\mu''}^2$ if i belongs to μ-th group, j belong to μ'-th group and k belongs μ''-th group, $\mu \neq \mu' \neq \mu'' = 1,2,\ldots,m$.

If n, the number of factors in a group, is less than 3, then condition $D_1(ii)$ will not be necessary for GDTORD when factors are divided into m groups of n factors each. In that case the solution of the problem will be much simpler than for n greater than or equal to 3.

Condition $D_1(ii)$ will be satisfied if $2b = 2(r - \lambda_1)$

Condition $D_1(i)$ will be satisfied if $2b = 5(r - \lambda_1)$

Condition D will be satisfied if $2b = 6(r - \lambda_1)$.

Again for μ-th group,

$$\Sigma\, x_{iu}^2\, x_{ju}^4 \gtrless 3\,\Sigma\, x_{iu}^2\, x_{ju}^2\, x_{ku}^2 \qquad 2b \lessgtr 7(r-\lambda_1)$$

$$\Sigma\, x_{iu}^6 \gtrless 5\,\Sigma\, x_{iu}^2\, x_{ju}^4 \qquad 2b \lessgtr 5(r-\lambda_1)$$

$$\Sigma\, x_{iu}^4 \gtrless 3\,\Sigma\, x_{iu}^2\, x_{ju}^2 \qquad 2b \lessgtr 6(r-\lambda_1)$$

<u>Case I</u>. $2b > 7(r-\lambda_1)$.

In this case $\Sigma\, x_{iu}^2\, x_{ju}^4 < 3\,\Sigma\, x_{iu}^2\, x_{ju}^2\, x_{ku}^2$, $\Sigma\, x_{iu}^2 < 5\,\Sigma\, x_{iu}^2\, x_{ju}^4$ and also $\Sigma x_{iu}^4 < 3\,\Sigma\, x_{iu}^2\, x_{ju}^2$. To satisfy the conditions D, $D_1(i)$ and $D_1(ii)$, we add points in such a manner that $\Sigma\, x_{iu}^2\, x_{ju}^2\, x_{ku}^2$ remains unchanged but $\Sigma\, x_{iu}^2\, x_{ju}^2$, $\Sigma\, x_{iu}^2\, x_{ju}^4$, $\Sigma\, x_{iu}^4$ and Σx_{iu}^6 are increased. For this we proceed as follows :

Construct

$$N^{(1):v \times 4tm\binom{n}{2}} = \begin{bmatrix} N_1^{(1)} & 0 & \ldots & 0 \\ 0 & N_2^{(1)} & \ldots & 0 \\ \ldots & \ldots & \ldots & \ldots \\ 0 & 0 & \ldots & N_m^{(1)} \end{bmatrix}$$

where

$$N_\mu^{(1):n \times 4t\binom{n}{2}} = \begin{bmatrix} N_{\mu i}^{(1)} & \ldots & \ldots & N_{\mu t}^{(1)} \end{bmatrix};$$

$$N_{\mu i}^{(1):n \times 4\binom{n}{2}} = \begin{bmatrix} b_{\mu i} & b_{\mu i} & -b_{\mu i} & -b_{\mu i} \\ b_{\mu i} & -b_{\mu i} & b_{\mu i} & -b_{\mu i} \end{bmatrix} \otimes N, \qquad \begin{matrix} i = 1,2,\ldots,t \\ \mu = 1,2,\ldots,m \end{matrix}$$

and $N^{n \times \binom{n}{2}}$ is the incidence matrix of the BIBD with parameters $(n, \binom{n}{2}, 2, n-1, 1)$.

Also construct

$$N^{(2)}: v \times 2sv = \begin{bmatrix} N_1^{(2)} & 0 & \cdots & 0 \\ 0 & N_2^{(2)} & \cdots & 0 \\ \cdots & \cdots & \cdots & \cdots \\ 0 & 0 & & N_m^{(2)} \end{bmatrix}$$

where

$$N_\mu^{(2)}: n \times 2sn = [N_{\mu 1}^{(2)} \; \cdots \; \cdots \; N_{\mu s}^{(2)}],$$

$$N_{\mu j}^{(2)}: n \times 2n = (c_{\mu j}, -c_{\mu j}) \otimes I_n, \quad j = 1,2,\ldots,s$$
$$\mu = 1,2,\ldots,m$$

For $\{b.2^{p+1} + 4tm\binom{n}{2} + 2sv\}$ points given by the columns of $\left[N^* \,\middle|\, N^{(1)} \,\middle|\, N^{(2)}\right]$ conditions A and C_1(ii) remain unchanged. The other conditions are as follows :

$B(i) : \Sigma x_{iu}^2 = b.2^p.a_\mu^2 + 2 \sum_j c_{\mu j}^2 + 4(n-1) \sum_i b_{\mu i}^2$ if i belongs to μ-th group.

$B(ii) : \Sigma x_{iu}^4 = b.2^p.a_\mu^4 + 2 \sum_j c_{\mu j}^4 + 4(n-1) \sum_i b_{\mu i}^4$ if i belongs to μ-th group.

$(B_1) : \Sigma x_{iu}^6 = b.2^p.a_\mu^6 + 2 \sum_j c_{\mu j}^6 + 4(n-1) \Sigma b_{\mu i}$ if i belongs to μ-th group.

$C : \Sigma x_{iu}^2 x_{ju}^2 = (b-2r+2\lambda_1).2^p.a_\mu^4 + 4 \Sigma b_{\mu i}^4$ if i and j both belong to μ-th group

$= (b-2r+2\lambda_2).2^p.a_\mu^2 a_{\mu'}^2$, if i belongs to μ-th and j belongs to μ'-th group.

$C_1(ii) : \Sigma x_{iu}^2 x_{ju}^4 = (b-2r+2\lambda_1).2^p.a_\mu^6 + 4 \Sigma b_{\mu i}^6$ if i and j both belong to μ-th group

$= (b-2r+2\lambda_2).2^p.a_\mu^2 a_{\mu'}^4$, if i belongs to μ-th and j belongs to μ'-th group, $\mu \neq \mu'$.

Conditions D, D_1(i) and D_1(ii) will be satisfied if

$D_1(ii) : (b-2r+2\lambda_1).2^p.a_\mu^6 + 4 \Sigma b_{\mu i}^6 = 3(b-3r+3\lambda_1).2^p.a_\mu^6$.

$D_1(i) : b.2^p.a_\mu^6 + 2 \Sigma c_{\mu j}^6 + 4(n-1) \Sigma b_{\mu i}^6 = 5\{(b-2r+2\lambda_1).2^p a_\mu^6 + 4 \Sigma b_{\mu i}^6\}$.

$D : b.2^p.a_\mu^6 + 2 \Sigma c_{\mu j}^4 + 4(n-1) \Sigma b_{\mu i}^4 = 3\{(b-2r+2\lambda_1).2^p.a_\mu^4 + 4 \Sigma b_{\mu i}^4\}$.

Conditions E and E_1 may automatically be satisfied, condition E can be satisfied by adding some central points, if necessary.

Case II. $2b = 7(r-\lambda_1)$.

In this case, $\Sigma x_{iu}^2 x_{ju}^4 = 3\Sigma x_{iu}^2 x_{ju}^2 x_{ku}^2, \Sigma x_{iu}^6$ and also $\Sigma x_{iu}^4 < 3\Sigma x_{iu}^2 x_{ju}^2$. So $D_1(ii)$ is automatically satisfied. To satisfy the conditions D, $D_1(i)$, we add points in such a manner that Σx_{iu}^2, Σx_{iu}^4 and Σx_{iu}^6 are increased and others remain unchanged. For this with the design points given by N^*, we add the design points given by $N^{(2)}$. For the design points of $[N^*, N^{(2)}]$, all the conditions of GDTORD except D and $D_1(i)$ are satisfied and D, $D_1(i)$ will be satisfied if $c_{\mu j}$'s are such that

$D_1(ii) : b.2^p.a_\mu^6 + 2\Sigma c_{\mu j}^6 = 5(b-2r+2\lambda_1).2^p.a_\mu^6$.

$D : b.2^p.a_\mu^4 + 2\Sigma c_{\mu j}^4 = 3(b-2r+2\mu_1).2^p.a_\mu^4$.

Case III. $2b < 7(r-\lambda_1)$.

In this case, $\Sigma x_{iu}^2 x_{ju}^4 > 3\Sigma x_{iu}^2 x_{ju}^2 x_{ku}^2$. So condition $D_1(ii)$ is not satisfied. To satisfy $D_1(ii)$, we have to add points in such a manner that $\Sigma x_{iu}^2 x_{ju}^2 x_{ku}^2$ is increased and which indirectly increases $\Sigma x_{iu}^2 x_{ju}^2, \Sigma x_{iu}^2$ etc. So to satisfy the conditions D, $D_1(i)$ and $D_1(ii)$, we proceed in the following manner :

Construct

$$N^{(3)}: v \times \omega.m.2^q = \begin{bmatrix} N_1^{(3)} & 0 & \cdots & 0 \\ 0 & N_2^{(3)} & \cdots & 0 \\ \cdots & \cdots & \cdots & \cdots \\ 0 & 0 & \cdots & N_m^{(3)} \end{bmatrix}$$

where

$$N_\mu^{(3)}: n \times \omega.2^q = (N_{\mu 1}^{(3)}, \ldots, N_{\mu\omega}^{(3)}), \quad \mu = 1,2,\ldots,m$$

and $N_{\mu k}^{(3)}: n \times 2^q$ is the design matrix of $\frac{1}{2^{n-q}}$ replicate of 2^n-experiment with the level combinations $\pm d_{\mu k}$ so that no effect or interaction less than seven factors are confounded.

For $\{b.2^{p+1} + 4tm\binom{n}{2} + 2sv + \omega.m.2^q\}$ points given by the columns of $[N^*|N^{(1)}|N^{(2)}|N^{(3)}]$, all the conditions except D, $D_1(i)$ and $D_1(ii)$ are satisfied. They are as follows :

$B(i) : \Sigma x_{iu}^2 = b.2^p.a_\mu^2 + 2\Sigma c_{\mu j}^2 + 4(n-1)\Sigma b_{\mu i}^2 + 2^q \sum_k d_{\mu k}^2$ if i belongs to μ-th group.

B(ii) : $\Sigma\, x_{iu}^4 = b.2^p.a_\mu^4 + 2\,\Sigma\, c_{\mu j}^4 + 4(n-1)\,\Sigma\, b_{\mu i}^4 + 2^q \sum_k d_{\mu k}^4$ if i belongs to μ-th group.

B_1 : $\Sigma\, x_{iu}^6 = b.2^p.a_\mu^6 + 2\,\Sigma\, c_{\mu j}^6 + 4(n-1)\,\Sigma\, b_{\mu i}^6 + 2^q \sum_k d_{\mu k}^6$ if i belongs to μ-th group.

C : $\Sigma\, x_{iu}^2\, x_{ju}^2 = (b-2r+2\lambda_1).2^p.a_\mu^4 + 4\,\Sigma\, b_{\mu i}^4 + 2^q.\,\Sigma\, d_{\mu k}^4$ if i and j both belong to μ-th group

$= (b-2r+2\lambda_1).2^p.a_\mu^2\, a_{\mu'}^2$, if i belongs to μ-th and j belongs to μ'-th group, $\mu \neq \mu'$.

C_1(i) : $\Sigma\, x_{iu}^2\, x_{ju}^4 = (b-2r+2\lambda_1).2^p.a_\mu^6 + 4\,\Sigma\, b_{\mu i}^6 + 2^q\,\Sigma\, d_{\mu k}^6$ if i and j belong to the μ-th group

$= (b-2r+2\lambda_2).2^p.a_\mu^2\, a_{\mu'}^4$, if i belongs to μ-th group and j belongs to μ'-th group, $\mu \neq \mu'$.

C_1(ii) : $\Sigma\, x_{iu}^2\, x_{ju}^2\, x_{ku}^2 = (b-3r+3\lambda_1).2^p.a_\mu^6 + 2^q\,\Sigma\, d_{\mu k}^6$ if all the three factors i,j and k belong to μ-th group

$= (b-3r+\lambda_1+2\lambda_2).2^p.a_\mu^4\, a_{\mu'}^2$, if i and j both belong to μ-th group and k belongs to the μ'-th group $\mu \neq \mu' = 1,2,\ldots,m$

$= (b-3r+3\lambda_2).2^p.a_\mu^2\, a_{\mu'}^2\, a_{\mu''}^2$ if the three factors i,j and k belong to three different groups μ-th, μ'-th and μ''-th, $\mu \neq \mu' \neq \mu'' = 1,2,\ldots,m$.

Conditions D, D_1(i) and D_1(ii) will be satisfied if $b_{\mu i}$'s, $c_{\mu j}$'s and $d_{\mu k}$'s are chosen such that

D(ii) : $(b-2r+2\lambda_1).2^p.a_\mu^6 + 4\,\Sigma\, b_{\mu i}^6 + 2^q.\,\Sigma\, d_{\mu k}^6 = 3\{(b-3r+3\lambda_1).2^p\, a_\mu^6 + 2^q\,\Sigma\, d_{\mu k}^6\}$.

D_1(i) : $b.2^p.a_\mu^6 + 2\,\Sigma\, c_{\mu j}^6 + 4(n-1)\,\Sigma\, b_{\mu i}^6 + 2^q\,\Sigma\, d_{\mu k}^6 = 5\{(b-2r+2\lambda_1).2^p.a_\mu^6 + 4\,\Sigma\, b_{\mu i}^6 + 2^q\Sigma d_{\mu k}^6\}$

D : $b.2^p.a_\mu^4 + 2\,\Sigma\, c_{\mu j}^4 + 4(n-1)\,\Sigma\, b_{\mu i}^4 + 2^q\,\Sigma\, d_{\mu k}^4 = 3\{(b-2r+2\lambda_1).2^p.a_\mu^4 + 4\,\Sigma\, b_{\mu i}^4 + 2^q\,\Sigma\, d_{\mu k}^4\}$.

Example. $N_2^{12 \times 12}$ is the incidence matrix of the GD design with the parameters $v = 12$, $b = 12$, $r = 5$, $k = 5$, $\lambda_1 = 1$, $\lambda_2 = 2$, $n^* = 3$, $m^* = 4$ and $\bar{N}_2^{12 \times 12}$ is the incidence matrix of the complementary design. Here $k = 5$, $v-k = 7$. $N_{31}^{5 \times 32}$ is the design matrix of 2^5-experiment with the level combinations ± 1.

$$N_{2(1)}^{12 \times 384} = N_{31} \otimes N_2.$$

$N_{32}^{7 \times 64}$ is the design matrix of $\frac{1}{2}$ replicate of 2^7-experiment with the level combinations ± 1 and the identity equation is $I = 1234567$.

$$N_{2(2)}^{12 \times 768} = N_{32} \otimes \overline{N}_2$$

$$N_2^{*12 \times 1536} = [N_{2(1)} | N_{2(1)} | N_{2(2)}].$$

Take

$$N_1' = (a_1\ a_1\ a_1\ a_2\ a_2\ a_2\ a_3\ a_3\ a_3\ a_4\ a_4\ a_4)^{1 \times 12}$$

$$N^{*12 \times 1536} = N_1 \otimes N_2^*.$$

The 12 factors are divided into 4 groups consisting of 3 factors each. The four groups of factors are obtained corresponding to four groups of treatment of GD design.

Here $2b < 7(r-\lambda_1)$. So we proceed as in Case III of method 3.1. Here we do not take $N^{(1)}$. Take

$$N^{(2):12 \times 24} = \begin{bmatrix} N_1^{(2):3 \times 6} & 0 & 0 & 0 \\ 0 & N_2^{(2):3 \times 6} & 0 & 0 \\ 0 & 0 & N_3^{(2):3 \times 6} & 0 \\ 0 & 0 & 0 & N_4^{(2):3 \times 6} \end{bmatrix}$$

where

$$N_u^{(2):3 \times 6} = (c_\mu, -c_\mu)^{1 \times 2} \otimes I_3$$

Now $s = 1$ and take

$$N^{(3):12 \times 64} = \begin{bmatrix} N_1^{(3):3 \times 16} & 0 & 0 & 0 \\ 0 & N_2^{(3):3 \times 16} & 0 & 0 \\ 0 & 0 & N_3^{(3):3 \times 16} & 0 \\ 0 & 0 & 0 & N_4^{(3):3 \times 16} \end{bmatrix}$$

where

$$N_\mu^{(3):3 \times 16} = \left[N_{\mu 1}^{(3):3 \times 18} \,\middle|\, N_{\mu 2}^{(3):3 \times 8} \right], \ \mu = 1,2,3,4.$$

$N_{\mu k}^{(3)}$ is the design matrix of 2^3-experiment with the level combinations $\pm d_{\mu k}$, $k = 1,2;\ \mu = 1,2,3,4.$

For 1624 design points given by the columns of $[N^*|N^{(2)}|N^{(3)}]$, condition A is satisfied. The other conditions are as follows :

B(i) : $\Sigma x_{iu}^2 = 12.64a_\mu^2 + 2c_\mu^2 + 2^3(d_{\mu 1}^2 + d_{\mu 2}^2)$ if i belongs to μ-th group, $\mu = 1,2,3,4$.

B(ii) : $\Sigma x_{iu}^2 = 12.64a_\mu^4 + 2c_\mu^4 + 2^3(d_1^4 + d_{\mu 2}^4)$ if i belongs to μ-th group, $\mu = 1,2,3,4$.

B_1 : $\Sigma x_{iu}^4 = 12.64a_\mu^6 + 2c_\mu^6 + 2^3(d_{\mu 1}^6 + d_{\mu 2}^6)$ if i belongs to μ-th group, $\mu = 1,2,3,4$.

C : $\Sigma x_{iu}^2 x_{ju}^2 = 4.64a_\mu^4 + 2^3(d_{\mu 1}^4 + d_{\mu 2}^4)$ if i and j belong to the same group.

$= 4.64\, a_\mu^2 a_{\mu'}^2$, if i and j belong to two different groups.

C_1(i) : $\Sigma x_{iu}^2 x_{ju}^4 = 4.64a_\mu^6 + 2^3(d_{\mu 1}^6 + d_{\mu 2}^6)$ if i and j belong to the same group

$= 4.64\, a_\mu^2 a_{\mu'}^4$, if i and j belong to two different groups.

C_1(ii) : $\Sigma x_{iu}^2 x_{ju}^2 x_{ku}^2 = 2^3(d_{\mu 1}^6 + d_{\mu 2}^6)$ if the factors belong to the same group

$= 2.64\, a_\mu^4 a_{\mu'}^2$, if i and j both belong to the μ-th group and k belongs to the μ'-th group, $\mu \neq \mu' = 1,2,3,4$

$= 3.64\, a_\mu^2 a_{\mu'}^2 a_{\mu''}^2$ if the three factors belong to three different groups μ-th, μ'-th and μ''-th, $\mu \neq \mu' \neq \mu'' = 1,2,3,4$.

If we take $C_\mu^2 = 8a_\mu^2$ and $d_\mu^2 = 2a_\mu^2$, then conditions D, D_1(i) and D_1(ii) are satisfied.

By the above conditions we have for the factors μ-th group

$$\Sigma x_{iu}^2 = 816\, a_\mu^2 \qquad \Sigma x_{iu}^4 = 960\, a_\mu^4 \qquad \Sigma x_{iu}^6 = 1920\, a_\mu^6$$

$$\Sigma x_{iu}^2 x_{ju}^2 = 320\, a_\mu^4 \quad \Sigma x_{iu}^2 x_{ju}^4 = 384\, a_\mu^6 \quad \Sigma x_{iu}^2 x_{ju}^2 x_{ku}^2 = 128\, a_\mu^6.$$

For these 1624, non-singularity condition is satisfied. Thus we have GDTORD on 12 factors in four groups of three factors each in 1624 points.

It is to be noted that the minimum number of design points for TORD on 12 factors is 3224.

4.2 We have also developed methods for construction of GDTORD from PBIB designs and Balanced Block Designs (Adhikary (1965)).

REFERENCES

1. B. Adhikary, On the properties and construction of Balanced Block Designs with variable replications, Calcutta Statist. Assoc. Bull.,14(1965), 36-64.

2. B. Adhikary, A note on Restricted Kronecker Product method of constructing statistical designs, Calcutta Statist. Assoc. Bull., 21(1972), 193-196.

3. B. Adhikary and B.K. Sinha, On Group Divisible Rotatable Designs, Calcutta Statist. Assoc. Bull., 25(1976), 79-93.

4. B. Adhikary and R.N. Panda, Restricted Kronecker Product method of constructing second order rotatable designs, Calcutta Statist. Assoc. Bull., 26(1977),61-78.

5. R.C. Bose, W.H. Clatworthy and S.S. Shrikhande, Tables of Partially Balanced Designs with two associate classes, North Carolina Agricultural experimental station, Technical Bulletin No. 107, 1954.

6. G.E.P. Box and J.S. Hunter, Multifactor experimental designs for exploring response surfaces, 1957.

7. M.N. Das and A. Dey, Group Divisible Rotatable Designs, Ann. Inst. Stat. Math., 19(1967), 331-347.

8. A. Dey and A.K. Nigam, Group Divisible Rotatable Designs - Some further consideration, Ann. Inst. Stat. Math., 20(1968), 477-481.

9. A.M. Herzberg, Cylindrically Rotatable Designs, Ann. Math. Stat., 37(1966), 242-247.

10. A.M. Herzberg, Cylindrically Rotatable Designs of types 1,2 and 3, Ann. Math. Stat., 38(1967), 167-176.

11. R. Meed and D.J. Pike, A review of response surface methodology from a biometric view point, Biometrics, 31(1975), 803-851.

SOME PATH-LENGTH PROPERTIES OF GRAPHS AND DIGRAPHS

W. G. BRIDGES
The University of Wyoming*
Laramie, Wyoming 82071 U.S.A.

1. INTRODUCTION

Various path-length properties for a graph or directed graph are expressible in terms of matrix equations involving the (0,1) adjacency matrix, A, of the graph. Perhaps the most common example is that of strongly regular graphs where an equation of the form : $A^2 + aA = dI + \lambda J$ (I, the identity; J, the matrix of all ones) captures the "strong-regularity" assumption. Here the matrix A is symmetric with $a_{ii} = 0$ and, for $\lambda \neq 0$, regularity comes from the implied equation $AJ = JA$ or by inspection of the main diagonal entries. But the assumption that for any 2 <u>distinct</u> vertices the number of paths of length 2 joining them depends only on whether they are joined or not is reflected directly in the more general equation $A^2 + aA = D + \lambda J$ where D is a diagonal (not necessarily scalar) matrix. Moreover this equation reflects the directed analogue to strongly regular graphs - for any 2 distinct vertices i and j the number of (directed) paths of length two from i to j depends on whether or not i is joined <u>to</u> j. Here regularity cannot be deduced from the matrix equation even if loops are excluded, i.g.,

$$A = \begin{bmatrix} 0 & 1 & 1 & 1 \\ 1 & 0 & 1 & 0 \\ 1 & 0 & 0 & 1 \\ 1 & 1 & 0 & 0 \end{bmatrix}$$

corresponding to the cone over a directed 3-cycle. If loops are allowed there are, in fact, infinitely many examples of "non-regular, strongly regular di-graphs". But the non-regular examples have been completely classified [2,21] and several other path length properties have been shown to force regularity (in and out) for digraphs through

---

* This article was prepared while the author was visiting the Math-Stat. Division of I.S.I., Calcutta and the Statistics Department, The University of Indore, Indore.

the study of the general matrix equation $f(A) = D + \lambda J$ where $f(x)$ is a polynomial, D is diagonal and $\lambda \neq 0$. These so called polynomial digraphs will be discussed in the next section. In particular this study answers affirmatively a question of H.J. Ryser - Does a digraph on more than 2 vertices such that there are λ dipaths of length k from any vertex to any other vertex have to be regular for $\lambda \neq 0$, $k > 2$? (All non-regular examples for $k = 2$ were found by Ryser [21]).

Replacing the symmetry of the adjacency matrix by normality - in degree = out degree for every vertex and, for $i \neq j$, the number of vertices joined to both i and j is the same as the number of vertices to which both are joined - gives some additional results which we mention in Section 3.

In Section 4 we describe a situation in which "regularizing" an ordinary graph via the addition of "multi-loops" brings matric techniques into play. This is in the study of geodetic graphs of diameter two or equivalently the Moore Geometries of W. Kantor [13] used in the classification problem for rank three groups.

In Section 5 we discuss briefly the determination of cyclic strongly regular graphs via association scheme methods and finally in Section 6 we discuss briefly a family of 3-eigenvalue ordinary graphs associated with some generalized design problem and Section 7 discusses the directed analogue of the Moore graph problem.

2. POLYNOMIAL DIGRAPHS

If $G_i (i = 1,\dots,t)$ are digraphs (loops allowed), by the complementary direct sum (c.d.s.), $\mathrm{cds}(G_i)_{i=1}^t$, of the G_i, we mean the digraph obtained by adding to the disjoint union of the G_i all edges (V_r, V_s) when $V_r \in G_i$, $V_s \in G_j$ and $i \neq j$. Thus if A_i is the adjacency matrix of G_i the adjacency matrix of the c.d.s. is

$$J - \sum_{i=1}^{t} \oplus (j - A_i) \qquad \dots \qquad (2.1)$$

where $\Sigma \oplus$ is the usual matrix direct sum and the J-matrices have appropriate sizes. If $f(x)$ is a monic polynomial with $f(0) = 0$ we say the digraph G is an $f(x)$-graph if

$$f(A) = D + \lambda J \qquad \dots \qquad (2.2)$$

where A is the adjacency matrix of G, D is diagonal and $\lambda \neq 0$. (Should $\lambda = 0$ we say G is a degenerate $f(x)$-graph). Recall that G is regular means in and out regular or $AJ = JA = rJ$.

Theorem 2.1 [2]. If G is an $f(x)$-graph G is the c.d.s. of no more than degree $(f(x))$ regular $f(x)$-graphs (possibly degenerate).

If, in the above theorem G is the c.d.s. of the regular $f(x)$-graphs G_i ($i=1, \ldots, t$) with G_i on n_i vertices, regular of degree r_i we define the $t \times t$ parameter matrix $R = (r_{ij})$ by

$$r_{ij} = n_j \quad \text{for } i \neq j$$

$$r_{ii} = r_i .$$

That is, R is obtained from the block decomposition (2.1) by replacing each block by its row sum. We then have also

Theorem 2.2 [2]. If G is an $f(x)$-graph with parameter matrix R so that (2.2) holds $f(R) - \lambda R$ is a scalar matrix.

These results are obtained by considering the (0,1) centralizer of $D + \lambda J$ in (2.2) and then applying an appropriate similarity transformation to the form (2.1). See [2] for details. While the bound, degree $(f(x))$, on the number of valences an $f(x)$-graph may have is in some cases attained, Theorem 2.2 implies that for some scalar, b, $f(x) - \lambda x - b$ has t distinct roots (the distinct eigenvalues of R). Thus e.g., one may show if $f(x)$ is an even polynomial with non-negative coefficients an $f(x)$-graph has at most 2 valences. The case of x^2-graphs with regard to regularity was settled by Ryser [21].

Theorem 2.3 (Ryser). An x^2-graph is either regular or one of the following :

a)

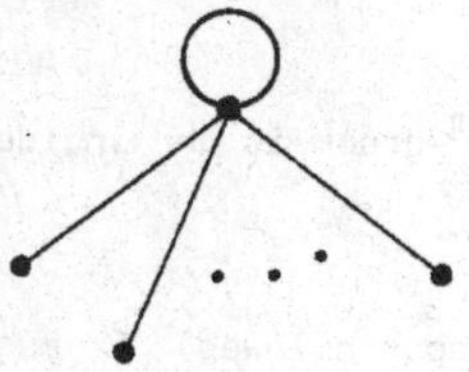

(looped claw)

b)

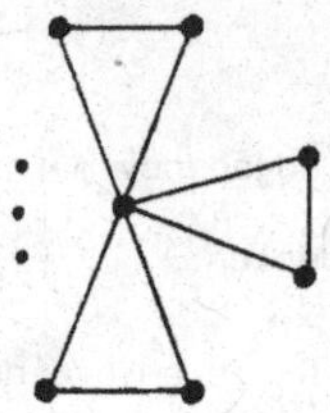

(pinwheel)

or,

c)

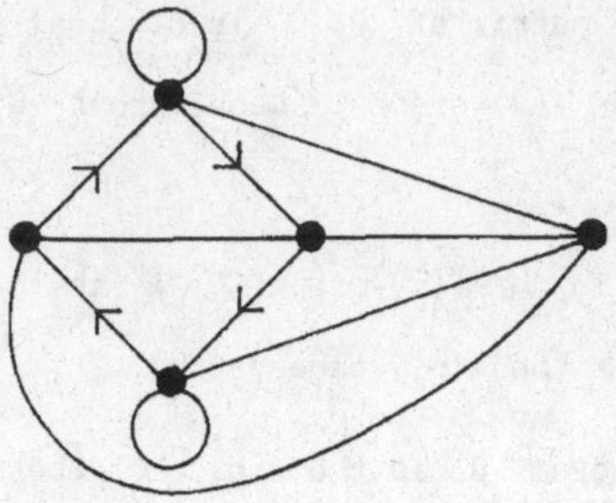

(where edges without arrows are doubly directed).

The result mentioned in the introduction stated in these terms is

Theorem 2.4 [3,8]. An x^k-graph on more than 2 vertices is regular if $k > 2$.

The case of regular x^k-graphs has been extensively considered by Lam [15-18], Lam and VanLint [19], Ryser [21], and Bridges and Mena [8]. We will say a bit more on this in the next section. The classification of the non-regular $f(x)$-graphs with degree $(f(x)) = 2$ is in [2]. We remark, in conclusion, that we know of no graph-theoretic proof of Theorem 2.4 for the case of ordinary graphs.

3. NORMAL AND CYCLIC DIGRAPHS

Recall (Section 1) that a digraph G is called normal if some (then any) adjacency matrix is normal. It is called cyclic if some incidence matrix is circulant ($a_{i,j} = a_{i+1,j+1}$) or equivalently for a suitable vertex labelling the full cycle $(1,2,\ldots,n)$ is an automorphism. If G is regular on n vertices of degree d, we call G non-trivial if $2 \leq d \leq n - 2$.

Theorem 3.1 (Lam [18]). There are no non-trivial cyclic x^2-graphs.

This result appears in [18] with a short proof in [8] where the following are also proven :

Theorem 3.2. The adjacency matrix of a normal x^m-graph is the incidence matrix of a symmetric block design.

Theorem 3.3. If G is an ordinary x^{2m}-graph (loops allowed) G must be an x^2-graph.

Theorem 3.2 raises the further question whether this block design must have a polarity and m must be even. We conjecture this must be so.

Eigenvalue restrictions may have strong consequences for normal digraphs. For example

<u>Theorem 3.4.</u> If A is the adjacency matrix of a connected normal digraph having 3 eigenvalues then A is either symmetric or the skew ($A+A^t = J \pm I$) incidence matrix of a Hadamard symmetric block design.

Here in the symmetric case even regularity does not follow as we discuss further in Section 6. Another result along these lines in which one assume the number of paths of length two from vertex i to vertex j depends (in a special way) on the four possible adjacency arrangements, between i and j occurs in [4] where again skew Hadamard designs crop up. We reserve for Section 5 a discussion of cyclic digraphs with 3 eigenvalues.

4. GEODETIC, DIAMETER TWO GRAPHS

These (ordinary) graphs are those for which there is a unique 2-path between any pair of unjoined vertices (thus diameter two with unique shortest paths). The basic structure of such graphs exhibited in [13,23] is as follows.

<u>Theorem 4.1.</u> If G is a geodetic graph of diameter two one of the following holds :

(a) G is the cone over a disjoint union of cliques.

(b) G is regular (and then strongly regular).

(c) G has two valences, say $a > b$, the total number of vertices is $ab + 1$ and the points of valence b form an anticlique. The maximal cliques in G are either edges joining points of different valence (and all such edges are maximal) or are of size $a - b + 2$.

The classification of possibilities (c) remains unsolved. The known examples [5,13,20] are :

I. The birdcage $\underline{\underline{B}}_n$ on $2n + 1$ vertices $1,2,\dots,n, 1',\dots,n',\infty$ joined as follows :

$$i \text{ joins } i' \ (i = 1,\dots,n)$$
$$i \text{ joins } j \ (i,j = 1,\dots,n,\ i \neq j)$$
$$\infty \text{ joins } i' \ (i = 1,\dots,n)$$

Here $a = n$, $b = 2$.

II. The affine graph AG_n with vertices the points and lines of an affine plane of order n. The edges are of two types - a point joins a line if they are incident and two lines are joined if they are parallel. Here $a = 2n - 1$, $b = n + 1$.

III. The projective graph $\widehat{PG}_n$ obtained from AG_n by adding an $(n+1)$-clique, joining each added point to one complete parallel class of lines. Here $a = 2n$, $b = n + 1$.

IV. The polar projective graph $\widehat{PG}_n$. If π is a projective plane of order n with a polarity we may label points $\{P_i\}$ and lines $\{\ell_j\}$ so that $P_j \in \ell_i \leftrightarrow P_i \in \ell_j$. The vertices of $\widehat{PG}_n$ are then $1,2,\dots,n^2+n+1$ with i joined to j if $i \neq j$ and $P_i \in \ell_j$. Here $a = n + 1$, $b = n$. (These correspond to the polar Moore geometries in [13]).

A direct consequence of Theorem 4.1 is

Theorem 4.2. If $A = \begin{bmatrix} 0 & N \\ N^t & X \end{bmatrix}$ is the adjacency matrix of a graph meeting conditions (c) in Theorem 4.1 then

$$Z = \begin{bmatrix} (a-b)I & N \\ N^t & X \end{bmatrix}$$

satisfies $ZJ = aJ$ and $Z^2 = (a-b-1)Z + (a-1)I + J$.

This shows examples IV correspond to $a - b = 1$ [10,5] and consequences for other classes are explored in [5]. See also [23,24].

5. CYCLIC, STRONGLY REGULAR GRAPHS

A graph is called cyclic [13,25] if some adjacency matrix is a circulant. That is equivalent to the existence of an automorphism permuting the vertices in single cycle. The classic examples of cyclic strongly regular graphs are the so called Paley

Graphs built as follows. For a prime $P \equiv 1 \pmod 4$ join two residues i and j (mod P) whenever $i - j$ is a non-zero quadratic residue. Here the graph is regular of degree $(\frac{P-1}{2})$ and unjoined vertex pairs have $(\frac{P-1}{4})$ common joins while joined vertex pairs have $(\frac{P-5}{4})$ common joins.

Call a strongly regular graph on n vertices Paley-type if it is of degree $(\frac{n-1}{2})$ and the number of common joins is $(\frac{n-1}{4})$ or $(\frac{n-5}{4})$ for unjoined and joined vertex pairs respectively. Then from [14] :

Theorem 5.1. A Paley-type cyclic strongly regular graph is a Paley Graph, and from [22] :

Theorem 5.2. A strongly regular graph is either of Paley-type or has rational eigenvalues.

Through a study of the algebra of $n \times n$ rational circulant matrices with rational eigenvalues viewed as the (rational) Bose-Mesner algebra of an appropriate association scheme, the possible rational spectra of cyclic graphs is explored in [11] where the following is established.

Theorem 5.3. There are no non-trivial cyclic strongly regular graphs with rational eigenvalues.

Combining these results the case of cyclic strongly regular graphs is settled.

Theorem 5.4. The only non-trivial cyclic strongly regular graphs are the Paley Graphs.

Actually the Paley construction in the case $P \equiv 3 \pmod 4$ gives a skew ($A+A^t = J-I$) cyclic, digraph with 3 eigenvalues (which carries a Hadamard symmetric block design). There is an analogue of Theorem 5.1 (also in [14]) for this case and the net result is that the Paley construction yields all non-trivial 3 eigenvalues cyclic digraphs.

6. MULTIPLICATIVE GRAPHS

As noted in Section 3 a three eigenvalue ordinary graph need not be regular (e.g., $K_{m,n}$). Indeed the general analogue of the strongly regular graph equation for the adjacency matrix A is readily seen to be

$$A^2 + aA = dI + \alpha\alpha^t$$

where $\alpha = (a_1, a_2, \ldots, a_n)^t > 0$ is a Perron eigenvector of A (assuming connectivity). If the eigenvalues of A are $r > s > t$ here $a = s + t$ and should $a = 0$ the matrix A is also the incidence matrix of a uniform multiplicative design [6,7]. That is, we have the analogue for multiplicative designs of the (v,k,λ)-graphs or $G_2(d)$-graphs related to symmetric block designs.

A few of these non-regular <u>multiplicative graphs</u> are known. One of the exceptional graphs with least eigenvalue -2 on 22 vertices has spectrum $14^{(1)}$, $2^{(7)}$, $-2^{(14)}$ (the superscripts denoting multiplicities) and adjacency matrix

$$A = \begin{bmatrix} 0 & N \\ N^t & B \end{bmatrix},$$

where

$$B = \begin{bmatrix} J-I & J-I \\ J-I & J-I \end{bmatrix}$$

and N is the 8×14 incidence matrix of the Hadamard 3-design. The valences are 7 and 16. An infinite family of such graphs may be developed as follows. Take a $(\lambda^3+2\lambda^2, \lambda^2+\lambda, \lambda)$-graph, G, [1] and add an additional vertex joined to all its vertices. The resulting graph (the <u>cone</u> over G) has eigenvalues $\lambda(\lambda+2)$, $\pm\lambda$. The base graphs G are constructed in [1] for λ a prime power. There are, it happens, only three further possibilities for a multiplicative graph which is a cone (some vertex joined to all others). These are on 46, 97 and 289 vertices. The first two of these are constructed in [10] and the third which is parametrically related to a (288, 42, 6) is unknown.

7. DIRECTED MOORE GRAPHS

A digraph, regular of degree d and having diameter k can have at most $n = \sum_{i=0}^{k} d^i$ vertices. If this bound is attained one easily argues there are no cycles of length k or less and there is a unique path of length less than $k + 1$ between any two distinct vertices. Then the adjacency matrix A will satisfy

$$\sum_{i=0}^{k} A^i = J.$$

Thus the eigenvalues of A will be d and say $x_1, x_2, \ldots, x_{n-1}$ where $x_i^{k+1} = 1$. Since trace $(A^j) = 0$ for $j = 1,2,\ldots,k$, $d^j + \sum_{i=1}^{n-1} x_i^j = 0$ for $j = 1,2,\ldots,k$. Then

$$-d = \sum_{i=1}^{n-1} X_i = \sum_{i=1}^{n-1} \overline{X}_i = \sum_{i=1}^{n-1} X_i^k = -d^k$$

so G is either a complete digraph or a directed cycle [9].

REFERENCES

1. R.W. Ahrens and G. Szekeres, On a combinatorial generalization of the 27 lines associated with a cubic surface, J. Austral. Math. Soc., 10(1969), 485-492.
2. W.G. Bridges, The polynomial of a non-regular digraph, Pacific J. Math., 38 (1971), 325-342.
3. W.G. Bridges, The regularity of X^3-graphs, J. Combinatorial Theory, 12(1972), 174-176.
4. W.G. Bridges, A class of normal (0,1)-matrices, Can. J. Math., 25(1973), 621-626.
5. W.G. Bridges, D.M. Finucane and R.A. Mena, On graphs with unique 2-paths for unjoined vertices, J.C.I.S.S., 2(1977), 12-19.
6. W.G. Bridges and R.A. Mena, Multiplicative designs I : the reducible cases, J. Combinatorial Theory (A), 27(1979), 69-84.
7. W.G. Bridges and R.A. Mena, Multiplicative designs II : uniform normals and related structures, to appear in J. Combinatorial Theory (A).
8. W.G.Bridges and R.A. Mena, X^k-digraphs, to appear in J. Combinatorial Theory (B).
9. W.G.Bridges and Sam Toueg, On the impossibility of directed Moore graphs, to appear in J. Combinatorial Theory (B).
10. W.G.Bridges and R.A.Mena, Multiplicative cones - a family of three eigenvalue graphs, (submitted).
11. W.G.Bridges and R.A.Mena, Rational circulants with rational spectra and cyclic strongly regular graphs, to appear in Ars Combinatoria.
12. B. Elspas and J. Turner, Graphs with circulant adjacency matrices, J. Combinatorial Theory, 9(1970), 297-307.
13. W.M. Kanter, Moore geometries and rank 3 groups having $\mu = 1$, Quarterly J. Math. Oxford, 28(1977), 309-328.
14. J.B. Kelly, A characteristic property of quadratic residues, Proc. Amer. Math. Soc., 5(1954), 38-46.
15. Clement W.H. Lam, A generalization of cyclic difference sets I, J. Combinatorial Theory (A), 19(1975), 51-65.
16. Clement W.H. Lam, A generalization of cyclic difference sets II, J. Combinatorial Theory (A), 19(1075), 177-191.
17. Clement W.H. Lam, On some solutions of $A^k = dI + \lambda J$, J. Combinatorial Theory (A), 23(1977), 140-147.

18. Clement W.H. Lam, Rational 9-circulants satisfying the matrix equation $A^2 = dI + \lambda J$, Thesis, Calif. Inst. of Tech., 1974.

19. Clement W.H. Lam and J.H. VanLint, Directed graphs with unique paths of fixed length, J. Combinatorial Theory (B), 24(1978), 331-337.

20. Ho-Jin Lee, A note on geodetic graphs of diameter two and their relation to orthogonal latin squares, J. Combinatorial Theory (B), 22(1977), 165-167.

21. H.J. Ryser, A generalization of the matrix equation $A^2 = J$, Linear Algebra and Appl., 3(1970), 451-460.

22. J.J. Seidel, Strongly regular graphs with (-1,1,0) adjacency matrix having eigenvalue 3, Linear Algebra and Appl., 1 (1968), 281-298.

23. J.G. Stemple, Geodetic graphs of diameter two, J. Combinatorial Theory, 17(1974), 266-280.

24. J.G. Stemple and M.E. Watkins, On planar geodetic graphs, J. Combinatorial Theory, 4(1968), 101-117.

25. M. Tuero, A contribution to the theory of cyclic graphs, Matrix and Tensor Quarterly (1961), 74-80.

2-2 PERFECT GRAPHIC DEGREE SEQUENCES

P. D. CHAWATHE
Department of Mathematics
University of Bombay, Bombay 400 098

1. INTRODUCTION

We consider only finite undirected graphs without loops and multiple edges. Behzad and Chartrand [1] defined a graph with at least two vertices to be quasiperfect, if there are precisely two vertices with same degree. They also characterised quasi-perfect graphs completely. In [2], these ideas were generalised further by V.N. Bhat, who defined a graph to be 3-perfect, if there are precisely three vertices with the same degree and among all other vertices no two vertices have the same degree.

In this paper, we pursue these ideas further :

Definition 1. A sequence $\pi : d_1 \geq d_2 \geq \dots \geq d_p$ of non-negative integers, $p \geq 4$, whose sum is even and where $p - 1 \geq d_1$, is said to be 2-2 perfect, if there exist two integers k and ℓ, $1 \leq k,\ell \leq p$ with $\{k,k+1\} \cap \{\ell, \ell+1\} = \phi$ such that $d_k = d_{k+1}$, $d_\ell = d_{\ell+1}$, $d_k \neq d_\ell$ and all other d_r's are mutually distinct and distinct from d_ℓ and d_k. If $d_k > d_\ell$, we call d_ℓ the lower repeated degree and d_k the higher repeated degree.

Definition 2. A graph G of order $p \geq 4$ is said to be 2-2 perfect, if its degree sequence π is 2-2 perfect. In this case, the lower and the higher repeated degrees of π are called the lower and the higher repeated degrees, respectively, of the graph G.

The purpose of this paper is to characterise graphical 2-2 perfect degree sequences. For this we divide 2-2 perfect degree sequences into three types as follows :

Definition 3. If G is a 2-2 perfect graph with degree sequence $\pi : d_1 \geq d_2 \geq \dots \geq d_p$, then we say that G, as well as π, is of type I,II or III according as $d_1 = p-1$, $d_p = 0$ or $p-2 \geq d_1 \geq d_p \geq 1$ respectively.

If G is a 2-2 perfect type I graph, then G has no isolated point. Moreover, there exists precisely one integer m, $1 \leq m \leq p-2$, which is not the degree of any vertex of G. The integer m is called the missing degree of G as well as the missing degree of the degree sequence of G. The 2-2 perfect type I degree sequence of order p with higher repeated degree j, lower repeated degree i and missing degree m is called the (p,j,i,m) degree sequence; a graph with such a degree sequence is called a (p,j,i,m) graph. To characterise 2-2 perfect degree sequences of types I and II, it is enough to characterise those of type I, since the complement of a 2-2 perfect type II graph is a 2-2 perfect type I graph.

A 2-2 perfect type III graph or a 2-2 perfect type III degree sequence has no "missing degrees" besides 0 and $p-1$. The 2-2 perfect type III degree sequence of order p with higher repeated degree j and the lower repeated degree i is called the (p,j,i) degree sequence; a graph with such a degree sequence is called a (p,j,i) graph.

Throughout this paper, p denotes an integer > 4, a denotes the integer $[\frac{p-1}{2}]$ and b denotes the integer $[p/2]$. We shall frequently make use of the following result by Erdős and Gallai [3].

Theorem. A sequence $\pi : d_1 \geq d_2 \geq \dots \geq d_p$ of non-negative integers whose sum is even and where $p-1 \geq d_1$, is graphic if and only if

$$\sum_{k=1}^{r} d_k \leq r(r-1) + \sum_{k=r+1}^{p} \min(r,d_k) \qquad \dots \qquad (1)$$

for every integer r, $1 \leq r \leq p-1$.

2. 2-2 PERFECT TYPE III DEGREE SEQUENCES

We first deal with 2-2 perfect type III degree sequences. We prove

Proposition 1. If $j < a-i$, then the (p,j,i) degree sequence is not graphic.

Proof. We consider two cases.

Case (1). The order p is even, say $p = 2t$. Then $a = t-1$ and $j < t-1-i$. It follows that $j < t-1$ and $i < t-1$. In this case the degree sequence is

$d_1 = 2t-2$, $d_2 = 2t-3$, ..., $d_t = t-1$, ..., $d_{2t-1-j} = j$, $d_{2t-j} = j$, ..., $d_{2t-i} = i$, $d_{2t-i+1} = i$, ..., $d_{2t} = 1$. With $r = t-1$, the left hand expression in (1) becomes $\frac{1}{2}(t-1)(3t-2)$ and the right hand expression in (1) becomes $\frac{1}{2}(t-1)(3t-4) + (i+j) < \frac{1}{2}(t-1)(3t-2)$. Thus (1) is not satisfied for $r = t-1$ and the (p,j,i) degree sequence is not graphic.

Case (2). The order p is odd, say $p = 2t+1$. In this case, $a = t$ and $j < t$ and $i < t$. The degree sequence is $d_1 = 2t-1$, $d_2 = 2t-2, \ldots, d_t = t, \ldots$, $d_{2t-j} = j$, $d_{2t-j+1} = j$, ..., $d_{2t-i+1} = i$, $d_{2t-i+2} = i$, ..., $d_{2t+1} = 1$. With $r = t$, the left hand expression in (1) becomes $\frac{1}{2}t(3t-1)$ and the right hand expression in (1) becomes $\frac{3}{2}t(t-1) + (i+j) < \frac{1}{2}t(3t-1)$. Thus (1) is not satisfied for $r = t$ and the (p,j,i) degree sequence is not graphic.

Corollary 1. The (p,j,i) degree sequence is not graphic, if $j > 2(p-1) - a - i$.

Proof. We know that a degree sequence $d_1 \geq d_2 \geq \ldots \geq d_p$ is graphic iff its complementary degree sequence $p-1-d_p \geq p-1-d_{p-1} \geq \ldots \geq p-1-d_1$ is graphic. The complementary degree sequence of the (p,j,i) degree sequence is the $(p,p-1-i,p-1-j)$ degree sequence. The result follows by applying the Proposition 1 to this complementary degree sequence.

Corollary 2. For $p = 4t+1, 4t+2, 4t+3, 4t+4$, the (p,j,i) degree sequence, with $j \leq t$, is not graphic.

Proof. Since $i < j$, so $i+j < 2j \leq 2t \leq a$, i.e., $j < a-i$ and the degree sequence is not graphic.

Corollary 3. The (p,j,i) degree sequence is not graphic if

(i) $p = 4t+1$ and $i \geq 3t$,

(ii) $p = 4t+2$ and $i \geq 3t+1$,

(iii) $p = 4t+3$ and $i \geq 3t+2$,

(iv) $p = 4t+4$ and $i \geq 3t+3$.

Proof. The result follows by applying Corollary 3 to the complementary degree sequence, wich is the $(p, p-1-i, p-1-j)$ degree sequence.

Proposition 2. If $j > a+i$, then the (p,j,i) degree sequence is not graphic.

Proof. If p is even, say $p = 2t$, it can be verified that (1) is not satisfied for $r = t$ and if p is odd, say, $p = 2t+1$, then it can be verified that (1) is not satisfied for $r = t+1$.

Theorem 1. The (p,j,i) degree sequence with $i \leq a$ is graphic iff the following three conditions are satisfied :

(i) the sum of degrees is even,

(ii) $i+1 \leq j \leq p - 2$,

(iii) $a-i \leq j \leq a + i$.

Proof. Let the (p,j,i) degree sequence be graphic. Then conditions (i) and (ii) are trivially satisfied. By Propositions 1 and 2, condition (iii) is satisfied.

We prove the sufficiency by induction on p. For $p \leq 6$, it it easy to verify that if the (p,j,i) degree satisfies conditions (i),(ii) and (iii) then it is graphic. Let now $p > 6$ and the result be true for degree sequences of order $< p$. We shall use the "laying off" technique described in [4]. We consider two cases : (1) p is even, say $p = 2t$ and (ii) p is odd, say $p = 2t+1$.

Case 1. $p = 2t$. Then $a = t-1$. The degree sequence is $d_1 = 2t-2$, $d_2 = 2t-3$, ..., $d_{2t-1-j} = j$, $d_{2t-j} = j$, ..., $d_{2t-i} = i$, $d_{2t-i+1} = i$, ..., $d_{2t} = 1$. We divide this into the following subcases.

I. $i < 2t-2-j$. If we lay off a vertex of degree i, we get the $(2t-1, 2t-2-i, j)$ degree sequence which has the $(2t-1, 2t-2-j, i)$ degree sequence as its complementary degree sequence; the latter is graphic by induction hypothesis.

II. $i = 2t-2-j$. If we lay off a vertex of degree i, we get the 3-perfect type III degree sequence of order $2t-1$ and special degree j. This is graphic by Corollary 6 of [2].

III. $i = 2t-1-j$. If we lay off a vertex of degree i, we get the $(2t-1, j, j-1)$ degree sequence. Its complementary degree sequence - the $(2t-1, i, 2t-2-j)$ degree sequence is graphic by induction hypothesis.

IV. $i = 2t-j$. If we lay off a vertex of degree i, then we get the 3-perfect type III degree sequence of order $2t-1$ with special degree $j-1$, which is graphic by Corollary 6 of [2].

V. $i > 2t-j$. If we lay off a vertex of degree i, then we get the $(2t-1, j-1, 2t-i-1)$ degree sequence. Its complementary degree sequence - the $(2t-1, i-1, 2t-j-1)$ degree sequence is graphic by induction hypothesis.

Case 2. $p = 2t+1$. Then $a = t$. The degree sequence in this case is $d_1 = 2t-1$, $d_2 = 2t-2, \ldots, d_{2t-j} = j, d_{2t-j+1} = j, \ldots, d_{2t-i+1} = i, d_{2t-i+2} = i, \ldots, d_{2t+1} = 1$. We divide this into the following subcases.

VI. $i < 2t-j-1$. When we lay off a vertex of degree i, we get the $(2t, 2t-i-1, j)$ degree sequence. Its complementary degree sequence is graphic by induction hypothesis.

VII. $i = 2t-j-1$. When we lay off a vertex of degree i, we get the 3-perfect type III degree sequence of order $2t$ with special degree j. This is graphic by Corollary 9 of [2].

VIII. $i = 2t-j$. If we lay off a vertex of degree i, then we get the $(2t, j, j-1)$ degree sequence. Its complementary degree sequence - the $(2t, i, 2t-1-j)$ degree sequence is graphic by induction hypothesis.

IX. $i = 2t-j+1$. If we lay off a vertex of degree i, then we get the 3-perfect type III degree sequence of order $2t$ with special degree $j-1$ and this is graphic by Corollary 9 of [2].

X. $i > 2t-j+1$. If we lay off a vertex of degree i, then we get the $(2t, j-1, 2t-i)$ degree sequence. Its complementary degree sequence - the $(2t, i-1, 2t-j)$ sequence is graphic by induction hypothesis.

The result now follows by induction.

Theorem 2. The (p,j,i) degree sequence with $i > a$ is graphic iff the following three conditions are satisfied :

(i) the sum of degrees is even,

(ii) $1 \leq i \leq j - 1$,

(iii) $j - a \leq i \leq 2(p-1) - j - a$.

Proof. The result follows by applying Theorem 1 to the complementary degree sequence, which is the (p, p-1-i, p-1-j) degree sequence.

3. 2-2 PERFECT TYPE I DEGREE SEQUENCES

We now study type I degree sequences. We prove

Lemma 1. If the (p,j,i,m) degree sequence is graphic, then $m \neq b$.

Proof. We prove the result by induction on p. If $p = 4$, the only graphic 2-2 perfect type I degree sequence is the (4,3,2,1) sequence; if $p = 5$, the only graphic 2-2 perfect type I degree sequences are (5,2,1,3), (5,4,3,1) and (5,3,2,1) degree sequences. In all these cases, $m \neq b$. Let now $p > 5$ and the result be true for degree sequences of order $< p$. Consider the (p,j,i,m) degree sequence. If $j = p-1$ and the sequence is graphic, then we must have $m = 1 \neq b$. If $j < p-1$, by laying off the vertex of degree $p-1$ and removing the resulting isolated vertex, we see that the (p,j,i,m) degree sequence is graphic iff the (p-2, j-1, i-1, m-1) degree sequence is graphic. By induction hypothesis, $m-1 \neq [\frac{p-2}{2}] = b - 1$. Hence $m \neq b$.

Lemma 2. Let the (p,j,i,m) degree sequence be graphic. If $m > b$, then $j < m$.

Proof. The result follows by induction on $k = p - m$.

Lemma 3. Let the (p,j,i,m) degree sequence be graphic. If $m < b$, then $i > m$.

Proof. The result follows by induction on m.

Theorem 3. The (p,j,i,m) degree sequence with $m < b$ is graphic iff

(i) the sum of the degrees is even,

(ii) $i+1 \leq j \leq p - 1$ and

(iii) $j \in [b - (i-m), b + (i-m)]$ if $i \leq b$ and

$i \in [(j+m) - b, 2p - (j+m) - b]$ if $i > b$,

where $[x,y]$ denotes the closed interval from x to y.

Proof. We prove the result by induction on m. Let $m = 1$. If $j \neq p-1$, then by laying off the vertex of degree $p-1$, we get the (p-1, j-1, i-1) degree sequence.

By Theorems 1 and 2, the (p-1, j-1, i-1) degree sequence is graphic iff the conditions (i), (ii) and (iii) are satisfied.

If $j = p-1$, then by laying off the vertex of degree $p - 1$, we get the quasiperfect degree sequence $p-2, p-3, \ldots, i-1, i-1, \ldots, 2,1$. By [1], this is graphic iff besides (i) and (ii), we have $i - 1 = [\frac{p-1}{2}]$, i.e., iff (i), (ii) and (iii) are satisfied. Thus the result is true for $m = 1$. Let now $2 \le m < b$. By Lemma 3, $i > m \ge 2$. Since $m \ne 1$, $j \ne p - 1$. By laying off the vertex of degree $p-1$ and removing the resulting isolated vertex, we get a (p-2, j-1, i-1, m-1) degree sequence. The result now follows by induction hypothesis.

Theorem 4. The (p,j,i,m) degree sequence with $m > b$ is graphic iff

(i) the sum of the degrees is even,

(ii) $i + 1 \le j \le p - 1$, and

(iii) $j \in [a - (i+m-p+1), a + (i+m-p+1)]$ if $i \le a$ and
$i \in [b - (m-j), b + (m-j)]$ if $i > a$.

Proof. We prove the result by induction on $k = p - m$. If $k = 2$, $m = p - 2$. If $i = 1$, then by laying off the vertex of degree $p-1$ and removing the resulting two isolated vertices, we get the quasiperfect degree sequence

$$p-4, p-5, \ldots, j-1, j-1, \ldots, 2,1.$$

By [1], this is graphic iff $j - 1 = [\frac{p-3}{2}]$, i.e., iff conditions (i),(ii) and (iii) are satisfied. If $i > 1$, then by laying off the vertex of degree $p-1$ and removing the resulting isolated vertex, we get the (p-2, j-1, i-1) degree sequence. This is graphic iff conditions (i),(ii) and (iii) are satisfied (Theorem 1 and 2). Thus the result is true for $k = 2$.

Let now $k = p - m > 2$. Since $m \ne p - 2$, $i \ne 1$. By Lemma 2, $j < m < b$, i.e., $j \ne p = 1$. By laying off the vertex of degree $p - 1$ and removing the resulting isolated vertex, we get the (p-2, j-1, i-1, m-1) degree sequence. For this degree sequence, $k' = (p-2) - (m-1) = p-m-1 = k - 1$. Applying induction hypothesis to the (p-2, j-1, i-1, m-1) degree sequence, we get the required result.

ACKNOWLEDGEMENTS

The author wishes to express her sincere thanks to Vasanti Bhat for her interest in this work and for the stimulating discussions throughout the preparation of this paper. Thanks are also due to the referee for many useful suggestions.

REFERENCES

1. M. Behzad and G. Chartrand, No graph is perfect, Amer. Math. Monthly, 74(1967), 962-963.

2. V. Bhat, Characterization of 3-perfect graphical degree seq., Proc. Ind. Nat. Sci. Acad, 41(1975), 228-244.

3. P. Erdös and T. Gallai, Graphs with prescribed degrees of vertices (Hungarian), Mat. Lapok, 11(1960), 264-274.

4. D.J. Kleitman and D.L. Wang, Algorithms for constructing graphs and digraphs with given valences and factors,Discrete Math., 6(1973), 79-88.

CHARACTERIZATION OF FORCIBLY OUTERPLANAR GRAPHIC SEQUENCES

S. A. CHOUDUM
School of Mathematical Sciences
Madurai Kamaraj University, Madurai 625 021

ABSTRACT

Graphic sequences every realization of which is outerplanar are characterized.

<u>Theorem</u>. A graphic sequence $\pi : d_1 \geq d_2 \geq \ldots \geq d_p$ is forcibly outerplanar iff π satisfies one of the following conditions.

(1) $d_2 \leq 2$.

(2) $d_2 \geq 3$ and $d_4 \leq 1$.

(3) $d_2 \geq 3$, $d_4 = 2$ and $d_5 \leq 1$.

(4) $\pi = (4,4,4,2,2,2)$.

(5) $d_1 = p - 1$, $d_2 = d_3 = 3$ and $d_4 = d_5 = 2$.

(6) $d_1 = p - 1$, $d_3 = 3$ and $d_3 = d_4 = d_5 = 2$.

INTRODUCTION

By a graph we mean a finite, undirected graph without loops or multiple edges. We refer to Harary [2] for terminology.

Let π be a graphic sequence and let P be an invariant property. π is said to be potentially P if there exists a realization of π having the property P. π is forcibly P if every realization of π has property P.

In this paper we characterize forcibly outerplanar graphic sequences. Our proof is algorithmic and hence it is different from that of Rao's Theorem [4] which characterizes forcibly planar graphic sequences. In fact, our proof can be straightaway modified to give an algorithmic proof of Rao's Theorem.

<u>Laying Off Algorithm and Some Simple Observations</u>

A graph G is called an outerplanar graph if G can be embedded in the plane so that every vertex of G lies on the exterior region. A graph H is said to be homeomorphic from a graph G if H can be obtained from G by a finite sequence of subdivisions. Chartrand and Harary [1] have characterized outerplanar graphs as those graphs which fail to contain subgraphs homeomorphic from K_4 or $K_{2,3}$. Therefore it follows that if $\pi : d_1 \geq \ldots \geq d_p$ is a potentially nonouterplanar graphic sequence, then either (i) $d_4 \geq 3$ or (ii) $d_4 \leq 2$ but $d_5 \geq 2$ and $d_2 \geq 3$.

We shall call a graphic sequence $\pi : d_1 \geq \ldots \geq d_p$ a Type I sequence if $d_4 \geq 3$ and a Type 2 sequence if $d_4 \leq 2$ but $d_5 \geq 2$ and $d_2 \geq 3$. The following lemma is immediate.

Lemma 1. If π is a potentially nonouterplanar graphic sequence, then π is either a Type 1 sequence or a Type 2 sequence.

We repeatedly use an algorithm called laying off Algorithm of Kleitman and Wang [3] to realize a graph from a graphic sequence. We give it now.

If π is a graphic sequence, let $\pi(d_j)$ or $\pi(j)$ be the truncated sequence obtained from π by laying off d_j and then subtracting 1 from each of the largest d_j terms in $\pi - \{d_j\}$.

The terms in $\pi(j)$ from which 1 has been subtracted are called depleted terms and the corresponding vertices in a realization of $\pi(j)$ are called depleted vertices.

Now define

$$\pi(j_1,\ldots,j_k) = \pi(j_1,\ldots,j_{k-1})(j_k).$$

A realization G of π can be constructed from a realization H of $\pi(j_1,\ldots,j_k)$, $k \geq 1$ as follows : set $G_k = H$. Join the vertex v_{j_i} to the depleted vertices of $\pi(j_1,j_2,\ldots,j_i)$ to get G_{i-1}, $i = k, k-1,\ldots,2,1$ and define $G_o = G$.

Henceforth G always refers to a realization of π on the vertex set $V(G) = \{v_1,v_2,\ldots,v_p\}$ as constructed above.

The following propositions which are crucial are obvious.

Proposition 1. If G_k is a realization of $\pi(j_1,\ldots,j_k)$, $k \geq 1$, containing a subgraph K and if G_i is a realization of $\pi(j_1,\ldots,j_i)$, $i = k-1,\ldots,2,1$, constructed as above, then G_i contains K. In particular if $\pi(j_1,\ldots,j_k)$ is potentially nonouterplanar then π is also potentially nonouterplanar.

Proposition 2. Suppose $\pi : d_1 \geq \ldots \geq d_p$ is a graphic sequence. If a d_i is depleted in each of the truncated sequences $\pi(j_1), \pi(j_1,j_2),\ldots,\pi(j_1,j_2,\ldots,j_k)$ then the vertices $v_{j_1}, v_{j_2},\ldots,v_{j_k}$ are all adjacent to v_i in G.

For convenience, we treat a graphic sequence $\pi : d_1,\ldots,d_p$ as a p-tuple $(d_1,\ldots,d_p)$. We denote $\pi(j)$ by a p-tuple with its jth co-ordinate a star (*) and other co-ordinates as defined earlier.

Throughout this paper t denotes 1 or 2 and ε denotes 0 or -1. Unless otherwose stated π is a graphic sequence and is always in nonincreasing order. The length $\ell(\pi)$ of π is the number of strictly positive integers in π.

CHARACTERIZATION

Without loss of generality, we assume that $d_p \geq 1$ in $\pi : d_1 \geq \ldots \geq d_p$.

Theorem 1. If $\pi^* : d_1 \geq \ldots \geq d_p$ is a Type 1 sequence, then π^* is potentially nonouterplanar.

We first prove the following lemma.

Lemma 2. If a graphic sequence π has a realization G, which contains a subgraph H (of Figure 1), then π has a realization which contains a subgraph homeomorphic from K_4, that is π is potentially nonouterplanar.

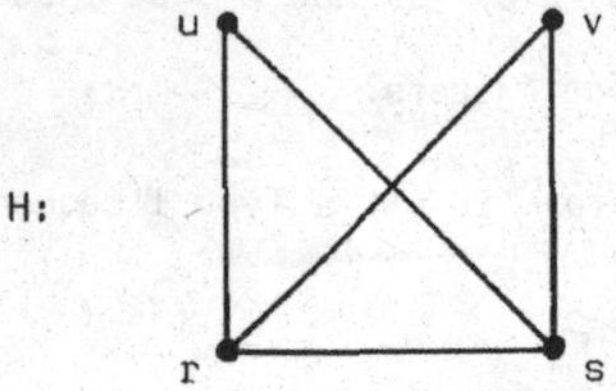

FIGURE 1

Proof. If $(u,v) \in E(G)$, then G contains a K_4 and we are through. Therefore suppose that $(u,v) \notin E(G)$. Since $\deg(u) \geq 3$ and $\deg(v) \geq 3$, there exist vertices u' and v' such that $(u,u') \in E(G)$ and $(v,v') \in E(G)$. If $u' = v'$, then the induced subgraph $\langle \{r,s,u,v,u'\} \rangle$ contains a subgraph homeomorphic from K_4. Therefore assume that $u' \neq v'$. If $(u',v') \in E(G)$, then $< \{r,s,u,v,u',v'\} >$ contains a subgraph homeomorphic from K_4. If $(u',v') \notin E(G)$, then let G' be the graph obtained from G, by deleting the edges (u',u'), (v,v') and adding the edges (u,v), (u',v'). Then G' is a realization of π and contains a K_4, namely $< \{r,s,u,v\} >$. This proves Lemma 2.

Proof of Theorem 1. In view of the above lemma, Theorem 1 is proved if we construct a realization of π containing a subgraph H as described in the lemma. We construct such a realization by the following algorithm.

Step 0. Set $\pi^* = \pi$ and go to Step 1 with π.

Step 1. Lay off d_p from π and go to Step 2 with $\pi(p)$. (If $\pi(p)$ is not in non-increasing order, order it so and also delete the 0's in $\pi(p)$ and then go to Step 2).

Step 2. (i) If $\ell(\pi(p)) \geq 4$ and $\pi(p)$ is a Type 1 sequence, then go to Step 1, with $\pi(p)$ setting $\pi(p) = \pi$.

(ii) If $\ell(\pi(p)) = 3$, then go to Step 3 with π.

(iii) If $\ell(\pi(p)) \geq 4$ and if $\pi(p)$ is not a Type 1 sequence, then go to Step 4 with π.

Step 3. π is a Type 1 sequence with $\ell(\pi) = 4$. Therefore, $\pi = (3,3,3,3)$ and K_4 is a realization of π. Hence if G is a realization of π^* constructed from π, then G contains a K_4.

Step 4. Since $\ell(\pi(p)) \geq 4$ and $\pi(p)$ is not a Type 1 sequence, it follows that $\ell(\pi) \geq 5$ and $d_p \leq 3$. We consider 3 cases.

Case 1 : $d_p = 1$. Since $\pi(p)$ is not a Type 1 sequence we have $d_1 = 3$ and $d_5 \leq 2$ in . Therefore,

$$\pi = (3,3,3,3,t,\ldots,t,1).$$

Let

$$\pi(4) = (2,2,2,*,t,\ldots,t,1),$$

$$\pi(4,3) = (1,1,*,*,t,\ldots,t,1).$$

Since first and second co-ordinates are depleted in $\pi(4)$ and $\pi(4,3)$, G contains the subgraph H, described in Lemma 2.

Case 2 : $d_p = 2$. Since $\pi(p)$ is not a Type 1 sequence and $d_p = 2$ we have $d_6 = 2$ and $d_2 = 3$. Moreover, if $d_5 = 3$, then $d_1 = 3$, in which case however π is not graphic. Therefore

$$\pi = (d_1,3,3,3,2,2,\ldots,2).$$

Let

$$\pi(4) = (d_1-1,2,2,*,2,2,\ldots,2),$$

$$\pi(4,1) = (*,1,1,*,2+\varepsilon,\ 2+\varepsilon,\ldots,2+\varepsilon).$$

Again G contains H.

Case 3 : $d_p = 3$. It follows that $d_3 = 3$ and $p \leq 7$. In fact π is one of the following.

(1) $\pi_1 = (5,3,3,3,3,3)$ (2) $\pi_2 = (3,3,3,3,3,3)$

(3) $\pi_3 = (4,3,3,3,3)$ (4) $\pi_4 = (3,3,3,3)$

Now $C_5 + K_1$, $K_{3,3}$, $C_4 + K_1$ and K_4 are nonouterplanar realizations of π_1, π_2, π_3 and π_4 respectively and each of these graphs contains a subgraph homeomorphic from K_4.

This completes the proof of Theorem 1.

<u>Theorem 2</u>. If $\pi : d_1 \geq \ldots \geq d_p$ is a Type 2 sequence different from the following sequences, then π is potentially nonouterplanar.

(1) $\pi = (4,4,4,2,2,2)$,

(2) $d_1 = p-1$, $d_2 = d_3 = 3$ and $d_4 = d_5 = 2$,

(3) $d_1 = p-1$, $d_2 = 3$ and $d_3 = d_4 = d_5 = 2$.

<u>Proof</u>. We first state the following lemma; its proof is similar to that of Lemma 2 and is therefore omitted.

<u>Lemma 3</u>. If a graphic sequence $\pi : d_1 \geq \ldots \geq d_p$ has a realization G which contains a subgraph K (of Figure 2) such that there exist vertices w_2' and $w_5' \notin \{w_1,w_2,w_3,w_4,w_5\}$ in K with $(w_2,w_2') \in E(K)$ and $(w_5,w_5') \in E(K)$, then π has a nonouterplanar realization. In particular, if

(i) $\deg_G(w_5) \geq 2$, $(w_5,w_4) \notin E(K)$, $(w_5,w_3) \notin E(K)$ and

(ii) $\deg_G(w_2) \geq 3$, $(w_2,w_1) \notin E(K)$,

then π has a nonouterplanar realization.

K:

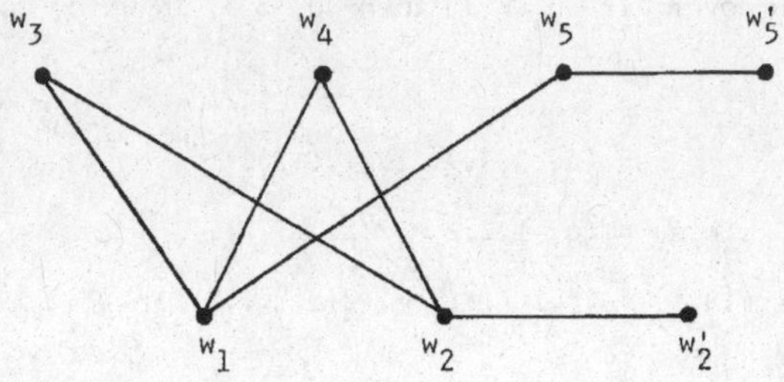

FIGURE 2

In view of Lemma 3, Theorem 2 is proved if we construct a realization of π containing a $K_{2,3}$ or a subgraph K as described in the Lemma. We consider several cases.

Case 1 : $\pi = (d_1,d_2,2,2,2,t,\ldots,t)$ where d_2 3. Let

$$\pi(5) = (d_1-1,d_2-1,2,2,*,t,\ldots,t),$$
$$\pi(5,4) = (d_1-2,d_2-2,2,*,*,t,\ldots,t).$$

We now consider two subcases as follows.

Case 1.1: $d_2 \geq 4$. Then let $\pi(5,4,3) = (d_1-3,d_2-3,*,*,*,t,\ldots,t)$. Clearly now G contains a $K_{2,3}$.

Case 1.2: $d_2 = 3$. By the hypothesis of the theorem, $d_1 \leq p-2$. Now let $\pi(5,4,1) = (*,1,1,*,*,t+\varepsilon,\ldots,t+\varepsilon)$. Clearly then G contains a subgraph K of Lemma 3, in which $(v_2,v_1),(v_3,v_4),(v_3,v_5) \notin E(K)$.

Case 2 : $\pi = (d_1,d_2,d_3,2,2,t,\ldots,t)$ where $d_3 \geq 3$. Using the graphicness of π it is not difficult to prove that $\ell(\pi) \geq 6$. Let $\pi(5) = (d_1-1,d_2-1,d_3,2,*,t,\ldots,t)$. We now consider five subcases as follows.

Case 2.1 : $d_2 = 3$, $d_6 = 1$. The nonouterplanar graph given in Figure 3 is a realization of π since $d_1 \equiv (p-3) \pmod 2$.

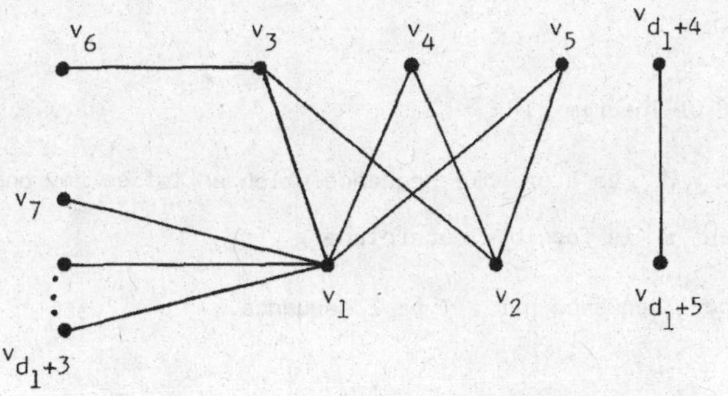

FIGURE 3

Case 2.2 : $d_2 = 3$, $d_6 = 2$. By the hypothesis in the theorem $d_1 \leq p-2$. Then let

$$\pi(5,3) = (d_1-2,1,*,2,*,1,t,\ldots,t),$$
$$\pi(5,3,1) = (*,1,*,1,*,1+\varepsilon,t+\varepsilon,\ldots,t+\varepsilon).$$

Clearly then G contains a subgraph K of Lemma 3.

Case 2.3 : $d_2 \geq 4$ and $d_1 \geq d_3+1$. Let

$$\pi(5,2) = (d_1-2,*,d_3-1,1,*,t+\varepsilon,\ldots,t+\varepsilon),$$
$$\pi(5,2,4) = (d_1-3,*,d_3-1,*,*,t+\varepsilon,\ldots,t+\varepsilon),$$
$$\pi(5,2,4,1) = (*,*,d_3-2,*,*,t+2\varepsilon,\ldots,t+2\varepsilon).$$

Clearly G contains a $K_{2,3}$.

Case 2.4 : $d_2 \geq 4$, $d_1 = d_3$ and $d_p = 1$. Let

$$\pi(5,p) = (d_1-1,d_1-1,d_1-1,2,*,t,\ldots,t,*),$$
$$\pi(5,p,4) = (d_1-2,d_1-2,d_1-1,*,*,t,\ldots,t,*),$$
$$\pi(5,p,4,3) = (d_1-3,d_1-3,*,*,*,t+\varepsilon,\ldots,t+\varepsilon,*).$$

Again G contains a $K_{2,3}$.

Case 2.5 : $d_2 \geq 4$, $d_1 = d_3$ and $d_p = 2$. By the Erdös-Gallai criteria for the graphicness of π, we have $12 \leq 3d_1 \leq 6 + 2(p-3) = 2p$. Now if $p = 6$, then $\pi = (4,4,4,2,2,2)$. So by the hypothesis of the theorem $p \geq 7$, hence $d_1 \leq [\frac{2p}{3}] \leq p-3$. Now let

$$\pi(5,3) = (d_1-2,d_1-2,*,1,*,t,\ldots,t,2),$$
$$\pi(5,3,p) = (d_1-3,d_1-3,*,1,*,t,\ldots,t,*).$$

Again G contains a $K_{2,3}$.

This completes the proof of Theorem 2.

Theorem 3. Let $\pi : d_1 \geq \ldots \geq d_p$ be a graphic sequence which satisfies any one of the following conditions. Then π is forcibly outerplanar.

(1) π is neither a Type 1 sequence nor a Type 2 sequence.

(2) $\pi = (4,4,4,2,2,2)$.

(3) $d_1 = p-1$, $d_2 = d_3 = 3$ and $d_4 = d_5 = 2$.

(4) $d_1 = p-1$, $d_2 = 3$ and $d_3 = d_4 = d_5 = 2$.

Proof. If π satisfies (1), then π is forcibly outerplanar by Lemma 1.

If π satisfies (2), then π is unigraphic and its realization is outerplanar.

If π satisfies (3), it follows that $\pi(G-v_1) = (2,2,1,1,t-1,\ldots,t-1)$ since $d_2 = p-1$. Hence G has no subgraph homeomorphic from $K_{2,3}$ and π is forcibly outerplanar.

If π satisfies (4), π is forcibly outerplanar as above.

Theorem 4. A graphic sequence $\pi : d_1 \geq \ldots \geq d_p$ is forcibly outerplanar iff satisfies one of the following conditions.

(1) $d_2 \leq 2$.

(2) $d_3 \geq 3$ and $d_4 \leq 1$.

(3) $d_2 \geq 3$, $d_4 = 2$ and $d_5 \leq 1$.

(4) $\pi = (4,4,4,2,2,2)$.

(5) $d_1 = p-1$, $d_2 = d_3 = 3$ and $d_4 = d_5 = 2$.

(6) $d_1 = p-1$, $d_2 = 3$ and $d_3 = d_4 = d_5 = 2$.

This theorem follows from Theorems 1,2,3 and the fact that π is (1),(2) or (3) iff π is neither a Type 1 sequence nor a Type 2 sequence.

ACKNOWLEDGEMENTS

We express our deep sense of gratitude to the referee for simplifying the proof of Theorem 2, to Dr.S.B. Rao for suggesting the problem and to Messrs. T.V.S.Jagannathan, S. Ramachandran and P. Thangaraju for the discussions we had regarding the proofs.

REFERENCES

1. G. Chartrand and F. Harary, Planar permutation graphs, Ann. Inst. H. Princare, Sec. B, 3(1967), 433-438.

2. F. Harary, Graph Theory, Addison-Wesley, Mass. 1969.

3. D.J. Kleitman and D.L. Wang, Algorithms for constructing graphs and digraphs with given valences and factors, Disc. Math., 6(1973), 79-88.

4. S.B. Rao, Characterization of forcibly planar degree sequences, I.S.I. Tech. Report No. 36/78.

CHARACTERIZATION OF POTENTIALLY SELF-COMPLEMENTARY, SELF-CONVERSE DEGREE-PAIR SEQUENCES FOR DIGRAPHS

PRABIR DAS
Indian Statistical Institute
203 B. T. Road, Calcutta 700 035

ABSTRACT

In this paper we characterize all finite sequences of ordered pairs of non-negative integers which form the outdegree-indegree sequence of (a) a self-complementary digraph and (b) a self-converse digraph.

1. INTRODUCTION AND DEFINITIONS

For definitions and notations we follow Harary [6]. All digraphs considered here are finite and without multiple arcs or loops. If D is a digraph then its vertex (arc) set is denoted by V(D) (A(D)) and the outdegree (indegree) of a vertex u of D is denoted by r(u) (s(u)). We also denote (r(u), s(u)) by d(u) and call it the degree-pair of u. The complement of D contains an arc uv if and only if it is not in D. If there is an isomorphism σ from a digraph D to its complement $\overline{D}$ (converse D^c) then D is said to be self-complementary (self-converse) and σ is said to be a complementing permutation or c.p. (antiautomorphism or a.a.) of D.

Let $\pi = (r_1,s_1), (r_2,s_2), \ldots, (r_p,s_p)$ be a finite sequence of ordered pairs of non-negative integers. We say that π is a digraphic (respectively potentially self-complementary (p.s.c.); potentially self-converse) degree-pair sequence if there exists a digraph (respectively s.c. digraph; self-converse digraph) D with $V(D) = \{v_1,v_2, \ldots, v_p\}$ such that $r(v_i) = r_i$, $s(v_i) = s_i$ for i, $1 \leq i \leq p$. Such a digraph D is then called a realization (respectively s.c. realization; self-converse realization) of π and then π is also said to be digraphic (respectively p.s.c.; potentially self-converse).

In this paper characterizations for p.s.c. and potentially self-converse degree-pair sequences are obtained. Digraphic degree-pair sequences have been characterized in Fulkerson [5]. However, the result is stated in a more direct and useful form on

p. 405 of Chen [1]. Also degree-pair sequences for self-converse tournaments have been characterized in Eplett [4]. For undirected graphs p.s.c. degree sequences have been characterized in Clapham and Kleitman [3] and Clapham [2] and degree sequences that are graphic with every realization a s.c. graph have been characterized in Rao [8].

We now introduce some definitions and notations to proceed towards the characterizations.

Let $\pi = (r_1,s_1), (r_2,s_2), \ldots, (r_p,s_p)$. Then

$$\bar{\pi} = (p-1-r_1,\ p-1-s_1), \ldots, (p-1-r_p,\ p-1-s_p),$$

$$\pi^c = (s_1,r_1), \ldots, (s_p,r_p), \text{ and } \ell(\pi) = p.$$

Let $\pi' = (r'_1,s'_1), \ldots, (r'_p,s'_p)$. Then $\pi = \pi'$ if there is a permutation ψ of $\{1,2,\ldots,p\}$ such that $(r_i,s_i) = (r'_{\psi(i)},\ s'_{\psi(i)})$ for i, $1 \leq i \leq p$.

Henceforth, if π is digraphic then all realizations D of π are considered on the vertex set $V(\pi) = \{v_1,v_2,\ldots,v_{\ell(\pi)}\}$ with $d(v_i) = (r_i,s_i)$ for i, $1 \leq i \leq \ell(\pi)$. When realizations of any other degree-pair sequence π' are also considered on $V(\pi)$ (i.e., $V(\pi') = V(\pi)$) then for $v \in V(\pi)$ we use the corresponding notations $r(v)(\pi')$, $s(v)(\pi')$, $d(v)(\pi')$.

An R-ordering (RO) of π is an ordering of $V(\pi)$ given by a bijection $f : V(\pi) \to \{1,2,\ldots,\ell(\pi)\}$ such that if $f(v_i) = f(v_j) + 1$ then $r_j \geq r_i$ and, furthermore, if $r_j = r_i$ then $s_j \geq s_i$.

Similarly an S-ordering (SO) is defined with r and s interchanged in the above.

Note that for any π there exists an RO and this may be taken, without loss of generality, to be given by the indices.

If D is a digraph, with $V(D) = \{x_1,x_2,\ldots,x_p\}$ say, then $\pi(D) = (r(x_1), s(x_1)), (r(x_2), s(x_2)), \ldots, (r(x_p), s(x_p))$.

Let $X,Y \subset V(D)$, then $D[X,Y]$ is the digraph defined by the following : $V(D[X,Y]) = X \cup Y$, $A(D[X,Y]) = \{uv \in A(D) : u \in X,\ v \in Y\}$. We denote $D[X,X]$ by $D[X]$. Note that $D[X,Y]$ can be described unambiguously by merely writing

$A(D[X,Y]) = \{u_1v_1, \ldots\}$.

If $X \subset V(D)$, then $D - X = D[V(D)-X]$.

If $L \subset A(D)$, then $D - L$ is the digraph defined by the following :

$$V(D-L) = V(D),\ A(D-L) = A(D) - L.$$

A digraph D_1 is said to be a subgraph of a digraph D if $V(D_1) \subset V(D)$ and $A(D_1) \subset A(D)$.

An antidirected alternating trail $I = (x_1 x_2 \ldots x_{2a})$ with respect to a digraph D is a sequence of vertices of D such that for all odd i,j, $i \neq j$, $1 \leq i$, $j \leq 2a$ we have $x_i x_{i+1} \in A(D)$, $x_i x_{i-1} \notin A(D)$, $(x_i, x_{i+1}) \neq (x_j, x_{j+1})$, $(x_i, x_{i-1}) \neq (x_j, x_{j-1})$ where the subscripts are taken modulo 2a. By writing $D \rightarrow I \rightarrow H$ we mean that I is an antidirected alternating trail with respect to D and that we obtain H from D by deleting the arcs $x_i x_{i+1}$ and adding the arcs $x_i x_{i-1}$, i odd and $1 \leq i \leq 2a$. Then clearly H is a digraph and for $v \in V(D) = V(H)$ we have $d(v)(\pi(D)) = d(v)(\pi(H))$. Hence $\pi(H) = \pi(D)$.

2. POTENTIALLY SELF-COMPLEMENTARY DEGREE-PAIR SEQUENCES

We begin with some canonical realizations of digraphic degree-pair sequences.

Lemma 2.1. Let π be digraphic. Let x,u,v be three distinct elements of $V(\pi)$. If there exists a realization D of π in which $vx \in A(D)$, $ux \notin A(D)$ and either (i) $r(u) > r(v)$ or (ii) $r(u) = r(v)$ and $s(u) \geq s(v)$ holds then there exists a realization D_1 in which $ux \in A(D_1)$, $vx \notin A(D_1)$ and if $y \neq v$, $yx \in A(D)$ then $yx \in A(D_1)$.

Proof. Now if there exists $w \in V(\pi)$, $w \neq v$ such that $uw \in A(D)$, $vw \notin A(D)$ then let $I = (vxuw)$ and $D \rightarrow I \rightarrow H$. Note that in obtaining H no arc joined to x, except for vx, is deleted from D. Hence H serves for D_1.

So suppose such a w does not exist. Then (i) cannot hold and therefore (ii) holds. As $r(u) = r(v)$ and there is no such w so $uv \in A(D)$ and $vu \notin A(D)$. As $s(u) \geq s(v)$ so there exists a $z \in V(\pi)$ such that $zu \in A(D)$, $zv \notin A(D)$. Then let $I = (vxuvzu)$, $D \rightarrow I \rightarrow H$. Again in obtaining H no arc joined to x, except for vx, is deleted from D. Hence H serves for D_1. The lemma is proved.

Repeated use of Lemma 2.1 gives the following result, earlier obtained by Kleitman Wang [7].

Theorem 2.1. Let π be digraphic. Let $x \in V(\pi)$ and $X \subseteq (V(\pi) - \{x\})$ such that $|X| = s(x)$ and for all $u \in X$, $v \in (V(\pi) - (\{x\} \cup X))$ either (i) $r(u) > r(v)$ or (ii) $r(u) = r(v)$ and $s(u) \geq s(v)$ holds. Then there exists a realization D of π where each vertex of X is joined to x.

A particular choice of X yields the following

Corollary 2.1. Let π be digraphic and let $x \in V(\pi)$. Then there exists a realization D_x of π where x is joined from the first s(x) vertices, other than x, in any fixed RO of π.

Henceforth any realization obtained as in Corollary 2.1 is denoted as D_x, and if the degree-pair sequence be different from π, say π', then we denote it as $D_x(\pi')$.

Let $\pi_x = \pi(D_x - \{ux : ux \in A(D_x)\})$. Then π_x is digraphic, $s(x)(\pi_x) = 0$ and if $y \in (V(\pi) - \{x\})$ then $s(y)(\pi_x) = s(y)$.

In the rest of this section till Theorem 2.2 we take π to be digraphic, $\ell(\pi) = 2m$, $\pi = \overline{\pi}$ and $V(\pi) = \{v_1, v_2, \ldots, v_{2m}\}$. Also we assume that the indices give an RO of π and refer to it as the RO of π. Since $\pi = \overline{\pi}$, we have $d(v_i) + d(v_{2m+1-i}) = (2m-1, 2m-1)$ for i, $1 \leq i \leq 2m$ and we refer to each such unordered pair of vertices v_i, v_{2m+1-i} as a conjugate pair, $1 \leq i \leq 2m$.

In the above context we have the following lemma, whose proof is omitted.

Lemma 2.2. Let x,y be a conjugate pair. Then exactly one of the following holds : (1) $yx \in A(D_x)$; (2) $xy \in A(D_y)$.

Let $V(\pi_x) = V(\pi)$. We choose an RO f_x of π_x such that if

$$i < j,\ d(v_i)(\pi_x) = d(v_j)(\pi_x)$$

then $f_x(v_i) < f_x(v_j)$. We refer to f_x as the RO of π_x in the following

Lemma 2.3. Let x,y be a conjugate pair and suppose $yx \in A(D_x)$. Then there exists a realization $D(xy)$ of π in which x is joined from the first $s(x)$ vertices, other than x, in the RO of π and y is joined from the first $s(y)$ vertices, other than x and y, in the RO of π_x.

Proof. We take D_x to have $D_y(\pi_x)$ as a subgraph. This can be done as $s(x)(\pi_x) = 0$. If $xy \notin A(D_x)$ then we let $D(xy) = D_x$ and we are done.

Suppose $xy \in A(D_x)$, that is $xy \in A(D_y(\pi_x))$.

CLAIM : We can get a realization D_1 of π in which $xy \notin A(D_1)$ and $\{ux \in A(D_1)\} = \{ux \in A(D_x)\}$.

From Lemma 2.2 we get $xy \notin A(D_y)$, but $xy \in A(D_x)$. So there exists $u \neq x,y$ such that $uy \notin A(D_x)$ and either (i) $r(u) > r(x)$ or (ii) $r(u) = r(x)$ and $s(u) \geq s(x)$. Now if there exists $w \neq x$ such that $xw \notin A(D_x)$ and $uw \in A(D_x)$ (note then $w \neq y$) then let $I = (xyuw)$, $D_x \to I \to H$. H serves for D_1 as no arcs joined to x or y, except for xy, is deleted from D_x.

So suppose there is no such w. Then (i) cannot hold and so (ii) holds. Since $r(u) = r(x)$ so $ux \in A(D_x)$ and $xu \notin A(D_x)$ else we can find such a w. But now as $s(u) \geq s(x)$ so there exists z such that $zu \in A(D_x)$, $zx \notin A(D_x)$. (Note then $z \neq x,y$).

Now we consider two cases.

Case I. $zy \notin A(D_x)$. Let $I = (xyzu)$, $D_x \to I \to H$. Again as before we see that H serves for D_1.

Case II. $zy \in A(D_x)$. Now if there exists $a \neq x,z$ such that $ua \in A(D_x)$, $za \notin A(D_x)$ (so $a \neq y$) then let $I = (xyuazu)$ and $D_x \to I \to H$. As before this H serves for D_1.

Now as $ux \in A(D_x)$ but $zx \notin A(D_x)$ so either (1) $r(u) > r(z)$ or (2) $r(u) = r(z)$ and $s(u) \geq s(z)$. If there is no such a then (1) cannot hold and so (2) holds. As $r(u) = r(z)$ we have $uz \in A(D_x)$. Now if there exists $b \neq z$ such that $bu \in A(D_x)$, $bz \notin A(D_x)$ (so $b \neq x$) then let $I = (buxyuz)$, $D_x \to I \to H$ and this H serves for D_1.

If there is no such b we have $xz \notin A(D_x)$ as $xu \notin A(D_x)$ and $s(u) \geq s(z)$. Let $I = (xyuz)$, $D_x \to I \to H$ and this H serves for D_1.

Hence the claim is proved.

Let $D_2 = D_1 - \{ux \in A(D_1)\}$. Then $\pi(D_2) = \pi_x$ and $xy \notin A(D_2)$. So using Lemma 2.1 we can get a realization D_3 of π_x where y is joined from highest $s(y)$ vertices, except for y and x, in RO of π_x. Then we get $D(xy)$ as follows. $V(D(xy)) = V(\pi)$, $A(D(xy)) = A(D_3) \cup \{ux \in A(D_1)\}$. Clearly $D(xy)$ satisfies the conditions of the lemma. Hence the lemma is proved.

We now make the following observation. The RO chosen for π_x implies that if $v_ix, v_iy \in A(D(xy))$ then for $j \leq i$ such that $v_j \neq x,y$ we have $v_jx, v_jy \in A(D(xy))$ and if $v_iy, v_ky \in A(D(xy))$ then for j, $i \leq j \leq k$ such that $v_j \neq x$ we have $v_jy \in A(D(xy))$.

Let $\pi^*_{xy} = \pi(D(xy) - \{uv \in A(D(xy)) : v = x \text{ or } y\})$. Then clearly π^*_{xy} is digraphic. As $s(x) + s(y) = 2m - 1$ and $xy \notin A(D(xy))$, $yx \in A(D(xy))$ so there are $2m - 2$ arcs in $D(xy)$ joined to x and y from the $2m - 2$ vertices of $V(\pi) - \{x,y\}$. This along with the above observation shows that if u,v be a conjugate pair different from x,y then $d(u)(\pi^*_{xy}) + d(v)(\pi^*_{xy}) = (2m-3, 2m-1)$. Also $r(x)(\pi^*_{xy}) = r(x)$, $r(y)(\pi^*_{xy}) = r(y) - 1$ and $s(x)(\pi^*_{xy}) = s(y)(\pi^*_{xy}) = 0$.

Now fix an SO f_2 of π^*_{xy} such that if $i < j$ and $d(v_i)(\pi^*_{xy}) = d(v_j)(\pi^*_{xy})$ then $f_2(v_i) < f_2(v_j)$. Then let G_2 be a realization of π^*_{xy} in which x is joined to the first $r(x)$ vertices in the ordering f_2.

Let $\pi^*_3 = \pi(G_2 - \{xu \in A(G_2)\})$. Now fix an SO f_3 of π^*_3 such that if $f_2(u) < f_2(v)$ and $d(u)(\pi^*_3) = d(v)(\pi^*_3)$ then $f_3(u) < f_3(v)$. Then let G_3 be a realization of π^*_3 in which y is joined to the first $r(y) - 1$ vertices in the ordering f_3.

Let $\pi_4 = \pi(G_3 - \{yu \in A(G_3)\})$. Then $d(x)(\pi_4) = d(y)(\pi_4) = (0,0)$.

Let $\pi_{xy} = \pi(G_3 - \{x,y\})$. Then $\ell(\pi_{xy}) = 2m - 2$ and π_{xy} is digraphic. Let $V(\pi_{xy}) = V(\pi) - \{x,y\}$.

Let $G(xy)$ be the following digraph. $V(G(xy)) = V(\pi)$, $A(G(xy)) = A(G_3) \cup \{xu \in A(G_2)\} \cup \{uv \in A(D(xy)) : v = x$ or $y\}$. Clearly $G(xy)$ is a realization of π.

Now we make the following assertion.

ASSERTION 1: π_{xy} is digraphic. Let u,v be a conjugate pair different from x,y then $d(u)(\pi_{xy}) + d(v)(\pi_{xy}) = (2m-3, 2m-3)$. Hence $\pi_{xy} = \bar{\pi}_{xy}$ and $|A(G(xy)[\{x,y\}, \{u,v\}])| = |A(G(xy)[\{u,v\}, \{x,y\}])| = 2$.

The assertion follows from the choice of f_2 and f_3 and the fact that in $G(xy)$ there are $2m - 2$ arcs joined from x and y to $(V(\pi) - \{x,y\})$, which has $2m - 2$ vertices.

Hence if u,v be a conjugate pair different from x,y then we can get a realization of π_{xy}, denoted as $G(uv)(\pi_{xy})$, in which u,v play the roles x,y played in $G(xy)$.

We now make some definitions and a series of assertions, which will lead to the next lemma.

Let x,y and u,v be two different conjugate pairs. Then x,y is said to be properly joined to u,v and u,v is said to be properly joined from x,y in a realization G of π if $\pi(G[\{x,y\}, \{u,v\}]) = (2,0), (0,0), (0,1), (0,1)$ or $(1,0), (1,0), (0,2), (0,0)$.

A conjugate pair x,y is said to be properly removable if it is both properly joined to and from each other conjugate pair in some realization of π.

ASSERTION 2 : If x,y is properly removable, say in realization G of π, then if H is a s.c. realization of $\pi_1 = \pi(G - \{x,y\})$ such that H has a c.p. σ consisting only of disjoint transpositions then H^* defined by $V(H^*) = V(H) \cup \{x,y\}$, $A(H^*) = A(H) \cup \{uv \in A(G) : \{u,v\} \cap \{x,y\} \neq \phi\}$ is a s.c. realization of π with c.p. $(xy)\sigma$.

Let a,b be a conjugate pair different from x,y. Then by Assertion 1 we know that $s(a) + s(b) - s(a)(\pi_{xy}) - s(b)(\pi_{xy}) = 2$. Hence either (i) $s(a) - s(a)(\pi_{xy}) = s(b) - s(b)(\pi_{xy}) = 1$ or (ii) without loss of generality $s(a) - s(a)(\pi_{xy}) = 2$ and

$s(b) - s(b)(\pi_{xy}) = 0$. So if x,y is not joined properly to a,b in $G(xy)$ then (ii) cannot hold and (i) holds. Hence, then, $\pi(G(xy)[\{x,y\}, \{a,b\}]) = (1,0),(1,0),(0,1),(0,1)$ and we can take without loss of generality $A(G(xy)[\{x,y\}, \{a,b\}]) = \{xa,yb\}$.

So now if there be two distinct conjugate pairs a,b and c,d to which x,y is not properly joined in $G(xy)$ then we have $A(G(xy)[\{x,y\}, \{a,b,c,d\}]) = \{xa,yb,xc,yd\}$ which can be changed to $\{xa,xb,yc,yd\}$ to give a realization in which x,y is properly joined to both a,b and c,d. The same can be done if x,y is not properly from a,b and c,d in $G(x,y)$. Thus proceeding in this way we can get a realization $H(xy)$ of π in which x,y is not properly joined to or from at most one conjugate pair and such that $\pi(H(xy) - \{x,y\}) = \pi(G(xy) - \{x,y\}) = \pi_{xy}$. Also we can similarly get $H(uv)(\pi_{xy})$ from $G(uv)(\pi_{xy})$.

A conjugate pair x,y is called a

T1 pair in π' if $s(u)(\pi') = s(v)(\pi') + 1$ and $r(v)(\pi') > r(u)(\pi')$,

T2 pair in π' if $r(u)(\pi') = r(v)(\pi') + 1$ and $s(v)(\pi') > s(u)(\pi')$,

T3 pair in π' if $r(u)(\pi') = s(u)(\pi') = r(v)(\pi') + 1 = s(v)(\pi') + 1$,

T pair in π' if it is a T1 or T2 or T3 pair in π' where $\{u,v\} = \{x,y\}$.

Note that a conjugate pair may be both T1 and T2 in π'.

Using these definitions, Assertion 1 and considering the relative positions of a and b, where a,b are a conjugate pair, in the different orderings encountered while obtaining $G(xy)$ we get the following immediately.

ASSERTION 3 : If x,y is joined properly in $G(xy)$ to a,b which is T1 in π then $\pi(G(xy)[\{x,y\}, \{a,b\}]) = (2,0), (0,0), (0,1), (0,1)$. If x,y is joined properly in $G(xy)$ from c,d which is T2 in π then

$$\pi(G(xy)[\{c,d\}, \{x,y\}]) = (1,0), (1,0), (0,2), (0,0).$$

ASSERTION 4 : If there is no T1 pair in π but there is one T1 pair a,b in π^*_{xy} then a,b is T3 in π.

ASSERTION 5: A T1 (respectively T2) pair different from x,y remains T1 (respectively T2) in π_{xy}.

ASSERTION 6: A T3 pair in π different from x,y is either T3 or T1 (and T2) in π_{xy}.

ASSERTION 7: If in G(xy) x,y is not properly joined from conjugate pair a,b then a,b is T2 in π.

ASSERTION 8: If in G(xy) x,y is not properly joined to conjugate pair a,b then a,b is either T1 or T3 in π.

ASSERTION 9: If a,b is T2 in π and x,y is not properly joined from conjugate pair u,v in H(xy), then we can take $\{u,v\} = \{a,b\}$.

ASSERTION 10: If c,d is T1 in π and x,y is not properly joined to conjugate pair p,q in H(xy), then we can take $\{p,q\} = \{c,d\}$.

Note that we can make Assertions 9 and 10 simultaneously.

<u>Lemma 2.4</u>. If π is digraphic, $\pi = \overline{\pi}$, $\ell(\pi) = 2m$ where $m \geq 3$ then there is a conjugate pair which is properly removable.

<u>Proof</u>. Suppose there is at most one T pair in π. Let x,y be the unique T pair if it exists, else let x,y be any conjugate pair. Then by Assertions 7 and 8 we get that x,y is properly removable in G(xy).

So now we may assume that there are at least two T pairs in π. We deal with this in four cases, which are exhaustive.

<u>Case I</u>. There exist a T1 and a distinct T2 pair in π.

Let a,b be T1 and c,d be a distinct T2 pair in π. As $m \geq 3$ we can get a third conjugate pair x,y and construct H(xy). Using Assertions 9 and 10 we have the following subcases if x,y is not properly removable in H(xy).

Subcase (i) : x,y is properly joined from all conjugate pairs but x,y is not properly joined to a,b only.

Now c,d is T2 in π_{xy} by Assertion 5. So let $d(c)(\pi_{xy}) = (m-1,s_1)$ and $d(d)(\pi_{xy}) = (m-2,s_1')$ where $s_1' > s_1$. Then c is in the first m - 1 places in any

RO of π_{xy} and as $s(d)(\pi_{xy}) \geq m - 1$ so we have $cd \in A(D_d(\pi_{xy}))$. So each of c,d has $m - 2$ arcs joined to the vertices of $(V(\pi) - \{x,y,c,d\})$ in $G(dc)(\pi_{xy})$ and as there are exactly $m - 2$ conjugate pairs in $(V(\pi) - \{x,y,c,d\})$ so we see that c,d is not joined properly to a,b.

We take $H(xy)$ to have $G(dc)(\pi_{xy})$ as a subgraph. Hence we have $A(H(xy)[\{x,y,c,d\}, \{a,b\}]) = \{xa,yb,ca,db\}$ which can be changed to $\{xa,ya,cb,db\}$ to give x,y properly removable in above realization, H_1 say.

Subcase (ii): x,y is properly joined to all conjugate pairs but x,y is not properly joined from c,d only.

This is the converse of Subcase (i).

Subcase (iii): x,y is not properly joined to a,b only and from c,d only.

First we proceed as in Subcase (i) to get x,y properly joined to a,b and hence to all other conjugate pairs in H_1. Let $\pi' = \pi(H_1 - \{uv \in A(H_1) : \{u,v\} \cap \{x,y\} \neq \phi\})$. Clearly π' is digraphic and $\pi' = \overline{\pi}'$. Also $d(c)(\pi_{xy}) = d(c)(\pi')$ and $d(d)(\pi_{xy}) = d(d)(\pi')$. Hence c,d is T2 in π'. Also if u,v be a conjugate pair, different from x,y then $d(u)(\pi') + d(v)(\pi') = (2m-3,2m-3)$.

We may also assume without loss of generality that $d(a) = (r',m-1)$ and $d(b) = (r,m)$ where $r' > r$. So now we have $d(a)(\pi') = (r_1',m-3)$, $d(b)(\pi') = (r_1,m)$ where $r_1' > r_1$.

We now take H_1 to have $H(ab)(\pi')$ as a subgraph. So if a,b is not joined properly from some conjugate pair in $H(ab)(\pi')$ then we can take that to be c,d by Assertion 9. Thus $A(H_1[\{c,d\}, \{a,b,x,y\}]) = \{ca,db,cx,dy\}$ which can be changed to $\{cx,cy,da,db\}$ to give x,y properly removable in this realization H_2.

So we suppose a,b is joined properly from all other conjugate pairs in $(V(\pi) - \{x,y\})$ in H_1. Thus a,b is properly joined from all other conjugate pairs in $V(\pi)$ in H_1.

Now as c,d is T2 in π' and a,b is joined properly from c,d so $A(H_1[\{c,d\}, \{a,b\}]) = \{ca,da\}$ or $\{cb,db\}$ as it cannot be $\{ca,cb\}$ or $\{da,db\}$.

Since $s(b)(\pi') - s(a)(\pi') = 3$ so we can take $A(H_1[\{c,d\}, \{a,b\}]) = \{cb,db\}$.

Now in $H(ab)(\pi')$ if a,b is not joined properly to some other conjugate pair u,v (u,v may be same as c,d) in $(V(\pi) - \{x,y\})$ then $A(H_1[\{a,b\}, \{u,v\}]) = \{au,bv\}$. Also $A(H_1[\{x,y\}, \{a,b\}]) = \{xa,ya\}$ as H_1 was constructed thus in Subcase (i) and $A(H_1[\{c,d\}, \{x,y\}]) = \{cx,dy\}$ as x,y is not properly joined from c,d. Also $A(H_1[\{a,b\}]) = \{ab\}$ or $\{ba\}$.

If $A(H_1[\{a,b\}]) = \{ba\}$ then let $I = (cxdbauba)$, $H_1 \to I \to H_2$. Then x,y is properly joined from c,d and as no other arcs joined to or from x,y are altered so x,y is properly removable.

If $A(H_1[\{a,b\}]) = \{ab\}$, let $I_1 = (xacb)$, $I_2 = (yadb)$, $I_3 = (cxdabvab)$ and $H_1 \to I_1 \to H_2 \to I_2 \to H_3 \to I_3 \to H_4$. Then x,y is properly removable in H_4.

So we suppose in H_1 a,b is properly joined to all other conjugate pairs in $V(\pi) - \{x,y\}$. Also a,b is properly joined to x,y in H_1 as x,y is not properly joined from only c,d. Hence a,b is properly removable.

Now we give the last three cases without the proofs as the strategy in each case is similar to that of Case I and the details, though different, are lengthy.

Case II. There are no T2 or T3 pairs in π (respectively no T1 or T3 pairs in π by considering π^c).

Case III. There exist a T1 and a distinct T2 pair in π.

Case IV. Otherwise.

Now we give the characterization of p.s.c. degree-pair sequences in the following.

Theorem 2.2. π is p.s.c. if and only if (i) π is digraphic, (ii) $\pi = \overline{\pi}$ and (iii) whenever $\ell(\pi)$ is even, say $\pi = (r_1,s_1), \ldots, (r_{2m},s_{2m})$ with $r_i \geq r_{i+1}$ for $i = 1,2, \ldots, 2m-1$, then $\sum_{i=1}^{m} (r_i+s_i) \equiv m \pmod{2}$. Furthermore, if π is p.s.c. then there exists a s.c. realization of π which has a c.p. consisting only of disjoint transpositions and fixed points.

Proof. To prove the necessity, let D be a s.c. digraph and $\pi = \pi(D)$. Then obviously π is digraphic and $\pi = \overline{\pi}$. Now suppose $\ell(\pi) = 2m$. We assume $V(D) = \{v_1,v_2,\ldots,v_{2m}\}$

with $r(v_i) \geq r(v_{i+1})$ for $i = 1,2, \ldots, 2m-1$. Let $d(v_i) = (r_i,s_i)$ for $i = 1,2, \ldots, 2m$. Let $A = \sum_{i=1}^{m} (r_i+s_i)$ and $B = \sum_{i=1}^{m} 1 = m$. Thus it is enough to show that $A + B$ is even.

Now of all the c.p. of D choose one, say σ, which has the maximum number of disjoint cycles. If the length of any of the disjoint cycles of σ is not of the form 2^k then it is of the form pq where $p(>1)$ is odd. Then σ^p is also a c.p. of D and has at least $p - 1$ more disjoint cycles than σ, a contradiction. Hence the length of each disjoint cycle of σ is of the form 2^k.

The total contribution from the vertices of a cycle of length 2^p, $p > 1$, to A and B is even as there are an even number of vertices in that cycle with same degree-pair. Now the total contribution from the vertices of the transpositions to $A + B$ can be broken up into the following parts. Let $(v_a v_b)$ be a transposition with $a \leq m$. The contribution from $A(D[\{v_a,v_b\}])$ to A is 1 and from v_a to B is 1; hence the contribution to $A + B$ is even. Then the number of arcs joined from (to) v_a to (from) the vertices of a cycle of length 2^p, $p > 1$, is even and contributes to A. Let $(v_c v_d)$ be another transposition with $c \leq m$. Then the number of arcs between v_a and $\{v_c,v_d\}$ added the number of arcs between v_c and $\{v_a,v_b\}$ is even (as v_a,v_b is properly joined to and from v_c,v_d) and contributes to A. Hence $A + B$ is even.

The sufficiency is proved, by induction, by considering the following two cases : $\ell(\pi)$ is even and $\ell(\pi)$ is odd. In each case our induction hypothesis is that if π satisfies the conditions of the theorem and is of length k then there is a s.c. realization of π with a c.p. consisting only of disjoint transpositions with at most one fixed point. The reduction is achieved with the help of Lemma 2.4. The details are omitted here.

3. POTENTIALLY SELF-CONVERSE DEGREE-PAIR SEQUENCES

We straightaway give the characterization for potentially self-converse degree-pair sequences for digraphs in the following

Theorem 3.1. π is potentially self-converse if and only if π is digraphic and $\pi = \pi^c$.

Proof. The necessity is obvious.

To prove the sufficiency, let π be digraphic and $\pi = \pi^c$. Let f_1 be a fixed RO of π. As $\pi = \pi^c$, we can get a permutation φ of $V(\pi)$ consisting only of disjoint transpositions and fixed points such that for $u,v \in V(\pi)$ if we have $d(u) = d(v)$, $f_1(u) = f_1(v)+1$ then $f_1(\varphi(v)) = f_1(\varphi(u))+1$ and for $u \in V(\pi)$ we have $(r(u), s(u)) = (s(\varphi(u)), r(\varphi(u)))$.

Now let $x \in V(\pi)$ and $y = \varphi(x)$ be such that $s(x) \geq s(y)$. We construct D_x in which x is joined from the first $s(x)$ vertices, except for x, in the RO f_1 of π. Let $B = \{u : ux \in A(D_x)\}$ and $\pi_1 = \pi(D_x - \{ux \in A(D_x)\})$. Note that then $s(x)(\pi_1) = 0$ and $r(y)(\pi_1) = |B - \{y\}|$. Let $V_1 = V(\pi) - \{x,y\}$. Then for $u \in V_1$, $s(u)(\pi_1) = s(u)$. Let $C = \{\varphi(u) : u \in (B - \{y\})\}$.

It can be seen that if $\varphi(u) \in C$ and $\varphi(v) \in (V_1 - C)$ then either $s(\varphi(u))(\pi_1) > s(\varphi(v))(\pi_1)$ or $s(\varphi(u))(\pi_1) = s(\varphi(v))(\pi_1)$ and $r(\varphi(u))(\pi_1) \geq r(\varphi(v))(\pi_1)$ holds.

As $|C| = r(y)(\pi_1)$ we see by Theorem 2.1 that we can get a realization, say D_1, of π_1 in which y is joined to all the vertices of C.

Now if φ has no fixed points, then $x \neq y$ and (xy) is a transposition of φ. Let $\pi_2 = \pi(D_1 - \{yu \in A(D_1)\})$. Then π_2 is digraphic and $\pi_2 = \pi_2^c$. Let $V(\pi_2) = V(\pi)$. Now we can get an RO f_2 of π_2 such that we still have for $v \in V(\pi_2)$, $r(u)(\pi_2) = s(\varphi(u))(\pi_2)$ and $s(u)(\pi_2) = r(\varphi(u))(\pi_2)$ and for $u,v \in V(\pi_2)$ if $d(u)(\pi_2) = d(v)(\pi_2)$ and $f_2(u) = f_2(v) + 1$ then $f_2(\varphi(v)) = f_2(\varphi(u)) + 1$. Hence we can repeat the previous construction with roles of y and x interchanged to get π^* with π^* digraphic, $\pi^* = (\pi^*)^c$ and $d(x)(\pi^*) = d(y)(\pi^*) = (0,0)$. Let $V(\pi^*) = V(\pi)$. Now if there is a self-converse realization of π^* with an a.a. φ then by adding all the arcs deleted in the foregoing we can get a self-converse realization of π with a.a. φ.

Let $\varphi = (xy)(x_1y_1) \ldots (x_my_m)$ where all the transpositions are disjoint. Thus taking $(x_1y_1), (x_2y_2), \ldots, (x_my_m)$ successively and repeating the above construction each time we get a canonical self-converse realization H of π with a.a. φ when φ has only disjoint transpositions.

Now let φ have $k(>0)$ distinct fixed points, say, $z_1, z_2, \ldots, z_k$. Let

$d(z_i) = (a_i, a_i)$ with $a_i > a_j$ if $i < j$. As π is digraphic so $\ell(\pi) > k$ and hence there is at least one transposition in φ.

Now we take $x = y = z_i$ for a fixed i, $1 \leq i \leq k$ and construct first D_x and then D_1 as above. Let $\pi_1 = \pi(D_1 - \{x\})$, $V_1 = V(\pi) - \{x\}$ and $A_1 = \{uz_i \in A(D_x)\} \cup \{z_i u \in A(D_1)\}$. Let φ_1 be the restriction of φ to V_1. Then for $u \in V_1$ we have $r(u)(\pi_1) = s(\varphi_1(u))(\pi_1)$ and $s(u)(\pi_1) = r(\varphi_1(u))(\pi_1)$. Thus $\pi_1 = \pi_1^c$ and π_1 is digraphic.

Now if for two distinct fixed points of φ_1 (say z_j and z_ℓ) we have $d(z_j)(\pi_1) = d(z_\ell)(\pi_1)$ then we have without loss of generality $j = \ell + 1$ as $d(z_r)(\pi_1) = d(z_r)$ or $d(z_r) - (1,1)$ for $r \neq i$, $1 \leq r \leq k$. Also then if $r \leq \ell$ we have $d(z_r)(\pi_1) = d(z_r) - (1,1)$ and if $r \geq \ell + 1$ we have $d(z_r)(\pi_1) = d(z_r)$. Hence there can be at most one such ℓ with $d(z_\ell)(\pi_1) = d(z_{\ell+1})(\pi_1)$. If there be one such then choose that ℓ, else choose any ℓ.

Now we can get an RO f' of π_1 such that for $u,v \in V_1 - \{z_\ell\}$ if we have $d(u)(\pi_1) = d(v)(\pi_1)$ and $f'(u) = f'(v) + 1$ then $f'(\varphi_1(v)) = f'(\varphi_1(u)) + 1$. Thus we can repeat the above constructions to get A_2, π_2, V_2 and φ_2 similarly. After k such stages we finally get φ_k to consist only of disjoint transpositions, $\pi_k = \pi_k^c$ and π_k is digraphic. Thus, as already shown, we can get a self-converse realization H_k of π_k with a.a. φ_k. Define H as follows $V(H) = V(\pi)$, $A(H) = A(H_k) \cup \bigcup_{i=1}^{k} A_i$. Then H is a self-converse realization of π with a.a. φ.

This completes the proof of the theorem.

Further, if we assume that $\pi = (r_1,s_1), \ldots, (r_p,s_p)$ with $r_i + s_i = p - 1$ for $i = 1,2, \ldots, p$ then we get the following corollary (a result earlier obtained in Eplett [4]).

<u>Corollary 3.1</u>. Let $\pi = (r_1,s_1), \ldots, (r_p,s_p)$ be such that $r_i \geq r_{i+1}$ for $i = 1,2, \ldots, p-1$. Then π has a realization which is a self-converse tournament if and only if π is digraphic, $r_i + s_i = p - 1$ and $r_i + r_{p+1-i} = p - 1$ for $i = 1,2, \ldots, p$.

Proof. Clearly $\pi = \pi^c$. Then as $r_i + s_i = p - 1$ for $i = 1,2, \ldots, p$ it can be seen that at every stage of the construction of the canonical self-converse realization H, given in the proof of Theorem 3.1, if $uv \in A(H)$ then $vu \notin A(H)$. Hence H is a self-converse tournament.

ACKNOWLEDGEMENT

The author wishes to thank Dr. S.B. Rao for his valuable suggestions regarding the problems solved in this paper and also for his constant help and encouragement.

REFERENCES

1. W.K. Chen, Applied Graph Theory, North-Holland Publishing Company, Amsterdam, London, 1971.
2. C.R.J. Clapham, Potentially self-complementary degree sequences, J. Combinatorial Theory, Ser. B, 20(1976), 75-79.
3. C.R.J. Clapham and D.J. Kleitman, The degree sequences of self-complementary graphs, J. of Combinatorial Theory, Ser. B, 20(1976), 67-74.
4. W.J.R. Eplett, Self-converse tournaments, Canadian Math. Bull., Vol.22, No.1 (1979), 23-27.
5. D.R. Fulkerson, Zero-one matrices with zero trace, Pacific J. Math.,10(1960), 831-836.
6. F. Harary, Graph Theory, Addison Wesley, New York, 1969.
7. D.J. Kleitman and D.L. Wang, Algorithms for constructing graphs and digraphs with given valences and factors, Discrete Math., 6(1973), 79-88.
8. S.B. Rao, Characterization of forcibly self-complementary degree sequences, Submitted for publication.

SET-RECONSTRUCTION OF CHAIN SIZES IN A CLASS OF FINITE TOPOLOGIES

SHAWPAWN KUMAR DAS
10 Raja Dinendra Street, Calcutta 700 009

ABSTRACT

A finite T_0 topology, or an acyclic transitive digraph, partitions its underlying point set uniquely into certain ordered subsets called chains, and the size of a chain is the number of points in it. This paper shows that if a T_0 topology, or an acyclic transgraph, satisfies a prescribed condition then, for any i, the number of chains with size i is set-reconstructible.

1. THE PROBLEM AND ITS MOTIVATION

1.1 Notation and Definitions

A topology $\underline{T}$ is a family of sets, called the open sets of $\underline{T}$ satisfying the following three conditions :

(1) $\phi \varepsilon \underline{T}$.

(2) The union of the members of each subfamily of $\underline{T}$ is a member of $\underline{T}$.

(3) The intersection of any two members of $\underline{T}$ is a member of $\underline{T}$.

The point-set of $\underline{T}$, denoted $P(\underline{T})$, is the union of all the open sets of $\underline{T}$. If $S \subseteq P(\underline{T})$ then $(\underline{T}:S)$ denotes the topology $\{S \cap A | A \varepsilon \underline{T}\}$; $(\underline{T}:S)$ is known as the topology induced by $\underline{T}$ on S. $\underline{T}$ is a T_0 topology if for every pair of distinct points, there exists an open set which contains one of these points but not the other. Moreover, for each $x \varepsilon P(\underline{T})$,

$$\underline{T}^*(x) = \cap\{A | x \varepsilon A \varepsilon \underline{T}\}.$$

The notation $y \underline{T} x$ indicates that $y \varepsilon \underline{T}^*(x) - x$. Two topologies $\underline{T}$ and $\underline{U}$ are homeomorphic, written $\underline{T} \simeq \underline{U}$, provided there exists a bijection f from $P(\underline{T})$ onto $P(\underline{U})$ such that for each $S \subseteq P(\underline{T})$, $f(S) \varepsilon \underline{U}$ iff $S \varepsilon \underline{T}$.

A digraph $\underline{G}$ is an ordered pair $(P(\underline{G}), L(\underline{G}))$ where $P(\underline{G})$ is a set of points, and $L(\underline{G})$ is a set of lines or odered pairs of distinct points. (It is recommended that the line (y,x) be visualized as $y \leftarrow x$). The notation $y \underline{G} x$ indicates that (y,x) is a line of $\underline{G}$. If $S \subseteq P(\underline{G})$, then $(\underline{G}:S)$ denotes the digraph $(S, L(\underline{G}) \cap (S \times S))$, which is the digraph induced by $\underline{G}$ on S. $\underline{G}$ is a transgraph (abbreviation for transitive digraph) provided $z \underline{G} y$ and $y \underline{G} x$ imply $z \underline{G} x$. An acyclic transgraph $\underline{G}$ can be characterized as a transgraph in which $y \underline{G} x$ precludes $x \underline{G} y$. For each $x \in P(\underline{G})$,

$$\underline{G}^*(x) = x + \{y \in P(\underline{G}) \mid y \underline{G} x\}.$$

Two digraphs $\underline{G}$ and $\underline{H}$ are isomorphic, written $\underline{G} \cong \underline{H}$, provided there exists a bijection f from $P(\underline{G})$ to $P(\underline{H})$ such that for each pair of points, $x, y \in P(\underline{G})$, $f(x)\underline{H}f(y)$ iff $x \underline{G} y$.

The subsequent material is a unified treatment of T_0 topologies as well as acyclic transgraphs. Depending on the reader's preference or inclination, a structure is to be interpreted as either a T_0 topology or an acyclic transgraph. All structures are assumed to have finite point-sets.

For any set X, $|X|$ denotes the cardinality of X.

Let $\underline{S}$ be a structure and suppose that $x \in P(\underline{S})$. Then

$$\underline{S}^+(x) = x + \{y \in P(\underline{S}) \mid y \underline{S} x\} + \{y \in P(\underline{S}) \mid x \underline{S} y\},$$

$$(S:x^*) = (\underline{S}:\underline{S}^*(x)),$$

$$(\underline{S}:x^+) = (\underline{S}:\underline{S}^+(x)),$$

and

$$\underline{S} - x = (\underline{S} : P(\underline{S}) - x).$$

If $x \in A \subseteq P(\underline{S})$ then x is a maximal point of A (relative to $\underline{S}$) provided $x \underline{S} y$ does not hold for any $y \in A$. x is a maximal point of $\underline{S}$ provided x is a m ximal point of $P(\underline{S})$. The level of x in $\underline{S}$ is denoted $\underline{S}^{\cdot}(x)$ and is defined as follows : If $\underline{S}^*(x) = x$ then $\underline{S}^{\cdot}(x) = 1$, otherwise $\underline{S}^{\cdot}(x) = \max\{n \mid$ there exists a sequence of points $p_1, \ldots, p_n$ such that $x = p_1$, $\underline{S}^*(p_n) = p_n$ and $p_{i+1} \underline{S} p_i$, $1 \leq i < n\}$. For each i, $L_i(\underline{S})$ denotes the set of level i points of $\underline{S}$. $M(\underline{S})$ denotes the set of maximal points of $\underline{S}$, and $M_i(\underline{S}) = M(\underline{S}) \cap L_i(\underline{S})$. The level distribution (resp., maximal point distribution) of $\underline{S}$ is the sequence of nonnegative integers $n_1, \ldots, n_i, \ldots$ where $n_i = |L_i(\underline{S})|$ (resp., $n_i = |M_i(\underline{S})|$). The level of $\underline{S}$, denoted $^{\cdot}\underline{S}$, is defined to be the maximum value of i for which $L_i(\underline{S}) \neq \phi$.

A chain of $\underline{S}$ of size m, $m \geq 1$, is a sequence $[x_1,\ldots,x_m]$ of m distinct points satisfying the following requirements (a) - (d) :

(a) If $\underline{S}^*(x_1) - x_1 = \underline{S}^*(y)$ for some $y \in P(\underline{S})$, then there exists a $z \in P(\underline{S})$ such that $z \neq x_1$ and $\underline{S}^*(z) - z = \underline{S}^*(y)$.

(b) If $\underline{S}^*(y) - y = \underline{S}^*(x_m)$ for some $y \in P(\underline{S})$, then there exists a $z \in P(\underline{S})$ such that $z \neq y$ and $\underline{S}^*(z) - z = \underline{S}^*(x_m)$.

Moreover, if $m > 1$ and $1 \leq i < m$ then :

(c) $\underline{S}^*(x_{i+1}) - x_{i+1} = \underline{S}^*(x_i)$, and

(d) $\{y \in P(\underline{S}) | \underline{S}^*(y) - y = \underline{S}^*(x_i)\} = x_{i+1}$.

It can be verified that the set of chains of $\underline{S}$ constitutes a uniquely defined partition of $P(\underline{S})$. Chains were introduced in [2], and used in [3] to analyse $T_0 + T_5$ topologies. The chain distribution of $\underline{S}$ is the sequence of nonnegative integers $n_1,\ldots,n_i, \ldots$ where n_i is the number of chains of $\underline{S}$ with size i.

If x_1 and x_2 are distinct points of $\underline{S}$ then $x_1 \underline{S}^{\not{c}} x_2$ indicates that the condition :

$x_i \underline{S} x_j$ and there does not exist a third point y such that $x_i \underline{S} y$ and $y \underline{S} x_j$

holds either for $(i,j) = (1,2)$ or for $(i,j) = (2,1)$. $\underline{S}$ is a tree structure provided :

(a) $\underline{S}$ is connected, i.e., given any pair of distinct points x and y there is a sequence of distinct points $p_1,\ldots,p_n$ such that $x = p_1$, $y = p_n$ and for each i, $1 \leq i < n$, either $p_i \underline{S} p_{i+1}$ or $p_{i+1} \underline{S} p_i$.

(b) There does not exist a sequence of distinct points $x_1,\ldots,x_n$, with $n \geq 3$, such that $x_1 \underline{S}^{\not{c}} x_n$ and $x_i \underline{S}^{\not{c}} x_{i+1}$, $1 \leq i < n$.

Let $\underline{\underline{A}}$ be a set of structures. Then the number $\not{n}(\underline{\underline{A}};\underline{S})$ is defined by :

$$\not{n}(\underline{\underline{A}};S) = |\{\underline{X} \in \underline{\underline{A}} | \underline{X} \simeq \underline{S}\}|.$$

If $\underline{\underline{B}}$ is also a set of structures then $\underline{\underline{A}}$ Ⓡ $\underline{\underline{B}}$ means that $\not{n}(\underline{\underline{A}};\underline{X}) = \not{n}(\underline{\underline{B}};\underline{X})$ for any structure $\underline{X}$.

A structure $\underline{R}$ is a <u>set-reconstruction</u> of $\underline{S}$ provided the following holds : to each $x \in P(\underline{R})$ there corresponds a $y \in P(\underline{S})$ such that $\underline{R} - x \simeq \underline{S} - y$ and, conversely, to each $x \in P(\underline{S})$ there corresponds a $y \in P(\underline{R})$ such that $\underline{S} - x \simeq \underline{R} - y$. $\underline{S}$ is <u>set-reconstructible</u> if $\underline{S} \simeq \underline{R}$ whenever $\underline{R}$ is a set-reconstruction of $\underline{S}$. A structural parameter of $\underline{S}$ is set-reconstructible if the parameter has the same value for every set-reconstruction of $\underline{S}$.

1.2 <u>The set-reconstruction problem</u>

For a general survey of reconstruction activity see the review paper by Bondy and Hemminger [1].

The set-reconstruction conjecture for structures enquires whether every structure with at least 4 points is set-reconstructible. It is known [5] that tree structures with at least 4 points are set-reconstructible and, as yet, there is no known example of a non-set-reconstructible structure with 4 or more points. The following proposition (est blished in [5]) is used in the proof of tha main theorem.

<u>Proposition 1</u>. Let $\underline{A}$ be a structure with at least 4 points, and suppose that $\underline{B}$ is a set-reconstruction of $\underline{A}$. Then

(a) $\underline{A}$ and $\underline{B}$ have identical level distributions.

(b) $\underline{A}$ and $\underline{B}$ have identical maximal point distributions.

(c) $\{(\underline{A}:x^+) | x \in L_i(\underline{A})\} \, Ⓣ \, \{(\underline{B}:x^+) | x \in L_i(\underline{B})\}, \quad 1 \leq i \leq \cdot\underline{A}$.

McKay [6] has announced the use of computers in reconstructing small graphs. Very recently, the author has embarked on a project involving the application of digital computers in the set-reconstruction of structures with a small number of points. The procedure which is being used for computer-coding a structure is the one developed in [4]. This coding procedure makes direct use of the chain c ncept. In this context, the question : "Is the chain distribution a set-reconstructible parameter?" naturally arises. The answer is important, for if it is in the affirmative then a significant reduction in c mputer execution time (for set-reconstructing a given structure) can possibly be achieved in most cases. As yet, the question as not been settled completely. However, an affirmative answer does exist when a certain restruction is imposed and the object of this paper is to establish the following preliminary result :

Theorem. Let $\underline{T}$ be a structure with at least 4 points such that $\cdot\underline{T} = k \geq 2$ and $M_{k-1}(\underline{T}) = \phi$. Then the chain-distribution of $\underline{T}$ is set-reconstructible. In other words, if $\underline{U}$ is a set-reconstruction of $\underline{T}$, then $\underline{T}$ and $\underline{U}$ have identical chain-distributions.

The above Theorem is proved in the next section. Observe that if $\cdot\underline{T} = 1$ then the chain-distribution is trivially set-reconstructible. For in this case $\underline{T}$ has $|P(\underline{T})|$ chains, and the size of each chain is 1.

2. PROOF OF THE THEOREM

Let $\underline{\underline{H}}$ be any set of mutually nonhomeomorphic (or nonisomorphic) structure such that for each $x \in P(\underline{T})$ there is some $\underline{H} \in \underline{\underline{H}}$ such that $\underline{T} - x \simeq \underline{H}$ and, conversely, for each $\underline{H} \in \underline{\underline{H}}$ there is some $x \in P(\underline{T})$ such that $\underline{H} \simeq \underline{T} - x$. Then it is enough to prove that $\underline{\underline{H}}$ determines the chain-distribution of $\underline{T}$. Let

$\underline{n} = n_1, \ldots, n_i, \ldots$ be the level distribution of $\underline{T}$, and

$\underline{m} = m_1, \ldots, m_i, \ldots$ the maximal point distribution of $\underline{T}$.

It is a consequence of Proposition 1(a and b) that $\underline{\underline{H}}$ determines both $\underline{n}$ and $\underline{m}$. It follows that $\underline{\underline{H}}$ determines whether or not $k \geq 2$ and $M_{k-1}(\underline{T}) = \phi$; therefore, for the rest of the proof, it will be assumed that $\underline{T}$ does possess these properties. By Proposition 1(c), a set of structures, henceforth denoted by the symbol $\underline{\underline{A}}$, can be constructed (purely from the information contained in $\underline{\underline{H}}$) such that $\underline{\underline{A}} \circledT \{(\underline{T}:x^*) \mid x \in L_k(\underline{T})\}$. The proof will be completed by the construction of a reconstruction algo--rithm which determines the chain distribution of $\underline{T}$ using $\underline{\underline{H}}, \underline{\underline{A}}, \underline{n}$ and $\underline{m}$ as inputs. This algorithm is described in Section 2.2, and some necessary notation and definitions are compiled in Section 2.1.

2.1 Notation and Definitions

Let $\underline{S}$ be a structure, and suppose that $\cdot\underline{S} = r \geq 2$.

A point y is a <u>support</u> of a point x (in $\underline{S}$) provided $y \,\underline{S}\, x$ and $\underline{S}^\cdot(y) = \underline{S}^\cdot(x) - 1$. x is a <u>type</u> 1 (resp., <u>type</u> 2) point of $\underline{S}$ provided $\underline{S}^\cdot(x) = r$ and x has exactly 1 (resp., at least 2) supports in $\underline{S}$. For any point a, let $\bar{a} = \{b \in P(\underline{S}) | a$ is a support of $b\}$. Then a point x is a <u>type</u> 1 (res. <u>type</u> 2) support point of $\underline{S}$ if

$\underline{S}^{\cdot}(x) = r - 1$ and each point in $\bar{x}$ is a type 1 (res., type 2) point of $\underline{S}$; x is a type 3 support point provided $\underline{S}^{\cdot}(x) = r - 1$ and $\bar{x}$ contains at least one type 1 and at least one ty e 2 point. A *(i) set of $\underline{S}$, where $i \geq 2$, is a subset A of $L_r(\underline{S})$ such that $|A| = i$, each point in A is a type 1 point and, moreover, the same point of $L_{r-1}(\underline{S})$ is the support of each point in A ; this common support is called a *(i) support point.

$\underline{S}$ is a class 1 structure if it has exactly one maximal point and this maximal point is a type 1 point of $\underline{S}$. If $\underline{S}$ is a class 1 structure, x is the maximal point of $\underline{S}$, and y is the support of x then the notation ${}^{o}\underline{S}$ is defined by :

$${}^{o}\underline{S} = |\underline{S}^*(y)|,$$

and $\underline{S}$ is a class 2 structure if $\underline{S}^*(x) = \underline{S}^*(y) + x$.

The quantity $\Sigma\, |\underline{S}^*(x)|$, where the summation extends over all $x \in M(\underline{S})$, is denoted by $m(\underline{S})$.

If x is a point of $\underline{S}$ then $(\underline{S};x)$ is used to denote the chain distribution of a structure $\underline{R}$ characterized by the following properties : $P(\underline{R}) = P(\underline{S})$ + exactly one additional point, say y, $\underline{R}^*(y) = \underline{S}^*(x) + y$, and $\underline{R}^*(z) = \underline{S}^*(z)$ for all $z \neq y$.

Let $\underline{e} = e_1, \ldots, e_i, \ldots$ be a sequence of integers. If u,v are integers, with $u \geq 1$, then $(\underline{e} : u,v)$ denotes the sequence

$$\underline{f} = f_1, \ldots, f_i, \ldots$$

where $f_u = e_u + v$ and $f_i = e_i$ for all $i \neq u$. If $a_1, \ldots, a_n, b_1, \ldots, b_n, n \geq 2$, are integers, with the restriction that each $a_i \geq 1$, then the notation $(\underline{e} : a_1,b_1; \ldots ; a_n,b_n)$ indicates the sequence $(\underline{h} : a_n,b_n)$, where $\underline{h} = (\underline{e} : a_1,b_1; \ldots ; a_{n-1},b_{n-1})$. Finally, $(\underline{S} : a_1,b_1; \ldots ;a_n,b_n)$ is defined to mean $(\underline{e} : a_1,b_1; \ldots ;a_n,b_n)$ where $\underline{e}$ = chain distribution of $\underline{S}$.

Let $\underline{\underline{X}}$ be a set of structures. Then $\underline{S}$ is said to possess the $\langle\underline{\underline{X}}\rangle$ property if there exists a structure $\underline{Z} \in \underline{\underline{X}}$ for which the following conditions hold: $\not{n}(\underline{\underline{X}};\underline{Z}) = \not{n}(\{(\underline{S}{:}x^*)\,|\,x \in L_r(\underline{S})\};\underline{Z}) + 1$, $\underline{Z}$ has exactly one maximal point and, denoting this maximal point by z, $\underline{Z}^*(z) - z$ contains at least two maximal points (relative to $\underline{Z}$).

2.2 Reconstruction Algorithm

In the END statements, $\underline{s}$ denotes the chain distribution of $\underline{I}$.

START 1. Set $\underline{\underline{G}} = \{\underline{H} \in \underline{\underline{H}} \mid \text{the level distribution of } \underline{H} \text{ is } (\underline{n} : k,-1)\}$. If there exists a class 1 structure in $\underline{\underline{A}}$ then go to 2, otherwise select any structure in $\underline{\underline{G}}$, identify it as $\underline{N}$, and go to 24.

2. Set $\underline{\underline{F}} = \{\underline{H} \in \underline{\underline{G}} \mid M_{k-1}(\underline{H}) = \phi\}$. If $\underline{\underline{F}} = \phi$ then set $\underline{\underline{Z}} = \underline{\underline{A}}$, $\underline{\underline{W}} = \underline{\underline{G}}$, and go to 13, otherwise go to 3.

3. Set $\underline{\underline{K}} = \{\underline{H} \in \underline{\underline{F}} \mid \underline{H}$ has the <$\underline{\underline{A}}$> property$\}$. If $\underline{\underline{K}} = \phi$ then go to 4, otherwise identify any structure in $\underline{\underline{K}}$ as $\underline{N}$ and go to 24.

4. For each i, $i \geq 2$, set $\underline{\underline{P}}_i = \{\underline{H} \in \underline{\underline{H}} \mid \text{the level distribution of } \underline{H} \text{ is } (\underline{n} : k,-i\ ;\ k-1,i-1)\}$. If $\underline{\underline{P}}_i = \phi$ for all i then go to 12, otherwise go to 5.

5. Set $t = \max\{i \mid \underline{\underline{P}}_i \neq \phi\}$. If $t = 2$ then go to 6, otherwise proceed as follows : For each $\underline{H} \in \underline{\underline{F}}$ define $f(\underline{H})$ = number of $*(t)$ sets of $\underline{H}$. Select, and identify as $\underline{N}$, any structure in $\underline{\underline{F}}$ for which $f(\underline{N}) = \min\{f(\underline{H}) \mid \underline{H} \in \underline{\underline{F}}\}$. Now go to 24.

6. Set $\underline{\underline{M}} = \{\underline{H} \in \underline{\underline{F}} \mid \underline{H}$ possesses at least one $*(2)$ set$\}$. If $\underline{\underline{M}} = \phi$ then go go to 8, otherwise go to 7.

END 7. Select any $\underline{M} \in \underline{\underline{M}}$. Select any $*(2)$ support point of $\underline{M}$, and let u denote the size of the chain $\underline{M}$ which contains this support point. For each $\underline{H} \in \underline{\underline{F}}$ define $f(\underline{H}) = |\{x \in L_k(\underline{H}) \mid \text{the chain of } \underline{H} \text{ which contains } x \text{ is of size } u+1\}|$. Select any $\underline{N} \in \underline{\underline{F}}$ such that $f(\underline{N}) = \max\{f(\underline{H}) \mid \underline{H} \in \underline{\underline{F}}\}$. Then $\underline{s} = (\underline{N} : 1,2\ ;\ u,1\ ;\ u+1,-1)$.

8. Identify the structure in $\underline{\underline{F}}$ by the symbol $\underline{R}$. If $n_k = 2$ then denote the point in $L_{k-1}(\underline{R})$ by x and go to 11, otherwise go to 9.

9. If $\underline{R}$ possesses at least one type 2 point then go to 10. Otherwise proceed as follows : Select a structure $\underline{X}$ from $\underline{\underline{A}}$ such that $\not{n}(\underline{\underline{A}};\underline{X}) = \not{n}(\{(\underline{R}:x^*) \mid x \in L_k(\underline{R})\};\ \underline{X}) + 1$. Select a structure $\underline{X}'$ from $\underline{\underline{A}}$ such that $\underline{X}' \simeq \underline{X}$ but $\underline{X}' \neq \underline{X}$. Let $\underline{\underline{Z}} = \underline{\underline{A}} - \{\underline{X},\underline{X}'\}$, and $\underline{\underline{W}} = \{\underline{H} \in \underline{\underline{G}} \mid \underline{H}$ has a $*(2)$ set$\}$. Now go

to 13.

10. $\underline{\text{If}}$ $\underline{R}$ has a type 3 support point $\underline{\text{then}}$ identity this point as x and go to 11. $\underline{\text{Otherwise}}$ identify as $\underline{N}$ any structure in $\underline{\underline{G}}$ such that the number of type 2 points of $\underline{N}$ = (number of type 2 points of $\underline{R}$) - 1, and go to 24.

END 11. $\underline{s} = (\underline{R};x)$.

12. Select any structure from $\underline{\underline{F}}$ and identify it as $\underline{X}$. $\underline{\text{If}}$ $\underline{X}$ has a type 1 support point $\underline{\text{then}}$ select, and identify as $\underline{\underline{Z}}$, any subset of $\underline{\underline{A}}$ such that $\underline{\underline{Z}}$ Ⓡ $\{(\underline{X}:x^*)|x$ is a type 1 point of $\underline{X}$ with a type 1 support$\}$, set $\underline{\underline{W}} = \{\underline{H} \in \underline{\underline{G}} | \Sigma_1 |\underline{H}^*(x)| = \Sigma_2 |\underline{X}^*(x)|$, where Σ_1 (resp., Σ_2) denotes summation over all x such that x is a type 2 point of $\underline{H}$ (resp., $\underline{X}$)$\}$ and go to 13; $\underline{\text{otherwise}}$ go to 16.

13. $\underline{\text{If}}$ there exists a class 2 space in $\underline{\underline{Z}}$ $\underline{\text{then}}$ go to 14, $\underline{\text{otherwise}}$ proceed as follows :

Set $h = \min\{{}^{0}\underline{H}|\underline{H}$ is a class 1 structure in $\underline{\underline{Z}}\}$. Identify as $\underline{N}$ any structure in $\underline{\underline{W}}$ such that $|M_{i-1}(\underline{N})| = 1$ and $|\underline{N}^*(x)| = h$, where x denotes the point in $M_{i-1}(\underline{N})$. Now go to 24.

14. Set $m = \max\{m(\underline{H}) | \underline{H} \in \underline{\underline{H}}$ and $|M(\underline{H})| = \sum_{i=1}^{\infty} m_i\}$. Identify any class 2 structure in $\underline{\underline{Z}}$ as $\underline{Z}$, and identify the maximal point of $\underline{Z}$ as z. $\underline{\text{If}}$ $|\underline{\underline{Z}}| = 1$ $\underline{\text{then}}$ set

$$\underline{\underline{E}} = \{\underline{H} \in \underline{\underline{W}} \Big| |M_{k-1}(\underline{H})| = 1 \text{ and } m(\underline{H}) = m\}$$

and go to 15. $\underline{\text{Otherwise}}$ set

$$\underline{\underline{E}} = \{\underline{H} \in \underline{\underline{W}} \Big| |M_{k-1}(\underline{H})| = 1, m(\underline{H}) = m \text{ and } \{(\underline{H}:x^*) \Big| x \in L_k(\underline{H})\} \text{ Ⓡ } \underline{\underline{A}} - Z\}$$

and go to 15.

END 15. Select any structure $\underline{N} \in \underline{\underline{E}}$ such that $(\underline{N}:x^*) \simeq \underline{Z} - z$; here x denotes the point in $M_{k-1}(\underline{N})$. Then $\underline{s} = (\underline{N};x)$.

16. $\underline{\text{If}}$ $\underline{X}$ has a type 1 point $\underline{\text{then}}$ go to 17, $\underline{\text{otherwise}}$ go to 18.

END 17. Select any type 1 point of $\underline{X}$ and let u denote the size of the chain of $\underline{X}$ which contains this point. For each $\underline{H} \in \underline{\underline{F}}$ define

$f(\underline{H}) = |\{x \in L_k(\underline{H}) | x$ is a type 1 point of $\underline{H}$ and the length of the chain of $\underline{H}$ which contains x is $u\}|$.

Select any $\underline{N} \in \underline{\underline{F}}$ such that $f(\underline{N}) = \min\{f(\underline{H}) | \underline{H} \in \underline{\underline{F}}\}$. Then $\underline{s} = (\underline{N} : u,1 ; u-1,-1)$.

18. If $\underline{X}$ has exactly one type 2 point then go to 19. Otherwise select any structure in $\underline{\underline{G}}$ which has a type 3 support point, identify this structure as $\underline{N}$, and go to 24.

19. If there exists a structure, hereby identified as $\underline{N}$, in $\underline{\underline{G}}$ such that $M_{k-1}(\underline{N})$ contains at least 2 points then go to 24, otherwise go to 20.

20. Let $\underline{Y}$ and $\underline{Z}$ denote the two structures in $\underline{\underline{A}}$, $\underline{Y}$ being the class 2 structure. Label, arbitrarily, the two points in $L_{k-1}(\underline{Z})$ as z_1 and z_2; label the point in $L_k(\underline{Z})$ as z. Set $a = {}^{o}\underline{Y}$, $b_1 = |\underline{Z}^*(z_1)|$, $b_2 = |\underline{Z}^*(z_2)|$, $c = |\underline{Z}^*(z_1) \cap \underline{Z}^*(z_2)|$ and $d = |\underline{Z}^*(z) - (z + \underline{Z}^*(z_1) \cup \underline{Z}^*(z_2))|$. If $a-c+d = 1$ then go to 21. Otherwise proceed as follows : For each $\underline{H} \in \underline{\underline{G}} - \underline{X}$ define $f(\underline{H}) = |\underline{H}^*(h_1)| + |\underline{H}^*(h_2)|$, where h_1 denotes the point in $L_k(\underline{H})$ and h_2 the point in $M_{k-1}(\underline{H})$. Identify as $\underline{N}$ the structure in $\underline{\underline{G}} - \underline{X}$ for which $f(\underline{N}) = \min|f(\underline{H}) | \underline{H} \in \underline{\underline{G}} - \underline{X}\}$. Now go to 24.

21. If $b_1 = b_2$ then put $e = b_1$ and go to 22; otherwise put $e = b_i$, where i is such that $b_i \neq a$, and go to 22.

22. If $e-c+d = 1$ then go to 23. Otherwise proceed as follows : For each $\underline{H} \in \underline{\underline{G}} - \underline{X}$ define $f(\underline{H}) = |\underline{H}^*(h)|$ where h is the point in $L_k(\underline{H})$. Identify as $\underline{N}$ the structure in $\underline{\underline{G}} - \underline{X}$ for which $f(\underline{N}) = \min\{f(\underline{H}) | \underline{H} \in \underline{\underline{G}} - \underline{X}\}$. Now go to 24.

END 23. $\underline{s} = (\underline{Z} : 1,-1 ; 2,1)$.

END 24. $\underline{s} = (\underline{N} ; 1,1)$.

A demonstration of the fact that the Algorithm indeed reconstructs $\underline{s}$ is rather lengthy. The author therefore omits it in the interests of brevity, with the hope of publishing a more complete version elsewhere. However, the following comments should prove to be useful to those interested in trying to understand the logic involved.

Comment 1. In each END statement, the structure involved in the expression for $\underline{s}$ is homeomorphic (or isomorphic) to $\underline{T} - x$ for some $x \in L_k(\underline{T})$.

Comment 2. Suppose $x \in L_k(\underline{T})$. Let $\not{C}(\underline{T})$ and $\not{C}(\underline{T} - x)$ denote the set of chains of $\underline{T}$ and $\underline{T} - x$ respectively.

(a) If $\underline{T}^*(x) - x \neq \underline{T}^*(y)$ for any $y \in L_{k-1}(\underline{T})$ then the 1-term sequence [x] is a (size 1) chain of $\underline{T}$ and $\not{C}(\underline{T} - z) = \not{C}(\underline{T}) - [x]$.

(b) Suppose $\underline{T}^*(x) - x = \underline{T}^*(y)$ for some y in $L_{k-1}(\underline{T})$. If there exists a z, $z \neq x$, such that $\underline{T}^*(z) - z = \underline{T}^*(y)$ then $[x] \in \not{C}(\underline{T})$ and $\not{C}(\underline{T} - x) = \not{C}(\underline{T}) - [x]$. If no such z exists then x and y are in the same chain of $\underline{T}$ and, denoting this chain by $[a_1,\ldots,a_m]$ (where $a_{m-1} = y$ and $a_m = x$),

$$\not{C}(\underline{T} - x) = (\not{C}(\underline{T}) - [a_1,\ldots,a_m]) + [a_1,\ldots,a_{m-1}].$$

REFERENCES

1. J.A. Bondy and R.L. Hemminger, Graph Reconstruction - A survey, J. Graph Theory, 1(1977),

2. S.K. Das, A Partition of Finite T_0 Topologies, Canad. J. Math., 25(1973), 1137-1147.

3. S.K. Das, On the structure of Finite $T_0 + T_5$ Spaces, Canad. J.Math., 25(1973), 1148-1158.

4. S.K. Das, A Machine Representation of Finite T_0 Topologies. Journal of the ACM., 24(1977), 676-692.

5. S.K. Das, Some studies in the Theory of Finite Topologies, Doctoral thesis submitted to the University of Calcutta, 1979.

6. B.D. McKay, Computer Reconstruction of Small Graphs, J. Graph Theory, 1(1977), 281-283.

CHARACTERIZATION OF FORCIBLY BIPARTITE SELF-COMPLEMENTARY BIPARTITIONED SEQUENCES

T. GANGOPADHYAY
Indian Statistical Institute
203 B. T. Road, Calcutta 700 035

ABSTRACT

A bipartitioned graph is an ordered pair (G,P) where G is a bipartite graph and P = {A,B} is a bipartition of G. The bipartite complement of (G,P) is the bipartitioned graph $(\overline{G}(P),P)$ where $\overline{G}(P)$ is the complement of G with respect to the complete bipartite graph $K_{|A|,|B|}$. If $G \simeq \overline{G}(P)$ then (G,P) is called bipartite self-complementary. In this paper we characterise all those bipartitioned sequences $\pi = (d_1,\ldots,d_m|e_1,\ldots,e_n)$ such that π is graphic and all realisations of π are bipartite self-complementary.

1. INTRODUCTION AND MAIN RESULT

All graphs in this paper are finite, undirected and without loops or multiple edges. For a graph G, the symbols V(G), E(G), $N_G(v)$, $d_G(v)$ respectively denote the vertex set of G, the edge set of G, the neighbourhood of a vertex v in G and the degree of a vertex v in G. For terms and notation undefined here we refer to Bondy and Murty [1].

Let G be a graph with $V(G) = \{v_1,\ldots,v_p\}$. Then the sequence of non-negative integers $\pi(G) = (d_1,\ldots,d_p)$ where $d_i = d_G(v_i)$ is called the <u>degree sequence</u> of G. Conversely a sequence π of nonnegative integers is said to be <u>graphic</u> if there is a graph G such that $\pi(G) = \pi$. In this case G is called a <u>realisation</u> of π. Let P be any invariant property of a graph. A sequence π is said to be <u>forcibly</u> P if π is graphic and every realisation of π has property P.

A graph G is said to be <u>bipartite</u> if there exist A,B such that $A \cup B = V(G)$, $A \cap B = \phi$ and A and B are independent sets in G. Such a partition {A,B} is called a bipartition of G. A <u>bipartitioned graph</u> is a pair (G,P) where G is a bipartite graph and P is a bipartition of G. A complete bipartite graph is a bipartitioned graph(G,P) in which each vertex in A is adjacent to all vertices in B. Throughout this paper, (G,P) will denote a bipartitioned graph and A,B will denote the sets of P.

The bipartite complement of a bipartitioned graph (G,P) is defined to be the bipartitioned graph $(\overline{G}(P),P)$ where $V(\overline{G}(P)) = V(G)$ and $E(\overline{G}(P)) = \{uv | u \in A, v \in B, uv \notin E(G)\}$. A bipartitioned graph (G,P) is said to be bipartite self-complementary (bipsc) if $G \simeq \overline{G}(P)$.

Let (G,P) be bipsc. A bipartite complementing permutation (bipcp) of (G,P) is an isomorphism between G and $\overline{G}(P)$. We denote by $\underline{\underline{C}}((G,P))$ the class of all bipcp's of the bipsc graph (G,P). If (G,P) is a bipartitioned graph which is not bipsc then we define $\underline{\underline{C}}((G,P))$ to be the empty set. A cycle of a bipcp is said to be pure if it permutes only vertices belonging to a single set of P and is said to be mixed otherwise. Let $\underline{\underline{C}}_p((G,P)) = \{\sigma \in \underline{\underline{C}}((G,P)) |$ all cycles of σ are pure$\}$ and $\underline{\underline{C}}_m((G,P)) = \{\sigma \in \underline{\underline{C}}((G,P)) |$ all cycles of σ are mixed$\}$.

We now state two propositions for the proof of which the reader is referred to [3]. These have also been proved independently by Quinn [5].

Proposition 1. If (G,P) is a disconnected bipsc graph then $\underline{\underline{C}}_p((G,P)) \neq \phi$.

Proposition 2. If (G,P) is a connected bipsc graph then $\underline{\underline{C}}((G,P)) = \underline{\underline{C}}_p((G,P)) \cup \underline{\underline{C}}_m((G,P))$. Further if $\sigma \in \underline{\underline{C}}_m((G,P))$ and τ is a cycle of σ then the length of τ is a multiple of 4 and τ takes vertices alternately from A and B.

Given a bipartitioned graph (G,P), if $A_1 \subseteq A$ and $B_1 \subseteq B$ then $G[A_1|B_1]$ denotes the bipartitioned graph (H,Q) where $H = G[A_1 \cup B_1]$ and the sets of Q are A_1 and B_1. Further if $A_1 = \{u_1,\ldots,u_s\}$ and $B_1 = \{v_1,\ldots,v_t\}$ then we will write $G[u_1,\ldots,u_s|v_1,\ldots,v_t]$ to mean $G[A_1|B_1]$. Finally $\overline{G}[A_1|B_1]$ will be used to denote the bipartite complement of $G[A_1|B_1]$.

If (G,P) is a bipartitioned graph, let $A = \{u_1,\ldots,u_m\}$ and $B = \{v_1,\ldots,v_n\}$ where $d_G(u_1) \geq \ldots \geq d_G(u_m)$ and $d_G(v_1) \geq \ldots \geq d_G(v_n)$. Let $d_i = d_G(u_i)$ and $e_j = d_G(v_j)$, then bipartitioned sequence $\pi((G,P)) = (d_1,\ldots,d_m|e_1,\ldots,e_n)$ is called the degree

<u>sequence</u> of (G,P).

If $A = \{u_1,\dots,u_m\}$ and $B = \{v_1,\dots,v_n\}$, then we say that $S = (u_1,\dots,u_m|v_1,\dots,v_n)$ is an <u>ordering</u> of (G,P). The bipartitioned graph (G,P) with the ordering $(u_1,\dots,u_m|v_1,\dots,v_n)$ is said to be a realisation of the bipartitioned sequence $\pi = (d_1,\dots,d_m|e_1,\dots,e_n)$ if $d_G(u_i) = d_i$ and $d_G(v_j) = e_j$ for all i and j. We also say that (G,P) is a realisation of π if (G,P) with some ordering S, is a realisation of π. A bipartitioned sequence π is said to be <u>graphic</u> if there is a realisation of π. Further, following Koren [9] π is said to be <u>unigraphic</u> if given any two realisations (G,P) and (H,P) of π, there is an isomorphism σ from G onto H such that $\sigma(B) = B$.

<u>Henceforth π will denote the bipartitioned sequence</u> $(d_1,\dots,d_m|e_1,\dots,e_n)$ <u>where</u> $n \geq d_1 \geq \dots \geq d_m \geq 0$ and $m \geq e_1 \geq \dots \geq e_n \geq 0$.

A graphic bipartitioned sequence π is said to be <u>potentially bipsc</u> if there exists at least one bipsc realisation (G,P) of π. A bipartitioned sequence π is said to be <u>forcibly bipsc</u> if π is graphic and every realisation of π is bipsc. Potentially bipsc bipartitioned sequence have been earlier characterised in Gangopadhyay [2]. In this paper we will characterise forcibly bipsc bipartitioned sequences. This characterisation is in terms of the following conditions C1 and C2 on π :

$$C1 : \begin{cases} d_i + d_{m+1-i} = n, & 1 \leq i \leq m, \\ e_j + e_{n+1-j} = m, & 1 \leq j \leq n. \end{cases}$$

$$C2 : \begin{cases} m = n \text{ is even,} \\ d_i + e_{m+1-i} = m, & 1 \leq i \leq m, \\ d_{2i-1} = d_{2i}, & 1 \leq i \leq \frac{m}{2}. \end{cases}$$

Our characterisation also uses the characterisation of forcibly self-complementary sequences as obtained by Rao [6] and the characterisation of unigraphic bipartitioned sequences as obtained by Koren [4].

Finally, if $w_1, w_2, \dots, w_{2k}$ are distinct vertices of a graph G and if $w_i w_{i+1}$ (with $w_{2k+1} = w_1$) is an edge of G or not according as i is odd or even, then by an <u>interchange along</u> $(w_1,\ldots,w_{2k},w_1)$ we mean removing the edges $w_i w_{i+1}$ for odd i and adding the edges $w_i w_{i+1}$ for even i.

<u>Henceforth, given</u> $\pi = (d_1,\ldots,d_m|e_1,\ldots,e_n)$, <u>we denote</u> $\frac{m}{2}$ <u>by</u> s <u>iff</u> m <u>is even and</u> $\frac{n}{2}$ <u>by</u> t <u>if</u> n <u>is even. Moreover, if</u> C2 <u>holds then we denote</u> $\frac{m}{2} = \frac{n}{2}$ <u>by</u> t. We note here that if π satisfies C1 and $d_i = d_{m+1-i}$ for some i, then n is even and so t is well-defined. Similarly, if π satisfies C1 and $e_j = e_{n+1-j}$ for some j, then s is well-defined.

<u>In what follows, we assume without loss of generality that</u> π <u>satisfies the conditions</u> (I) - (III) <u>given below since, if any one of</u> (I) - (III) <u>is violated by</u> π <u>then</u> $\tilde{\pi} = (e_1,\ldots,e_n|d_1,\ldots,d_m)$ <u>satisfies all of</u> (I) - (III).

(I) If $d_1 > d_m$ then $e_1 > e_n$.

(II) If some $e_j = \frac{m}{2}$, then some $d_i = \frac{n}{2}$.

(III) If $d_1 > d_m$, $e_1 > e_n$, some $d_i = \frac{n}{2}$ and some $e_j = \frac{m}{2}$, then $d_p - n + q \geq e_q - m + p$ where $p = \max\{i|d_i > t\}$ and $q = \max\{j|e_j > s\}$.

We are now ready to state the main theorem of this paper as

<u>Theorem 1</u>. A bipartitioned sequence $\pi = (d_1,\ldots,d_m|e_1,\ldots,e_n)$ (with the above assumptions (I) - (III), which can be made without loss of generality), is forcibly bipsc iff $\sum_{i=1}^{m} d_i = \sum_{j=1}^{n} e_j$ and π satisfies one of the following four conditions :

(1) C2 holds and the sequence $\pi' = (d_1 + 2t-1,\ldots,d_{2t} + 2t-1, e_1,\ldots,e_{2t})$ is forcibly self-complementary.

(2) C1 holds, $d_1 = d_m$, $e_1 = e_n$ and either $\min(s,t) \leq 2$ or, $\min(s,t) = 3$ and $\max(s,t) \leq 4$.

(3) C1 holds, $d_1 = d_m$ and if k is the number of e_j's in π which are equal to zero, then either $t - k \leq 2$, or $\pi^o \stackrel{\text{def}}{=} ((t-k)^m|e_{k+1},\ldots,e_{2t-k})$ is one of the following bipartitioned sequences :

$\pi_1 = (3^6|3^6)$, $\pi_2 = (4^6|3^8)$, $\pi_3 = (3^8|4^6)$,

$\pi_4 = ((t-k)^2|1^{2(t-k)})$, $\pi_5 = ((t-k)^4|2^{2(t-k)})$,

$\pi_6 = ((t-k)^m|(m-1)^{t-k}, 1^{t-k})$,

$\pi_7 = ((t-k)^4|3, 2^{2(t-k-1)}, 1)$, $\pi_8 = (3^{2s}|2s-1, s^4, 1)$.

(4) C1 holds and if p is the number of d_i's greater than $\frac{n}{2}$ and q the number of e_j's greater than $\frac{m}{2}$, then $0 < p \leq \frac{m}{2}$ and $0 < q \leq \frac{n}{2}$. Further if h is the number of e_j's in π which are not less than $m-p$, then

(a) n is even,

(b) $\sum_{i=1}^{p} d_i = (n-h)p + \sum_{j=n-h+1}^{n} e_j$,

(c) $\sum_{j=1}^{h} e_j = h(m-p) + \sum_{i=m-p+1}^{m} d_i$,

(d) either $p = \frac{m}{2}$

or $t - h \leq 2$

or $\pi^+ \overset{\text{def}}{=} ((t-h)^{m-2p}|e_{h+1}-p, \ldots, e_{2t-h}-p)$

is one of $\pi_1 - \pi_8$ given in (3) above, with t replaced by $t-h$ and k replaced by 0,

(e) the bipartitioned sequence $\pi^* = (d_1-n+h, \ldots, d_p-n+h|e_{n-h+1}, \ldots, e_n)$ is unigraphic.

The proof of Theorem 1 is lengthy and we split it up into several sections. In Section 2, we prove certain preliminary lemmas which will be frequently used in the main body of the proof. In Section 3, we prove the necessity part of the theorem and finally, the sufficiency part of the theorem is proved in Section 4.

In some figures in this paper we <u>represent a set of vertices by a single vertex</u> for convenience and <u>give the size of the set by its side</u>, with the following understanding :

If xy is an edge in the figure then x (or every vertex of x in case x is a set) is adjacent to y (or every vertex of y in case y is a set). Further the vertices on the left (right) are labelled $u_1,\dots,u_m(v_1,\dots,v_n)$ from top to bottom.

2. PRELIMINARIES

In this section we present a few preliminary lemmas which will be used frequently in the course of proving Theorem 1. Of these only Lemma 9 is proved and we refer to [3] for proofs of the other lemmas.

Lemma 0. If (G,P) is a realisation of π with $\underline{\underline{C}}_p((G,P)) \neq \phi$, then π satisfies C1.

Lemma 1. If π is a potentially bipsc bipartitioned sequence then π satisfies either C1 or C2.

Lemma 2. If π is a bipartitioned sequence not satisfying C2, and (G,P) is a bipsc realisation of π, then $\underline{\underline{C}}_p((G,P)) \neq \phi$.

Lemma 3. Let $\pi = (d_1,\dots,d_m|e_1,\dots,e_n)$ be a forcibly bipsc bipartitioned sequence not satisfying C2 and let (G,P) with the ordering $S = (u_1,\dots,u_m|v_1,\dots,v_n)$ be a realisation of π. Let i,j be integers such that $1 \leq i \leq [\frac{m+1}{2}]$ and $1 \leq j \leq [\frac{n+1}{2}]$ and let $A_1 = \{u_i,u_{i+1},\dots,u_{m+1-i}\}$, $B_1 = \{v_j,v_{j+1},\dots,v_{n+1-j}\}$. If (1) $i = 1$ or $d_{i-1} > d_i$ and (2) $j = 1$ or $e_{j-1} > e_j$, then the bipartitioned sequence $\pi^* = \pi(G[A_1|B_1])$ is forcibly bipsc.

Lemma 4. If π satisfies C1 and $e_1 = e_n$ then π is graphic.

Lemma 5. If (G,P) with the ordering $S = (u_1,\dots,u_m|v_1,\dots,v_4)$ is any realisation of $\pi = (2^m|\, m-\alpha,\, m-\beta,\, \beta,\, \alpha)$ where $1 \leq \alpha \leq \beta \leq \frac{m}{2}$, then $\underline{\underline{C}}_p((G,P))$ contains an element σ such that $\sigma(v_j) = v_{5-j}$, $1 \leq j \leq 4$.

Lemma 6. If (G,P) with the ordering $S = (u_1,\dots,u_{2s}|v_1,\dots,v_4)$ is any realisation of $\pi = (2^{2s}|s^4)$, then $\underline{\underline{C}}_p((G,P))$ contains an element σ satisfying $\sigma(v_j) = v_j$ for all j.

Lemma 7. If (G,P) with the ordering $S = (u_1,\dots,u_{2s}|v_1,\dots,v_4)$ is any realisation of $\pi = (3,2^{2s-2},1|s^4)$ then $\underline{\underline{C}}_p((G,P))$ contains an element σ such that $\sigma(u_1) = u_{2s}$ and $\sigma(u_{2s}) = u_1$.

Lemma 8. Let $\pi = (d_1,\dots,d_m|e_1,\dots,e_n)$ be graphic and i any integer such that $1 \leq i \leq m$. Then there is a bipartitioned graph (G,p) and an ordering

$S = (u_1,\dots,u_m|v_1,\dots,v_n)$ such that (G,P) with the ordering S is a realisation of π and u_i is adjacent to $v_1,\dots,v_{d_i}$ in G.

Lemma 9. If n is even and $\pi = (d_1,\dots,d_m|e_1,\dots,e_n)$ is a graphic bipartitioned sequence satisfying C1, then there is a bipartitioned graph (G,P) and an ordering $S = (u_1,\dots,u_m|v_1,\dots,v_{2t})$ such that (G,P) with S is a realisation of π and u_iv_j is an edge of G for all i,j, $1 \leq i \leq [\frac{m+1}{2}]$ and $1 \leq j \leq t$.

Proof. We prove the lemma by induction on m.

If $m = 1$, then $\pi = (t|1^1,0^t)$ and any realisation (G,P) of π proves the theorem.

If $m = 2$, then $\pi = (d_1, 2t-d_1|2^r, 1^{2t-2r}, 0^r)$, where $0 \leq r \leq t$. Let (G,P) be the bipartitioned graph with $A = \{u_1,u_2\}$, $B = \{v_1,\dots,v_{2t}\}$ and

$$E(G) = \{u_1v_j|1 \leq j \leq d_1\} \cup \{u_2v_j|1 \leq j \leq r \text{ or } d_1+1 \leq j \leq 2t-r\}.$$

Clearly then (G,P) with the ordering $S = (u_1,u_2|v_1,\dots,v_{2t})$ is the required realisation of π.

We now assume the lemma for $m-2$ and prove it for m when $m \geq 3$. For convenience we will take $e_0 = m$ and $e_{2t+1} = 0$ in what follows. Let r be the number of e_j's in $\{e_1,\dots,e_{2t}\}$ such that $e_j - e_{2t+1-j} \geq 2$. Then since $e_1 \geq \dots \geq e_{2t}$, it follows that $0 \leq r \leq t$. Also by C1 we have $e_r > e_{r+1}$. Now let

$$\pi^o = (d_2,\dots,d_m|e_1^o,\dots,e_{2t}^o)$$

where

$$e_j^o = \begin{cases} e_j - 1 & \text{if } 1 \leq j \leq d_1 \\ e_j & \text{otherwise.} \end{cases}$$

By Lemma 8 (with $i = 1$) it easily follows that π^o is graphic.

Now let $k = \min\{r,d_m\}$. We will then show that $C = \{e_1^o,\dots,e_k^o, e_{d_1}^o+1,\dots,e_{2t-k}^o\}$ is the set of the largest d_m elements in $D = \{e_1^o,\dots,e_{2t}^o\}$. By C1, $|C| = d_m$. So let $\alpha = \min C$ and $\beta = \max(D-C)$. We will then prove that $\alpha \geq \beta$.

First let $k < d_m$. Then $\alpha = \min\{e_k-1, e_{2t-k}\}$ and $\beta = \max\{e_{k+1}-1, e_{2t-k+1}\}$. Also $k = r$ and by the definition of r, we have $e_k - e_{2t-k+1} \geq 2$ and

$e_{k+1} - e_{2t-k} \leq 1$. It easily follows now that $\alpha \geq \beta$.

Next let $k = d_m$. Then $C = \{e_1^o,\ldots,e_{d_m}^o\}$ and $\alpha = e_{d_m} - 1$. Also $\beta = \max\{e_{d_m+1} - 1, e_{d_1+1}\}$. Since $r \geq d_m$ we have $e_{d_m} - e_{d_1+1} \geq 2$ and it easily follows that $\alpha \geq \beta$.

Thus C is the set of the d_m largest elements in D. Since π^o is graphic, it follows from Lemma 8 (applied to π^o with $e_1^o,\ldots,e_{2t}^o$ rearranged in non-increasing order) that

$$\pi^* = (d_2,\ldots,d_{m-1}|e_1^*,\ldots,e_{2t}^*)$$

is graphic, where

$$e_j^* = \begin{cases} e_j^o - 1 = e_j - 2 & \text{if } 1 \leq j \leq k \\ e_j^o = e_j - 1 & \text{if } k+1 \leq j \leq d_1 \\ e_j^o - 1 = e_j - 1 & \text{if } d_1+1 \leq j \leq 2t-k \\ e_j^o = e_j & \text{if } 2t-k+1 \leq j \leq 2t. \end{cases}$$

Clearly then

$$e_j^* + e_{2t+1-j}^* = m-2 \quad \text{for} \quad 1 \leq j \leq t \qquad \ldots (1)$$

We next show that $\gamma \geq \delta$ where $\gamma = \min\{e_1^*,\ldots,e_t^*\}$ and $\delta = \max\{e_{t+1}^*,\ldots,e_{2t}^*\}$. Now $\gamma = \min\{e_k-2, e_t-1\}$ and $\delta = \max\{e_{t+1} - 1, e_{2t-k+1}\}$. Since $e_r > e_{r+1}$, we have

$$e_k - 2 \geq e_r - 2 \geq e_{r+1} - 1 \geq e_{t+1} - 1.$$

Also $e_{2t-r} > e_{2t-r+1}$ and so

$$e_t - 1 \geq e_{2t-r} - 1 \geq e_{2t-r+1} \geq e_{2t-k+1}.$$

Further $e_k - 2 \geq e_{2t-k+1}$ since $r \geq k$. It now easily follows that $\gamma \geq \delta$.

Now let θ be a permutation of $\{1,2,\ldots,t\}$ such that $e_{\theta(1)}^* \geq \ldots \geq e_{\theta(t)}^*$. Extend θ to a permutation ϕ of $\{1,\ldots,2t\}$ by defining $\phi(j) = 2t+1-\theta(2t+1-j)$ if $t+1 \leq j \leq 2t$. Let

$$\pi^{**} = (d_2,\ldots,d_{m-1}|e_1^{**},\ldots,e_{2t}^{**})$$

where

$$e_j^{**} = e_{\phi(j)}^*, \; 1 \leq j \leq 2t.$$

Then clearly π^{**} is a rearrangement of π^* and so is graphic. Also $e_1^{**} \geq \ldots \geq e_t^{**}$ by definition of θ, $e_t^{**} \geq e_{t+1}^{**}$ since $\gamma \geq \delta$, and $e_{t+1}^{**} \geq \ldots \geq e_{2t}^{**}$ by (1). Thus $e_1^{**} \geq \ldots \geq e_{2t}^{**}$. Since π satisfies C1, it follows from (1) that π^{**} also satisfies C1. Hence by induction hypothesis, there exists a bipartitioned graph (H,Q) and an ordering $S' = (u_2,\ldots,u_{m-1}|v_{\phi(1)},\ldots,v_{\phi(2t)})$ such that (H,Q) with S' is a realisation of π^{**} and $u_i v_{\phi(j)}$ is an edge in H whenever $2 \leq i \leq [\frac{m+1}{2}]$ and $1 \leq j \leq t$. Then (H,Q) with the ordering $(u_2,\ldots,u_{m-1}|v_1,\ldots,v_{2t})$ is a realisation of π^*. Also since $\{\phi(1),\ldots,\phi(t)\} = \{1,\ldots,t\}$, it follows that $u_i v_j$ is an edge in H whenever $2 \leq i \leq [\frac{m+1}{2}]$ and $1 \leq j \leq t$.

Now construct a bipartitioned graph (G,P) from (H,Q) by adding two new vertices u_1 and u_m and joining u_1 to $v_1,\ldots,v_{d_1}$, and u_m to $v_1,\ldots,v_k$, $v_{d_1+1},\ldots,v_{2t-k}$. Then clearly (G,P) with the ordering $S = (u_1,\ldots,u_m|v_1,\ldots,v_{2t})$ is a realisation of π and $u_i v_j$ is an edge in G whenever $1 \leq i \leq [\frac{m+1}{2}]$ and $1 \leq j \leq t$. This completes the induction and the lemma is proved.

Lemma 10. If π is graphic, m = n and $d_i = e_i$ for all i, then there exists a bipartitioned graph (G,P) and an ordering $S = (u_1,\ldots,u_m|v_1,\ldots,v_m)$ such that each non-trivial component G_h of G has an automorphism σ_h with $\sigma_h(A \cap V(G_h)) = B \cap V(G_h)$.

Finally, in the following lemma, which may be proved using Lemma 10, we show that unigraphicness in bipartitioned sequences is equivalent to an apparently weaker condition.

Lemma 11. A graphic bipartitioned sequence $\pi = (d_1,\ldots,d_m|e_1,\ldots,e_n)$ is unigraphic iff for any two realisations (G,P) and (H,P) of π, G is isomorphic to H.

3. PROOF OF NECESSITY

In this section we establish the necessity in Theorem 1. So let π be forcibly bipsc. Then π is graphic and so $\sum_{i=1}^{m} d_i = \sum_{j=1}^{n} e_j$. We now prove the necessity by showing that if π does not satisfy (1), then π satisfies one of conditions (2) - (4). So let π not satisfy (1). Then either π does not satisfy C2 or $\pi' = (d_1+2t-1,\ldots, d_{2t}+2t-1, e_1,\ldots,e_{2t})$ is not forcibly self-complementary.

We first prove that π satisfies C1. If π does not satisfy C2, then since π is also potentially bipsc, it follows by Lemma 1 that π satisfies C1. Suppose now π satisfies C2 and π' is not forcibly self-complementary. Let G' be a non-self-complementary realisation of π'. Let u_i be the vertex of G' having degree $d_i + 2t - 1$ and v_j be the vertex of G' having degree e_j, $1 \leq i, j \leq 2t$. Let $A = \{u_1,\ldots,u_{2t}\}$ and $B = \{v_1,\ldots,v_{2t}\}$. Then clearly

$$\sum_{i=1}^{2t} d_{G'}(u_i) = 2t(2t-1) + \sum_{j=1}^{2t} e_j.$$

Hence, it follows that $G'[A] = K$ and $G'[B] = \overline{K}$. Consider the bipartitioned graph (G,P) where G is the graph obtained from G' by removing all edges within A, and the sets of P are A and B. Clearing (G,P) with the ordering $S = (u_1,\ldots,u_{2t} | v_1,\ldots,v_{2t})$ is a realisation of π. Sincc π is forcibly bipsc, it follows that (G,P) is bipsc. Suppose now $\underline{\underline{C}}_p((G,P)) = \phi$. Then by Proposition 1, G is connected, and so by Proposition 2, there is an element σ in $\underline{\underline{C}}((G,P))$ such that $\sigma(A) = B$. It can be easily verified that σ also acts as an isomorphism between G' and $\overline{G}'$ and so G' is self-complementary, a contradiction. Hence $\underline{\underline{C}}_p((G,P)) \neq \phi$. By Lemma 0, it now follows that π satisfies C1.

We now consider three cases as follows :

<u>Case 1</u>. $d_1 = d_m$ and $e_1 = e_n$.

<u>Case 2</u>. $d_1 = d_m$ and $e_1 > e_n$.

<u>Case 3</u>. $d_1 > d_m$.

Clearly, these three cases are exclusive and exhaustive. We will now prove that if Case (x) holds, then π satisfies condition (x+1) of Theorem 1, x = 1,2,3.

<u>Case 1</u>. $d_1 = d_m$ and $e_1 = e_n$. Then $\pi = (t^{2s} | s^{2t})$. We assume without loss of generality that $s \leq t$. We will then prove that π satisfies (2) by constructing a non-bipsc realisation of π if $s \geq 3$ and $t \geq 5$ or $s = t = 4$.

First let $s \geq 3$ and $t \geq 5$. Then it can be verified that the bipartitioned graph (G,P) given in Figure 1, is a non-bipsc realisation of π.

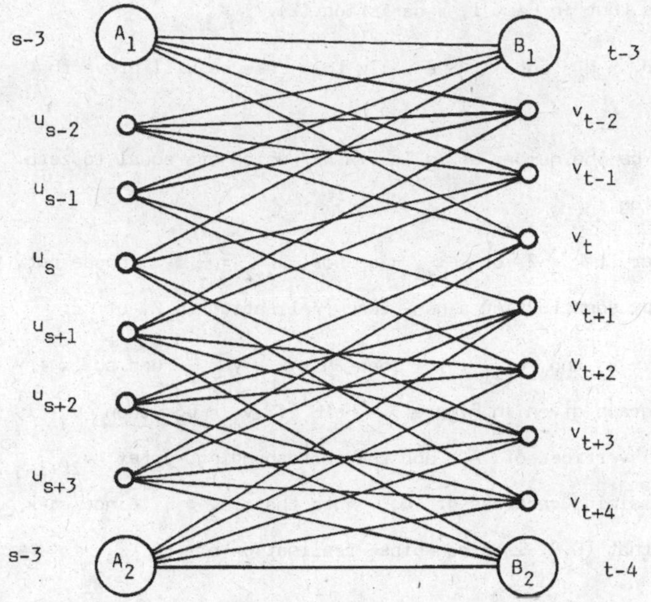

FIGURE 1

Next let $s = t = 4$. Then it can be verified that the bipartitioned graph (G,P) given in Figure 2 is a non-bipsc realisation of $\pi = (4^8|4^8)$.

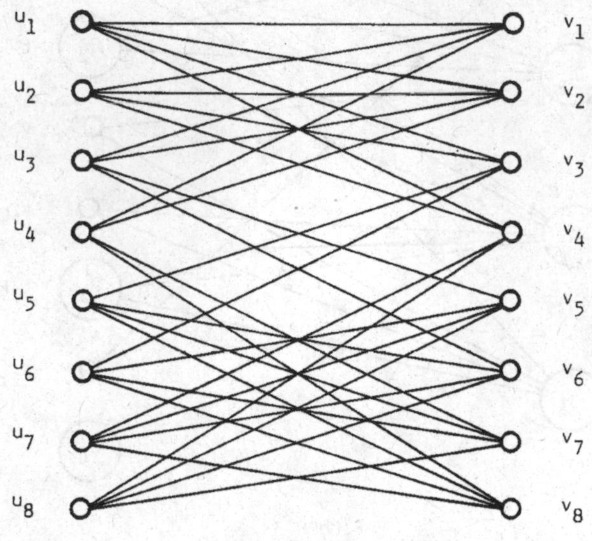

FIGURE 2

This proves that in Case 1, π satisfies (2).

Case 2. $d_1 = d_m$ and $e_1 > e_n$. In this case we will prove that π satisfies (3).

So let k be the number of e_j's in π which are equal to zero. We first prove the following :

1°. Either $t-k \leq 2$ or $e_{k+1} = e_t$ or $e_{k+2} = \frac{m}{2}$. Suppose not, then we obtain a contradiction by constructing a non-bipsc realisation (G,P) of π.

Let $e_{2t-k} = x$ and $e_{t+1} = y$. Then $0 < x < y \leq \frac{m}{2}$ and so $x + y < m$. Now take (G,P) to be the graph given in Figure 3. Here if $v_j \in B_2$ then v_j is joined to the first $e_j - (m-y)$ vertices of A_3 and the corresponding vertex v_{2t+1-j} of B_3 is joined to the remaining vertices of A_3. Note that $B_2 \neq \phi$ since $t-k \geq 3$. It can now be verified that (G,P) is a non-bipsc realisation of π.

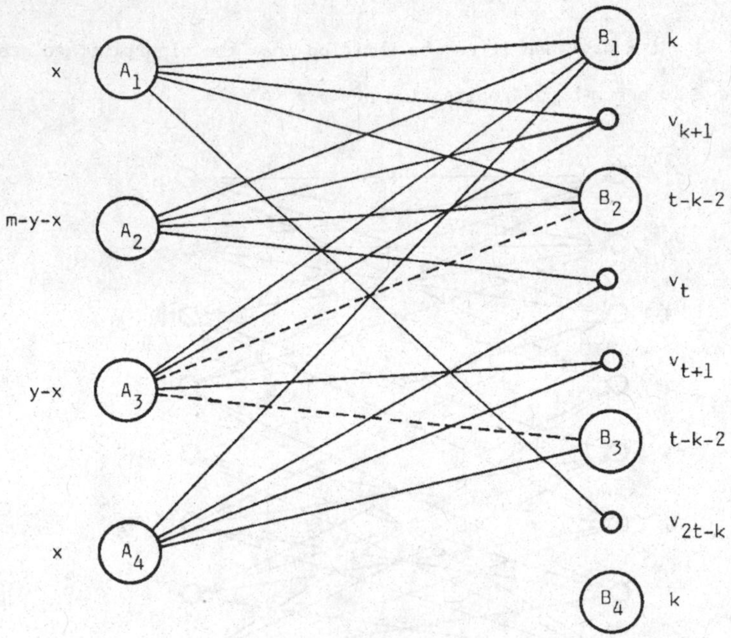

FIGURE 3

Now to prove that π satisfies (3), let $t - k \geq 3$ and $\pi^o = ((t-k)^m|$ $e_{k+1},\ldots,e_{2t-k})$. If (G,P) with the ordering $S = (u_1,\ldots,u_m|v_1,\ldots,v_{2t})$ is any realisation of π then we note that $\pi^o = \pi(G[A|v_{k+1},\ldots,v_{2t-k}])$. Hence by Lemma 3, π^o is forcibly bipsc. We now consider several cases.

Case 2(a). $e_{k+1} = e_t = \frac{m}{2}$. Then $\pi^o = ((t-k)^{2s}|s^{2(t-k)})$. Now since π^o is forcibly bipsc and $t - k \geq 3$, we have by Case 1 that π^o is one of $\pi_1, \pi_2, \ldots, \pi_5$.

Case 2(b). $e_{k+1} = e_t > \frac{m}{2}$. Then $e_{k+1} = m-x$ for some x, $1 \leq x < \frac{m}{2}$. We now prove that $\pi^o = \pi_6$ by constructing in Figure 4 a non-bipsc realisation (G,P) of π if $x > 1$.

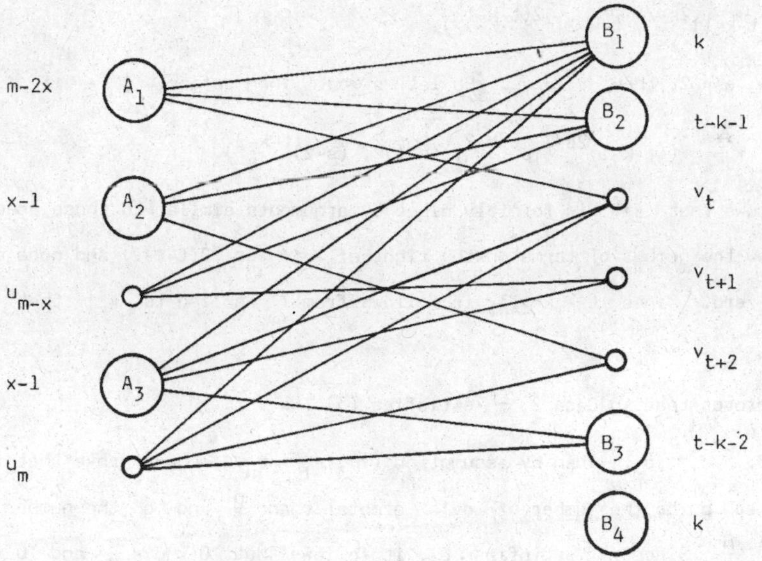

FIGURE 4

Case 2(c). $e_{k+1} > e_t$. Then by 1°, we have $e_{k+2} = \frac{m}{2}$ and so $\pi^o = ((t-k)^{2s}|$ $2s-\alpha, s^{2(t-k-1)}, \alpha)$ for some α, $1 \leq \alpha \leq s - 1$.

We will now prove that π^o is π_7 or π_8. For this define

$$\pi^* = ((t-k)^{\alpha}, (t-k-1)^{2s-2\alpha}, (t-k-2)^{\alpha}|s^{2(t-k-1)}).$$

Then by lemma 4, π^* is graphic. Let (H,Q) with the ordering $S^* = (u_1,\ldots,u_{2s}|$ $v_2,\ldots,v_{2(t-k)-1})$ be a realisation π^*. Get a new bipartitioned graph (G,P) from (H,Q) by adding two new vertices, say v_1 and $v_{2(t-k)}$, and joining v_1 to $u_{\alpha+1},\dots,u_{2s}$ and joining $v_{2(t-k)}$ to $u_{2s-\alpha+1},\dots,u_{2s}$. Clearly now (G,P) with the ordering $(u_1,\dots,u_{2s}|v_1,\dots,v_{2(t-k)})$ is a realisation of π^o. Now π^o is forcibly bipsc and does not satisfy C2, $2s-\alpha > s$, and $\pi^* = \pi(G[u_1,\dots,u_{2s}|v_2,\dots,v_{2(t-k)-1}])$, hence by Lemma 3, π^* is forcibly bipsc. Hence

$$(s^{2(t-k-1)}|(t-k)^{\alpha},\ (t-k-1)^{2s-2\alpha},\ (t-k-2)^{\alpha})$$

is also forcibly bipsc, and so by 1° applied to this sequence we have : either $s \leq 2$ or $\alpha = 1$ (note that the number of terms or the right is $2s$ and none of these is zero). Now $s \leq 2$ implies $s = 2$ and $\alpha = 1$, hence we alsways have $\alpha = 1$. Thus $\pi^* = (t-k,\ (t-k-1)^{2s-2},\ t-k-2|s^{2(t-k-1)})$.

If now $s = 2$, then $\pi^o = \pi_7$. So let $s \geq 3$. Then define

$$\pi^{**} = ((t-k-1)^{2s-2}|s^{t-k-2},\ (s-1)^2,\ (s-2)^{t-k-2}).$$

Then it follows that π^{**} is forcibly bipsc by arguments similar to those used above for π^*. Now the number of terms on the right of π^{**} is $2(t-k-1)$ and none of these is equal to zero. Since $t - k \geq 3$, it follows from 1° applied to π^{**} that $t - k = 3$ and $\pi^o = \pi_8$.

This proves that in Case 2, π satisfies (3).

<u>Case 3</u>. $d_1 > d_m$. Then by assumption (1), $e_1 > e_n$. We now prove that π satisfies (4). Let p be the number of d_i's greater than $\frac{n}{2}$ and q the number of e_j's greater than $\frac{m}{2}$. Since π satisfies C1, it follows that $0 < p \leq \frac{m}{2}$ and $0 < q \leq \frac{n}{2}$. Also let h be the number of e_j's which are not less than $m - p$. Then we will prove that π satisfies the conditions (a) - (e) of (4).

If n is odd then by C1, $e_{\frac{n+1}{2}} = \frac{m}{2}$ and so by assumption (II), some d_i equals $\frac{n}{2}$, a contradiction. This proves (a).

Next we prove (b) and (c) together. This is done in several steps as follows :

Let (G,P) with the ordering $S = (u_1,\dots,u_m|v_1,\dots,v_{2t})$ be any realisation of π. Since π is forcibly bipsc, (G,P) is bipsc. Define

$A_1 = \{u_i | 1 \leq i \leq p\}$, $A_2 = \{u_i | p+1 \leq i \leq [\frac{m+1}{2}]\}$,

$A_3 = \{u_i | [\frac{m+1}{2}] + 1 \leq i \leq m-p\}$, $A_4 = \{u_i | m-p+1 \leq i \leq m\}$,

$B_1 = \{v_j | 1 \leq j \leq q\}$, $B_2 = \{v_j | q+1 \leq j \leq t\}$,

$B_3 = \{v_j | t+1 \leq j \leq 2t-q\}$ and $\{B_4 = \{v_j | 2t-q+1 \leq j \leq 2t\}$.

Also let $B_{11} = \{v_j | 1 \leq j \leq h\}$ and $B_{12} = B_1 - B_{11}$, $B_{41} = \{v_j | 2t-q+1 \leq j \leq 2t-h\}$ and $B_{42} = B_4 - B_{41}$. We note that if $B_2 \neq \phi$ then by assumption (II), $|A_2| = |A_3| > 0$ and so $p < \frac{m}{2}$. We will now show that $B_{11} \subseteq B_1$. Let $v \in B_{11}$. If $B_2 \neq \phi$, then $p < \frac{m}{2}$ and $d(v) \geq m-p \geq \frac{m}{2}$. If $B_2 = \phi$, then $d(v) \neq \frac{m}{2}$, but $d(v) \geq m-p \geq \frac{m}{2}$. Thus, in either case $d(v) > \frac{m}{2}$ and $v \in B_1$. Hence $B_{11} \subseteq B_1$. Similarly it can be proved that $B_{42} \subseteq B_4$.

We now show that there exists an element σ of $\underline{\underline{C}}((G,P))$ such that $\sigma(A_1 \cup B_1) = A_4 \cup B_4$ and $\sigma(A_2 \cup A_3 \cup B_2 \cup B_3) = A_2 \cup A_3 \cup B_2 \cup B_3$. If $m = n$ then any element of $\underline{\underline{C}}((G,P))$ will do. If $m \neq n$ then π does not satisfy C2, hence by Lemma 2, $\underline{\underline{C}}_p((G,P)) \neq \phi$ and we take σ to be any element of $\underline{\underline{C}}_p((G,P))$.

From the result proved in the preceding paragraph it follows that $G[A_1|B_1] \simeq \overline{G}[A_4|B_4]$ and $G[A_2 \cup A_3|B_2 \cup B_3] \simeq \overline{G}[A_2 \cup A_3|B_2 \cup B_3]$. Now by Lemma 9, we can choose (G,P) such that $G[A_1 \cup A_2|B_1 \cup B_2] = K$. Throughout the rest of Case 3, we let (G,P) be such a graph. Since every vertex in A_2 (resp. B_2) has degree t (resp. $\frac{m}{2}$), it follows that $G[A_2|B_3 \cup B_4] = \overline{K}$ and $G[A_3 \cup A_4|B_2] = \overline{K}$.

Choose now a σ in $\underline{\underline{C}}((G,P))$ with the properties given above. Then it follows that $G[A_4|B_4] = \overline{K}$ since $G[A_1|B_1] = K$. Also if $B_2 \neq \phi$, then since $G[A_2 \cup A_3|B_2 \cup B_3] \simeq \overline{G}[A_2 \cup A_3|B_2 \cup B_3]$ and the former has at most $2(s-p)(t-q)$ edges whereas the latter has at least $2(s-p)(t-q)$ edges, it follows that $G[A_3|B_3] = K$ and $\sigma(A_2 \cup B_2)$ is either $A_2 \cup B_3$ or $A_3 \cup B_2$.

We now prove that $G[A_1|B_3] = K$. We may take $B_3 \neq \phi$ (hence B_2, A_2 and A_3 non-empty) since otherwise the claim is vacuously true. If $\sigma(A_2 \cup B_2) = A_2 \cup B_3$, then $\overline{G}[A_2 \cup A_4|B_3 \cup B_4] \simeq G[A_1 \cup A_2|B_1 \cup B_2] = K$, so $G[A_4|B_3] = \overline{K}$. Since every vertex of B_2 has degree s it follows that $G[A_1|B_3] = K$. So let

$\sigma(A_2 \cup B_2) = A_3 \cup B_2$. Then we can prove as above that $G[A_3|B_1] = K$. Thus v_q is joined to all vertices in $A - A_4$, hence $e_q \geq 2s - p$, so by assumption (III), $d_p \geq 2t - q$. If possible, let $u \in A_1$ and $v \in B_3$ be non-adjacent. Since $d(u) \geq 2t - q$ it follows that there exists $v' \in B_4$ adjacent to u and since $d(v) = s$ it follows that there exists $u' \in A_4$ adjacent to v. Now if H is the graph obtained from G by an interchange along (u,v',u',v,u), then (H,P) with the ordering S is a realisation of π, $H[A_1|B_1] = K$ and $H[A_4|B_4] \neq \overline{K}$, a contradiction. This proves that $G[A_1|B_3] = K$.

Since the degree of every vertex in B_3 is s, it follows that $G[A_4|B_3] = \overline{K}$.

Next we show that $G[A_3|B_{11}] = K$. If possible, let $u \in A_3$ and $v \in B_{11}$ be non-adjacent. Since $d(u) = t$, it follows that there exists $v' \in B_4$ adjacent to u and since $d(v) \geq m-p$, it follows that there exists $u' \in A_4$ adjacent to v. Now by an interchange along (u,v',u',v,u), we arrive at a contradiction. Hence $G[A_3|B_{11}] = K$. By similar arguments it can be shown that $G[A_3|B_{42}] = \overline{K}$, $G[A_1|B_{41}] = K$ and $G[A_4|B_{12}] = \overline{K}$.

Summing up, we obtain $G[A_1|B-B_{42}] = K$, $G[A_4|B-B_{11}] = \overline{K}$, $G[A-A_4|B_{11}] = K$ and $G[A-A_1|B_{42}] = \overline{K}$. From this (b) and (c) follow immediately.

We now prove that (d) holds. If $p = \frac{m}{2}$, then we are done. So let $p < \frac{m}{2}$. If $t - h \leq 2$ then (d) holds. So let $t - h \geq 3$ and

$$\pi^+ = ((t-h)^{m-2p}|e_{h+1} - p,\ldots,e_{2t-h} - p).$$

Clearly $\pi^+ = \pi(G[A_2 \cup A_3|B-B_{11}-B_{42}])$. We now prove that π^+ is forcibly bipsc. If π does not satisfy C2, then this claim follows by Lemma 3. So let π satisfy C2. Then $m = n$ and $d_i = e_i$ for all i, so $p = q$. Since $G[A_1|B-B_{42}] = K$, it follows that $d_p \geq n - q = m - p$, hence $e_q \geq m - p$ and $B_{11} = B_1$. Thus $\pi^+ = \pi(G[A_2 \cup A_3|B_2 \cup B_3])$. Let (G^+,Q) with the ordering $S^+ = (u_{p+1},\ldots,u_{m-p}|v_{q+1},\ldots,v_{2t-q})$ be any realisation of π^+. Let (H,P) be the graph obtained from (G,P) by replacing $G[A_2 \cup A_3|B_2 \cup B_3]$ by (G^+,Q). Then (H,P) with the ordering S is a realisation of π and hence (H,P) is bipsc. Hence, as shown earlier, there is a $\sigma \in \underline{\underline{C}}((H,P))$ such that $\sigma(A_2 \cup A_3 \cup B_2 \cup B_3) = A_2 \cup A_3 \cup B_2 \cup B_3$. It now follows that (G^+,Q) is bipsc with the restriction of σ to $A_2 \cup A_3 \cup B_2 \cup B_3$ as a bipcp. Thus π^+ is forcibly bipsc. Since $t - h \geq 3$, it now follows from Cases 1 and 2 (applied to π^+) that π^+ is one of $\pi_1 - \pi_8$ with t replaced by $t-h$ and k replaced by zero. This proves that (d) holds.

Finally, to prove that (e) holds, let

$$\pi^* = (d_1-n+h,\ldots,d_p-n+h|e_{n-h+1},\ldots,e_n).$$

Note that $\pi^* = \pi(G[A_1|B_{42}]) = \pi(\overline{G}[A_4|B_{11}])$. Since π does not satisfy condition (1) of Theorem 1, it follows that either π does not satisfy C2 or, π satisfies C2 and $\pi' = (d_1+2t-1,\ldots,d_{2t}+2t-1,\ e_1,\ldots,e_{2t})$ is not forcibly self-complementary. We accordingly consider two cases.

<u>Case 3(a).</u> π does not satisfy C2. Let (G_1,P_1) and (G_2,P_2) be two realisations of π^*. Let (H,P) be the graph obtained from (G,P) by replacing $G[A_1|B_{42}]$ by (G_1,P_1) and $G[A_4|B_{11}]$ by $(\overline{G}_2(P_2),P_2)$. Then (H,P) is a realisation of π and so is bipsc. But π does not satisfy C2, so $\underline{\underline{C}}_p((H,P))$ contains an element σ. Clearly $\sigma(A_1) = A_4$ and $\sigma(B_{42}) = B_{11}$. Hence

$$G_1 = H[A_1|B_{42}] \cong \overline{H}[A_4|B_{11}] = G_2.$$

Thus any two realisations of π^* are isomorphic, hence by Lemma 11, it follows that π^* is unigraphic. Thus (e) holds in this case.

<u>Case 3(b).</u> π satisfies C2 and $\pi' = (d_1+2t-1,\ldots,d_{2t}+2t-1,\ e_1,\ldots,e_{2t})$ is not forcibly self-complementary. We now prove that (e) holds by assuming that π^* is not unigraphic and obtaining a contradiction.

We first show that if (G^*,P^*) is a realisation of π^* then there is a $\sigma^* \in \underline{\underline{C}}((G^*,P^*))$ such that $\sigma^*(A^*) = B^*$ where A^* and B^* are the sets of P^*. Since π^* is not unigraphic, by Lemma 11, there exists another realisation (H^*,P^*) of π^* such that $G^* \ncong H^*$. Now let (H,P) be the graph obtained from (G,P) by replacing $G[A_1|B_{42}]$ by (G^*,P^*) and $G[A_4|B_{11}]$ by $(\overline{H}^*(P^*),\ P^*)$. Clearly (H,P) is a realisation of π and so is bipsc. If now $\underline{\underline{C}}_p((H,P))$ contains an element σ then $\sigma(A_1) = A_4$ and $\sigma(B_{42}) = B_{11}$, hence $G^* \simeq H^*$, a contradiction. Thus $\underline{\underline{C}}_p((H,P)) = \phi$. Hence by

Propositions 1 and 2, it follows that $\underset{=}{C}((H,P))$ contains an element σ such that $\sigma(A) = B$. Now since π satisfies C2 it follows that $m = n$, $d_i = e_i$ for all i, and so as before we have $B_{11} = B_1$ and $B_{42} = B_4$. Since $m = n$, it also follows that $\sigma(A_1) = B_4$ and $\sigma(B_4) = A_1$. Now the restriction of σ to $A_1 \cup B_4$ serves as the required σ^*.

Let now G' be any realisation of π'. Let u_i (resp. v_i) be the vertex with degree $d_i + 2t - 1$ (resp. e_i), $i = 1,\ldots,2t$. Also define $A_1,\ldots,A_4$, $B_1,\ldots,B_4$ as before and let $A = \bigcup_{i=1}^{4} A_i$, $B = \bigcup_{j=1}^{4} B_j$. If now $G'[A] \neq K$ or $G'[B] \neq \overline{K}$, then

$$\sum_{i=1}^{2t} (d_i+2t-1) < 2t(2t-1) + \sum_{i=1}^{2t} e_i,$$

a contradiction since $d_i = e_i$ for all i. Thus $G'[A] = K$ and $G'[B] = \overline{K}$. Let $P = \{A,B\}$ and (G_1,P) the bipartitioned graph obtained from G' by deleting the edges in A. Then (G_1,P) is a realisation of π, and it follows from (b) and (c) that

$$G_1[A_1|B-B_4] = K, \quad G_1[A-A_1|B_4] = \overline{K}.$$

$$G_1[A-A_4|B_1] = K, \quad G_1[A_4|B-B_1] = \overline{K}.$$

Hence $\pi(G_1[A_1|B_4]) = \pi^* = \pi(\overline{G}_1[A_4|B_1])$. So by the result proved in the preceding paragraph, there exist $\sigma_1 \in \underset{=}{C}(G_1[A_1|B_4])$ such that $\sigma_1(A_1) = B_4$ and $\sigma_2 \in \underset{=}{C}(\overline{G}_1[A_4|B_1])$ such that $\sigma_2(A_4) = B_1$. Now consider the permutation σ of $A \cup B$ defined by

$$\sigma(x) = \begin{cases} \sigma_1(x) & \text{if } x \in A_1 \cup B_4, \\ \sigma_2^{-1}(x) & \text{if } x \in A_4 \cup B_1, \\ v_{2t+1-i} & \text{if } x = u_i \in A_2 \cup A_3, \\ u_j & \text{if } x = v_j \in B_2 \cup B_3. \end{cases}$$

It is easy to see that σ is an isomorphism between G' and $\overline{G}'$. Hence G' is self-complementary and π' is forcibly self-complementary. This contradiction proves that (e) holds in this case.

Thus in Case 3, π satisfies (4) and the proof of necessity in Theorem 1 is complete.

4. PROOF OF SUFFICIENCY

In this section we establish the sufficiency in Theorem 1. So let $\pi = (d_1,\ldots,d_m|e_1,\ldots,e_n)$ be a bipartitioned sequence satisfying $\sum_{i=1}^{m} d_i = \sum_{j=1}^{n} e_j$ and at least one of conditions (1) - (4). We will prove that π is forcibly bipsc. We divide this proof into four cases.

Case 1. π satisfies (1). As π satisfies C2, we have $m = n = 2t$. We first prove that π is graphic and $d_{2t} > 0$. Since π' is forcibly self-complementary, it is also graphic. Let G' be a realisation of π'. Let $A = \{u_1,\ldots,u_{2t}\}$ and $B = \{v_1,\ldots,v_{2t}\}$, where u_i has degree $d_i + 2t - 1$ and v_i has degree e_i in G', $1 \leq i \leq 2t$. Then clearly

$$\sum_{i=1}^{2t} d_{G'}(u_i) = 2t(2t-1) + \sum_{i=1}^{2t} e_i.$$

Hence it follows that $G'[A] = K$ and $G'[B] = \overline{K}$. Consider the bipartitioned graph (G,P) where G is the graph obtained from G' by deleting all edges within A and the sets of P and A and B. Then (G,P) with the ordering $S = (u_1,\ldots,u_{2t}|v_1,\ldots,v_{2t})$ is a realisation of π and so π is graphic. If now $d_{2t} = 0$, then by C2, $e_1 = 2t$ and so π is not graphic, a contradiction. Hence $d_{2t} > 0$.

We next prove that any realisation of π is bipsc. Let (G,P) with the ordering $S = (u_1,\ldots,u_{2t}|v_1,\ldots,v_{2t})$ be any realisation of π. Let G' be the graph obtained from G by joining every pair of distinct vertices in A by an edge. Then G' is a realisation of π'. Since π' is forcibly self-complementary, it follows that G' is self-complementary. Let σ be a complementing permutation of G'. Now if $\sigma(u_i) = u_j$, then since $d_{2t} > 0$, it follows that

$$4t \leq d_i + d_j + 4t - 2 = d_{G'}(u_i) + d_{G'}(u_j) = 4t-1,$$

a contradiction. Hence $\sigma(A) = B$ and $\sigma(B) = A$. It now follows that σ is also an isomorphism between G and $\overline{G}(P)$. Thus (G,P) is bipsc and π is forcibly bipsc.

Case 2. π satisfies (2). By Lemma 4, it follows that π is graphic. Without loss of generality we assume that $s \leq t$. If then follows that π is one of $(t^2|1^{2t})$, $(t^4|2^{2t})$, $(3^6|3^6)$ and $(4^6|3^8)$.

If $\pi = (t^2|1^{2t})$ and (G,P) with the ordering $S = (u_1,u_2|v_1,\ldots,v_{2t})$ is a realisation of π, then without loss of generality one can take

$$E(G) = \{u_1v_j,\ u_2v_{t+j} | 1 \leq j \leq t\}.$$

Clearly (G,P) is bipsc and $\sigma = (u_1u_2) \prod_{j=1}^{2t} (v_j) \in \underline{\underline{C}}_p((G,P))$.

If $\pi = (t^4|2^{2t})$, then it follows by Lemma 6 that every realisation (G,P) of π is bipsc with $\underline{\underline{C}}_p((G,P)) \neq \phi$.

If $\pi = (3^6|3^6)$, then one can verify that π has exactly six non-isomorphic realisations (G,P) and each of these is bipsc with $\underline{\underline{C}}_p((G,P)) \neq \phi$. For a complete list of these realisations and a complementing permutation for each of them the reader is referred to [3].

If $\pi = (4^6|3^8)$, then one can verify that π has exactly twenty non-isomorphic realisations (G,P), and each of these is bipsc with $\underline{\underline{C}}_p(G,P)) \neq \phi$. For a complete list of these realisations and a complementing permutation for each of them the reader is referred to [3].

This proves that π is forcibly bipsc in this case.

<u>Case 3.</u> π satisfies (3). Then $\pi = (t^m|m^k, e_{k+1},\ldots,e_{2t-k}, 0^k)$. By Lemma 4 (applied to π with $d_1,\ldots,d_m$ and $e_1,\ldots,e_{2t}$ interchanged) we get that π is graphic. Let (G,P) with the ordering $S = (u_1,\ldots,u_m|v_1,\ldots,v_{2t})$ be any realisation of π. Then clearly u_iv_j is an edge whenever $1 \leq i \leq m$ and $1 \leq j \leq k$. If now $t - k = 0$ then

$$\sigma = \prod_{i=1}^{m} (u_i) \prod_{j=1}^{m} (v_j\ v_{2t+1-j}) \in \underline{\underline{C}}_p((G,P))$$

and π is forcibly bipsc. So let $t - k \geq 1$. Then let $(G^o,P^o) = G[u_1,\ldots,u_m|v_{k+1},\ldots,v_{2t-k}]$. If we now prove that (G^o,P^o) is bipsc with $\sigma^o \in \underline{\underline{C}}_p((G^o,P^o))$, then

$$\sigma = \sigma^o \prod_{j=1}^{k} (v_j\ v_{2t+1-j}) \in \underline{\underline{C}}_p((G,P))$$

and it follows that π is forcibly bipsc. Thus it remains to prove that $\underline{\underline{C}}_p((G^o,P^o)) \neq \phi$.

First let $t - k = 1$. Then without loss of generality we may take $E(G^o) = \{u_i v_t | 1 < i < e_t\} \cup \{u_i v_{t+1} | e_t + 1 \leq i \leq m\}$. Clearly then

$$\sigma^o = \prod_{i=1}^{m} (u_i)\ (v_t\ v_{t+1}) \in \underset{=}{C}_p((G^o,P^o)).$$

Next let $t - k = 2$. Then $\pi((G^o,P^o) = (2^m | e_{t-1}, e_t, m-e_t, m-e_{t-1})$. Hence by Lemma 5, $\underset{=}{C}_p((G^o,P^o)) \neq \phi$.

Finally let $t - k \geq 3$. Then $\pi((G^o,P^o)) = \pi^o$ and by (3), π^o is one of $\pi_1 - \pi_8$. If π^o is one of $\pi_1 - \pi_5$, then as proved in Case 2, $\underset{=}{C}_p((G^o,P^o)) \neq \phi$.

Let now $\pi^o = \pi_6 = ((t-k)^m | (m-1)^{t-k}, 1^{t-k})$. Let $B_1 = \{v_{k+1},\ldots,v_t\}$ and $B_2 = \{v_{t+1},\ldots,v_{2t-k}\}$. For $1 \leq i \leq m$, let B_{1i} be the set of all vertices of B_1 not adjacent to u_i and B_{2i} the set of all vertices of B_2 adjacent to u_i. Then since $|B_1| = |B_2| = t - k$ and $d_{G^o}(u_i) = t - k$, it follows that $|B_{1i}| = |B_{2i}|$. Also since the degree of every vertex of B_1 is $m - 1$ in G^o, it follows that B_{1i} and B_{1h} are disjoint if $i \neq h$. Similarly, B_{2i} and B_{2h} are disjoint if $i \neq h$. Further $B_1 = \bigcup_{i=1}^{m} B_{1i}$ and $B_2 = \bigcup_{i=1}^{m} B_{2i}$. Now if σ^o is any permutation such that $\sigma^o(u_i) = u_i$, $\sigma^o(B_{1i}) = B_{2i}$ and $\sigma^o(B_{2i}) = B_{1i}$ for $i = 1,\ldots,m$, then $\sigma^o \in \underset{=}{C}_p((G^o,P^o))$.

Next let $\pi^o = \pi_7 = ((t-k)^4 | 3, 2^{2(t-k-1)}, 1)$. Then by Lemma 7, $\underset{=}{C}_p((G^o,P^o)) \neq \phi$.

Finally let $\pi^o = \pi_8 = (3^{2s} | 2s-1, s^4, 1)$. Let $(H,Q) = G^o[u_1,\ldots,u_{2s} | v_{k+2},\ldots,v_{2t-k-1}]$. Note that $t - k = 3$ and so $v_{k+2} = v_{t-1}$ and $v_{2t-k-1} = v_{t+2}$.

First let v_{t-2} and v_{t+3} have disjoint neighbourhoods in G^o. Then $\pi((H,Q)) = (2^{2s} | s^4)$. Now let A_{ij} be the set of all vertices adjacent to both v_{t-2+i} and v_{t-2+j} in H and $n_{ij} = |A_{ij}|$, $1 \leq i \neq j \leq 4$. Without loss of generality we also assume that the vertex adjacent to v_{t+3} in G^o belongs to A_{34}. Now,

$$\sum_{i \neq j} n_{ij} = d_H(v_{t-2+j}) = s,\ j = 1,\ldots,4. \qquad \ldots \quad (2)$$

Summing (2) over all j and using the fact that $n_{ij} = n_{ji}$, we get

$$\sum_{j=1}^{4} \sum_{i<j} n_{ij} = 2s. \qquad \ldots \quad (3)$$

Subtracting the equations (2) corresponding to $j = 1$ and $j = 4$ from the equation (3) we get $n_{14} = n_{23}$. Now any permutation σ^o of $V(G^o)$ satisfying

$\sigma^{o}(A_{14}) = A_{23}$,

$\sigma^{o}(u) \quad = u$ if $u \in A_{12} \cup A_{13} \cup A_{24} \cup A_{34}$,

$\sigma^{o}(v_j) \quad = v_{2t+1-j}$, $t - 2 \leq j \leq t + 3$,

is an element of $\underset{=}{C}_p((G^o,P^o))$.

Next let some u_i be adjacent to both v_{t-2} and v_{t+3} in G^o. Without loss of generality we assume that in G^o, u_1 is not adjacent to v_{t-2} and u_{2s} is adjacent to v_{t+3}. Then $\pi((H,Q)) = (3, 2^{2s-2}, 1|s^4)$. Now by Lemma 7, $\underset{=}{C}_p((H,Q))$ contains an element σ such that $\sigma(u_1) = u_{2s}$ and $\sigma(u_{2s}) = u_1$. It now follows that $\sigma(v_{t-2}\ v_{t+3}) \in \underset{=}{C}_p((G^o,P^o))$.

Thus we have shown that $\underset{=}{C}_p((G^o,P^o)) \neq \phi$ if $t - k \geq 1$. As explained before, this proves that π is forcibly bipsc in Case 3.

<u>Case 4</u>. π satisfies (4). Then $n = 2t$. Let $A_1 = \{u_1,\ldots,u_p\}$, $A_2 = \{u_{m-p+1},\ldots,u_m\}$, $B_1 = \{v_1,\ldots,v_h\}$ and $B_2 = \{v_{2t-h+1},\ldots,v_{2t}\}$. We then prove the following

<u>Claim</u> : (G,P) with the ordering $S = (u_1,\ldots,u_m|v_1,\ldots,v_{2t})$ is a realisation of π iff

(i) $G[A_1|B-B_2] = K$, $G[A-A_1|B_2] = \overline{K}$,

(ii) $G[A-A_2|B_1] = K$, $G[A_2|B-B_1] = \overline{K}$,

(iii) $G[A_1|B_2]$ with the ordering $(u_1,\ldots,u_p|v_{2t-h+1},\ldots,v_{2t})$ as well as $\overline{G}[A_2|B_1]$ with the ordering $(u_m,\ldots,u_{m-p+1}|v_h,\ldots,v_1)$ is a realisation of π^*,

(iv) $G[A-A_1-A_2|B-B_1-B_2]$ with the ordering $(u_{p+1},\ldots,u_{m-p}|v_{h+1},\ldots,v_{2t-h})$ is a realisation of π^+.

The 'if part' of the claim is trivial. To prove the 'only if part', let (G,P) with the ordering $S = (u_1,\ldots,u_m|v_1,\ldots,v_{2t})$ be a realisation of π. Then (i) follows by (b), (ii) follows by (c), and (iii) and (iv) follow from (i) and (ii). This proves the claim.

Now a graph (G,P) satisfying (i) - (iv) above exists since by (e), π^* is graphic and by Lemma 4, π^+ is graphic. By the claim proved above, such a graph is a realisation of π and so π is graphic.

Next let (G,P) with the ordering $S = (u_1,\ldots,u_m|v_1,\ldots,v_{2t})$ be a realisation of π. Then (i) - (iv) of the above claim hold. Also since π^* is unigraphic it follows from (iii) that there exists an isomorphism σ^* from $G[A_1|B_2]$ to $\overline{G}[A_2|B_1]$ such that $\sigma^*(A_1) = A_2$ and $\sigma^*(B_2) = B_1$.

If now $p = \frac{m}{2}$ then no d_i is $\frac{n}{2}$, hence by assumption (II), no e_j is $\frac{m}{2}$, so $q = \frac{n}{2} = h$. Thus $A = A_1 \cup A_2$ and $B = B_1 \cup B_2$. It is easy to see that the permutation σ defined by

$$\sigma = \begin{cases} \sigma^* & \text{on } A_1 \cup B_2 \\ \sigma^{*-1} & \text{on } A_2 \cup B_1 \end{cases}$$

is an element of $\underline{\underline{C}}_p((G,P))$.

Next let $p < \frac{m}{2}$. If now $t - h = 0$ then $B = B_1 \cup B_2$ and the permutation σ defined by

$$\sigma(x) = \begin{cases} \sigma^*(x) & \text{if } x \in A_1 \cup B_2 \\ \sigma^{*-1}(x) & \text{if } x \in A_2 \cup B_1 \\ x & \text{if } x \in A - A_1 - A_2 \end{cases}$$

is an element of $\underline{\underline{C}}_p((G,P))$. So let $p < \frac{m}{2}$ and $t - h > 0$. Then by (d), $\pi(G[A - A_1 - A_2|B - B_1 - B_2]) = ((t-h)^{m-2p}|e_{h+1}-p,\ldots,e_{2t-h}-p)$ satisfies condition (3) of Theorem 1 with t replaced by $t - h$ and k replaced by 0. Hence by Case 3, it follows that $\underline{\underline{C}}_p(G[A-A_1-A_2|B-B_1-B_2])$ contains an element σ^+. Now the permutation σ defined by

$$\sigma = \begin{cases} \sigma^* & \text{on } A_1 \cup B_2 \\ \sigma^{*-1} & \text{on } A_2 \cup B_1 \\ \sigma^+ & \text{on } (A-A_1-A_2) \cup (B-B_1-B_2) \end{cases}$$

is an element of $\underline{\underline{C}}_p((G,P))$.

Thus π is forcibly bipsc in Case 4, and sufficiency in Theorem 1 is established.

ACKNOWLEDGEMENTS

The author wishes to express his gratitude to Dr. S.B. Rao who personally scrutinized each step in this paper in its genesis. He is also grateful to Dr. A.R. Rao and Dr. S.B. Rao for their critical comments and helpful suggestions during the presentation of this paper. He is also obliged to the referee for his comments and instructions.

REFERENCES

1. J.A. Bondy and U.S.R. Murty, Graph Theory with Applications, Macmillan, 1976.

2. T. Gangopadhyay, Characterization of potentially bipartite self-complementary bipartitioned sequences, Tech. Report No.4/79, ISI, Calcutta, submitted for publication to Discrete Math.

3. T. Gangopadhyay, Studies in multipartite self-complementary graphs, Ph.D. Thesis, submitted to ISI, May 1980.

4. M. Koren, Pairs of sequences with a unique realisation by bipartite graphs, J. Combinatorial Theory (B), 21(1976), 224-234.

5. S.J. Quinn, Factorization of complete bipartite graph into two isomorphic subgraphs, Lecture notes in Mathematics, 748, Springer Verlag, Berlin. Combinatorial Mathematics VI:Proc. Armidale, Australia 1978, Edited by A.F. Horadam and W.D. Wallis, 98-111.

6. S.B. Rao, Characterisation of forcibly self-complementary degree sequences, submitted for publication to Discrete Math.

A GRAPH THEORETICAL RECURRENCE FORMULA FOR COMPUTING THE CHARACTERISTIC POLYNOMIAL OF A MATRIX*

M. K. GILL**
Department of Mathematics
Indian Institute of Technology
Powai, Bombay 400 076

ABSTRACT

In this paper, a recurrence formula for computing the characteristic polynomial of a graph due to A.J. Schwenk is generalised to arbitrary networks, and some useful reductions of this formula are cited.

RESULTS

The purpose of this note is to give a graph theoretical recurrence formula for computing the characteristic polynomial $\Phi(M)$ of a square matrix M.

For terminology in digraph theory and graph theory, we refer the reader to [6,7] respectively.

Given a square matrix $M = (m_{ij})$ of reals of order n, its characteristic polynomial $\Phi(M)$ is defined by

$$\Phi(M) = \det(\lambda I - M) = \sum_{i=0}^{n} a_i(M)\, \lambda^{n-i} \qquad \ldots \quad (1)$$

where $\det(A)$ denotes the determinant of the matrix A, λ is a complex variable, I is a unit matrix of order n, and $a_i(M)$ is the coefficient of λ^{n-i}.

For directed network D_M associated with M and for the adjacency matrix $A(D)$ associated with the digraph D we refer to [2,7] respectively.

A linear subgraph D' of the directed network D is a subgraph in which every node has exactly one incoming arc and exactly one outgoing arc [4,5,7,8]. The set of linear subgraphs of order k in D will be denoted by $\Psi_k(D)$, where $\Psi_o(D)$ consists

---

* The work was done when the author was at Mehta Research Institute, Allahabad.

** Research supported by the Council of Scientific and Industrial Research, New Delhi, and partially by Government of India Research Project No. HCS/DST/409/76.

of the empty digraph K_o, i.e., the digraph having no nodes.

For any square matrix M of order n, the coefficient $a_i(M)$ of λ^{n-i}, $i = 0,1,2,\ldots,n$, in (1) is known to have the graph theoretical expression

$$a_i(M) = \sum_{L\varepsilon\Psi_i(D_M)} (-1)^{c(L)} w(L) \qquad \ldots \quad (2)$$

where $c(L)$ denotes the number of components in L, $w(L)$ is the product of the weights of the arcs belonging to L and $a_o(M) = 1$,[1].

When M is a symmetric matrix, (2) reduces to

$$a_i(M) = \sum_{L\varepsilon\Psi_i(G_M)} (-1)^{m(L)} 2^{c(L)} w(L),\ i = 0,1,2,\ldots,n \qquad \ldots \quad (3)$$

where G_M is the weighted graph associated with M, $m(L)$ is the number of components in L, $c(L)$ is the number of cycles having at least three nodes and $w(L)$ is as defined in [9]. Further, if M is the adjacency matrix $A(G)$ of a graph G of order n, then writing $a_i(A(G))$ as $a_i(G)$, the expression in (3) reduces to

$$a_i(G) = \sum_{L\varepsilon\Psi_i(G)} (-1)^{m(L)} 2^{c(L)},\quad i = 0,1,2,\ldots,n. \qquad \ldots \quad (4)$$

In [10], the following recurrence relation for computing $\Phi(A(G)) = \Phi(G)$ for a graph G is given :

$$\Phi(G) = \lambda\Phi(G-x) - \sum_{y \text{ adj } x} \Phi(G-x-y) - 2 \sum_{Z\varepsilon C(x)} \Phi(G-V(Z)), \qquad \ldots \quad (5)$$

where $G-A$, A being a set of points, has usual meaning and $C(x)$ is the set of cycles of G containing x. In [3], the relation (5) was further generalized to <u>sigraphs</u> (signed graphs which are undirected networks in which weight of any line is either 1 or -1) as follows : If S is a sigraph, then

$$\Phi(S) = \lambda\Phi(S-x) - \sum_{y \text{ adj } x} \Phi(S-x-y) - 2 \sum_{Z\varepsilon C^+(x)} \Phi(S-V(Z)) + 2 \sum_{Z\varepsilon C^-(x)} \Phi(S-V(Z)) \qquad \ldots \quad (6)$$

where $C^+(x)(C^-(x))$ is the set of positive (negative) cycles in S that contain x. This was achieved by first extending (4) to sigraphs S of order n as follows :

$$a_i(S) = \sum_{L\varepsilon\Psi_i(S)} (-1)^{m(L)} 2^{c(L)} \prod_{Z\varepsilon C(L)} s(Z),\quad i = 0,1,2,\ldots,n, \qquad \ldots \quad (7)$$

where $C(L)$ is the set of cycles of L and $s(Z) = 1$ or -1 according to whether

Z contains an even or odd number of negative lines (accordingly Z is then referred to as being positive or negative as in the definitions of $C^+(x)$ and $C^-(x)$ above).

Now, we are ready to state the main result of this paper.

<u>Theorem</u>. For any square matrix $M = (m_{ij})$ of reals of order n, the characteristic polynomial $\Phi(M)$ of M, written in the form (1), is given by the recurrence formula

$$\Phi(D_M) = \lambda\Phi(D_M - x) - \sum_{Z\varepsilon C^+(x)} w(Z)\,\Phi(D_M - V(Z)) + \sum_{Z\varepsilon C^-(x)} |w(Z)|\,\Phi(D_M - V(Z)) \quad \ldots \quad (8)$$

where x is an arbitrarily chosen node of the network D_M, and $|r|$ is the absolute value of r.

<u>Proof</u>. We present a one-to-one correspondence between those linear subgraphs of D_M contributing to $a_i(D_M)$, the coefficient of λ^{n-i} in $\Phi(D_M)$, on the left hand side (LHS) of (8) and those contributing to one of the terms on the right hand side (RHS) of (8). For any linear subgraph L of D_M, we have two possibilities with respect to the given node x :

<u>Case I</u>. If $x \notin L$, Let L' be the same linear subgraph, only now viewed as a linear subgraph of $D_M - x$.

<u>Case II</u>. If $x \varepsilon L$, let $L' = L - V(Z)$ viewed as a linear subgraph of $D_M - V(Z)$. Since L is a collection of mutually disjoint cycles, we have two subcases:

<u>Subcase (i)</u>. $x \varepsilon Z$ for some $Z \varepsilon C^+(L)$, the set of positive cycles of L.

<u>Subcase (ii)</u>. $x \varepsilon Z$ for some $Z \varepsilon C^-(L)$, the set of negative cycles of L.

Clearly this establishes a one-to-one correspondence between the terms on the LHS and those on the RHS of (8). Moreover, if L contributes an amount t to the coefficient of λ^{n-i} in $\Phi(D_M)$ on the LHS of (8), we show that L' also contributes t on the RHS of (8) in each case. Here, note that by a positive (negative) subgraph D' of a network D with weight function w we mean that $w(D') > 0$ $(w(D') < 0)$, where $w(D)$ is the weight of the subgraph D which is defined as the product of the weights of the lines belonging to D.

<u>Case I</u> : In this case, $L' = L$, so L' contributes an amount

$$(-1)^{c(L')}\, w(L') = (-1)^{c(L)}\, w(L) = t$$

to the coefficient $a_i(D_M-x)$ of $\lambda^{(n-1)-i}$ in $\Phi(D_M-x)$ or an amount t to the coefficient of $\lambda^{(n-1)-(i-1)} = \lambda^{n-i}$ in $\lambda\Phi(D_M-x)$ on the RHS of (8).

Case II :

Subcase (i) : In this case, $L' = L - V(Z)$ and $Z \in C^+(x)$, so L' contributes an amount

$$(-1)^{c(L')} w(L') = (-1)^{c(L)-1} w(L)/w(Z) = -t/w(Z)$$

to the coefficient $a_{i-v(Z)}(D_M-V(Z))$ of $\lambda^{(n-v(Z))-(i-v(Z))} = \lambda^{n-i}$ in $\Phi(D_M-V(Z))$, where $v(Z)$ is the number of nodes in Z. Hence L' contributes an amount t to the coefficient of λ^{n-i} in $-w(Z)\,\Phi(D_M-V(Z))$.

Subcase (ii) : In this case, $L' = L - V(Z)$ and $Z \in C^-(x)$, so L' contributes an amount

$$(-1)^{c(L')} w(L') = -(-1)^{c(L)-1} w(L)/|w(Z)|$$

to the coefficient of λ^{n-i} in $\Phi(D_M-V(Z))$. Hence L' contributes an amount t to the coefficient of λ^{n-i} in $|w(Z)|\Phi(D_M-V(Z))$.

This completes the proof of the theorem.

The formula (8) may be written in the following form :

$$\Phi(D_M) = \lambda\Phi(D_M-x) - \sum_{Z\in C(x)} w(Z)\,\Phi(D_M-V(Z)) \qquad \ldots (9)$$

where $C(x)$ is the set of cycles of D_M containing x. But we prefer the form of $\Phi(D_M)$ given in (8) due to its computational advantage. We can minimise computation by choosing x so that either $C^+(x) = \phi$ or $C^-(x) = \phi$. If no such x exists then we may choose x such that $|C(x)|$ is a minimum.

The author acknowledges financial assistance provided by Department of Atomic Energy, Bombay at the final stage.

REFERENCES

1. B.D. Acharya, A graph theoretical expression for the characteristic polynomial of a matrix, Proc. Nat. Acad. Sci. (India), 49 (Sec.A) (1979).

2. W.K. Chen, Applied Graph Theory : Graphs and Electrical Networks, North-Holland, Amsterdam, 1975.

3. M.K. Gill and B.D. Acharya, A recurrence formula for computing the characteristic polynomial of a sigraph, J. Comb. Infor. Sys. Sci., 5(1) (1980), 1-5.

4. F. Harary, A graph theoretical method for complete reduction of a matrix with a view toward finding its eigenvalues, J. Math. Physics, 38(1959), 104-111.

5. F. Harary, The determinant of the adjacency matrix of a graph, SIAM Rev., 4(3)(1962), 202-210.

6. F. Harary, R.Z. Norman, and D. Cartwright, Structural models : An introduction to the theory of directed graphs, Wiley, 1965.

7. F. Harary, Graph Theory, Addison-Wesley, Reading, Mass., 1972.

8. D. König, Theorie der endlichen und unendlichen graphen, Leipzig, 1936 (Reprinted New York, 1950).

9. M.J. Rigby, R.B. Mallion, and A.C. Day, Comment on a graph theoretical description of heteroconjugated molecules, Chemical Physics Letters, 51(1) (1977), 178-182.

10. A.J. Schwenk, Computing the characteristic polynomial of a graph, Springer-Verlag Lecture Notes in Mathematics, Vol.406 (1974), 153-172.

A NOTE CONCERNING ACHARYA'S CONJECTURE ON A SPECTRAL MEASURE OF STRUCTURAL BALANCE IN A SOCIAL SYSTEM*

MUKHTIAR KAUR GILL**
Department of Mathematics
Indian Institute of Technology
Powai, Bombay 400 076

ABSTRACT

In this note, we establish a theorem concerning the common polynomials of the cospectral classes of signed graphs on a given graph in which all the cycles are of the same length and pass through a single point. This theorem is observed to give a doubly infinite class of graphs serving as counterexamples to a recent conjecture on a certain number associated with a cospectral class of unbalanced signed graphs on a given graph.

INTRODUCTION

Generally, we follow Harary [8] for terminology in graph theory. However, we recall some definitions here in order to make this note easily readable.

A <u>signed graph</u> (briefly, a <u>sigraph</u>) is an ordered pair $S = (S^u, s)$ where S^u is a graph, called the <u>underlying graph of</u> S, and $s : E(S^u) \to \{-1,1\}$ is a function, called the <u>weight function</u>, from the line set $E(S^u)$ of S^u into the two-element set $\{-1,1\}$; in other words, it is a graph in which each line is assigned the number 1 or -1 (cf. Chartrand [3]). The lines which are assigned the weight 1 are called <u>positive</u> and those that are assigned -1 are called <u>negative</u>. We denote by $E^+(S)$ and $E^-(S)$ the sets of positive and negative lines of S respectively. We regard every graph as a sigraph in which each line is positive.

Given any sigraph S of order n, its <u>adjacency matrix</u> $A(S) = (a_{ij})$ is an n by n (0,1,-1)-matrix defined by

---

* This work was done when the author was at Mehta Research Institute of Mathematics and Mathematical Physics, Allahabad.

** The research partially supported by Department of Atomic Energy, Bombay.

$$a_{ij} = \begin{cases} 0 & \text{if } v_i v_j \notin E(S) \\ 1 & \text{if } v_i v_j \in E^+(S) \\ -1 & \text{if } v_i v_j \in E^-(S). \end{cases} \qquad \ldots \quad (1)$$

Clearly, A(S) is a symmetric matrix in which all diagonal elements are zero.

The characteristic polynomial of S is defined to be the characteristic polynomial of its adjacency matrix A(S), and is denoted by $\Phi(S)$. We write it in the form,

$$\Phi(S) = \det(\lambda I - A(S)) = \sum_{i=0}^{n} a_i(S)\lambda^{n-i} \qquad \ldots \quad (2)$$

where n is the order of S, λ is a complex variable, I is the $n \times n$ unit matrix, and $a_i(S)$ is the coefficient of λ^{n-i} in the expansion of the determinant $\det(\lambda I-A(S))$ of the matrix $\lambda I - A(S)$. By the spectrum of S we mean the set of roots of $\Phi(S)$ (i.e., the set of eigenvalues of A(S)).

A sigraph is said to be positive if it contains an even number of negative lines; this definition can be extended to the subgraphs of the sigraph S as well, because every subgraph S' of S is a sigraph in its own right with the weight function s of S restricted to E(S'), the line set of S'.

A sigraph S is said to be balanced (cf. Harary [6], Cartwright and Harary [2]) if every cycle of S is positive. A number of criteria for balance in a signed digraph may be found in [7] (Ch.13). These criteria hold for sigraphs as well and may be found in [3] (Ch.8).

Recently, the following surprising spectral characterization of structural balance in a social system was discovered.

Theorem 1 (Acharya [1]). A sigraph S is balanced if and only if $\Phi(S) = \Phi(S^u)$.

Two sigraphs S_1 and S_2 are said to be cospectral if they have the same spectrum. Thus, Theorem 1 states that S is balanced if, and only if, it is cospectral with S^u.

In [1], a new measure β_o of the degree of balance in a 'social system' (i.e., a sigraph), suggested by Theorem 1, was proposed as follows :

$$\beta_o(S) = 1 - \frac{T(S)}{n-2} \qquad \ldots \qquad (3)$$

where $T(S)$ is the number of indices i for which $a_i(S) \neq a_i(S^u)$, and n is the order of sigraph S. Furthermore, the following conjecture was stated :

Conjecture(Acharya [1]). Let S_1 and S_2 be two sigraphs such that $S_1^u = S_2^u$ and $\beta_o(S_1) = \beta_o(S_2)$. Then S_1 and S_2 are cospectral.

In this note, we shall disprove this conjecture. Towards this end it is sufficient to deal with $T(S)$ since n is fixed once the graph is given.

A GENERAL CLASS OF COUNTEREXAMPLES

Now, we give a general class of sigraphs for which the above conjecture fails. In general, let $H = H(k;m)$ denote the graph satisfying the following conditions :

(i) H has exactly k cycles.

(ii) All the cycles of H have the same length m.

(iii) All the cycles of H pass through a single point.

Now choose any r of the k cycles. We may do this in $\binom{k}{r}$ ways. In each selection of r cycles, make all of them negative. Hence there are $\binom{k}{r}$ sigraphs having r negative cycles when the manner in which they have been made negative is disregarded. Since all the cycles of H are of the same length, we prove from the following Lemma 1 that all the sigraphs on H having exactly r negative cycles are cospectral.

Lemma 1 (Gill and Acharya [4]). Let S be a sigraph and x be a point given in S. Then

$$\Phi(S) = \lambda\Phi(S-x) - \sum_{y \text{ adj } x} \Phi(S-x-y) - 2\Big[\sum_{Z\varepsilon C^+(x)} \Phi(S-V(Z)) - \sum_{Z\varepsilon C^-(x)} \Phi(S-V(Z))\Big] \qquad \ldots \qquad (4)$$

where for a set A of points of S the notation $S - A$ ($S-y$ if $A = \{y\}$) means, as usual, the sigraph obtained from S by deleting all the points of A together with the lines of S having a point in A, $C^+(x)$ and $C^-(x)$ denote respectively the sets of positive and negative cycles in S containing the point x.

Let H_i, $i = 1,2,\ldots,\binom{k}{r}$, be the sigraphs on H having exactly r negative cycles. Using (4) the characteristic polynomial $\Phi(H_i)$, $i = 1,2,\ldots,\binom{k}{r}$, comes out to be

$$\Phi(H_i) = \lambda[\Phi(P_{m-2})]^k - 2k\,\Phi(P_{m-3})\,[\Phi(P_{m-2})]^{k-1} - 2(k-2r)\,[\Phi(P_{m-2})]^{k-1} \qquad \ldots (5)$$

where P_t denotes the parth of length t and r is same for each i. Hence all the sigraphs on H having exactly r negative cycles are cospectral.

So, of all the sigraphs on H having exactly r negative cycles it is sufficient to choose just one sigraph to represent the whole class and there are thus $k + 1$ cospectral classes of sigraphs on H. We label them $\alpha_o, \alpha_1, \alpha_2, \ldots, \alpha_k$ such that α_j consists of the sigraphs on H in which exactly j cycles are negative. Now, using Lemma 1 we establish that the graphs $H(k;m)$ are precisely the required counterexamples to the conjecture.

We are now ready to state the main result of this note.

Theorem 2. On the graph $H = H(k;m)$ there are $k + 1$ cospectral classes of signed graphs, say $\alpha_o, \alpha_1, \alpha_2, \ldots, \alpha_k$ where α_j consists of the sigraphs on H in which exactly j cycles are negative. Let Φ_j denote the common characteristic polynomial of the sigraphs in the cospectral class α_j, and for $j \neq 0$ let T_j denote the number of indices i for which $a_i(\alpha_j) \neq a_i(\alpha_o)$, where $a_i(\alpha_j)$ is the coefficient of λ^{n-i} in Φ_j. Then $T_1 = T_2 = T_3 = \ldots = T_k$.

Proof. The first part is already established above. Using (4) the characteristic polynomial Φ_j, for $j = 0,1,2,\ldots,k$, comes out to be

$$\Phi_j = \lambda[\Phi(P_{m-2})]^k - 2k[\Phi(P_{m-2})]^{k-1}\{\Phi(P_{m-3}) + 1\} + 4j[\Phi(P_{m-2})]^{k-1} \qquad \ldots (6)$$

where P_t denotes the path of length t. It is clear that the first two terms in (6) remain same for all $j = 0,1,2,\ldots,k$, while for $j \neq 0$ the coefficient of every power of λ in $[\Phi(P_{m-2})]^{k-1}$, whenever it is nonzero, differs from the corresponding coefficient in Φ_o, and obviously this number is fixed for all $j \neq 0$, Thus it follows that $T_1 = T_2 = T_3 = \ldots = T_k$, and the proof is complete.

A COUNTEREXAMPLE TO THE CONJECTURE

Clearly, because of Theorem 1, one cannot hope to produce counterexamples to the above conjecture in the class of balanced sigraphs on a given graph. Therefore, both the sigraphs S_2 and S_3 serving as counterexamples to the conjecture must be unbalanced. In Figure 1, we have a graph $G = S_1$ such that $s_1(e) = 1$ for

$G = S_1$:

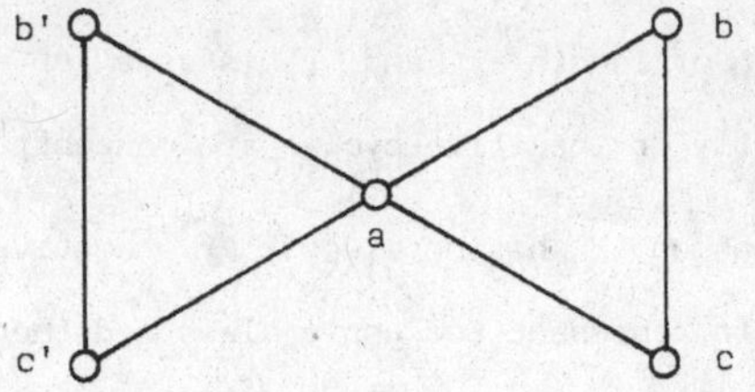

Figure 1

every $e \in E(S_1)$, i.e., the sign of every edge of S_1 is 1. In other words $S_1 \in \alpha_o$. Two sigraphs $S_2 \in \alpha_1$ and $S_3 \in \alpha_2$ on S_1 are as follows

$$s_2(e) = \begin{cases} -1 & \text{if } e = (b,c) \\ 1 & \text{otherwise} \end{cases}$$

and

$$s_3(e) = \begin{cases} -1 & \text{if } e = (b,c) \text{ or } (b',c') \\ 1 & \text{otherwise.} \end{cases}$$

It is easy to check that

$$\begin{aligned} \Phi(S_1) &= \lambda^5 - 6\lambda^3 - 4\lambda^2 + 5\lambda + 4 \\ \Phi(S_2) &= \lambda^5 - 6\lambda^3 + 5\lambda \\ \Phi(S_3) &= \lambda^5 - 6\lambda^3 + 4\lambda^2 + 5\lambda - 4 \end{aligned} \quad \ldots (7)$$

so that $T(S_2) = T(S_3) = 2$ which implies that $\beta_o(S_2) = \beta_o(S_3) = 1 - \frac{2}{3} = \frac{1}{3}$. However, we have $\Phi(S_2) \neq \Phi(S_3)$. Thus the conjecture is false. We note from the catalogue of sigraphs on less than six points (cf. Gill [5]) that this is a pair of sigraphs having smallest order which puts down the above conjecture.

ACKNOWLEDGEMENT

The author expresses her sincere thanks to Dr. B.D. Acharya of Mehta Research Institute of Mathematics and Mathematical Physics, Allahabad for his guidance in preparing this paper as also for the facilities he provided to her under the Department of Science and Technology (Government of India) Research Project No. HCS/DST/409/76 entitled 'Investigations on the Applications of Sigraphs in Behavioural Sciences'.

REFERENCES

1. B.D. Acharya, A spectral criterion for cycle-balance in networks and its consequences, J. Graph Theory, 3(4) (1979).

2. D. Cartwright and F. Harary, Structural balance : A generalization of Heider's theory, Psych. Rev., 63(1956), 277-293.

3. G.T. Chartrand, Graphs as mathematical models, Prindle, Weber and Schmidt(S), Inc., Boston, Mass., 1977.

4. M.K. Mill and B.D. Acharya, A recurrence formula for computing the characteristic polynomial of a sigraph, J. Comb. Infor. Sys. Sci., 5(1) (1980), 1-5.

5. M.K. Gill, The catalogue of sigraphs of orders less than six, their characteristic polynomials and their spectra (under preparation).

6. F. Harary, On the notion of balance of a signed graph, Mich. Math. Journal, 2(1953), 143-146.

7. F. Harary, R.Z. Norman and D. Cartwright, Structural Models : An Introduction to the Theory of Directed Graphs, Wiley, 1965.

8. F. Harary, Graph Theory, Addison Wesley, Reading, Mass., 1972.

ON PERMUTATION-GENERATING STRINGS AND ROSARIES

HANSRAJ GUPTA*
Honorary Professor of Mathematics
Panjab University, Chandigarh

1. INTRODUCTION

In a note, probably still awaiting publication, S.P. Mohanty and Daljit Rao had given examples of permutation-generating strings for $n = 4, 5, 6$ and 7 with $(n-1)^2 + 3$ elements in each string. I give one example of each from their work :

n	String	Type
4	1234 123 1423 1	(4 3 3 2)
5	12345 1234 1523 14235 1	(5 4 4 3 3)
6	1234526 13425 16324 15236 14523 1	(6 6 5 4 4 3)
7	1234567 123456 172345 167234 156723 1456723 1	(7 6 6 5 5 5 5)

This means that all the permutations of the first n natural numbers occur in the string as subsequences. By the type of the string we mean the sequence of frequencies of the elements $1,2,3,\ldots,n$ in the string. Strings of other types also exist with the same number of elements as in the examples. It was believed that the strings with $(n-1)^2 + 3$ elements for $n > 2$, were the shortest.

I felt that if we were to write the elements of a sequence in order on the circumference of a circle (to be called a rosary hereafter), the permutations could be generated by a shorter sequence. Two cases arise :

(i) when we are permitted to move only in the counter-clock-wise direction;

(ii) when we can move in the counter-clock-wise as also in the clock-wise direction.

In locating a permutation in these cases, one can start from any point on the circle, but it is not permissible to turn back or to cross the point of start.

* Address for correspondence : 402 Mumfordganj, Allahabad 211 002.

I conjectured that in case (i), the shortest rosary will have $[n^2/2]$ elements, in case (ii) only $[(3n^2 + 4)/8]$. These two conjectures have been verified for values of n from 2 to 6, but it has not yet been possible to prove them, not even to show that at least one rosary will exist in each case with these many elements.

We give the results below :

Case (i)

n	Rosary	Type
2	12	(1 1)
3	1232	(1 2 1)
4	14234 324	(1 2 2 3)
5	12345 43215 45	(2 2 2 3 3)
6	12345 26154 31251 625	(4 4 2 2 4 2)

Case (ii)

2	12	(1 1)
3	123	(1 1 1)
4	1234 23	(1 2 2 1)
5	12345 2345	(1 2 2 2 2)
6	12345 63126 4365 (S.P. Khare)	(2 2 3 2 2 3)

2. RESULTS

In this section, we give some positive results for rosaries of the second kind mentioned in the preceding section. Writing $(a_1, a_2, \ldots, a_k)_j$ for the juxtaposition of j copies of the sequence in the bracket, we in fact prove that the rosaries given by

$$1(2,3,\ldots,n)_m\ (2,3,\ldots,n-1)_1 \quad \text{for } n \text{ even}; \qquad \ldots \quad (1)$$

and by

$$1(2,3,\ldots,n)_m \quad \text{for } n \text{ odd}; \qquad \ldots \quad (2)$$

with $m = [(n-1)/2]$ in each case and $n > 2$, yield all the permutations of the first n natural numbers. These rosaries have

$$[n/2]\ (n-1) + h \qquad \ldots \quad (3)$$

elements where $h = 0$ or 1, according as n is even or odd.

The rosaries for $n = 6$ and 7 are given in Figure 1.

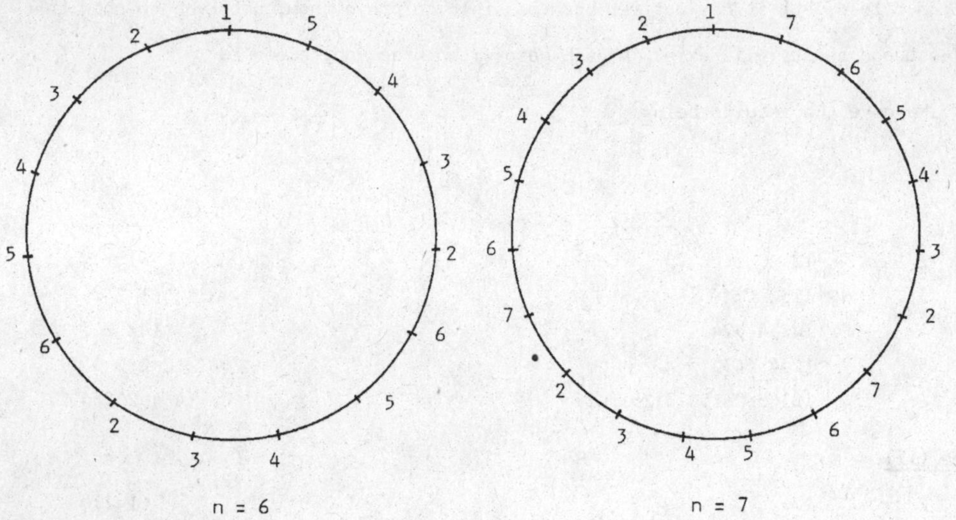

Figure 1

Proof. Let

$$a_1\, a_2 \ldots a_n \qquad \ldots \quad (4)$$

be any permutation of the first n natural numbers, $n > 2$. Since the elements of (1) and (2) are written in order in the counter-clock-wise direction round a circle and we can move from any point in the counter-clock-wise or the clock-wise direction with the restrictions mentioned earlier, we can assume without loss of generality, that in (4), $a_1 = 1$.

Now, in the permutation

$$a_2\, a_3 \ldots a_n \qquad \ldots \quad (5)$$

of $2,3,4,\ldots,n$, put a comma after every a_i for which $a_i > a_{i+1}$. This gives a partition of (5) into a number of sections say r_1. Next, start with the permutation in (5) anew, and put a comma after every a_j for which $a_j < a_{j+1}$. This gives another partition of (5) into a number of sections, r_2 say. Evidently, we have

$$r_1 + r_2 = n. \qquad \ldots \quad (6)$$

If $r_1 < r_2$, the permutation in (4) is located by moving from 1 in the counter-clock-wise direction, otherwise by moving in the clock-wise direction.

Take, for example the permutation

1 3 4 2 7 6 5 9 8

of the first 9 natural numbers. Then, the two partitions of 34276598 are

3 4, 27, 6, 59, 8

and

3, 42, 765, 98.

Here $r_1 = 5$, $r_2 = 4$. It will be readily seen that the permutation 134276598 can be located by moving from 1 in the clock-wise direction.

Here it is not claimed that our rosaries are the shortest, but they are easy to write.

For a string which can be read from left to right or right to left, our permutation will not necessarily have $a_1 = 1$. Using the same type of argument as for rosaries of the second kind, we can show that the permutation-generating string readable both ways is given by

$$(1,2,3,\ldots,n)_t \quad \text{with} \quad t = [(n+1)/2];\ n \geq 1.$$

For $n = 5$, such a string is

12345 12345 12345.

ENUMERATION OF LABELLED DIGRAPHS AND HYPERGRAPHS

MALATI HEGDE
Department of Applied Mathematics
Indian Institute of Science, Bangalore 560 012

M. R. SRIDHARAN
Department of Mathematics
Indian Institute of Technology, Kanpur 208 016

ABSTRACT

We present formulae for the number of labelled k-colored hypergraphs, labelled strongly k-colored hypergraphs, labelled connected hypergraphs, labelled even digraphs and labelled even hypergraphs.

We refer to [1] and [2] for the basic definitions. The degree of a vertex v of a hypergraph is the number of edges E_i of H such that v is in E_i. A hypergraph H is even if every vertex of H has even degree. In an even digraph the sum of indegree and outdegree of every vertex is even.

R.C. Read [4] has counted the number of k-colored graphs. Here we generalize this result by extending Read's method to obtain the formula for the number of labelled k-colored and strongly k-colored hypergraphs.

1. LABELLED k-COLORED HYPERGRAPHS

Let $p_1, p_2, \ldots, p_k$ be positive integers that form an ordered partition of p, so that

$$\sum_{i=1}^{k} p_i = p. \qquad \ldots (1.1)$$

We write $\{p\}$ for an arbitrary solution of (1.1). Let

$$\binom{p}{p_1, p_2, \ldots, p_k} \qquad \ldots (1.2)$$

denote the multinomial coefficient.

<u>Theorem 1</u>. The number $C_p(k)$ of k-colored hypergraphs of order p is

$$C_p(k) = \frac{1}{k!} \sum_{\{p\}} \binom{p}{p_1,\ldots,p_k} 2^{\sum_{m=2}^{p} \binom{p}{m} - \sum_{i=1}^{k} \sum_{m=2}^{p_i} \binom{p_i}{m}} \qquad \ldots \quad (1.3)$$

Proof. Let $C_p(k)$ be the number of k-colored labelled hypergraphs of order p. For a particular assignment of k colors the total number of k-colored labelled hypergraphs is $k!\,C_p(k)$.

Hence we suppose that the k-colors are fixed. Each solution of $\{p\}$ determines a k-part ordered partition of p. We seek the number of labelled hypergraphs with p_i vertices of ith color. The number of ways the labels can be selected for the vertices is the multinomial (1.2).

Now the possible number of subsets that contain at least two non-equivalent vertices, that is, vertices with different colors, is

$$\sum_{m=2}^{p} \binom{p}{m} - \sum_{i=1}^{k} \sum_{m=2}^{p_i} \binom{p_i}{m} \qquad \ldots \quad (1.4)$$

Since each of these subsets may or may not be an edge in forming labelled hypergraphs, the number of choices is

$$2^{\sum_{m=2}^{p} \binom{p}{m} - \sum_{i=1}^{k} \sum_{m=2}^{p_i} \binom{p_i}{m}} \qquad \ldots \quad (1.5)$$

Now from (1.2), (1.4) and (1.5), the formula for number of k-colored labelled hypergraphs of order p is given by (1.3) and the generating function for the number of k-colored labelled hypergraphs of order p is

$$C_p(x) = \frac{1}{k!} \sum_{\{p\}} \binom{p}{p_1,\ldots,p_k} (1+x)^{\sum_{m=2}^{p} \binom{p}{m} - \sum_{i=1}^{k} \sum_{m=2}^{p_i} \binom{p_i}{m}} \qquad \ldots \quad (1.6)$$

2. LABELLED STRONGLY k-COLORED HYPERGRAPHS

Theorem 2. The number $SC_p(k)$ of strongly k-colored labelled hypergraphs of order p is

$$SC_p(k) = \frac{1}{k!} \sum_{\{p\}} \binom{p}{p_1,\ldots,p_k} 2^{\Sigma p_1p_2+\Sigma p_1p_2p_3+\cdots+\Sigma p_1p_2\cdots p_k} \quad \ldots (2.1)$$

where $\Sigma\, p_1p_2\cdots p_m$ means that the summation is taken over all $\binom{k}{m}$ different choices of m sets from k different classes with $\binom{p_i}{2}$ zero if p_i is equal to 1.

Proof. Let $SC_p(k)$ be the number of strongly k-colored labelled hypergraphs of order p. For a particular assignment of k colors, the total number of strongly k-colored hypergraphs is $k!\, SC_p(k)$.

Hence we first consider the case where the k-colors are fixed. Each solution of $\{p\}$ determines a k-part ordered partition of p. We count the number of labelled strongly k-colored hypergraphs with p_i vertices of ith color. The number of ways the labels can be selected for the vertices is the multinomial coefficient (1.2). Now the possible number of subsets such that each edge has no equivalent vertices is

$$\Sigma\, p_1p_2 + \Sigma\, p_1p_2p_3 + \cdots + \Sigma\, p_1p_2p_3\cdots p_k \quad \ldots (2.2)$$

Since each of the subsets in (2.2) may or may not be an edge in forming labelled hypergraphs, we have

$$2^{\Sigma\, p_1p_2 + \Sigma\, p_1p_2p_3 + \cdots + \Sigma\, p_1p_2p_3\cdots p_k} \quad \ldots (2.3)$$

choices. From (2.2) and (2.3), the number of strongly k-colored labelled hypergraphs of order p is given by (2.1) and the generating function for the number of labelled strongly k-colored hypergraphs of order p is

$$SC_p(x) = \frac{1}{k!} \sum_{\{p\}} \binom{p}{p_1,\ldots,p_k} (1+x)^{\Sigma p_1p_2 + \Sigma p_1p_2p_3 + \cdots + \Sigma p_1p_2\cdots p_k} \quad \ldots (2.4)$$

3. LABELLED CONNECTED HYPERGRAPHS

Theorem 3. The number C_p of labelled connected hypergraphs without loops (that is, edges of size 1) is given by

$$C_p = 2^{2^p-p-1} - \frac{1}{p} \sum_{k=1}^{p-1} k\binom{p}{k} C_k\, 2^{2^{p-k}-(p-k)-1} \quad \ldots (3.1)$$

Proof. A rooted hypergraph has one of its vertices, called the root, distinguished from the others. Two rooted hypergraphs are isomorphic if there is a one-to-one mapping from the vertex set of one hypergraph to that of the other which preserves not only adjacency but also the roots. Now we can obtain the recursive formula as below.

Different rooted, labelled hypergraphs are obtained when a labelled hypergraph is rooted at each of its vertices. Hence the number of rooted labelled hypergraphs of order p is pH_p, where H_p is the number of labelled hypergraphs of order p. The number of labelled hypergraphs in which the root is a component of exactly k vertices is

$$kC_k \binom{p}{k} H_{p-k}. \qquad \ldots \quad (3.2)$$

Now summing from $k = 1$ to p

$$pH_p = \sum_{k=1}^{p} k\binom{p}{k} C_k H_{p-k}. \qquad \ldots \quad (3.3)$$

The number of labelled hypergraphs of order p without loops is

$$H_p = 2^{2^p-p-1}.$$

So the equation (3.3) reduces to

$$p.2^{2^p-p-1} = \sum_{k=1}^{p-1} k\binom{p}{k} C_k H_{p-k} + pC_p.$$

This completes the proof of the theorem.

4. LABELLED EVEN DIGRAPHS

Theorem 4. The number of labelled even digraphs of order p is $2^{(p-1)^2}$.

Proof. R.C. Reed [5] has enumerated labelled even graphs. Following Read's method, here we generalize the result.

Let L be the set of all labelled digraphs with p vertices and q edges. Consider any digraph D in L and arbitrarily assign $+1$ and -1 to the vertices. Let the sign of an edge be the product of the signs of the end vertices. The sign of the digraph D, denoted by $\sigma(D)$ is the product of the signs of the edges of D. There are 2^p ways in which the signs can be assigned to the labels of a given digraph. There are $\binom{p(p-1)}{q}$ labelled digraphs with p vertices and q edges.

Let α = sum of the indegrees of the vertices with negative sign and β = sum of the outdegrees of the vertices with positive sign. Then

$$\sigma(D) = (-1)^{\alpha+\beta} = (-1)^{a+b} \qquad \ldots \quad (4.1)$$

where

a = number of edges from positive to negative vertices,

b = number of edges from negative to positive vertices.

Now we can write

$$\sum_{D\varepsilon L} \{\sum_{S} (-1)^{\alpha+\beta}\} = \sum_{S} \{ \sum_{D\varepsilon L} (-1)^{a+b}\} \qquad \ldots \quad (4.2)$$

where S is the set of all possible assignments of -1 and +1 to the vertices, and there are 2^p elements in S.

Let us first consider the left hand side or L.H.S. of (4.2). If D is an even digraph then for all $v \varepsilon V$ net degree of v is even. This means that $\alpha+\beta$ is even if D is an even digraph. Hence

$$\sum_{S} \sigma(D) = 2^p$$

if D is even digraph. If D is not even, at least one vertex v has net degree odd. The sign $\sigma(D)$ is -1 if an odd number of odd vertices have negative sign. The allocations in S for which the label of v is positive and for which it is negative is equinumerous. Hence D contributes nothing L.H.S. of (4.2). Hence L.H.S. of (4.2) is 2^p times the number of even digraphs in L. Now we consider the right hand side of (4.2). Consider an allocation in S for which n vertices are positive and m = p-n are negative. There are $\binom{p}{n}$ such allocations. Consider the contribution to the R.H.S. of (4.2) from those digraphs for which a+b+d = q where d is the number of edges with positive sign. Observe that d edges can be chosen in $\binom{m(m-1)+n(n-1)}{d}$ different ways, a edges can be chosen in $\binom{nm}{a}$ different ways, and b edges can be chosen in $\binom{nm}{b}$ different ways. Thus the contribution to the R.H.S. of (4.2) from those digraphs for which a+b+d = q is

$$\sum_{a+b+d=q} \binom{p}{n} (-1)^{a+b} \binom{nm}{a}\cdot\binom{nm}{b}\binom{m(m-1)+n(n-1)}{d}. \qquad \ldots \quad (4.3)$$

This is the coefficient of x^q in the series

$$\binom{p}{n}(1-x)^{2mn}(1+x)^{m(m-1)+n(n-1)}. \qquad \ldots \quad (4.4)$$

Now summing over $n+m = p$ we obtain a series in which the coefficient of x^q is the R.H.S. of (4.2),

$$\sum_{n+m=p} \binom{p}{n}(1-x)^{2mn}(1+x)^{m(m-1)+n(n-1)} \qquad \ldots \quad (4.5)$$

$$= \sum_{n+m=p} \binom{p}{n}(1-x)^{2mn}(1+x)^{p(p-1)-2nm}. \qquad \ldots \quad (4.6)$$

Therefore the number of labelled even digraphs, denoted by $E(x)$, is given by

$$E(x) = \frac{1}{2^p}(1+x)^{p(p-1)} \sum_{n=0}^{p} \binom{p}{n}\left(\frac{1-x}{1+x}\right)^{2n(p-n)}. \qquad \ldots \quad (4.7)$$

The total number of labelled even digraphs of order p is obtained by putting $x = 1$ in (4.7). Hence, the number of labelled even digraphs of order p is given by $2^{(p-1)^2}$. This completes the proof of the theorem.

The counting series for labelled even digraphs with any number of vertices and edges is given by the formula

$$E(t,x) = \sum_{p=0}^{\infty} E(x)\frac{t^p}{p!}$$

with $p!$ being introduced because the vertices are labelled. The counting series for connected even digraphs is given by the series $\log E(t,x)$, and the number of connected even digraphs on p labelled vertices and k-edges is $p!$ times the coefficient of $t^p x^k$ in this series.

5. LABELLED EVEN HYPERGRAPHS

<u>Theorem 5</u>. The number of labelled even hypergraphs, without loops, of order p is 2^{2^p-2p-1}, $p \geq 3$.

<u>Proof</u>. As in Section 4, the signs are assigned to the vertices. The sign of an edge is the product of the signs of its vertices. The sign of the hypergraph is defined as in the case of digraphs. Now we have the identity

$$\sum_{H\varepsilon B}\{\sum_{S}\sigma(H)\} = \sum_{S}\{\sum_{H\varepsilon B}\sigma(H)\} \qquad \ldots \quad (5.1)$$

where B is the set of all labelled hypergraphs of order p, and

$$\sigma(H) = (-1)^a = (-1)^b \qquad \ldots \qquad (5.2)$$

where

a = sum of the degrees of the negative vertices,

b = number of negative edges.

Consider the L.H.S. of (5.1). If H is an even hypergraph then a is even for all possible allocations of signs to its vertices. Hence

$$\sum_S (-1)^a = 2^p$$

if H is an even hypergraph, and if H is not even then it contributes zero to L.H.S. of (5.1). Thus, L.H.S. of (5.1) is 2^p times the number of even hypergraphs in B. Now consider the R.H.S. of (5.1). Consider an allocation in S for which n vertices are positive and $m = p-n$ are negative. If there are k edges of negative sign they may occur in

$$f(k) = \binom{\sum_{i=0}^{n} \binom{n}{i} ({}^mC_3 + {}^mC_5 + \ldots) + \sum_{i=1}^{n} \binom{n}{i}\binom{m}{1}}{k}$$

different ways. The remaining $q-k$ edges can occur in

$$g(k) = \binom{\sum_{i=2}^{n} \binom{n}{i} + \sum_{i=0}^{n} \binom{n}{i} ({}^mC_2 + {}^mC_4 + \ldots)}{q-k}$$

different ways. Note that by our convention $\binom{n}{i}\binom{n}{j}$ is zero if $i + j > n$. Now summing from $k = 0$ to q we obtain

$$\sum_{k=0}^{q} (-1)^k f(k)\, g(k) \qquad \ldots \qquad (5.3)$$

as the contribution to R.H.S. of (5.1) for each allocation with a given n and m. This is the coefficient of x^q in

$$(1-x)^{\sum_{i=0}^{n} \binom{n}{i} ({}^mC_3+{}^mC_5+\ldots) + \sum_{i=1}^{n} \binom{n}{i}\binom{m}{1}} \; (1+x)^{\sum_{i=2}^{n} \binom{n}{i} + \sum_{i=0}^{n} \binom{n}{i}({}^mC_2+{}^mC_4+\ldots)}$$

and $\binom{n}{i}\binom{n}{j}$ is zero if $i + j > n$. $\qquad \ldots \qquad (5.4)$

Hence R.H.S. of (5.1) is the coefficient of x^q in

$$\sum_{n=0}^{p} \binom{p}{n} (1-x)^{\sum_{i=0}^{n} \binom{n}{i} ({}^mC_3 + {}^mC_5 + \dots) + \sum_{i=1}^{n} \binom{n}{i}\binom{m}{1}} \times$$

$$(1+x)^{\sum_{i=2}^{n} \binom{n}{i} + \sum_{i=0}^{n} \binom{n}{i} ({}^mC_2 + {}^mC_4 + \dots)} \qquad \dots \quad (5.5)$$

and this coefficient of x^q is 2^p times the number of even hypergraphs in B. Now the formula (5.5) is simplified as follows.

$$\sum_{n=0}^{p} \binom{p}{n} (1+x)^{2^n(1+{}^mC_2+\dots)-n-1} (1-x)^{2^n({}^mC_1+{}^mC_3+{}^mC_5+\dots)-m} .$$

Let

$$E^1(x) = \sum_{n=0}^{p} \binom{p}{n} (1+x)^{2^p-p-1} \left(\frac{1-x}{1+x}\right)^{2^n({}^mC_1+{}^mC_3+{}^mC_5+\dots)-m} \qquad \dots \quad (5.6)$$

Therefore the generating function for the number of labelled hypergraphs of order p is

$$E_h(x) = \frac{1}{2^p} E^1(x) \qquad \dots \quad (5.7)$$

Substituting $x = 1$ in (5.7), we have that the number of labelled even hypergraphs of order p is 2^{2^p-2p-1}. This completes the proof of the theorem.

The counting series for labelled even hypergraphs with any number of vertices and edges will be

$$A(t,x) = \sum_{p=0}^{\infty} E_h(x) \frac{t^p}{p!} \qquad \dots \quad (5.8)$$

and the counting series for connected even hypergraphs is given by its formal logarithm, that is, $\log A(t,x)$.

REFERENCES

1. C. Berge, Graphs and Hypergraphs, North Holland, Amsterdam, 1973.
2. F. Harary, Graph Theory, Addison-Wesley, Reading, Mass., 1972.
3. F. Harary and E.M. Palmer, Graphical Enumeration, Academic Press, New York, 1973.
4. R.C. Read, Euler graphs on labelled nodes, Can. J. Math., 14(1962), 482-486.
5. R.C. Read, On the number of self-complementary graphs and digraphs, J. Lond. Math. Soc., 38(1963), 99-104.

ANALYSIS OF A SPANNING TREE ENUMERATION ALGORITHM

R. JAYAKUMAR
K. THULASIRAMAN
Computer Centre
Indian Institute of Technology, Madras 600 036

1. INTRODUCTION

Enumeration of all the spanning trees of a graph is one of the widely studied graph problems. Several methods have been proposed for listing all the spanning trees of a graph. Minty's method [1] and Gabow and Meyer's method [2] are known to be very efficient. In 1968 Char [3] presented a conceptually simple and elegant algorithm. In this paper we present an analysis of Char's algorithm and report some interesting behaviour of this algorithm. For notations, we follow Harary [4].

2. CHAR'S ALGORITHM

Consider an undirected graph G. Let the vertices of G be denoted as $1,2,\ldots,n$. Consider any sequence $(i_1,i_2,\ldots,i_{n-1})$ with $1 \leq i_j \leq n$ whenever $1 \leq j \leq n-1$. Each such sequence may be considered as representing the following (not necessarily distinct) edges of G.

$$(1,i_1),\ (2,i_2),\ldots,(n-1,i_{n-1}).$$

Char's algorithm is based on the following result.

<u>Theorem 1</u> (Tree compatibility test). The sequence $(i_1,i_2,\ldots,i_{n-1})$ represents a spanning tree of G if and only if for each $j \leq n-1$ there exists a sequence of edges (chosen from among the set $(1,i_1), (2,i_2),\ldots,(n-1,i_{n-1})$) with (j,i_j) as the starting edge, which leads to a vertex $k > j$.

The sequences passing the tree compatibility test are called <u>tree sequences</u> and those failing the test are called <u>non-tree sequences</u>. We represent the sequence $(i_1,i_2,\ldots,i_{n-1})$ by an array DIGIT, where DIGIT(j) = i_j, $1 \leq j \leq n-1$. To start with, Char's algorithm selects a spanning tree T, called the <u>initial spanning tree</u>, of G and numbers the vertices of G so that the tree sequence (DIGIT(1), DIGIT(2),...,

DIGIT(n-1)) corresponding to T satisfies the property

$$\mathrm{DIGIT}(i) > i,\ 1 \leq i \leq n-1 \qquad \ldots \quad (1)$$

For a description of Char's algorithm, see [3]. If t_o and t denote respectively the numbers of non-tree sequences and tree sequences, then it can be shown that Char's algorithm is of complexity $O(m+n+n(t+t_o))$.

3. ANALYSIS OF CHAR'S ALGORITHM

We obtain, in this section, a characterisation of the non-tree subgraphs which correspond to the non-tree sequences generated by Char's algorithm and develop a formula for computing t_o.

Consider a connected undirected graph $G = (V,E)$, the spanning trees of which have been enumerated by Char's algorithm. Let the initial tree sequence be (REF(1),REF(2), ...,REF(n-1)). We know that $\mathrm{REF}(i) > i$, $1 \leq i \leq n-1$. Let $G_i = (V_i,E_i)$ be the initial spanning tree of G.

For any k, $2 \leq k \leq n-1$, consider a non-tree sequemce (DIGIT(1), DIGIT(2),..., DIGIT(k-1), DIGIT(k), DIGIT(k+1),...,DIGIT(n-1)) which fails the tree compatibility test at position k. In other words, this non-tree sequence is obtained when DIGIT(k) of the previous sequence is changed and it is found that there exists, in G, a sequence of edges, with the edge (k, DIGIT(k)) as the first edge, which leads to vertex k. Note that $\mathrm{DIGIT}(i) = \mathrm{REF}(i)$, $k+1 \leq i \leq n-1$. Thus the non-tree sequence is (DIGIT(1),DIGIT(2), ...,DIGIT(k-1), DIGIT(k), REF(k+1),...,REF(n-1)).

Let $\overline{G} = (\overline{V},\overline{E})$ be the subgraph of G corresponding to this non-tree sequence. Let $G_A = (V_A,E_A)$ be the subgraph of $\overline{G}$ such that

$$V_A = \{k+1,k+2,\ldots,n-1,\ \mathrm{REF}(k+1),\ \mathrm{REF}(k+2),\ldots,\mathrm{REF}(n-1)\}$$

and

$$E_A = \{(k+1,\ \mathrm{REF}(k+1)),\ (k+2,\ \mathrm{REF}(k+2)),\ldots,(n-1,\ \mathrm{REF}(n-1))\}.$$

Since $\mathrm{REF}(i) > i$, $k+1 \leq i \leq n-1$, we get

$$V_A = \{k+1,\ k+2,\ldots,n\}.$$

Note that G_A is a subgraph of the initial spanning tree G_i of G and there is a path in G_A from each one of the vertices k+1, k+2,...,n-1 to the vertex n.

So G_A is a connected subgraph of G_i.

Let $G_B = (V_B, E_B)$ be the subgraph of $\overline{G}$ such that

$$V_B = \{1,2,\dots,k, \text{DIGIT}(1), \text{DIGIT}(2),\dots,\text{DIGIT}(k)\}$$

and

$$E_B = \{(1,\text{DIGIT}(1)),(2,\text{DIGIT}(2)),\dots,(k,\text{DIGIT}(k))\}.$$

If, in $\overline{G}$, there is a path from vertex k to a vertex in V_A, then DIGIT(k) would have passed the tree compatibility test. Thus, in $\overline{G}$, there is no path from vertex k to any vertex in V_A. Let V_C be the subset of the vertex set $V_B - \{k\}$ such that in $\overline{G}$ there is a path from each vertex in V_C to a vertex in V_A, and let $E_C \subset \overline{E}$ be the set of all the edges in such paths. Let $G_C = (V_C, E_C)$. Now define the graphs G_1 and G_2 as follows :

$$G_1 = (V_1, E_1) = G_A \cup G_C,$$

$$G_2 = (V_2, E_2) = \overline{G} - G_1.$$

It can now be seen that a non-tree sequence which fails the tree compatibility test at position k corresponds to a subgraph of G which is of one of the following two types.

<u>Type 1</u>. Spanning subgraph G' of G such that

1. G' has exactly two components G_1 and G_2.
2. The edges (k+1,REF(k+1)), (k+2,REF(k+2)),...,(n-1,REF(n-1)) are in the component G_1 of G' and the vertex k is in the other component G_2.
3. There is exactly one circuit in G' and it passes through the vertex k.

<u>Type 2</u>. Spanning 2-tree T' of G such that

1. The edges (k+1,REF(k+1)), (k+2,REF(k+2)),...,(n-1,REF(n-1)) are in the component G_1 of T' and the vertex k is in the other component G_2.
2. The component of T' containing vertex k has at least two vertices.

Note that for the subgraphs of Type 1, the circuit passing through vertex k can be traversed in one of two directions and hence each subgraph of Type 1 corresponds to two different non-tree sequences. Similarly each spanning 2-tree T' of Type 2 corresponds to $d^{(k)}(T')$ distinct non-tree sequences where $d^{(k)}(T')$ is the degree of vertex k in T'.

From the above observations we get the following

Theorem 2. Let $G = (V,E)$ be a connected undirected graph. The number of non-tree sequences generated by Char's algorithm which fail the tree compatibility test at position k is equal to $2|S_1| + \sum_{T' \in S_2} d^{(k)}(T')$, where S_1 denotes the set of all spanning $G' \subseteq G$ of Type 1, S_2 denotes the set of all spanning 2-trees T' of Type 2, and $d^{(k)}(T')$ is the degree of vertex k in T'.

For any k, $2 \leq k \leq n-1$, let $\pi_k = (V_1, V_2)$ be a partition of the vertex set V of G where V_1 and V_2 satisfy the following :

$$\{k+1, k+2, \ldots, n\} \subseteq V_1, \qquad \ldots \quad (2)$$

$$k \in V_2, \qquad \ldots \quad (3)$$

$$|V_2| \geq 2. \qquad \ldots \quad (4)$$

Note that the vertex sets V_1 and V_2 of the subgraphs G_1 and G_2 defined in the previous section form a partition of V which satisfy the above properties.

Let

1. $G_a(\pi_k)$ be the subgraph of G induced by the vertex set V_1,
2. $G_a^{(s)}(\pi_k)$ be the graph obtained from $G_a(\pi_k)$ by coalescing the vertices $k+1, k+2, \ldots, n$ and
3. $G_b(\pi_k)$ be the subgraph of G induced by the vertex set V_2.

Further let

1. $t_a^{(s)}(\pi_k)$ and $t_b(\pi_k)$ be the number of spanning trees of $G_a^{(s)}(\pi_k)$ and $G_b(\pi_k)$ respectively, and
2. $d_b^k(\pi_k)$ be the degree of vertex k in $G_b(\pi_k)$.

Using Theorem 2, the number t_o of non-tree sequences generated by Char's algorithm can be obtained as

$$t_o = \sum_{k=2}^{n-1} \sum_{\pi_k} d_b^k(\pi_k) t_a^{(s)}(\pi_k) t_b(\pi_k) \qquad \ldots \quad (5)$$

Note that the vertices of G are numbered with respect to the initial spanning tree. Hence the subgraphs $G_a(\pi_k)$, $G_a^{(s)}(\pi_k)$ and $G_b(\pi_k)$ depend on the initial spanning tree. Hence the value of t_o depends on the choice of the initial spanning tree.

Using (5) we can show that $t_o \leq n^2 t$, so that the complexity of Char's algorithm is $O(m+n+n^3 t)$. However the inequality $t_o \leq n^2 t$ is not tight as will be seen from the results of the next section.

4. ANALYSIS OF CHAR'S ALGORITHM FOR SPECIAL CLASSES OF GRAPHS

In this section, we report the behaviour of Char's algorithm in the case of certain special classes of graphs.

Let $G^{(n-1)}$ denote the class of all n-vertex connected graphs in which there exists a vertex with degree $n-1$. Let $P_{n-1} + K_1$ be called an n-vertex ladder L_n, $C_{n-1} + K_1$ be called an n-vertex wheel W_n. Let $t_o(G)$ and $t(G)$ be the number of non-tree sequences and tree sequences generated by Char's algorithm when applied on a graph. In the following we assume that $K_{1,n-1}$ has been chosen as the initial spanning tree. The following results are based on [5] and the fact that the spanning trees of Ladders are alternate numbers in the Fibbonacci sequence.

Theorem 3. (i) For $G \in G^{(n-1)}$, $t_o(G) \leq t(G)$.

(ii) Complexity of Char's algorithm for all graphs in $G^{(n-1)}$ is $O(m+n+nt)$.

Theorem 4. (i) $t_o(L_n) = t(L_{n-1})$.

(ii) $\underset{n\to\infty}{\mathrm{Lt}} \dfrac{t_o(L_n)}{t(L_n)} = .382.$

Theorem 5. (i) $t_o(W_n) = t(L_n) + 1$.

(ii) $\underset{n\to\infty}{\mathrm{Lt}} \dfrac{t_o(W_n)}{t(W_n)} = 0.4472.$

Theorem 6. $t_o(K_n) = \sum_{n=2}^{n-1} \sum_{p=0}^{k-2} \binom{k-1}{p} (n-k)(n-k+p)^{p-1} (k-p-1)t(K_{k-p}).$

We conjecture that

$$\underset{n\to\infty}{\mathrm{Lt}} \frac{t_o(K_n)}{t(K_n)} = 1.$$

Yet another interesting result with respect to K_n is that $t_o(K_n)$ is independent of the choice of the spanning tree. This can be proved using equation (5).

Proofs of all the above results are given in [5].

5. COMPUTATIONAL EXPERIENCE

All the three algorithms due to Char, Minty, and Gabow and Myer have been programmed in PL/I and tested on a number of randomly generated graphs. In all the cases tested it has been found that Char's algorithm is superior to the other two algorithms, Char's algorithm becomes more and more efficient as the number of trees generated increases.

We have proved in Theorem 3 that $t_o(G) \leq t(G)$, if $G \in G^{(n-1)}$: Computational results suggest that $t_o \leq t$ even in the case of graphs which do not belong to $G^{(n-1)}$.

There is strong computational evidence to believe that Char's algorithm is a very efficient one. This would be so if we could prove that $t_o \leq t$ in the case of all graphs. This is an interesting open problem.

REFERENCES

1. G.J. Minty, A simple algorithm for listing all the spanning trees of a graph, IEEE Trans. Circuit Theory, CT-12(1965), 120.
2. H.N. Gabow and E.W. Myers, Finding all spanning trees of directed and undirected graphs, SIAM J. Comput., 7(1978), 280-287.
3. J.P. Char, Generation of trees, two-trees and storage of master forests, IEEE Trans. Circuit Theory, CT-15(1968), 128-138.
4. F. Harary, Graph Theory, Addison-Wesley, Reading, Mass., 1969.
5. R. Jayakumar, Analysis and Study of a Spanning Tree Enumeration Algorithm, M.S. Thesis, Computer Centre, I.I.T., Madras, 1980.

BINDING NUMBER, CYCLES AND COMPLETE GRAPHS

V. G. KANE
S. P. MOHANTY
Department of Mathematics
Indian Institute of Technology, Kanpur

ABSTRACT

In this paper we study the relationship between the binding number and the existence of cycles and complete subgraphs in a given graph. In particular, we prove the following results :

(i) If $\text{bind}(G) \geq c \geq 1$ and $n > 1 + c/(c-1)^2$, then G has a cycle of length 4.

(ii) If $\text{bind}(G) \geq 3/2$, $|V(G)| \geq 5$, then G has cycles of length 4 and 5.

(iii) If $\text{bind}(G) \geq r - 4/3$ (where r is an integer not less than 3) then G contains K_r.

1. TERMINOLOGY AND KNOWN RESULTS

In this paper, only finite graphs with neither loops nor multiple edges will be considered. The concept of the binding number of a graph was introduced by Woodall [7]. We reproduce below the definition of the binding number of a graph for convenience.

Let $V(G)$ denote the set of vertices of a graph G. Let $x \in V(G)$ and let $\Gamma(x)$ denote the set of all vertices of G adjacent to x. If $X \subseteq V(G)$, we write $\Gamma(X) = \bigcup_{x \in X} \Gamma(x)$. Let $\underline{\underline{F}} = \{X \subseteq V(G) \mid X \neq \phi$ and $\Gamma(X) \neq V(G)\}$. The binding number of G, denoted $\text{bind}(G)$, is defined as :

$$\text{bind}(G) = \min_{X \in \underline{\underline{F}}} \frac{|\Gamma(X)|}{|X|}$$

For other definitions, we refer the reader to [3].

In what follows, $n, \rho, \kappa, \beta, \chi$ denote the order, minimum degree, connectivity, point independence number, chromatic number of a graph respectively. We cite below some of the results proved earlier in the literature, which we use in this paper.

Theorem 1.1 (Andrasfai et. al. [1], Theorem 1.1). Let $r \geq 3$ be an integer. For any graph G with n vertices, at most two of the following properties can hold :

1. $K_r \not\subseteq G$,
2. $\rho > \frac{3r-7}{3r-4} n$,
3. $\chi(G) \geq r$.

Theorem 1.2 (Nash-Williams [5], Lemma 4). Let G be a graph with n vertices, $n \geq 3$. Let G contain no vertex of degree smaller than k, where k is an integer such that $k \geq \frac{1}{3}(n+2)$. Then G either has a Hamiltonian circuit or is separable or has $k+1$ independent vertices.

Theorem 1.3 (Woodall [7], Corollary 7.1). If $\text{bind}(G) \geq c$, then every vertex v has valency $\rho(v) \geq |G| \frac{(c-1)}{c} + \frac{1}{c}$. Thus, if G is a graph on n vertices with minimum valency ρ, then

$$c \leq \text{bind}(G) \leq \frac{n-1}{n-\rho}$$

that is,

$$\rho \geq n - \frac{n-1}{\text{bind}(G)} \geq n - \frac{n-1}{c}.$$

Theorem 1.4 (Woodall [7], Proposition 8). If $|G| = n(\geq 1)$ and the connectivity of G is $\kappa(\geq 0)$ (so that G is κ-connected but not $(\kappa+1)$-connected) then

$$\text{bind}(G) \leq \frac{n+\kappa}{n-\kappa} \quad \text{i.e.,} \quad \kappa \geq \frac{\text{bind}(G)-1}{\text{bind}(G)+1} n.$$

Theorem 1.5 (Woodall [7], Theorem 12(a)). If G is a graph on n vertices such that $\text{bind}(G) \geq 3/2$, then G has a Hamiltonian circuit.

2. CYCLES

Woodall [7] has conjectured that if $\text{bind}(G) \geq 3/2$, then G is pancyclic. In this section, we study the existence of cycles in a given graph, with reference to the above conjecture.

Lemma 2.1. If G is a graph with n vertices, $n \geq 3$ and $\text{bind}(G) > \frac{5(n-1)}{3n}$, then G has a triangle (a cycle of length 3).

Proof. Since $\frac{5(n-1)}{3n} > 1$ for $n \geq 3$, it follows from Proposition 2 of [7] that G is not bipartite. Now, if G does not have a triangle, then by Theorem 1.1, $\rho \leq \frac{2n}{5}$.

But by Theorem 1.3 $\text{bind}(G) \leq \frac{n-1}{n-\rho}$. So

$$\frac{5(n-1)}{3n} < \text{bind}(G) \leq \frac{n-1}{n-\rho} \quad \text{i.e., } \rho > \frac{2n}{5}$$

which is a contradiction. Hence the lemma follows.

Corollary 2.2. If $\text{bind}(G) \geq 5/3$, then G has a triangle.

In the following lemma, we give a proof of a remark on page 233 of [7] for completeness.

Lemma 2.3. If $\text{bind}(G) \geq \frac{1+\sqrt{5}}{2}$, then G has a triangle.

Proof. Let the binding number of G be c. If G does not have a triangle, then clearly $\rho \leq \beta$. We see that $\beta \leq \frac{n}{c+1}$ and by Theorem 1.3, $\frac{nc-n+1}{c} \leq \rho$. Hence

$$\frac{nc-n+1}{c} \leq \rho \leq \beta \leq \frac{n}{c+1}, \text{ so } c < \frac{1+\sqrt{5}}{2}$$

which is a contradiction. Hence the lemma follows (for the proof of $\beta \leq \frac{n}{c+1}$ the reader can refer to Theorem 3.1 which is independent of the present lemma).

Theorem 2.4. Let G be a graph with n vertices, $n \geq 4$. If $\text{bind}(G) \geq c > 1$, then G has a cycle of length 4, whenever $n > 1 + \frac{c}{(c-1)^2}$.

Proof. By Theorem 1.3, $\rho \geq \frac{nc-n+1}{c}$. Let v be a vertex of G of degree h. Let $\Gamma(v) = \{v_1, \ldots, v_h\}$. Assume that G does not have a cycle of length 4. Then $(\Gamma(v_i) - \{v\}) \cap (\Gamma(v_j) - \{v\}) = \phi$, for $i \neq j$. So

$$\left| \bigcup_{i=1}^{h} (\Gamma(v_i) - \{v\}) \cup \{v\} \right| = 1 + \sum_{i=1}^{h} |\Gamma(v_i) - \{v\}| \leq n,$$

or

$$1 - h + \sum_{i=1}^{h} |\Gamma(v_i)| \leq n.$$

But $\rho \leq h$ and $\rho \leq |\Gamma(v_i)|$, for $i = 1, \ldots, h$. So

$$\rho(\rho - 1) \leq n - 1.$$

Since $\rho \geq \frac{nc-n+1}{c}$ and $n \geq 4$, we get

$$(n - \frac{n-1}{c})(n - \frac{n-1}{c} - 1) \leq n-1,$$

or

$$n \leq 1 + \frac{c}{(c-1)^2},$$

which is a contradiction. Hence the theorem follows.

Corollary 2.5. Let G be a graph with n vertices, $n \geq 4$. If $\mathrm{bind}(G) \geq 3/2$, then G has a cycle of length 4.

Proof. By Theorem 2.4, if $\mathrm{bind}(G) \geq 3/2$ and $n > 7$ then G has a cycle of length 4. It can be easily verified, by using Theorem 1.5 and Theorem 1.3, that for $4 \leq n \leq 7$ also, G has a cycle of length 4.

Lemma 2.6. If $\mathrm{bind}(G) \geq 3/2$ and G does not have a triangle, then G has a cycle of length 5.

Proof. Let $|V(G)| = n$. By Theorem 1.3, $\rho \geq \frac{n+2}{3}$. Also, since $\mathrm{bind}(G) > 1$, G is not bipartite. So $\chi(G) \geq 3$ and G has an odd cycle. Let the length of the shortest odd cycle in G be t. Clearly, $t \geq 5$ and $\rho \leq \frac{2n}{t}$. Hence $\frac{n+2}{3} \leq \rho \leq \frac{2n}{t}$. So $t < \frac{6n}{n+2} < 6$, that is, $t = 5$.

Theorem 2.7. Let $|V(G)| = n \geq 5$. If $\mathrm{bind}(G) \geq \frac{3}{2}$, then G has a cycle of length 5.

Proof. If $n = 5,6$ or 7, then by using Theorem 1.3 and Theorem 1.5, it is easy to see that G has a cycle of length 5.

Let $n \geq 8$. Again, if G does not contain a triangle, by Lemma 2.6, G has a cycle of length 5.

Now let G contain a triangle. By Theorem 1.3, $\rho \geq \{\frac{n+2}{3}\} \geq 4$. Let a_1,a_2,a_3 be three vertices of G which are mutually adjacent. Let $A_i = \Gamma(a_i) - \{a_1,a_2,a_3\}$ for $i = 1,2,3$.

Case 1. $|A_1 \cap A_2| \geq 2$ and $|A_1 \cap A_3| \geq 1$.

Let $a_4 \in A_1 \cap A_3$ and $a_5 \in A_1 \cap A_2$, $a_5 \neq a_4$. Clearly $a_1a_4a_3a_2a_5a_1$ is a cycle of length 5.

Case 2. $|A_1 \cap A_2| = |A_1 \cap A_3| = 1$ and $A_1 \cap A_2 \neq A_1 \cap A_3$.

Let $a_4 \in A_1 \cap A_3$ and $a_5 \in A_1 \cap A_2$. Clearly $a_1a_4a_3a_2a_5a_1$ is a cycle of length 5.

Case 3. $|A_1 \cap A_2| = |A_1 \cap A_3| = 1$, $A_1 \cap A_2 = A_1 \cap A_3$.

Let $a_4 \in A_1 \cap A_2 = A_1 \cap A_3$. Let $B_i = \Gamma(a_i) - \{a_1,a_2,a_3,a_4\}$ for $i = 1,2,3,4$.

Since $\rho \geq 4$, $B_i \neq \phi$ for $i = 1,2,3,4$. We can assume that B_1,B_2,B_3,B_4 are pairwise disjoint, for otherwise G has a cycle of length 5.

Let $\Gamma(B_1) \cap B_j \neq \phi$ for some $j \in \{2,3,4\}$, then G has a cycle of length 5. If $\Gamma(B_1) \cap B_j = \phi$ for $j \in \{2,3,4\}$ then $\Gamma(B_1),B_2,B_3,B_4$ are pairwise disjoint and none of them intersects $\{a_2,a_3,a_4\}$. Hence

$$|\Gamma(B_1)| + |B_1| + |B_3| + |B_4| + 3 \leq n.$$

But $|B_j| \geq \rho - 3$ whenever $j \in \{1,2,3,4\}$ and $|\Gamma(B_1)\} \geq \rho$. So $4\rho \leq n+6$. But $\rho \geq \frac{n+2}{3}$. Hence $\{\frac{n+2}{3}\} \leq \rho \leq [\frac{n+6}{4}]$, i.e., $n = 10$ and $\rho = 4$.

Let G have 10 vertices. If $x \in \Gamma(B_1)$ then $x \in B_2 \cup B_3 \cup B_4$ also for otherwise degree of x is at most 3 which is a contradiction. Thus if $n = 10$, then also G has a cycle of length 5.

Case 4. $A_1 \cap A_2 = A_1 \cap A_3 = \phi$, $A_2 \cap A_3 \neq \phi$.

Let $a_4 \in A_2 \cap A_3$. Let $C_i = \Gamma(a_i) - \{a_2,a_3\}$ for $i = 1,2,3,4$. Clearly, if $\Gamma(C_1) \cap C_2 \neq \phi$ or $C_2 \cap C_4 \neq \phi$ or $C_4 \cap \Gamma(C_1) \neq \phi$, then G has a cycle of length 5. So let $\Gamma(C_1),C_2,C_4$ be mutually disjoint. None of $\Gamma(C_1),C_2$ or C_4 intersects $\{a_2,a_3\}$. So

$$|\Gamma(C_1)| + |C_2| + |C_4| + 2 \leq n.$$

But $|\Gamma(C_1)| \geq \rho$, $|C_2| \geq \rho - 1$ and $|C_4| \geq \rho - 2$. Hence $\rho+\rho-1+\rho-2+2 \leq n$, i.e., $\rho \leq \frac{n+1}{3}$ which is a contradiction.

Case 5. $A_1 \cap A_2 = A_2 \cap A_3 = A_3 \cap A_1 = \phi$.

If $\Gamma(A_i) \cap \Gamma(A_j) \neq \phi$ for $i \neq j$ and $i,j \in \{1,2,3\}$ then G has a cycle of length 5. Let $\Gamma(A_1)$, $\Gamma(A_2)$, $\Gamma(A_3)$ be mutually disjoint. Then $3\rho \leq \sum_{i=1}^{3} |\Gamma(A_i)| \leq n$ or $\rho \leq \frac{n}{3}$ which is a contradiction.

Case 6. $A_1 \cap A_j \neq \phi$, $A_1 \cap A_k = \phi$, $\{j,k\} = \{2,3\}$.

This case is easy and the theorem is proved.

Proposition 2.8. Let G be a graph with n vertices, $3 \leq n \leq 24$. If $\text{bind}(G) \geq 3/2$, then G is Hamiltonian.

Theorem 2.9. Let G be a graph with n vertices such that $\text{bind}(G) = c \geq \frac{1+\sqrt{5}}{2}$. Then G is Hamiltonian. (This result, though weaker than that of Woodall, is easy to prove).

Proposition 2.10. Let G be a graph with n vertices, $n \geq 3$. If $\text{bind}(G) > \frac{2(n-1)}{n+1}$, then G is pancyclic.

Proof. Clearly, G is not bipartite. By Theorem 1.3, we have

$$\frac{n-1}{n-\rho} \geq \text{bind}(G) > \frac{2(n-1)}{n+1} .$$

Hence $\rho > \frac{n-1}{2}$. So by a Corollary in [2], we have the required result.

Corollary 2.11. If $\text{bind}(G) \geq 2$, then G is pancyclic.

3. COMPLETE GRAPHS

Theorem 3.1. Let G be a graph with n vertices. Then $\text{bind}(G) \leq \frac{n}{\beta} - 1 \leq \chi - 1$.

Proof. Let S be an independent subset of $V(G)$ with β vertices. Clearly, $|\Gamma(S)| = n - \beta$. Hence $\text{bind}(G) \leq \frac{|\Gamma(S)|}{|S|} = \frac{n-\beta}{\beta} = \frac{n}{\beta} - 1$. But $\frac{n}{\beta} \leq \chi$. Hence $\text{bind}(G) \leq \frac{n}{\beta} - 1 \leq \chi - 1$.

If G is a complete r-partite graph $(r \geq 2)$ on r sets of equal size then the bounds given in the above theorem are attained. If $G = C_{2m}$, then also for every integer $m \geq 2$, the bounds in the above theorem are attained.

Theorem 3.2. Let $r \geq 3$ be an integer. If $\text{bind}(G) \geq r - 4/3$, then G contains K_r.

Proof. By Theorem 1.3, $r - 4/3 \leq \text{bind}(G) < \frac{n-1}{n-\rho}$. So $\rho > \frac{(3r-7)n}{(3r-4)}$. Also, by Theorem 3.1, $\chi(G) \geq r$. Hence by Theorem 1.1, G contains K_r.

Let $f(r) = \inf\{t \mid t$ is real and $\text{bind}(G) \geq t$ implies G contains $K_r\}$. By Theorem 3.2, $f(r) \leq r - 4/3$. Let $(r-2)^2 \leq n-1$. Consider $K_{r-1} \otimes K_n$ ($\otimes$ denotes the tensor product [6]). It is easy to see that $K_{r-1} \otimes K_n$ does not contain K_r. Also by Theorem 4.2 [4], $\text{bind}(K_{r-1} \otimes K_n) = r - 2$. Hence $f(r) \geq r-2$. It may be interesting to determine $f(r)$.

REFERENCES

1. B. Andrasfai, P. Erdös and V.T. Sos, On connection between chromatic number, maximal clique and minimal degree of a graph, Discrete Math., 8(1974), 205-218.

2. J.A. Bondy, Pancyclic graphs I, J. Comb. Theory Ser.B, 11(1971), 80-84.

3. F. Harary, Graph Theory, Addison-Wesley, Mass., 1969.

4. V.G. Kane and S.P. Mohanty, Product graphs and binding number, Communicated to J. Comb. Theory.

5. C. St. J.A. Nash-Williams, Edge disjoint Hamiltonian circuits in graphs with vertices of large valency, in : L. Mirsky, ed., Studies in Pure Mathematics (Papers presented to Richard Rado), Academic Press, New York, 1971.

6. A.T. White, On the genus of Product of Graphs, Recent Trends in Graph Theory, Springer Verlag, Lecture Notes in Mathematics, Vol.186(1971), 217-219.

7. D.R. Woodall, The binding number of a graph and its Anderson Number, J. Comb. Theory Ser.B, 15(1973), 225-255.

A CLASS OF COUNTEREXAMPLES TO A CONJECTURE ON DIAMETER CRITICAL GRAPHS

V. KRISHNAMOORTHY
Department of Mathematics
Madras Institute of Technology Campus
Perarignar Anna University of Technology, Madras 600 044

R. NANDAKUMAR
Department of Mathematics
Indian Institute of Technology, Madras 600 036

ABSTRACT

A graph is diameter critical if removal of any edge increases the diameter of the graph. In this note we construct an infinite family of counterexamples to a recent conjecture by Caccetta and Häggkvist [2] on the maximum number of edges in a diameter critical graph.

MAIN RESULTS

We consider finite simple graphs and follow Bondy and Murty [1] for general notation and terminology. A graph G is said to be 'diameter critical' if diameter of $G-e$ is greater than the diameter of G, for any edge e of G; it is also called k-critical if diameter of G is k.

In [2], Caccetta and Häggkvist conjecture that for $k \geq 3$, the maximum number of edges a k-critical graph can have is

$$2\left(\frac{\nu-r}{k+1}\right)^2 + \left(\frac{\nu-r}{k+1}\right)(k+r-2)$$

where $\nu \equiv r \bmod (k+1)$ and they also give a construction of a k-critical graph with this many edges on ν vertices. The construction is as follows.

Let $m = [\frac{\nu}{k+1}]$. The graph G is constructed by first taking m distinct paths $P^i = u_1^i u_2^i \ldots u_{k-1}^i$, $i = 1,2,\ldots,m$, and then by joining each u_1^i, $i = 1,2,\ldots,m$, to m new vertices and each u_{k-1}^i, $i = 1,2,\ldots,m$, to another set of $(m+r)$ new vertices. This graph contains

$$m(2m + k + r - 2)$$

edges, in terms of m.

Instead of taking m distinct paths of length $(k-2)$, we take $m+1$ such paths. Let one new vertex be joined to all u_1^i, $i = 1,2,\ldots,m+1$. Let all the other vertices be joined to u_{k-1}^i, $i = 1,2,\ldots,m+1$. The number of edges in this graph is

$$m(2m+k+r-2) + (3-k)m + r - 1.$$

This graph is an obvious counterexample to the above conjecture if

$$(3-k)m + r - 1 \geq 1.$$

To simplify we take $r = k$ and this inequality holds if (i) $k = 2$ or (ii) $k = 3$ or (iii) $k = 4$ and m is 1 or 2 or (iv) $k \geq 5$ and m is 1.

Similarly, if only $m-1$ paths of length $k-2$ are taken and the graph is constructed just as above, the graph contains

$$m(2m+k+r-2) + m(k-3) - 2k + 3 - r$$

edges. As before to get counterexamples, we set

$$m(k-3) - 2k + 3 - r \geq 1.$$

By taking $r = 0$, and $k = 3 + s$, we get

$$s(m-2) \geq 4.$$

This inequality holds if (i) $m \geq 6$ and $k \geq 4$ or (ii) $m = 5$ or 4 and $k \geq 5$ or (iii) $m = 3$ and $k \geq 7$.

We give below a graph given in [2] and the corresponding counterexample.

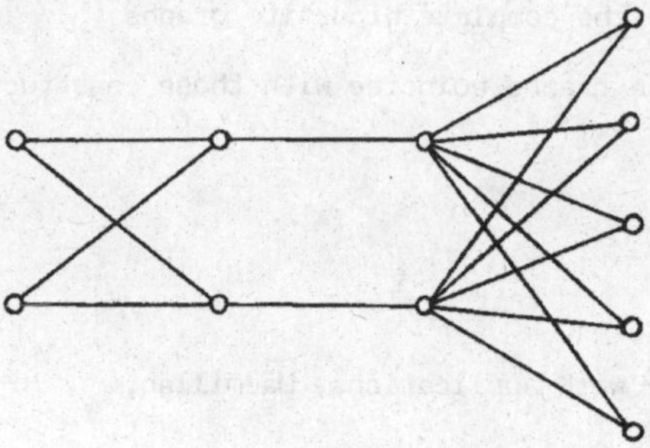

$\nu = 11;\ k = 3;\ \varepsilon = 16$

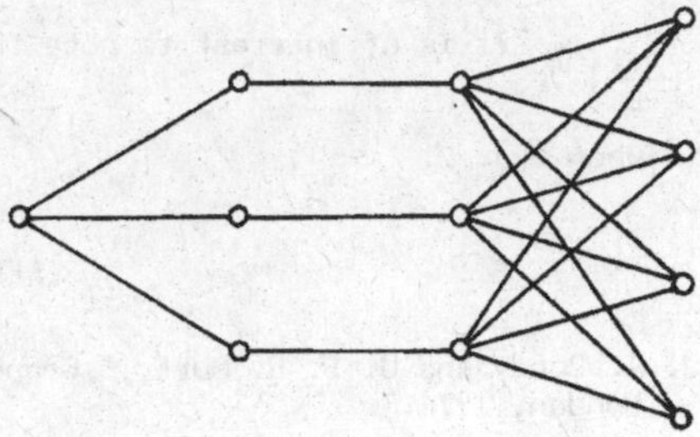

$\nu = 11;\ k = 3;\ \varepsilon = 18$

In general, if there are p paths of length $k-2$ in our construction, the number of edges is given by

$$\varepsilon = p(k-2) + [\nu - p(k-1)]\, p = p(\nu+k-2) - p^2(k-1),$$

for given ν and $k(\geq 2)$. Clearly p should satisfy

$$p(k-1) \leq \nu - 2 \qquad \ldots \qquad (1)$$

Our aim is to maximise ε subject to (1), where p is an integer.

Treating ε as a function of p, where p is real, ε attains its unique maximum at $p_o = \frac{\nu+k-2}{2(k-1)}$. Substituting $m = p_o + f$, we get

$$\varepsilon = \frac{(\nu+k-2)^2}{4(k-1)} - f^2(k-1).$$

Hence to maximise ε we minimise f, that is, take $m = (\frac{\nu+k-2}{2(k-1)})^* = p_o + f$ where $-\frac{1}{2} \leq f < \frac{1}{2}$ and a^* denotes the integer nearest to a, the smaller one being chosen in case of a tie. It can be shown that m satisfies (1).

We conjecture that the graphs given in our construction by taking $m = (\frac{\nu+k-2}{2(k-1)})^*$ realise the maximum number of edges in a k-critical graph, $k \geq 2$, with ν vertices.

When $k = 2$, we get $m = [\frac{\nu}{2}]$ and the graphs are the complete bipartite graphs $K_{[\frac{\nu}{2}][\frac{\nu+1}{2}]}$. It is of interest to note that these graphs coincide with those constructed in [2] when $k = 2$.

REFERENCES

1. J. A. Bondy and U. S. R. Murty, Graph Theory with applications, Macmillan, London, 1976.
2. Caccetta and Häggkvist, On diameter critical graphs, Discrete Math., 28(1979), 223-229.

RECONSTRUCTION OF A PAIR OF CONNECTED GRAPHS FROM THEIR LINE-CONCATENATIONS

V. R. KULLI
Department of Mathematics
Karnatak University, Post Graduate Centre, Gulbarga 585 105

N. S. ANNIGERI*
Department of Mathematics
Shri Sharanabasaveshwar College of Science, Gulbarga 585 103

ABSTRACT

Reconstruction of a pair of graphs from their line-concatenations is an open problem in [4]. We have shown in [2] that a large class of disconnected graphs are not reconstructible from their line-concatenations. In this paper an arbitrary pair of connected graphs are shown to be reconstructible from the class of their line-concatenations.

1. INTRODUCTION

The graphs considered will be finite, undirected, with no loops or multiple lines. Definitions and notation not given here may be found in Harary [1]. The original graph reconstruction conjecture [5] states that a graph with three or more points is uniquely reconstructible from the collection of its point deleted subgraphs. Sampathkumar [4] considered some variations of this conjecture, namely, reconstruction of a pair of graphs instead of one graph from some collection of graphs obtained from the two given graphs. We give its general formulation below.

Let $\underline{\underline{G}}$ be a class of graphs which contains an isomorphic image of each of its members. For any two point disjoint graphs G_1 and G_2, let $\underline{\underline{G}}(G_1,G_2)$ denote the class of all graphs $H \varepsilon \underline{\underline{G}}$ such that both G_1 and G_2 contain an induced subgraph $\simeq$ to H. Let $H \varepsilon \underline{\underline{G}}(G_1,G_2)$ and H_1,H_2 be induced subgraphs of G_1,G_2 respectively such that $H_1 = H_2 = H$. Given an isomorphism $f : V(H_1) \rightarrow V(H_2)$ from H_1 onto H_2, we denote by $G_1(H_1,f,H_2)G_2$, the graph obtained from G_1 and G_2 by coalescing the

* Research supported by the Teacher Fellowship, U.G.C., New Delhi, under F.I.P.

point $u \in V(H_1)$ with $f(u) \in V(H_2)$, for every $u \in V(H_1)$. We refer to this graph as an H-concatenation of G_1 and G_2. In particular, if $\underline{\underline{G}} = \{K_2\}$ then $H_1 = u_1v_1 = e_1 \in E(G_1)$ and $H_2 = u_2v_2 = e_2 \in E(G_2)$ and in this case we write $G_1(e_1,e_2)G_2$ for a K_2-concatenation (or line-concatenation) of G_1 and G_2, obtained by coalescing e_1 of G_1 with e_2 of G_2. Let $G = G_1(e_1,e_2)G_2$ and e be the line of G due to identification of lines e_1 and e_2. De-identification of line e of G is the process of dividing e to obtain two lines e' and e'' in order to get two point disjoint graphs G' and G'' such that $G'(e',e'')G'' = G$.

Given $H \in \underline{\underline{G}}(G_1,G_2)$, if h is the cardinality of the automorphism group $aut(H)$ of H and g_i denotes the number of induced subgraphs of G_i isomorphic to H, $i = 1,2$ then hg_1g_2 gives the total number of H-concatenations of G_1 and G_2. In the case of line-concatenations it is obvious to see that $h = 2$, $g_1 = q(G_1) = q_1$ and $g_2 = q(G_2) = q_2$ so that there are $\alpha = 2q_1q_2$ line-concatenations of G_1 and G_2 where $q(G)$ denotes the number of lines of the graph G. Also every line-concatenation $G_1(e_1,e_2)G_2$ contains $\beta = (q_1+q_2-1)$ lines. Since in the reconstruction problem, α and β are known, one can solve for q_1 and q_2. Note that, in some cases it may not be possible to determine directly, whether q_1 (or q_2) is the number of lines of G_1 or G_2.

By the $\underline{\underline{G}}$-concatenation deck of G_1 and G_2 we mean the collection $\underline{\underline{C}}(G_1,G_2) = \{G_1(H_1,f,H_2)G_2/H_1$ and H_2 are induced subgraphs of G_1 and G_2, $H_1 = H_2 = H \in \underline{\underline{G}}(G_1, G_2)$ and $f \in iso(H_1,H_2)\}$ where $iso(H_1,H_2)$ denotes the set of all isomorphisms from H_1 onto H_2. It is straightforward to see that $\underline{\underline{C}}(G_1,G_2) = \underline{\underline{C}}(G_2,G_1)$.

Given $\underline{\underline{G}}$, a pair (G_3,G_4) of graphs is called a $\underline{\underline{CG}}$-reconstruction of the given pair of graphs (G_1,G_2) if $\underline{\underline{C}}(G_3,G_4) = \underline{\underline{C}}(G_1,G_2)$. Further (G_1,G_2) is said to be $\underline{\underline{CG}}$-reconstructible if for every $\underline{\underline{CG}}$-reconstruction (G_3,G_4) of (G_1,G_2), one finds $(G_3,G_4) = (G_1,G_2)$. Suppose $\underline{\underline{G}} = \{H\}$. We write $\underline{\underline{CG}}$-reconstruction as simply $\underline{\underline{C}}H$-reconstruction.

Attempts on $\underline{\underline{C}}K_1$-reconstructibility of a pair of graphs are made in [2], [3] and [4]. We have produced in [2] infinitely many pairs of disconnected graphs which are not $\underline{\underline{C}}K_2$-reconstructible and some pairs of connected graphs are shown to be

$\underline{\underline{C}}K_2$-reconstructible. Our purpose here is to establish the $\underline{\underline{C}}K_2$-reconstructibility of any two connected graphs.

2. PRELIMINARIES

In the $\underline{\underline{C}}K_2$-reconstruction problem it is obvious that the point disjoint graphs considered are to be nontrivial. We restrict the graphs considered here to be connected. The following theorem from [2], is of use.

Theorem A. Let G_1 and G_2 be two connected graphs each of which contains at least one nonend line and one end line. Then (G_1,G_2) is K_2-reconstructible.

The following are the simple lemmas the proofs of which we omit.

Lemma 1. Let G_1 and G_2 be two point disjoint graphs. The graph G_1 is K_2 if and only if every $G_1(e_1,e_2)G_2$ is isomorphic with G_2 and either $q_1 = 1$ or $q_2 = 1$.

Lemma 2. Let $m = \max_{i,j}(\Delta(G_1(e_i,e_j)G_2))$. Both the grapho G_1 and G_2 are stars if and only if every $G_1(e_i,e_j)G_2$ with $\Delta(G_1(e_i,e_j)G_2) = m$, is a star.

By Lemma 1, for any graph G, K_2-reconstruction of (K_2,G) follows trivially. Hence we assume throughout, none of the graphs of a pair considered is K_2.

3. THE MAIN THEOREM

Before the main theorem we prove the following lemma.

Lemma 3. The pair of graphs $(G_1,G_2) = (K_{1,n},G)$ is $\underline{\underline{C}}K_2$-reconstructible, where $n \geq 2$ and G is any graph.

Proof. Note that $n = q_1$ or q_2. If G is also a star, by Lemma 2, $\underline{\underline{C}}K_2$-reconstruction of (G_1,G_2) follows. In fact, $(G_1,G_2) = (K_{1,q_1},K_{1,q_2})$.

Assume now G is not a star. Now G_1 and G_2 may be considered as :

(i) $G_1 = K_{1,n}$, $G_2 = G$ has no end lines. It is so only when all the line-concatenations contain the same number $(n-1)$ of end lines, all being incident at a point.

(ii) $G_1 = K_{1,n}$, $G_2 = G$ is not a star but has end lines. It will be the case only when every line-concatenation contains a point which is incident with at least

q_1-1 (or q_2-1) end lines and the set of all line-concatenations can be bifurcated into two subsets containing the line-concatenations with s-1 and s-2 end lines in the respective subsets, s being the total number of end lines in G_1 and G_2.

In either case let $m = \max_{i,j}(\Delta(K_{1,n}(e_i,e_j)G))$ and $H = K_{1,n}(e_i,e_j)G$ be such that $\Delta(H) = m$. Also let u be a point of H such that $\deg_H u = m$. Then one can see that u is the point due to identification of u_1 of $K_{1,n}$ and v_1 of G where $\deg_{K_{1,n}} u_1 = \Delta(K_{1,n})$ and $\deg_G v_1 = \Delta(G)$. Clearly each of u_1 and v_1 is one of the incident points of e_i and e_j respectively.

If $q_1 = q_2$ then $n = q_1$. de-identify a line of H incident with u so as to get one of the two graphs as K_{1,q_1}. Pair of graphs obtained is uniquely $(K_{1,n},G)$.

If $q_1 \neq q_2$, de-identify a line of H incident with u in such a way that one of the two graphs obtained is star $K_{1,n}$ where $n = q_1$ or q_2. Existence of a pair of graphs containing one of such stars is obvious. If n assumes single value q_1 or q_2 for some H, the pair obtained is $(K_{1,n},G)$. If it is not so, let (K_{1,q_1},G') and (K_{1,q_2},G'') be two pairs obtained. Now either $G' = G$ or $G'' = G$. The necessary condition for G' to be G is that u_1 and v_1 must be the points of maximum degrees in the respective graphs K_{1,q_1} and G'. If, only one of the pairs (K_{1,q_1},G') and (K_{1,q_2},G'') contains the points with the above property, the respective pair is uniquely $(K_{1,n},G)$. Otherwise without loss of generality, suppose $K_{1,n}$ and G' have respectively q_1 and q_2 lines such that $q_1 > q_2$ and $\Delta(G') = s$. Now $q_1 + s = m = q_2 + \Delta(G'')$. Therefore $\Delta(G'') = s + q_1 - q_2$. Obviously there exist line-concatenat-ions of (K_{1,q_2},G'') containing a point of degree $s + q_1 - q_2$. Now G' does not contain any point of degree $s + q_1 - q_2$, since $s + q_1 - q_2 > s$ and $\Delta(G') = s$. Also no line-concatenation of K_{1,q_1} and G' contains a point of degree $s + q_1 - q_2$. For, if there is some line-concatenation with a point of degree $s + q_1 - q_2$, then since $\Delta(G') = s$, $|E(G')| = q_2$ and G' is not star, we have $s - q_2 < 0$. Also there is a point of G' of degree i such that $1 \leq i < s$ and $i + q_1 - 1 = s + q_1 - q_2$. It is because a point of degree $s + q_1 - q_2$ in the line-concatenation can be obtained by identifying a point of degree i of G' with the only point of maximum degree q_1 of $K_{1,n}$ in the identification of two lines. Now $i - 1 = s - q_2 < 0$ that is

$i < 1$, which is impossible. Thus we conclude that if none of the given line-concatenations contains a point of degree $s + q_1 - q_2$ then $(K_{1,n},G) = (K_{1,q_1},G')$; otherwise $(K_{1,n},G) = (K_{1,q_2},G'')$.

Main Theorem. An arbitrary pair of connected graphs is $\underline{\subseteq}K_2$-reconstructible.

Proof. Determination of the connectedness of two graphs from their line-concatenations is simple, since two graphs are connected if and only if each line-concatenation of these graphs is connected. We now divide the class of pairs of connected graphs into mainly three subclasses depending on the end lines in each graph. Let (G_1,G_2) be a pair of graphs. The following are the three cases.

Case 1. Each of G_1 and G_2 contains at least one end line :

We now consider the following subcases.

Subcase 1.1. G_1 and G_2 both have at least one nonend line.

Subcase 1.2. Exactly one of G_1 and G_2 has nonend lines.

Subcase 1.3. Neither of G_1 and G_2 has nonend lines.

Graphs G_1 and G_2 belong to Subcase 1.1 if and only if it is always possible to collect the line-concatenations into 3 groups such that each of the line-concatenations of respective group contains t, $t-1$ and $t-2$ end lines. In fact t is the total number of end lines in G_1 and G_2. Then by Theorem A, (G_1,G_2) is $\underline{\subseteq}K_2$-reconstructible.

Pairs of graphs (G_1,G_2) belonging to Subcases 1.2 and 1.3 are recognized by Lemmas 3 and 2 respectively. Also by Lemma 3, (G_1,G_2) is $\underline{\subseteq}K_2$-reconstructible.

Case 2. Neither of G_1 and G_2 contains end lines :

It is so only when no line-concatenation contains end lines. Suppose $q_1 = q_2$. Then there is a line-concatenation containing exactly one line (particularly the line which can be shrunk into a cutpoint by its contraction), the de-identification of which gives a unique pair of graphs with equal number of lines. Such a line-concatenation exists particularly due to identification of two lines the incident points of which are not cutpoints in the respective graphs. (Let such lines be called x-lines).

The pair of graphs obtained here is (G_1,G_2).

Suppose now $q_1 \neq q_2$. Without loss of generality assume $q_1 > q_2$. Let $\underline{\underline{A}}$ be the collection of all nonisomorphic line-concatenations with the following property. Each $H \in \underline{\underline{A}}$ contains lines e_i's the de-identification of which yields pairs of graphs with q_1 and q_2 lines. Let $\underline{\underline{A}}_H$ be the set of pairs of graphs obtained by de-identifying lines e_i's of H as above. Obviously $(G_1,G_2) \in \bigcap \underline{\underline{A}}_H$ for each $H \in \underline{\underline{A}}$. We assert that for each $H \in \underline{\underline{A}}$, $\bigcap \underline{\underline{A}}_H = (G_1,G_2)$. For, if $(G_3,G_4) \in \bigcap \underline{\underline{A}}_H$ such that $(G_3,G_4) \neq (G_1,G_2)$ and $q(G_3) = q_1$, $q(G_4) = q_2$, then one can see that G_4 is a subgraph of G_1, and G_2 is a subgraph of G_3 such that G_1 and G_3 have lines e' and e'', the de-identification of which yields pairs of graphs (G_4,G') and (G_2,G') where G' contains $q_1 - q_2 + 1$ lines. Let H' be a line-concatenation of G_1 and G_2 obtained by coalescing a line of G_4 in G_1 distinct from e' and a line of G_2. Now H' does not contain any line, the de-identification of which can yield (G_3,G_4) unless $(G_3,G_4) = (G_1,G_2)$. That is $(G_3,G_4) \notin \underline{\underline{A}}_{H'}$, unless $(G_3,G_4) = (G_1,G_2)$. Thus $\underline{\underline{C}}K_2$-reconstruction of (G_1,G_2) is established.

<u>Case 3</u>. Exactly one of G_1 and G_2, say G_1, contains end lines and G_2 does not :

If G_1 is a star, then (G_1,G_2), by Lemma 3, is $\underline{\underline{C}}K_2$-reconstructible. Therefore we assume G_1 is not a star. Reconstruction of graphs G_1 and G_2 in this case follows from Lemma 3 and Case 2 of this theorem. Let $H = G_1(e_1,e_2)G_2$ contain minimum number, say n, of end lines among all line-concatenations. It is easy to see here that e_1 is an end line of G_1. Then G_1 has $n+1$ end lines and the number of line-concatenations with n end lines is $2m(n+1)$ where m is the number of lines of G_2. In the present problem, n and $2m(n+1)$ are known. Hence $q_2(= m)$ and q_1 are determined distinctly. Since G_2 has no end lines, it contains at least one x-line, for otherwise G_2 will not be a finite graph. Now choose a line e of H such that de-identification of e yields a pair of point disjoint graphs with q_1 and q_2 lines and the graph with q_2 lines contains no end lines. Also verify that if the line e of H is shrunk into a point by its contraction, then it is a cutpoint lying on minimum number of blocks among all H which have n end lines. Clearly the line e_1 of G_1 is an end line and e_2 of G_2 is an x-line. De-identify e in such a way that the graph with q_1 lines contains the end line e_1, that of q_2 lines contains e_2 as x-line. The pair of graphs obtained is uniquely (G_1,G_2). This completes the proof of the theorem.

REFERENCES

1. F. Harary, Graph Theory, Addison—Wesley, Reading, 1969.
2. V.R. Kulli and N.S. Annigeri, On reconstruction of a pair of graphs from their concatenations, 2nd Conference, Forum for Interdisciplinary Mathematics, Jaipur, 1978, presented.
3. S. Kundu and E. Sampathkumar, Reconstruction of a pair of trees from their concatenations, preprint.
4. E. Sampathkumar, Reconstruction of two graphs from their concatenations, preprint.
5. S.M. Ulam, A Collection of Mathematical Problems, Wiley (Interscience), New York, 1960.

ON DOMINATION RELATED CONCEPTS IN GRAPH THEORY

RENU LASKAR
Clemson University
Clemson, S.C. 29631, U.S.A.

H. B. WALIKAR
Karnatak Arts College
Dharwar 580 001, India

ABSTRACT

A survey of recent results on domination and related concepts in graph theory is presented.

1. INTRODUCTION

Let $G = (V,E)$ be an undirected graph with no loops and multiple edges. A set of vertices D in G is a dominating set if every vertex not in D is adjacent to at least one vertex in D. A set of vertices D is independent if no two vertices in D are adjacent.

The concepts of domination and independence have existed in literature for a long time. Ore [25] and Berge [5] are the first ones to discuss these concepts. There are many results in the literature discussing the theory of independent sets and the closely related topic of graph coloring. Quite surprisingly, there are few papers on domination, even though there is a wide variety of applications of this concept.

A survey of the literature reveals the following applications. Some of these are cited in Cockayne and Hedetniemi's elegant paper [9]. Ore [25] considers the problem of placing a minimum number of queens on a chessboard so that each square is controlled by at least one queen. Liu [21] discusses the application of dominance to communication links in a network, where a dominating set consists of a set of cities which, considered as transmitting stations, can transmit messages to every city in the network. In a similar vein, Berge [5] briefly considers the problem of keeping all points in a network under serveillance by a set of radar stations.

Domination is a concept which is known to coding theory. If we define a graph G for which V(G) is the set of n-dimensional vectors with coordinates chosen from {1,2,...,q} and for which E(G) is the set of pairs of vectors from V(G) which differ in one coordinate, then the sets of vectors which are (i) (n,q)-covering sets, (ii) single error correcting codes, or (iii) perfect covering sets, are all dominating sets of G with certain additional properties (see Kalbfleisch, Stanton, Horton [20]).

Further, if we take V(G) to be all of the (t-1)-dimensional subspaces of PG(n-1, q), the projective geometry of dimension n-1 over GF(q), and take E(G) to be the set of pairs of distinct elements from V(G) which have non-empty intersection, then the maximal (t-1) spreads are exactly the independent dominating sets of G as discussed by Drake and Freeman in [15].

Dominating sets also occur in applications of the work of Edmonds and Fulkerson [16] on clutters and blockers of clutters.

The 1977 paper [9] of Cockayne and Hedetniemi on domination theory aroused considerable interest among the researchers and since then some new concepts related to domination have been developed. Cockayne discusses some of these concepts in his survey article [14]. In this paper we survey the recent results. However, we do not claim completeness.

2. DOMINATION AND INDEPENDENT DOMINATION

A set of vertices D in G = (V,E) is a dominating set if every vertex not in D is adjacent to at least one vertex in D. A set of vertices D is independent if no two vertices in D are adjacent. The domination number $\gamma(G)$ is the minimum cardinality taken over all minimal dominating sets. The independent domination number i(G) is the minimum cardinality taken over all maximal independent sets. The open neighborhood of v, $N(v) = \{x | xv \in E\}$, and the closed neighborhood of v, $N[v] = N(v) \cup \{v\}$.

2.1 Basic properties.

Propositions 2.1.1. (Ore). If G is a graph without isolated vertices and D is a minimal dominating set, then V-D contains a minimal dominating set.

Proposition 2.1.2. (Berge). D is a maximal independent set if and only if D is an independent dominating set.

Proposition 2.1.3. Let H be a spanning subgraph of a graph G. Then, any dominating set in H is also a dominating set in G.

2.2 Bounds and related results

Berge [5] gives the following bounds for the domination number.

Proposition 2.2.1. If G is a graph having p vertices, q edges with maximum degree Δ, then

$$p - q \leq \gamma(G) \leq p - \Delta.$$

Proposition 2.2.2. (Vizing [28]). If G is a graph with p vertices and q edges, then $q \leq [(p-\gamma(G))(p-\gamma(G)+2)/2]$.

Proposition 2.2.3. (Walikar, Acharya and Sampathkumar [30]). For any graph G with p vertices and q edges, $\gamma(G) = p - q$ iff each component is a star.

Theorem 2.2.4. (Vizing [28]).

$$\gamma(G) \leq p + 1 - \sqrt{1+2q}.$$

Theorem 2.2.5. (Walikar, Sampathkumar and Acharya [29]).

$$\{\frac{p}{1+\Delta}\} \leq \gamma(G) \leq p - k(G)$$

where $\Delta(G)$ and $k(G)$ denote respectively the maximum degree and the vertex connectivity of G. The least integer greater than or equal to the real number r is denoted by $\{r\}$. Furthermore, $\gamma(G) = \frac{p}{1+\Delta}$ iff $V(G)$ can be partitioned into subsets V_1 and V_2 with $\gamma(G) = |V_1| \leq |V_2|$ satisfying all of the following conditions :

(i) V_1 is independent,

(ii) For $u \in V_2$, there exists unique $v \in V_2$ such that $N(u) \cap V_1 = \{v\}$,

(iii) $d(u) = \Delta$, for every $u \in V_1$.

2.3 Independent dominating sets

Clearly $\gamma(G) \leq i(G) \leq \beta_0(G)$, where $\beta_0(G)$ is the largest number of vertices in an independent set of G.

Allan and Laskar [2] have established a sufficient condition for $\gamma(G) = i(G)$ in terms of forbidden graphs.

Theorem 2.3.1. (Allan and Laskar). If G has no induced subgraphs isomorphic to $K_{1,3}$, then $\gamma(G) = i(G)$.

Since the line graph $L(G)$ does not have an induced subgraph $K_{1,3}$, we have

Corollary 2.3.2. $\gamma(L(G)) = i(L(G))$.

This extends the result due to Mitchell and Hedetniemi [23].

Corollary 2.3.3. (Mitchell and Hedetniemi). $\gamma(L(T)) = i(L(T))$, where T is any tree.

The middle graph of a graph G, denoted $M(G)$, is the intersection graph $\Omega(F)$ on $V(G) = \{v_1, v_2, \ldots, v_p\}$ where $F = \{\{v_1\}, \{v_2\}, \ldots, \{v_p\}\} \cup E(G)$.

The graph G^+ is defined as follows : add to $V = \{v_1, \ldots, v_p\}$ p vertices $u_1, u_2, \ldots, u_p$ different from the elements of V and from each other. Add the p edges $u_i v_i$, $i = 1,2,\ldots,p$ to E. The graph G^+ is the graph with vertex set $V \cup \{u_1, u_2, \ldots, u_p\}$ and edge set $E \cup \{u_1 v_1, u_2 v_2, \ldots, u_p v_p\}$. The following result is due to Hamada and Yoshimura [17].

Theorem 2.3.4. (Hamada and Yoshimura). $L(G^+)$ is isomorphic to $M(G)$.

This theorem and Corollary 2.3.2 give

Corollary 2.3.5. $\gamma(M(G)) = i(M(G))$, where $M(G)$ is the middle graph of G.

It may be pointed out that the condition of the Theorem 2.3.1 is not necessary, since $K_{1,3}$ itself has $\gamma = i = 1$.

A challenging question is to characterize those graphs G, for which $\gamma(G) = i(G)$.

Sumner and Moore [27] define a graph G to be domination perfect if $\gamma(H) = i(H)$, for every induced subgraph H of G. Clearly $K_{1,3}$-free graphs are domination perfect.

Bollabas and Cockayne [7] generalize the result of Allan and Laskar in the following.

Theorem 2.3.6. (Bollabas and Cockayne). If G has no induced subgraph isomorphic to $K_{1,k+1}$ $(k \geq 2)$, then $i(G) \leq \gamma(G)(k-1) - (k-2)$.

Bollabas and Cockayne [7] further proved that

Theorem 2.3.7. If G with p vertices has no isolated vertex, then

$$i(G) \leq p - \gamma(G) + 1 - \{\frac{p-\gamma(G)}{\gamma(G)}\}.$$

In [22], Meir and Moon proved that, for any tree T of order $p \geq 2$,

$$\gamma(T) + \beta_0(T) \leq p.$$

But this result is true for any graph G of order $p \geq 2$ without isolated vertices.

Proposition 2.3.8. For any graph G of order $p \geq 2$ without isolated vertices,

$$\gamma(G) + \beta_0(G) \leq p.$$

A set S of vertices of a graph G is said to be a transversal in G, if every edge in G has at least one end in S. The smallest number of vertices in any transversal of G is called the transversal number of G and is denoted by $\alpha_0(G)$. Clearly, every transversal is a dominating set, but the converse is not true.

Proposition 2.3.9. Let G be a graph without isolated vertices. Then the dominating set D is a transversal if, and only if $V-D$ is independent.

Proposition 2.3.10. For any graph G, without isolated vertices,

$$\gamma(G) \leq \alpha_0(G).$$

By a well known theorem of Gallai [18], $\alpha_0(G) + \beta_0(G) = p$, for any graph G of order p without isolated vertices.

Theorem 2.3.11. For any graph G of order $p \geq 2$ without isolated vertices, the following are equivalent :

(i) $\gamma(G) = \alpha_0(G)$

(ii) $\gamma(G) + \beta_0(G) = p$

(iii) there exists a minimum dominating set D such that $V-D$ is a maximal independent set.

3. DOMATIC NUMBER OF A GRAPH AND INDOMINABLE GRAPHS

3.1 Domatic number

A partition of $V(G)$ into dominating sets of G is called a D-partition. A set $L \subset V$ is a loose set if for $v \in V-L$, $N(v) \cap L \neq L$. Clearly L is a loose set of G iff it is a dominating set of $\overline{G}$, the complementary graph of G. A partition of V into loose sets of G is called an L-partition. The maximum order of a D-partition of G is called the domatic number of G and is denoted by $d(G)$. The maximum order of a L-partition of G is denoted $\ell(G)$.

Proposition 3.1.1. For any graph G, $\ell(G) = d(\overline{G})$.

Proposition 3.1.2. (Ore). $d(G) \geq 2$ iff G has no isolated vertex.

Proposition 3.1.3. (Cockayne and Hedetniemi [9]). For any graph G, $d(G) \leq \delta(G) + 1$, where δ is the minimum degree of G. Thus for any G with no isolated vertex,

$$2 \leq d(G) \leq \delta(G) + 1.$$

A graph is called (domatically) full if $d(G) = \delta(G) + 1$.

In their paper [9] Cockayne and Hedetniemi cite some of the following examples of full graphs. It is a challenging problem to characterize those graphs which are full or have domatic number 2.

Proposition 3.1.4. (Cockayne and Hedetniemi [9]).

(i) Any tree T with at least 2 vertices is full with $d(T) = 2$.

(ii) C_{3n} is full with $d(C_{3n}) = 3$.

Jaegar and Payan [19] consider the Nordhaus-Gaddum type results in the following theorem.

Theorem 3.1.5. (Jaegar and Payan). For any graph G with p vertices

$$\gamma(G) \cdot \gamma(\overline{G}) \leq p,$$

and

$$\gamma(G) + \gamma(\overline{G}) \leq p + 1.$$

A by-product of their elegant proof of the theorem is the following result.

Corollary 3.1.6. (Jaegar and Payan). $\gamma(\overline{G}) \leq d(G)$.

Allan and Laskar [3] have given an alternative proof of the above result which is algorithmic in nature.

Let $\varepsilon(G)$ be the maximum number of end edges in a spanning forest of G.

Theorem 3.1.7. (Nieminen [24]). For any graph G with p vertices

$$\gamma(G) + \varepsilon(G) = p.$$

Cockayne and Hedetniemi consider the Nordhaus-Gaddum type theorems on domatic numbers in the following :

Theorem 3.1.8. (Cockayne and Hedetniemi [9]).

$$d(G) + d(\overline{G}) \leq p + 1, \quad \text{where} \quad |V(G)| = p.$$

Theorem 3.1.9. (Walikar, Acharya and Sampathkumar [32]). For any graph G of order p, $\gamma(G) \cdot \gamma(\overline{G}) = p$ if, and only if the following two conditions hold simultaneously :

(i) $\gamma(G) \cdot d(G) = p$

(ii) $\gamma(\overline{G}) = d(G)$.

Theorem 3.1.10. (Walikar, Acharya and Sampathkumar [32]). Let T be a tree of order $p \geq 2$. Then the following are equivalent :

(i) $\gamma(T) \cdot \gamma(\overline{T}) = p$

(ii) $\gamma(T) \cdot d(T) = p$

(iii) $\gamma(T) = \frac{p}{2}$

(iv) $\gamma(T) = \beta_0(T)$

(v) $T = T_1^+$, for some tree T_1.

In [1], Acharya and Walikar introduced a new invariant called the star-partition number, denoted $\gamma^*(G)$ and defined as the smallest order of the partition of the vertex set $V(G)$ of a graph G in which each set induces a star in G. With the help of this parameter, they characterized the trees having unique minimum dominating sets by proving $\gamma(T) = \gamma^*(T)$, for any tree T.

In [31], Walikar extends this result to a triangle-free graph and thereby derives the above result as a corollary.

Theorem 3.1.11. For any triangle-free graph G,

$$\gamma(G) = \gamma^*(G).$$

Corollary 3.1.12. (Acharya and Walikar [1]). For any tree T,

$$\gamma(T) = \gamma^*(T).$$

3.2 Indominable graphs

G is indominable if its vertex set may be partitioned into independent dominating sets. K_n, $\overline{K}_n$, connected bipartite graphs, complete r-partite graphs are some of the examples of indominable graphs [11].

A graph is uniquely n-colorable if the chromatic number $X(G) = n$ and there exists a unique partition of $V(G)$ into n independent sets.

Proposition 3.2.1. (Cockayne and Hedetniemi [11]). If G is uniquely n-colorable, then G is indominable.

Benzaken and Hammer [6] define a graph G to be domistable if every minimal dominating set is independent. Further they consider "weighted" graphs and call a graph domishold if G has the property that there exist positive real numbers associated to their vertices so that S is dominating if and only if the sum of the corresponding "weights" of vertices of S exceeds a certain threshold θ. G is said to be a threshold graph if real nonnegative numbers can be associated to its vertices so that two vertices are adjacent iff the sum of their weights exceeds a certain threshold. Several characterizations of threshold graphs are given by Chvatal and Hammer[8].

A graph is called split if the vertex set can be partitioned into a complete subgraph and an independent set.

Theorem 3.2.2. (Benzaken and Hammer[6]). Every threshold graph is domishold and has the following properties :

(i) G is split;

(ii) G is domistable;

(iii) G is an interval graph.

4. DOMINATION AND IRREDUNDANCE

A set S V is an irredundant set if for each $w \in S$, $N[w] \not\subset \bigcup_{v \in S-\{w\}} N[v]$. The irredundance number $ir(G)$ is the minimum cardinality taken over all maximal irredundant sets.

Proposition 4.1. (Cockayne and Hedetniemi). X is a minimal dominated set iff X is an irredundant set and a dominating set.

Proposition 4.2. (Cockayne and Hedetniemi). If X is a minimal dominating set, then X is a maximal irredundant set.

Proposition 4.3. (Cockayne and Hedetniemi).

$$ir(G) \leq \gamma(G).$$

Theorem 4.4. (Allan and Laskar [3], Bollabas and Cockayne [7]). For any graph G,

$$\gamma(G) \leq 2ir(G) - 1.$$

Corollary 4.5. $\{\frac{\gamma(G)+1}{2}\} \leq ir(G) \leq \gamma(G) \leq 2ir(G) - 1.$

Let $\Gamma(G)$ denote the maximum cardinality taken over all minimal dominating sets and $\beta_0(G)$ denote the independence number of G, and $IR(G)$ denote the maximum cardinality taken over all maximal irredundant sets of vertices of G. The following results relating to $ir(G)$, $IR(G)$, $\gamma(G)$, $\Gamma(G)$, $i(G)$, $\beta_0(G)$ are due to Cockayne and Thomason [13].

Theorem 4.6. (Cockayne and Thomason). If G has p vertices and $\gamma(G) + IR(G) = p$, then $\beta_0(G) = \Gamma(G) = IR(G)$.

Theorem 4.7. (Cockayne and Thomason). If G has p vertices and no isolated vertex, then

$$i(G) + IR(G) \leq 2p + 2 - 2\sqrt{2p},$$

and if equality holds, then $\beta_0(G) = \Gamma(G) = IR(G)$.

Theorem 4.8. (Cockayne and Thomason). If G is bipartite, then $\beta_0(G) = \Gamma(G) = IR(G)$.

5. DOMINATION, TOTAL DOMINATION AND CONNECTED DOMINATION

5.1 Domination and total domination

A set D of vertices is a total dominating set if each vertex of V is adjacent to some vertex in D. The cardinality of the smallest total dominating set is called the total domination number of G, denoted $\gamma_t(G)$. This parameter is only defined for graphs without isolated vertices. Clearly $\gamma(G) \leq \gamma_t(G)$.

The following results are due to Cockayne and Hedetniemi [12].

Theorem 5.1.1. If G is a connected graph with more than 2 vertices, then

$$\gamma_t(G) \leq [\tfrac{2p}{3}].$$

Theorem 5.1.2. If G has p vertices and no isolates, then

$$\gamma_t(G) \leq p - \Delta(G) + 1.$$

Theorem 5.1.3. If G has p vertices, no isolates and $\Delta(G) < p - 1$, then

$$\gamma_t(G) + \gamma_t(\overline{G}) \leq p + 2$$

with equality if and only if G or $\overline{G} = mK_2$.

Total domatic number is defined in [12] analogously. The largest number of sets in a partition of V into total dominating sets is called the total domatic number of G and denoted $d_t(G)$.

Theorem 5.1.4. (Cockayne and Hedetniemi). If G has p vertices, no isolates and $\Delta(G) < p - 1$, then

$$d_t(G) + d_t(\overline{G}) \leq p - 1,$$

with equality if and only if G or $\overline{G} = C_4$.

Theorem 5.1.5. (Allan, Laskar, Hedetniemi [4]). If G is a graph with p vertices, such that each component has at least 3 vertices, then

$$i(G) + \gamma_t(G) \leq p.$$

Corollary 5.1.6. If G is connected with more than 2 vertices, then

$$\gamma(G) + \gamma_t(G) \leq p.$$

5.2 Domination and connected domination

If a dominating set D is such that the induced subgraph D has no isolates then it becomes a total dominating set. One can define analogously a connected, a cycle, or a path, etc., dominating set D if the induced subgraph D has the corresponding property. Along this line new results similar to those outlined in the above sections can be obtained.

A dominating set D is a connected dominating set if it induces a connected subgraph in G. The minimum cardinality taken over all connected dominating sets is called the connected domination number denoted by $\gamma_c(G)$.

Sampathkumar and Walikar consider the connected domination number and have the following results in [26].

Proposition 5.2.1. For any connected graph G with at least 3 vertices, $\gamma_c(G) \leq p - 2$.

Proposition 5.2.2. For any connected graph G with p vertices and q edges and with maximum degree Δ,

$$\{\frac{p}{\Delta+1}\} \leq \gamma_c(G) \leq 2q - p.$$

Theorem 5.2.3. If G is connected with at least 4 vertices, such that both G and $\overline{G}$ are connected,

$$\gamma_c(G) + \gamma_c(\overline{G}) \leq p(p-3)$$

and equality holds iff $G = P_3$.

Open Problems

We conclude this paper by raising the following structural characterization problems.

Problem 1. Characterize the classes of graphs for which the following relations hold.

(a) $\gamma(G) = \frac{p}{\Delta+1}$,

(b) $\gamma(G) = i(G)$,

(c) $\gamma(G) = \beta_0(G)$,

(d) $\gamma(G) = \gamma^*(G)$,

(e) $\gamma(G) \cdot \gamma(\overline{G}) = p$,

(f) $\gamma(G) \cdot d(G) = p$,

(g) $\gamma(G) + \beta_0(G) = p$.

Problem 2. Let Q_n denote the n-dimensional cube [18]. Calculate the domination number of Q_n.

REFERENCES

1. B.D. Acharya and H.B. Walikar, On the graphs having unique minimum dominating sets, Abstract No.2, Graph Theory Newsletter, 8(15), (1979), 1.
2. R.B. Allan and R. Laskar, On domination and independent domination numbers of a graph, Discrete Math., No.2, 23(1978), 73-76.
3. R.B. Allan and R. Laskar, On domination and some related topics in graph theory, Proc. of Ninth S.E. Conference on Combinatorics, Graph Theory and Computing, Utilitas Math., (1979), 43-56.
4. R.B. Allan and R. Laskar, S.T. Hedetniemi, On the total domination of a graph. To appear.
5. C. Berge, Theory of Graphs and Its Applications, Methuen, London, 1962.
6. C. Benzaken, and P.L. Hammer, Linear separation of dominating sets in graphs, Annals of Discrete Math., Vol.3, 1978.
7. B. Bollabas, and E.J. Cockayne, Graph theoretic parameters concerning domination, independence and irredundance, J. Graph Theory (to appear).
8. V. Chvatal, and P.L. Hammer, Set packing problems and threshold graphs, Annals of Discrete Math., Vol.1, 1977.
9. E.J. Cockayne and S.T. Hedetniemi, Towards a theory of domination in graphs, Networks, Fall 1977, 247-271.
10. E.J. Cockayne and S.T. Hedetniemi, Independence graphs, Proc. of Fifth S.E. Conference on Combinatorics, Graph Theory and Computing, Utilitas Math.,(1974), 471-491.
11. E.J. Cockayne and S.T. Hedetniemi, Disjoint independent dominating sets in graphs, Discrete Math., 15 (1976), 213-222.
12. E.J. Cockayne and S.T. Hedetniemi, Total domination in graphs. To appear.
13. E.J. Cockayne and A.G. Thomason, Contributions to the theory of domination, independence and irredundance in graphs, Research Report, University of Victoria, April 1979.

14. E.J. Cockayne, Domination of undirected graphs : A survey; in "Theory and Applications of Graphs", (Proc. International Conf., Western Michigan Univ., Kalamazoo, Michigan 1976), Lecture Notes in Math., 642 (1978), 141-147, Springer Verlag, Berlin.

15. D.A. Drake and J.W. Freeman, Partial t-spreads and group constructible (s,r,μ)-nets (preprint).

16. J. Edmonds and D.R. Fulkerson, Bottleneck extrema, J. Comb. Theory, 8(1970), 299-306.

17. T. Hamanda and I. Yoshimura, Traversability and connectivity of the middle graph of graph, Discrete Math., 14 (1976), 247-255.

18. F. Harary, Graph Theory, Addison Wesley, 1972.

19. F. Jaegar and C. Payan, Relations der type Nordhaus-Gaddum pour le nombre d'absorption d'un graphe simple, C.R. Acad. Sc. Series A, Paris, 274 (1972). 728-730.

20. J.G. Kalbflisch, R. Stanton, J.D. Horton, On covering sets and error correcting codes, J. Comb. Th., 11A (1971), 233-250.

21. C.L. Liu, Introduction to Combinatorial Mathematics, McGraw-Hill, N.Y., 1968.

22. A. Meir and J.W. Moon, Relations between packing and covering numbers of a tree, Pac. J. Math., 61 (1975), 225-233.

23. S. Mitchell and S.T. Hedetniemi, Independent domination in trees, Proc. Eighth S.E. Conference on Combinatorics, Graph Theory and Computing, (1977), 489-509.

24. J. Nieminen, Two bounds for the domination number of a graph, J. Inst. Math. Applics., 14 (1974), 183-187.

25. O. Ore, Theory of Graphs, Amer. Math. Soc. Colloq. Publ., 38, Providence, 1962.

26. E. Sampathkumar and H.B. Walikar, Some bounds on the connected domination number of a graph. (Preprint).

27. D. Sumnar and John I. Moore, Jr., Notices, AMS., Oct. 1979, A-569.

28. V.G. Vizing, A bound on the external stability number of a graph, Doklady A.N., 169, (1965), 729-731.

29. H.B. Walikar, E. Sampathkumar and B.D. Acharya, Two new bounds for the domination number of a graph. (Preprint).

30. H.B. Walikar, B.D. Acharya and E. Sampathkumar, Recent developments in the theory of domination in graphs and its applications, MRI, Lecture Notes in Math., No.1, 1978.

31. H.B. Walikar, On star-partition number of a graph. (Preprint).

32. H.B. Walikar, B.D. Acharya and E. Sampathkumar, On an extremal problem concerning a Nordhaus-Gaddum type result in the theory of domination in graphs. (Preprint).

THE LOCAL CENTRAL LIMIT THEOREM FOR STIRLING NUMBERS OF THE SECOND KIND AND AN ESTIMATE FOR BELL NUMBERS

V. V. MENON
Applied Mathematics Section
Institute of Technology
Banaras Hindu University
Varanasi 221 005

1. INTRODUCTION

The Stirling number of the second kind,

$$S(n,r) = \frac{r^n}{r!} \sum_{k=0}^{r} (-1)^k \binom{r}{k} \left(1 - \frac{k}{r}\right)^n = \frac{\Delta^r 0^n}{r!},$$

denotes the number of (i) distinct partitions of a set of n elements into r non-empty subsets, or (ii) equivalence relations with r equivalence classes, or (iii) Borel fields with r atoms on a set of size n. The Bell numbers B_n are defined as

$$B_n = \sum_{r=1}^{n} S(n,r).$$

We consider the random variable which assumes the value r with probability $B_n^{-1} S(n,r)$, $1 \leq r \leq n$. A CLT (central limit theorem) for this variable was established by Harper [1] who used an interesting property of the generating function to show that the variable is a sum of independent nonidentical variables. However, the local CLT (which yields estimates for the probabilities as well as the rate of convergence to the normal law) is difficult to obtain by his method. Much more difficult and stronger result is the series of further approximations (the so-called Edgeworth expansion), which is usually obtained by the saddle-point method applied to the characteristic function. In the present paper, we obtain such a series expansion and hence the local theorem, by an elementary argument. In addition, we demonstrate a simple method for estimating

---

AMS 1970 Subject Classification : Primary 05A10, 60F05; Secondary 05A17, 05A05.

Key words and phrases : Stirling numbers of the second kind, Bell numbers, local central limit theorem, asymptotic estimates, rate of convergence to normal law.

B_n, the mean (or, Σ $rS(n,r)$) and the variance (or, Σ $r^2S(n,r)$). One may compare the method of estimating B_n upto $O(n^{-2})$ used by Szekeres and Binet [3], which involves complex variable techniques.

An elementary method was illustrated in Menon [2] for asymptotic estimation of quantities like $S(n,r)$ which arise from the principle of inclusion and exclusion. For a certain range of values of r, which covers the maximum of $S(n,r)$, it was shown there that $n^{-1} \log\{S(n,r)(2\pi)^{1/2}\}$ is approximately $f(p) + g(p)$, where $p = r/n$, $g(p)$ is of smaller order of magnitude than $f(p)$, the maximum of $f(p)$ occurs at $p = \beta^{-1}$ where $\beta \exp \beta = n$, and the maximum p_o of $f(p) + g(p)$ occurs close to β^{-1}. The maximum of $S(n,r)$ was then located and estimated; and the occurrence of β is natural there.

In our analysis, the maximum of $S(n,r)$ occupies a prominent place. When we expand the above estimate $f(p) + g(p)$ in a Taylor series around its maximum p_o, the first derivative vanishes and the second derivative is negative because of concavity, so that by an obvious change of the variable, we obtain the local CLT and further approximations to any degree of accuracy. In this sense, the maximum is more natural than the mean for normalizing the variable to obtain the local CLT.

From the local theorem, B_n is easily estimated by approximating the sum by an integral involving the normal density: the trapezoidal rule or the Euler-McLaurin summation formula yield the result. Similarly for Σ $rS(n,r)$, etc.

The estimates obtained in the present paper can be carried to any degree of accuracy; we restrict ourselves to estimation upto and including the term involving n^{-2}, except the basic result (Theorem 1) which implies the Edgeworth series expansion for the probability function. The ratio of B_n to $\max S(n,r)$, and the estimate for B_n is found in Theorem 2; the local theorem with two different normalizations is obtained in Theorem 3 and Corollary 1; finally, the mean and variance are estimated in Theorem 4, and the local theorem with the usual normalization occurs as Corollary 2. The exact rates of convergence are obtained for all the local theorems. As a simple consequence, we find for instance that $S(n,r)$ is approximately normal over the range of r such that $r - r_n = o(n^{2/3}\beta^{-1})$, i.e., $n^{-2/3}(r-r_n) \log n \to o$, where $r_n \sim n\beta^{-1}$

is the location of the maximum of $S(n,r)$.

2. PRELIMINARIES

In [2] it was shown that

$$n^{-1}\log[S(n,r)(2\pi)^{\frac{1}{2}}] = F(p) + R = f(p) + g(p) + R, \qquad \dots (1)$$

where

$$p = r/n,\ f(p) = (1-p)\log np + p;$$

$$g(p) = -(2n)^{-1}\log np - p\exp(-p^{-1}) - (12n^2)^{-1}p^{-1} + [(2n)^{-1} - (8n^2)^{-1}p^{-3} + (3n^2)^{-1}p^{-2}]\exp(-p^{-1}) - [(1+p)/2 - 3(4n)^{-1}p^{-2}]\exp(-2p^{-1}) - [(p^{-1}/2) + (2/3) + (p/3)]\exp(-3p^{-1}); \qquad \dots (2)$$

$$R = O[n^{-4}(\log n)]. \qquad \dots (3)$$

The remainder R is of the above order for values of p such that $p \sim p_o$. The series (1) can be continued to any number of terms. We shall ignore R in what follows.

The maximum of $f(p)$ occurs at $p = \beta^{-1}$. The maximum p_o of $f(p) + g(p)$ is close to β^{-1}, and is easily shown to be estimated by the following series in powers of n^{-1},

$$p_o = \beta^{-1} - [n(\beta+1)]^{-1}(\beta+3/2) + [24n^2(\beta+1)^3]^{-1}\beta(12\beta^4+24\beta^3+14\beta^2+4\beta-1) + O(n^{-3}\beta^4), \qquad \dots (4)$$

where $\beta\exp\beta = n$.

The following theorem with the Lemma 1 contains our basic expansion.

Theorem 1. With the notation of (1),(2) and (4), denote the kth derivative of F by $F^{(k)}$, and write

$$\sigma^2 = [-F^{(2)}(p_o)]^{-1}. \qquad \dots (5)$$

Let $n \to \infty$, and consider the normalization

$$t = n^{\frac{1}{2}}(p-p_o)\sigma^{-1}. \qquad \dots (6)$$

Then, for any integer $m \geq 3$, if $t^{m+1}n^{-(m-1)/2} \to o$, we have

$$S(n,r) = \exp[nF(p_o)].\varphi(t).\exp[\sum_{k=3}^{m} a_k t^k n^{-(k-2)/2} + O(t^{m+1}n^{-(m-1)/2})], \dots (7)$$

where

$$\varphi(t) = (2\pi)^{-\frac{1}{2}} \exp(-t^2/2),$$

$$a_k = (k!)^{-1} \sigma^k F^{(k)}(p_o).$$

Proof. The function F in (1) is differentiable any number of times near its maximum p_o, and we obtain the Taylor series

$$nF(p) = n[F(p_o) + 2^{-1}(p-p_o)^2 F^{(2)}(p_o) + \sum_{k=3}^{m} (k!)^{-1}(p-p_o)^k F^{(k)}(p_o) +$$

$$+ \{(m+1)!\}^{-1} (p-p_o)^{m+1} F^{m+1}\{p_o+\theta(p-p_o)\}],$$

where $0 < \theta < 1$. The first derivative is missing because the maximum of F occurs at p_o. Consider now the remainder term. Note that $p_o \sim \beta^{-1}$, $\beta \sim \log n$, and as will be shown in the next lemma, $\sigma \sim \beta^{-1}$. Thus the condition $t^{m+1} n^{-(m-1)/2} \to o$ is the same as $(p-p_o)\sigma^{-1} n^{1/(m+1)} \to o$, which implies that $p \sim p_o$. The derivative in the remainder is therefore asymptotically equal to $F^{(m+1)}(p_o)$. However, the dominant part of $F^{(m+1)}(p_o)$ is $f^{(m+1)}(p_o)$ because $g^{(m+1)}(p_o)$ is of the order of $n^{-1} \beta^{2(m+1)}$. Since $f^{(2)}(p) = -p^{-2} - p^{-1}$, we find that $|f^{(m+1)}(p_o)| \sim m!\, \beta^{m+1}$. Hence the remainder term is asymptotically equal to $(m+1)^{-1}(p-p_o)^{m+1} \beta^{m+1}$, and therefore to $(m+1)^{-1}t^{m+1}n^{-(m+1)/2}$.

Now substitute for t in the above Taylor series, and multiply through by n to obtain (7). This proves the theorem.

It remains to evaluate the derivatives $F^{(k)}(p_o)$ using (4) and the Taylor series around $p = \beta^{-1}$. The values for $2 \leq k \leq 6$ are required in order to obtain the final result upto the term involving n^{-2}.

Lemma 1.

$$\sigma^2 = [\beta(\beta+1)]^{-1}[1-\{2n(\beta+1)^2\}^{-1}\beta(2\beta^3+6\beta^2+7\beta+2)+\{24n^2(\beta+1)^4\}^{-1}\beta^2(12\beta^6+24\beta^5-12\beta^4-48\beta^3-26\beta^2-10\beta+1)+O(n^{-3}\beta^6)];$$

$$3!a_3 = [\beta(\beta+1)]^{-3/2}[\beta^2(2\beta+1)-\{4n(\beta+1)^2\}^{-1}\beta^3(4\beta^5+8\beta^4-2\beta^3-16\beta^2-15\beta-2)+O(n^{-2}\beta^8)];$$

$$4!a_4 = -[\beta(\beta+1)]^{-2}[2\beta^3(3\beta+1)+\{n(\beta+1)^2\}^{-1}\beta^4(\beta^6-6\beta^5-15\beta^4-20\beta^3+21\beta^2+19\beta+2)+O(n^{-2}\beta^{10})];$$

$$5!a_5 = [\beta(\beta+1)]^{-5/2}[3!\beta^4(4\beta+1)+O(n^{-1}\beta^{10})];\quad 6!a_6 = -[\beta(\beta+1)]^{-3}[4!\beta^5(5\beta+1)+O(n^{-1}\beta^{12})];$$

In general,

$$(-1)^{k-1}\ k!a_k = (k-1)!\ [1+O(\beta^{-1})].$$

2. BELL NUMBERS

We now use the basic expansion (7) to estimate for large n the Bell number

$$B_n = \sum_{r=0}^{n} S(n,r). \qquad \ldots \quad (8)$$

The method is as follows. The sum is first restricted to a range of values of r around the maximum so that the basic expansion can be applied and the contribution outside the range is negligible, and the sum approximated by an integral related to the Normal density φ.

Choose two sequences $t_1 \to -\infty$ and $t_2 \to \infty$ from the possible values of t such that $n^{-1/2}t_i^3 \to o$, $i = 1,2$; for instance, let $|t_i| \sim n^{1/8}$, $(1 = 1,2)$ so that $n^{-1/2}|t_i|^3 \sim n^{-1/8}$, and $\exp(-t_i^2/2) = o(n^{-K})$, for any fixed K, and our expansions can be continued to any number of powers of n^{-1} without being affected by such terms.

Let K be any fixed positive integer, say $K = 4$. Consider $\Sigma\ S(n,r)$ over values of r such that $\{t < t_1$ or $t > t_2\}$. There are at most n terms in the sum and no term can exceed the term corresponding to $t = t_1$ or t_2, say t_2, because $S(n,r)$ increases upto its maximum and then decreases. Consequently the ratio of this sum to the term corresponding to $t = o$ is, as estimated by (7), less than $n(2\pi)^{-1/2}\ \exp(-t_2^2/2)\ \exp[O(n^{-1/8})]$, which is $o(n^{-K})$, and the ratio to the sum over $\{t_1 \leq t \leq t_2\}$ is still smaller. Therefore we may write

$$B_n = [\sum_{t_1}^{t_2} S(n,r)][1+o(n^{-K})]. \qquad \ldots \quad (9)$$

We have arrived in the range $\{t_1 \leq t \leq t_2\}$ where the basic expansion (7) is valid with any $m \geq 3$, because $t^{m+1}\ n^{-(m-1)/2} = (n^{-1/2}t^3)^{(m+1)/3}\ n^{-(m-2)/3} \to o$ for any m. We substitute (7) into the right side of (9) and use the Euler-McLaurin formula as follows: if y_i is the value of a function at equally spaced points $x_i (i = o,1,\ldots$ $\ldots,n-1)$ with step-length h, then

$$h \sum_{i=0}^{n-1} y_i = \int_{x_o}^{x_n} y(x)dx - \tfrac{1}{2}h(y_n-y_o) + 12^{-1}h^2(y_n'-y_o') - 720^{-1}h^3(y_n'''-y_o''') + \ldots .$$

In our case, the step-length for t is $n^{-1/2}\sigma^{-1} \sim n^{-1/2} \log n$, and the function is $\exp(-t^2/2)\exp[O(n^{-1/8})]$. Clearly, the correction terms involving $\varphi(t_1)$ and $\varphi(t_2)$ are $o(n^{-k})$, and the sum on the right side of (9) is approximated by the corresponding integral

$$\sum_{t_1}^{t_2} S(n,r) = (n^{1/2}\sigma)\exp[nF(p_o)][\int_{t_1}^{t_2} \varphi(t)R(t)dt][1+o(n^{-k})], \qquad \ldots (10)$$

where

$$R(t) = \exp[\sum_{k=3}^{m} a_k t^k n^{-(k-2)/2} + O(t^{m+1}n^{-(m-1)/2})].$$

It remains to estimate the integral on the right side of (10). Since we require the final estimates to the term containing n^{-2}, we choose $m = 6$. (For the term containing n^{-1}, we need $m = 4$). The evaluation of the integral is simplified by the following lemma.

Lemma 2. Let $-t_1 \sim t_2 = o(n^{1/6})$, and $t_2 \to \infty$ such that $\varphi(t_2) = o(n^{-3})$, where $\varphi(t) = (2\pi)^{-1/2}\exp(-t^2/2)$. Also let $a_k = O(1)$ for $3 \leq k \leq 8$. Then

$$\int_{t_1}^{t_2} t^k \varphi(t)dt = \int_{-\infty}^{\infty} t^k \varphi(t)dt + O(n^{-3});$$

$$\int_{t_1}^{t_2} \varphi(t)\exp[\sum_{k=3}^{8} a_k t^k n^{-(k-2)/2} + O(t^9 n^{-7/2})]dt$$

$$= 1 + n^{-1}(2^{-1}.5.3a_3^2+3a_4) + n^{-2}(24^{-1}.11.9.7.5.3a_3^4 + 2^{-1}.9.7.5.3a_3^2a_4 + 7.5.3a_3a_5 + 2^{-1}.7.5.3a_4^2 + 5.3a_6) + O(n^{-3}).$$

Proof. Integrating $t^k\varphi(t)$ over (t_2,∞) by parts repeatedly, the leading term is $t_2^{k-1}\varphi(t_2)$ and the other terms are smaller. The result is well known for $k = o,1$. Thus the integral over (t_2,∞), and similarly that over $(-\infty,t_1)$, is negligible. This establishes the first result.

For the second result, the second exponential can be expanded upto terms involving $n^{-5/2}$ and the remaining terms are $O(t^8n^{-3})$. Then a term-by-term integration, applying the first result, gives the corresponding moments of the standard normal distribution. The lemma is proved.

Now apply the Lemma 2 to the integral on the right side of (10) with $K = 3$, and the values of a_k as in Lemma 1. After some simplification we obtain the last expansion on the right side of (11) in the following theorem, in view of (9). Further simplification of (11) leads to (12).

<u>Theorem 2</u>. $B_n = \sum_{r=1}^{n} S(n,r) =$

$$= n^{1/2}\sigma \exp[nF(p_o)][1+(24n)^{-1}(\beta+1)^{-3}\beta(2\beta^2-4\beta-1)-(288n^2)^{-1}(\beta+1)^{-6}\beta^2(36\beta^8+96\beta^7- \\ -96\beta^6-1296\beta^5-2125\beta^4-944\beta^3+411\beta^2+484\beta+4^{-1}.71)+O(n^{-3}\beta^6)], \quad \ldots \quad (11)$$

where

$$\sigma = [\beta(\beta+1)]^{-1/2}[1-(4n)^{-1}(\beta+1)^{-2}\beta(2\beta^3+6\beta^2+7\beta+2)+(48n^2)^{-1}(\beta+1)^{-4}\beta^2(6\beta^6-12\beta^5 \\ -108\beta^4-186\beta^3-2^{-1}.271\beta^2-52\beta-5)+O(n^{-3}\beta^6)],$$

and

$$nF(p_o) = (2\pi)^{1/2} \max_r \log S(n,r),$$

$$= n(\beta-1+\beta^{-1})-(\beta/2)-1+(24n)^{-1}(\beta+1)^{-1}\beta(12\beta^2+22\beta+13)+(24n^2)^{-1}(\beta+1)^{-3}\beta^2(3\beta^5 \\ +23\beta^4+64\beta^3+77\beta^2+2^{-1}.79\beta+7)+O(n^{-3}\beta^6).$$

Equivalently,

$$B_n = (\beta+1)^{-1/2}\exp[n(\beta-1+\beta^{-1})-1][1-(24n)^{-1}(\beta+1)^{-3}\beta^2(2\beta^2+7\beta+10) \\ -(288n^2)^{-1}(\beta+1)^{-6}\beta^2(\beta^6+1205\beta^5+4^{-1}.9481\beta^4+1071\beta^3-455\beta^2-456\beta+2^{-1}.25)+O(n^{-3}\beta^6)]. \quad \ldots \quad (12)$$

3. THE LOCAL CENTRAL LIMIT THEOREM

Let us now consider the probabilities

$$B_n^{-1}\, S(n,r) \quad \ldots \quad (13)$$

for the transformed variable

$$t = n^{1/2}(p-p_o)\sigma^{-1} = (r-r_n)(n\sigma^2)^{-1/2}, \quad \ldots \quad (14)$$

where the maximum of $S(n,r)$ occurs at $r = r_n = n\,p_o$. We choose $m = 3$ in Theorem 1, and find using Lemma 1 that for $n^{-1/2}t^3 \to o$,

$$S(n,r) = \exp[nF(p_o)].\varphi(t).\exp[a_3 n^{-1/2}t^3+O(n^{-1}t^4)]$$

$$= \exp[nF(p_o)].\varphi(t).[1+6^{-1}(\beta+1)^{-3/2}\beta^{1/2}(2\beta+1)n^{-1/2}t^3+O(n^{-1}t^4)].$$

Divide now by the estimate (11) of Theorem 2, and simplify to obtain the following theorem.

Theorem 3 (Local CLT). With the notation of (14) and Theorem 2, let $n^{-1/2}t^3 \to o$ as $n \to \infty$. Then

$$[B_n^{-1}S(n,r)] \div [(n^{1/2}\sigma)^{-1}\varphi(t)] = 1-(24n)^{-1}(\beta+1)^{-3}\beta(2\beta^2-4\beta-1)$$
$$+6^{-1}(\beta+1)^{-3/2}\beta^{1/2}(2\beta+1)n^{-1/2}t^3+O(n^{-2}\beta^3+n^{-1}t^4). \quad \ldots (15).$$

The above normalization of the variable can be simplified, although the convergence will be slower. For example, a minor rearrangement of (15) yields the following

Corollary 1. Let

$$u = [\beta(\beta+1)/n]^{1/2}(r-n\beta^{-1}),$$

and let $n^{-1/2}u^3 \to o$ as $n \to \infty$. Then

$$[B_n^{-1}S(n,r)] \div [\{\beta(\beta+1)/n\}^{1/2}\varphi(u)]$$
$$= 1+(2n^{1/2})^{-1}\beta^{1/2}(\beta+1)^{-1/2}[3^{-1}(\beta+1)^{-1}(2\beta+1)u^3-(2\beta+3)u]-(24n)^{-1}(\beta+1)^{-3}\beta(12\beta^3$$
$$+35\beta^2+32\beta+14)+O(n^{-1}\beta^2u^2+n^{-3/2}\beta^3u^3+n^{-2}\beta^4).$$

4. ESTIMATE OF MEAN AND VARIANCE

We have seen above that the shape of the distribution (13) near its maximum $r = r_n = np_o$ is approximately normal and that almost all the probability is concentrated in the range specified by Theorem 3. Hence, approximately, the mean of the distribution is r_n and the variance is $n\sigma^2$. We shall now estimate these parameters upto the term containing n^{-1}. There is sufficient information in the previous sections to also find the terms containing n^{-2}.

We employ the natural method developed in the Section 2 for estimating the mean and variance as well, and thus obtain indirectly estimate for $B_n^{-1}\ B_{n+1}$ and $B_n^{-1}\ B_{n+2}$. As in (10), the sum is first restricted to the range $t_1 \leq t \leq t_2$ and then converted to an integral over $(-\infty,\infty)$. For instance, the mean is estimated as follows.

$$B_n^{-1}\sum_0^n (r-r_n)S(n,r) = B_n^{-1}\, n^{1/2}\sigma[\sum_{t_1}^{t_2} t\, S(n,r)][1+o(n^{-3})]$$

$$= B_n^{-1}(n^{1/2}\sigma)^2 \exp[nF(p_o)][\int_{t_1}^{t_2} t\,\varphi(t)R(t)dt][1+o(n^{-3})] \qquad \ldots (16)$$

$$= n^{-1/2}3a_3+n^{-3/2}(6^{-1}9.7.5.3a_3^3+7.5.3a_3a_4+5.3a_5)+O(n^{-5/2}),$$

where the Lemma 1 is now used. We thus obtain the following estimates.

<u>Theorem 4</u>. Let the random variable X assume the value r with probability $B_n^{-1}S(n,r)$, $1 \leq r \leq n$. Then the mean and variance of X are

$$E(X) = r_n+(\beta+1)^{-2}(\beta+1/2)-(24n)^{-1}(\beta+1)^{-5}\beta(12\beta^6+48\beta^5+72\beta^4+118\beta^3+10\beta^2-18\beta-1)+O(n^{-2}\beta^5);$$

$$= n\beta^{-1}-[2(\beta+1)^2]^{-1}(2\beta^2+3\beta+2)-(24n)^{-1}(\beta+1)^{-5}\beta^2(8\beta^3+70\beta^2+5\beta-12)+O(n^{-2}\beta^5).$$

$$V(X) = n\sigma^2[1+(4n)^{-1}(\beta+1)^{-3}\beta(8\beta^2+4\beta+1)+O(n^{-2}\beta^4)],$$

$$= n[\beta(\beta+1)]^{-1}[1-(2n)^{-1}(\beta+1)^{-3}\beta(2\beta^4+8\beta^3+11\beta^2+9\beta+2)+O(n^{-2}\beta^4)],$$

and the mode of X occurs at $r = r_n$ where

$$r_n = n\beta^{-1}-(\beta+1)^{-1}(\beta+3/2)+(24n)^{-1}(\beta+1)^{-3}\beta(12\beta^4+24\beta^3+14\beta^2+4\beta-1)+O(n^{-2}\beta^4).$$

Let us also examine the convergence to normal law with the usual normalization. In view of (7),(11) and Theorem 4, we have the following result.

<u>Corollary 2</u>. Let, for the variable X of Theorem 4,

$$z = [X-E(X)][V(X)]^{-1/2}.$$

Then, if $n^{-1/2}z^3 \to o$ as $n \to \infty$,

$$[B_n^{-1}S(n,r)]\div[\{V(X)\}^{-1/2}\varphi(z)] = 1+n^{-1/2}(\beta+1)^{-3/2}\beta^{1/2}(2\beta+1)[6^{-1}z^3-2^{-1}(\beta+1)z]$$

$$-(24n)^{-1}(\beta+1)^{-3}\beta(12\beta^4+36\beta^3+89\beta^2+38\beta+8)+O(n^{-1}\beta^2z^2+n^{-3/2}\beta^3z^3+n^{-2}\beta^4).$$

Of the normalizations in the Theorem 3 and the two corollaries, the fastest rate of convergence appears to be achieved by the normalization (14) of Theorem 3. The normalization in Corollary 1 is simple in form.

REFERENCES

1. L.H. Harper, Stirling behaviour is asymptotically normal, Ann. Math. Statist., 38(1967), 410-414.

2. V.V. Menon, On the maximum of Stirling numbers of the second kind, J. Comb. Theory, 15(1973), 11-24.

3. G. Szekeres and F.E. Binet, On Borel fields over finite sets, Ann. Math. Statist., 28(1957), 494-498.

A FAMILY OF HYPO-HAMILTONIAN GENERALIZED PRISMS

S. P. MOHANTY
DALJIT RAO
Department of Mathematics
Indian Institute of Technology
Kanpur 208 016

ABSTRACT

In this paper we construct a family of hypo-hamiltonian generalized prisms with $4k + 2$ vertices $k \neq 1,3$. This family gives us new cubic-hypohamiltonian graphs for $k > 5$.

INTRODUCTION

In this paper we consider undirected graphs and call the number of vertices of a graph G the order of G. A graph G is hamiltonian if it has a circuit passing through every vertex of G. It is hypo-hamiltonian if it is not hamiltonian but every vertex deleted subgraph $G-v$ is hamiltonian. As minimum degree in a hypo-hamiltonian graph is three, cubic hypo-hamiltonian graphs are minimally hypo-hamiltonian. These graphs have been obtained in [1] and [3]. Here we consider the existence of cubic hypo-hamiltonian graphs in the special context of permutation graphs.

The concept of permutation graph of a graph G was introduced in 1967 by Chartrand and Harary in [2]. Let G be a graph with $V(G) = \{1,2,\ldots,n\}$ and let $\pi = (\pi(1), \pi(2)\ldots \pi(n))$ be a permutation on $V(G)$. Let G_1 and G_2 be two vertex disjoint copies of G such that the vertex i of G is labelled a_i in G_1 and b_i in G_2. Then the permutation graph (G,π) of the graph G consists of the two graphs G_1, G_2 and n additional edges $\{a_i, b_{\pi(i)}\}_{i=1}^{n}$ joining them.

In [5] Klee constructed permutations π for which (c_n,π) $(n \geq 3)$ is non-hamiltonian. They are called generalized n-prisms there. He proved the existence of a non-hamiltonian generalized prism with $4k + 2$ vertices for all $k \geq 2$, $k \neq 3$. Here we prove that, in fact, we have hypo-hamiltonian generalized prisms in each case. In particular we get new cubic hypo-hamiltonian graphs of order $4k + 2$, $k > 5$.

This family gives us two new cubic hypo-hamiltonian graphs of order 30.

PRELIMINARIES

The permutation $\pi_{m,n}$ is given by

$$\pi_{m,n}(i) = \text{residue of } im(\text{mod } n) \text{ if } 1 \leq i \leq n-1$$

$$\pi_{m,n}(n) = n,$$

where m and n are relatively prime and $1 \leq m \leq n/2$.

The generalized Petersen graph $G(n,d)$ has $V(G(n,d)) = \{u_i, v_i\}_{i=1}^{n}$ and $E(G(n,d)) = \{u_i u_{i+1}, u_i v_i, v_i v_{i+d}\}_{i=1}^{n}$ where $1 \leq d \leq n-1$, $n \neq 2d$ all the suffixes are to be read modulo n. We have

Theorem 1. $G(n,m) \simeq (c_n, \pi_{m,n})$ if $(m,n) = 1$.

Proof. Let $a_1 a_2 \ldots a_n a_1$ and $b_1 b_2 \ldots b_n b_1$ be consecutive Labellings along the two cycles c_n. Let k be such that $km \equiv 1(\text{mod } n)$. The map $u_i \to b_i$ and $v_i \to a_{ki}$ is seen to be an isomorphism.

Robertson [7] and Bondy [1] have established that $G(n,2)$ is non-hamiltonian if and only if $n \equiv 5(\text{mod } 6)$. Hence, $(c_n, \pi_{2,n})$ is non-hamiltonian if and only if $n \equiv 5(\text{mod } 6)$.

Let $\pi_1 \in S_{n_1}$ and $\pi_2 \in S_{n_2}$. Then catenate $\pi = (\pi_1, \pi_2) \in S_{n_1+n_2}$ where $\pi(i) = \pi_1(i)$ for $1 \leq i \leq n_1$ and $\pi(i) = n_1 + \pi_2(i-n)$ for $n_1 < i \leq n_1+n_2$. The operation of catenation can be extended in an obvious way to an arbitrary finite sequence of permutations. If $\pi \in S_{n-1}$ then $(\pi,(1)) \in S_n$ is denoted by π'. If $\pi \in S_n$ and $\pi(n) = n$ then $\bar{\pi}$ denotes the restriction of π to $\{1,2,3,\ldots,n-1\}$.

Let G be a graph. A pair (a,b) of vertices of G is called 'good' in G if G has a spanning path with end points a and b. Similarly, $((a,b),(c,d))$ is called good in G if G has a spanning subgraph consisting of two vertex disjoint paths one having end points a and b and the other c and d. Klee [5] called a permutation $\pi \in S_n$ 'bad' if none of (a_1,b_1), (a_n,b_n), (a_1,b_n), (a_n,b_1), $((a_1,b_n), (a_n,b_1))$ and $((a_1,b_1), (a_n,b_n))$ is 'good' in the graph (P_n,π). Otherwise π is 'good'. He proved the following.

Theorem 2 (Klee [5]). If $\pi \in S_n (n \geq 3)$, $\pi(n) = n$ then (c_n, π) is non-hamiltonian if and only if $\bar{\pi}$ is bad.

Theorem 3 (Klee [5]). If $\pi_i \in S_{n_i}$ for $1 \leq i \leq k$ and the following three conditions hold then the catenate $\pi = (\pi_1, \pi_2, \ldots, \pi_k)$ is a bad permutation.

(a) for each i, either $n_i = 1$ or π_i is bad,

(b) there is an even number (possibly zero) of 1's among the n_i's, and

(c) $1 < n_1$, $1 < n_k$ and no two 1's among the n_i's appear consecutively.

Theorem 4 (Klee [5]). For odd $n \geq 3$ there exists a generalized n-prism not admitting a hamiltonian cycle if and only if n is neither 3 nor 7.

The graphs for $n \neq 11$ in Theorem 4 were obtained by taking $\pi_i = \bar{\pi}_{2,5} = (2413)$ for $n_i > 1$ in Theorem 3 suitably. For $n = 11$, the example of $(c_{11}, \pi_{2,11})$ was given.

The first graph $(c_5, (24135))$ in the sequence of graphs in Theorem 4 is hypo-hamiltonian as it is the well known Petersen graph. The second graph $(c_9, \bar{\pi}_{2,5}, \bar{\pi}_{2,5}, (1))$ is also hypo-hamiltonian as it is isomorphic to Sousselier's hypo-hamiltonian graph on eighteen vertices. This observation helped us to construct a new family.

We call $\pi \in S_n$ hypo-good if $\pi(n) = n$, (i) $\bar{\pi}$ is bad, (ii) both (a_1, a_n) and (b_1, b_n) are good in $(P_{n-1}, \bar{\pi})$, (iii) for each vertex v of (P_n, π) at least one of (a_1, b_n), (a_n, b_1), $((a_1, b_1), (a_n, b_n))$, $((a_1, b_n)\ (a_n, b_1))$ is good in $(P_{n-1}, \bar{\pi}) - v$.

Theorem 5. $\pi_{2,n}$ is hypo-good for all $n \equiv 5 \pmod 6$.

Proof. Let $n = 6m + 5$, $m \geq 0$. By Theorem 1, $(c_n, \pi_{2,n})$ is non-hamiltonian and hence $\bar{\pi}_{2,n}$ is a bad permutation by Theorem 2. Thus the first condition holds. Now we show that conditions (ii) and (iii) also hold.

Let $G = (P_{6m+4}, \bar{\pi}_{2,6m+5})$. The pairs (a_1, a_{6m+4}) and (b_1, b_{6m+4}) are good in G for $a_1\ a_2 \ldots a_{3m+2}\ b_{6m+4}\ b_{6m+3} \ldots b_1\ a_{3m+3}\ a_{3m+4} \ldots a_{6m+4}$ and $b_1\ a_{3m+3}\ a_{3m+4} \ldots a_{6m+4}\ b_{6m+3}\ b_{6m+2} \ldots b_2\ a_1 a_2 \ldots a_{3m+2}\ b_{6m+4}$ are spanning paths in G. This proves (ii).

Let v be any vertex of G. Now we show that one of (a_1, b_{6m+4}), (a_{6m+4}, b_1), $((a_1, b_1), (a_{6m+4}, b_{6m+4}))$ $((a_1, b_{6m+4}), (a_{6m+4}, b_1))$ is good in G-v. We introduce the following notation for the description of a path P in G. A sequence $(d_1, d_2, \ldots, \overleftarrow{d}_i,$

$d_{i+1},\ldots,d_t$) of positive integers d_i denotes a path P in G which follows the a-path P_1 and b-path P_2 in G alternately, the number of vertices in the successive intercepts being as prescribed by the sequence. A d_i with an arrow above it means P passes through d_i consecutive vertices of P_i, $i = 1,2$, in a leftward direction while a d_i without an arrow above means that the path P follows the P_i, $i = 1,2$ in a rightward direction for d_i vertices.

First we consider deletion of vertices from the a-path.

$G-a_1$: (b_1,a_{6m+4})-path is given by $(4,3,\ldots,3,4,3,3,\ldots,3,2)$ where 3 occurs 2m times after the first 4 and (2m-1) times after the second 4.

$G-a_2$: (a_1,b_{6m+4})-path is given by $(1,2,2,3,3,\overleftarrow{4},3,2,3,\overleftarrow{4},3,2,3,\overleftarrow{4},3,\ldots,2,3,\overleftarrow{4},3,1)$ where 2,3,4,3 group occurs (m-1) times.

$G-a_k, 3 \leq k \leq 3m+1$: (a_1,b_1)-path is given by $(k-1, \overleftarrow{2k-2})$ and (a_{6m+4},b_{6m+4})-path is given by $(\overleftarrow{3m-k+2},\overleftarrow{3},\overleftarrow{3m+2},6m-2k+3)$.

$G-a_{3m+2}$: (a_1,b_{6m+4})-path is given by $(3m+1, \overleftarrow{6m+2}, 3m+2,2)$.

The paths for remaining a_i, $3m+3 \leq i \leq 6m+4$ can be written out by symmetry.

Next we consider deletion of vertices from the b-path.

$G-b_1$: (a_1,b_{6m+4})-path is given by $(1,3,3,\ldots,3,2,\overleftarrow{2},\overleftarrow{3},\overleftarrow{3},\ldots,\overleftarrow{3},\overleftarrow{4},1)$ where 3 occurs 2m times and $\overleftarrow{3}$ occurs 2m-1 times.

The path for $G-b_{6m+4}$ can be written by symmetry.

$G-b_{2k}, 1 \leq k \leq 3m+1$: (a_1,b_1)-path is given by $(3m+k+2,\overleftarrow{2k-1})$ and (a_{6m+4},b_{6m+4})-path is given by $(\overleftarrow{3m-k+2}, 6m-2k+4)$.

$G-b_{2k+1}, 1 \leq k \leq 3m+1$: (a_1,b_1)-path is given by $(k,\overleftarrow{2k})$ and (a_{6m+4},b_{6m+4})-path is given by $(\overleftarrow{6m-k+4}, 6m-2k+3)$.

This completes the proof of Theorem 5.

Theorem 6. Let π_i be an n_i-permutation for $1 \leq i \leq k$ and let the following three conditions hold

(a) for each i, either $n_i = 1$ or π_i is a hypogood permutation,

(b) there is an even (possibly zero) number of 1's among the n_i's,

(c) $1 < n_1$, $1 < n_k$ and no two 1's among the n_i's appear consecutively.

Then the generalized prism (c_n,π') is hypo-hamiltonian where $\pi = (\pi_1,\pi_2,\ldots,\pi_k)$ and $n = \sum_{i=1}^{k} n_i+1$.

Proof. Let $G = (c_n,\pi')$. Since each π_i with $n_i > 1$ is hypogood, it is bad and hence $\pi = (\pi_1,\pi_2,\ldots,\pi_k)$ is a bad permutation by Theorem 3. Therefore G is non-hamiltonian by Theorem 2. We show below that every vertex-deleted subgraph $G-v$ of G is hamiltonian. We call the induced subgraph (P_{n_i},π_i) of G a 'block' if $n_i > 1$ and a 'single edge' if $n_i = 1$.

Let v be any vertex of G. It is either in a block or in a single edge. We consider these cases separately.

Case 1. Let $V = a_i(b_i)$ where a_ib_i is a single edge. Then starting from $b_i(a_i)$ follow the paths (ii) in the hypogood definition through each block and the single edges otherwise. This traces a hamiltonian cycle of $G-v$ since the number of single edges in $G-v$ is even.

Case 2. Let v be a vertex in a block $H = (P_{n_i},\pi_i)$. Let the end points of H be $\{a_t,b_t,a_m,b_m\}$. Let (a_t,b_m) or (a_m,b_t) be good in $H-v$. These paths can be extended to a hamiltonian cycle of $G-v$ by using the paths (ii) in the definition for the remaining blocks and single edges otherwise as $G-v$ has an odd number of single edges.

If $((a_t,b_t),(a_m,b_m))$ or $((a_t,b_m),(a_m,b_t))$ is good in $H-v$ then these paths can be extended to a hamiltonian cycle of $G-v$ by using the rest of the a-cycle for going from a_t to a_m and rest of the b-cycle for going from b_t to b_m. This completes the proof of the theorem.

Theorem 7. For odd $n \geq 3$, there exists a hypo-hamiltonian generalized n-prism if and only if n is neither 3 nor 7.

Proof. The theorem follows from Theorems 4,5 and 6.

Let the family of cubic hypo-hamiltonian graphs constructed in [1], namely $G(6m+5,2)$, $m \geq 0$ be denoted by A, the family constructed in [3] by B and the one constructed here in Theorems 5 and 6 by C. Clearly $A \subset C$. Now we examine how the families B and C are related.

The family B was constructed by using the special graphs called flip-flops defined as follows :

A flip-flop is a quintuple (G,a,b,c,d) where G is a graph and a,b,c,d are distinct vertices of G such that

(I) (a,d), (b,c) and ((a,d), (b,c)) are good in G,

(II) none of (a,b),(a,c),(b,d),(c,d), ((a,b),(c,d)) and ((a,c),(b,d)) is good in G,

(III) for every vertex u of G at least one of (a,c),(b,d), ((a,b),(c,d)), ((a,c),(b,d)) is good in G-u.

The order of flip-flop is the number of vertices in G. A flip-flop is called cubic if a,b,c,d have degree two in G and all other points have degree three.

Let $F_i = (G_i, a_i, b_i, c_i, d_i)$ be flip-flops. (F_1,F_2) denotes the quintuple (G_4,a_1,b_1,c_2,d_2) where G_4 is obtained by taking (disjoint) graphs G_1,G_2 and joining c_1 to b_2 and d_1 to a_2. (F_1,F_2,F_3) denotes the quintuple (G_5,a_1,b_1,c_3,d_3) where G_5 is obtained as follows : take (pairwise disjoint) graphs G_1,G_2,G_3, add four more vertices u_1,v_1,u_2,v_2 and ten more lines u_1v_1, u_2v_2, c_1u_1, u_1a_2, d_1v_1, v_1b_2, c_2u_2, u_2a_3, d_2v_2, v_2b_3. It was proved that (F_1,F_2) and (F_1,F_2,F_3) are flip-flops.

Let H be a graph and a,b,c,d be four distinct vertices of H. The graph G(H) based on H is obtained rom H by adding two more vertices u,v and five more lines uv, ua, ud, vb and vc. It was shown that if F is a flip-flop then G(F) is hypo-hamiltonian. The family B was constructed by using two flip-flops F_8 and F_{26} where $G(F_8)$ is the Petersen graph and $G(F_{26})$ is Coxeter's hypo-hamiltonian graph on 28 vertices.

<u>Theorem 8</u>. $G(F_{26})$ is not a generalized prism.

<u>Proof</u>. $G = G(F_{26})$ based on F_{26} is Coxeter's graph on 28 vertices. We take its description given in [5] namely,

$$V(G) = \{a_i, b_i, c_i, d_i\}_{i=1}^{7},$$

$$E(G) = \{a_i a_{i+1}, b_i b_{i+2}, c_i c_{i+3}, a_i d_i, b_i d_i, c_i d_i\}_{i=1}^{7}.$$

If possible let G be a generalized prism. So G consists of two vertex-disjoint 14 cycles C_1 and C_2 such that each vertex of C_1 is adjacent to exactly one vertex of C_2. Since there are seven vertices d_i in G, one of these two cycles say C_1 must contain four of these vertices d_i. Let these vertices be $d_{i_1}, d_{i_2}, d_{i_3}$ and d_{i_4}. Then C_1 is of the form

$$\ldots x_{i_1} d_{i_1} y_{i_1} \ldots x_{i_2} d_{i_2} y_{i_2} \ldots x_{i_3} d_{i_3} y_{i_3} \ldots x_{i_4}, d_{i_4} y_{i_4} \ldots$$

where $x, y \in \{a,b,c\}$. Since i_1, i_2, i_3, i_4 are distinct the 14-cycle C_1 can use at most two more suffices $i = 1,2,\ldots,7$. Therefore, there exists a suffix $j \in \{1,2,\ldots,7\}$ such that d_j and all its three adjacent vertices a_j, b_j, c_j lie in C_2. This is a contradiction. Hence $G(F_{26})$ is not a generalized prism.

<u>Theorem 9</u>. Let F_i be cubic flip-flops of order n_i for $1 \leq i \leq k$. Let $F = (F_1, F_2, \ldots, F_k)$. If $G(F)$ is a generalized prism, then so is each $G(F_i)$ where $i \in \{1,2,\ldots,k\}$.

<u>Proof</u>. Let $G(F)$ be a generalized prism and let C_1 and C_2 be the a-cycle and b-cycle of G respectively. Let $F_j = (G_j, a_j, b_j, c_j, d_j)$, $1 \leq j \leq k$ be any flip flop in F considered as an induced subgraph of $G(F)$ in the natural way. Then a_j, b_j, c_j, d_j are the only vertices of F_j having an adjacency in $G(F) - G_j$. Since these four vertices have only one adjacency each in $G(F) - G_j$ the intersections of C_1 and C_2 with G_j are connected graphs and hence form two vertex disjoint n_j-paths in G_j. Rest of the proof follows from the fact that $G(F)$ is a generalized prism.

<u>Theorem 10</u>. The graphs of B formed from F_8 alone are in C.

<u>Proof</u>. It is easy to exhibit isomorphism to see that $G(F_8)$, $G(F_8,F_8)$ and $G(F_8,F_8,F_8)$ are generalized prisms.

From the above three theorems we see that the only graphs in B which are generalized peisms are those formed from F_8 alone and all of these are in C. Thus the family C contains all generalized prisms present among the two earlier families A and B. As the family C contains many more graphs besides these, all of them are new cubic hypo-hamiltonian graphs of order $4k + 2$.

REFERENCES

1. J.A. Bondy, Variations on the Hamiltonian theorem, Canad. Math. Bull., 15(1972), 57-62.

2. G. Chartrand and F. Harary, Planar permutation graphs, Ann. Inst. Henri Poincare, Vol.III, no.4(1967), 433-438.

3. V. Chvatál, Flip flops in hypohamiltonian graphs, Canad. Math. Bull.,16(1973), 33-42.

4. J. Doyen and V. VanDiest, Hypohamiltonian graphs, Discrete Math., 13(1975), 225-236.

5. V. Klee, Which Generalised Prisms Admit H-circuits, Lecture Notes No. 303, Springer-Verlag, 173-179.

6. Daljit Rao, On some transversability problems in graph theory and combinatorics (Dissertation), I.I.T., Kanpur, 1977.

7. G.N. Robertson, Graphs under girth, valency and connectivity constraints (Dissertation), University of Waterloo, Canada, 1968.

GRAPHICAL CYCLIC PERMUTATION GROUPS

S. P. MOHANTY
M. R. SRIDHARAN
S. K. SHUKLA
Department of Mathematics
Indian Institute of Technology, Kanpur 208 016

ABSTRACT

A permutation group H acting on a set X is said to be graphical if there is a graph G such that Γ(G), the automorphism group of G, is identical to H. Characterisation of graphical permutation groups seems to be difficult. Kagno and Chao have shown that the group generated by a single m-cycle is not graphical. Here we study the group generated by a permutation such that it consists of disjoint cycles whose lengths are multiples of the length of one of its cycles. Our results are obtained by constructing certain graphs which we call generalised permutation graphs. We also study graphical cycle permutation groups of order p^m where p is a prime.

1. INTRODUCTION

A permutation group H acting on a set X is said to be graphical if there is a graph G such that Γ(G), the automorphism group of G, is identical to H. To find a necessary and sufficient condition for a permutation group to be graphical is one of the difficult problems related to groups and graphs. Kagno [1] and Chao [2] have shown that the group generated by a single m-cycle is not graphical. To begin with, in [3] we studied the group generated by a permutation consisting of disjoint cycles of some fixed length. In this paper, we study the group generated by a permutation such that it consists of disjoint cycles whose lengths are multiples of the length of one of its cycles. Our results are obtained by constructing certain graphs which we call generalised permutation graphs. We also study graphical cyclic permutation groups of order p^m where p is a prime.

2. PRELIMINARIES

We refer to [4] and [5] for the definitions and results not mentioned here.

By <u>cyclic form</u> of a permutation α on a set X, we mean its representation as a product of disjoint cycles (including cycles of length one).

Let G and H be two labeled graphs with labelling $u_1,u_2,\ldots,u_m$ and $u'_1,u'_2,\ldots,u'_n$. If m divides n, we define a <u>generalised α-permutation graph</u> $P^*_\alpha(G,H)$ for a permutation α acting on $Z_m = \{1,2,\ldots,m\}$ by

$$V(P^*_\alpha(G,H)) = V(G) \cup V(H)$$

and

$$E(P^*_\alpha(G,H)) = E(G) \cup E(H) \cup \{[u_i u'_j] : i \in Z_m, j \in Z_n \text{ and } j \equiv \alpha(i) (\text{mod } m)\}.$$

Let G and H be two labeled graphs with labelling $u_1,u_2,\ldots,u_m$ and $u'_1,u'_2,\ldots,u'_n$ respectively such that m divides n. If

$$A = \{\alpha_1,\alpha_2,\ldots,\alpha_r\}$$

be the set of disjoint permutations acting on Z_m, then we define a <u>generalized A-permutation graph</u>

$$P^*_A(G,H) = \bigcup_{\alpha_i \in A} P^*_{\alpha_i}(G,H).$$

3. GRAPHICAL CYCLIC PERMUTATION GROUPS

For the construction of graphs whose automorphism group is generated by a permutation, the cyclic form of which contains r cycles of lengths $m_1,m_2,\ldots,m_r$ such that m_1 divides m_i for $i \in Z_r$, we first consider the case when $r = 2$. It is easy to verify that for $m = 1,3,4$ and 5, this group is not graphical. However for $m = 2$, it is not the case. We prove the following theorem for $m > 5$.

<u>Theorem 1</u>. If $\sigma = (u_1u_2\ldots u_m)(u'_1u'_2\ldots u'_n)$ such that $m > 5$ and m divides n, then $\langle\sigma\rangle$ is graphical.

<u>Proof</u>. Let C_m and C_n be two cycles $u_1,u_2,\ldots,u_m$ and $u'_1,u'_2,\ldots,u'_n$ respectively. We construct a generalised A-permutation graph $P^*_A(\overline{C}_m,C_n)$ where

$$A = \{(12\ldots m)^i : i = 0,1,3\}$$

and claim that its automorphism group is $<\sigma>$. For simplicity, we denote $P^*_A(\overline{C}_m,C_n)$ by P^* only. For proving our claim, we first observe that $d(u_k) \geq m$ ($\neq 5$, by hypothesis) and $d(u_\ell) = 5$ in P^* for every $k \in Z_m$ and $\ell \in Z_n$. Therefore $V(\overline{C}_n)$ are invariant under $\Gamma(P^*)$ which implies that every $\hat{\beta} \in \Gamma(P^*)$ can be written as $\beta_1\beta_2$ such that β_1 acts on $V(\overline{C}_m)$ and β_2 acts on $V(C_n)$. Since $\overline{C}_m$ and C_n are induced subgraphs of P^*, we have $\beta_1 \in \Gamma(\overline{C}_m)$ and $\beta_2 \in \Gamma(C_n)$. Therefore, $\Gamma(P^*)$ is a subgroup of the direct product of the groups $\Gamma(\overline{C}_m)$ and $\Gamma(C_n)$ which are identical to Dihedral groups D_m and D_n respectively. Let us evaluate these β_1 and β_2 explicitly for a given $\beta \in \Gamma(P^*)$. Let

$$\varphi_1 = (u_1u_m)(u_2u_{m-1})(u_3u_{m-2})\ \ldots$$
$$\psi_1 = (u_1u_2\ \ldots\ u_m)$$
$$\varphi_2 = (u_1'u_n')(u_2'u_{n-1}')(u_3'u_{n-2}')\ \ldots$$
$$\psi_2 = (u_1'u_2'\ \ldots\ u_n').$$

Thus

$$\Gamma(\overline{C}_m) = <\varphi_1,\ \psi_1>$$

and

$$\Gamma(C_n) = <\varphi_2,\ \psi_2>.$$

We show that

$$\beta_1 = \psi_1^r \Rightarrow \beta_2 = \psi_2^{mk+r}$$

for some integer k such that $0 \leq k \leq s-1$ where $n = ms$ (we have already assumed that m divides n). We also show that

$$\beta_1 \neq \varphi_1\ \psi_1^r$$

for any $r \in Z_m$ and

$$\beta_2 \neq \varphi_2\ \psi_2^r$$

for any $r \in Z_n$. Hence $\beta_1\beta_2$ is always in $<\sigma>$.

First put $\beta_1 = \psi_1$. Let B_{u_k} denote the set of vertices of $V(C_n)$ adjacent to $u_k \in V(\overline{C}_m)$ in the graph P^*. So

$$B_{u_1} = \{u_1',u_2',u_4',u_{m+1}',u_{m+2}',u_{m+4}',\ldots,u_{m(s-1)+1}',u_{m(s-1)+2}',u_{m(s-1)+4}'\}$$

$$B_{u_2} = \{u_2',u_3',u_5',u_{m+2}',u_{m+3}',u_{m+5}',\ldots,u_{m(s-1)+2}',u_{m(s-1)+3}',u_{m(s-1)+5}'\}.$$

Since β_1 sends u_1 to u_2, β_2 sends the vertices in B_{u_1} to the vertices in the set B_{u_2}. But u_1' and u_2' are adjacent in C_n and u_4' is at a distance 2 from u_2', therefore β_2 sends u_1' to u_{mk+2}', u_2' to u_{mk+3}' and u_4' to u_{mk+5}' for some

$$k = 0,1,\ldots,m(s-1).$$

Hence β_2 also sends u_3' to u_{mk+4}', u_5' to u_{mk+6}' and so on. (Take the addition in subscript modulo m). Hence

$$\beta_2 = \psi_2^{mk+1}.$$

Similarly we can show that

$$\beta_2 = \psi_2^{mk+1} \Rightarrow \beta_1 = \psi_1.$$

Therefore, $\beta_1 = \psi_1^r$ if and only if $\beta_2 = \psi_2^{mk+r}$, i.e.,

$$\beta_1\beta_2 = (\psi_1\psi_2)^{mk+r} = \sigma^{mk+r}.$$

Now put

$$\beta_1 = \varphi_1 = (u_1u_m)(u_2u_{m-1}) \ldots$$

We know that

$$B_{u_m} = \{u_m', u_{m+1}', u_{m+3}', u_{2m}', u_{2m+1}', u_{2m+3}', \ldots, u_{ms}', u_1', u_3'\}.$$

Since β_1 sends u_1 to u_m and u_m to u_1, β_2 has to map B_{u_1} to B_{u_m} and B_{u_m} to B_{u_1}. Therefore β_2 sends u_1' to u_{km}', u_2' to u_{km+1}' and u_4' to u_{km+3}'. This is not possible because $u_1 \in B_{u_m}$ also and $u_{km}' \notin B_{u_1}$ but β_2 sends B_{u_m} to B_{u_1}. This implies that β_1 cannot be equal to φ_1. Similarly, we can prove that β_2 is not equal to φ_2 in any case. Since β_1 can take the value ψ_1^r and β_1 cannot be equal to φ_1, it follows that

$$\beta_1 \neq \varphi_1 \psi_1^r$$

for any $r \in Z_m$. Similarly

$$\beta_2 \neq \varphi_2 \psi_2^r$$

for any $r \in Z_n$. Thus we have shown that $\beta_1\beta_2 \in \langle\sigma\rangle$. But every permutation in $\langle\sigma\rangle$ preserves adjacency in P^*, therefore $\langle\sigma\rangle \subseteq \Gamma(P^*)$. Thus

$$\langle\sigma\rangle = \Gamma(P^*).$$

This completes the proof of Theorem 1.

Remark 1. What happens if we consider $P_A^*(C_m,C_n)$ instead of $P_A(\overline{C},C_n)$? Certainly some vertices of C_m may go to the vertices of C_n under some automorphism of $P_A(C_m,C_n)$. However if β is some automorphism of $P_A(C_m,C_n)$ such that $\beta = \beta_1 \cdot \beta_2$ where $\beta_1 \in \Gamma(C_m)$ and $\beta_2 \in \Gamma(C_n)$ then

$$\beta = \psi_1^r \psi_2^{mk+r} = \sigma^{mk+r}$$

for some $r \in Z_m$. We will use this fact in Theorem 2.

Now we consider the general case dealing with a permutation group generated by a permutation whose cyclic form contains $r (r > 2)$ cycles of lengths $m_1, m_2, \ldots, m_r$ such that m_1 divides m_i for $i \in Z_r$. For $m_1 = 1,2,3,4$ and 5, we lack the result at the moment. We assume that $m_1 > 5$.

Theorem 2. If the cyclic form of a permutation σ consists of r cycles of lengths $m_1, m_2, \ldots, m_r$ such that m_1 divides m_i for $i \in Z_r$ and $m_1 > 5$, then $\langle\sigma\rangle$ is graphical.

Proof. We construct a graph G such that

$$\Gamma(G) = \langle\sigma\rangle$$

which is based on the construction as in Theorem 1.

If the cyclic form of σ consists of r cycles of lengths $m_1, m_2, \ldots, m_r$, then it may happen that $m_i = m_j$ for some $i,j \in Z_r$. So rearrange σ leaving one cycle of length m_1 such that it consists of r_1 cycles of length n_1, r_2 cycles of length $n_2, \ldots$, and r_k cycles of length n_k. Clearly

$$\sum_{i=1}^{k} r_i = r - 1$$

and $n_1, n_2, \ldots, n_k$ are taking values from $m_1, m_2, \ldots, m_r$. Let us denote the class of r_i cycles of length n_i by

$$\{(u_1^{(r_i,j)}\ u_2^{(r_i,j)}\ \ldots\ u_{n_i}^{(r_i,j)}) : j \in Z_{r_i}\}$$

for $i \in Z_k$. Represent all such cycles by the graph consisting of the cycle :

$$C_{n_i}^{(r_i,j)} = u_1^{(r_i,j)}, u_2^{(r_i,j)}, \ldots, u_{n_i}^{(r_i,j)}$$

for $i \in Z_k$ and $j \in Z_{r_i}$.

Let $C_{m_1} = u_1,u_2,\ldots,u_{m_1}$ be the cycle of length m_1 which we had left in the process of rearrangement of cycles. For $i \in Z_k$, we define a graph

$$G^{r_i} = P^*_A(\overline{C}_{m_1}, C_{n_i}^{(r_i,1)}) \cup (\bigcup_{j=1}^{r_i-1} P^*_B(C_{n_i}^{(r_i,j)}, C_{n_i}^{(r_i,j+1)}))$$

where

$$A = \{(12\ldots m_1)^{\ell} : \ell = 0,1,3\}$$

and

$$B = \{(12\ldots n_j)^{\ell} : \ell = 0,1,3\}.$$

Construct a graph

$$G = \bigcup_{r=1}^{k} G^{r_i}.$$

In the following we prove that $\Gamma(G) = \langle\sigma\rangle$.

Since $\overline{C}_{m_1}$ is the induced subgraph of G^{r_i} for $i \in Z_k$, the degree of every vertex of G which is in $V(\overline{C}_{m_1})$ is greater than five. The same is true for the vertices of G which are in $\bigcup_{j=1}^{r_i-1} V(C_{n_i}^{(r_i,j)})$ for all $i \in Z_k$. We also know that the vertices of G which lie in $V(C_{n_i}^{(r_i,r_i)})$ have degree five and $C_{n_i}^{(r_i,r_i)}$ is mot isomorphic to $C_{n_j}^{(r_j,r_j)}$ for $i \neq j$. Hence for all $i \in Z_k$, $V(C_{n_i}^{(r_i,r_i)})$ is invariant under $\Gamma(G)$. Since the vertices of G which lie in $V(C_{n_i}^{(r_i,r_i)})$ are adjacent to those of $V(C_{n_i}^{(r_i,r_{i-1})})$, $V(C_{n_i}^{(r_i,r_i-1)})$ is also invariant under $\Gamma(G)$. By the same argument $V(C_{n_i}^{(r_i,r_i-1)})$, $V(C_{n_i}^{(r_i,r_i-3)})$,...,$V(C_{n_i}^{(r_i,1)})$ are invariant under $\Gamma(G)$. Therefore, $V(C_{n_i}^{(r_i,j)})$ is invariant under $\Gamma(G)$ for all $i \in Z_k$ and $j \in Z_{r_i}$. Then from Theorem 1 and Remark 1, it follows that if $\beta \in \Gamma(G)$ then

$$\beta = \beta_1.(\prod_{\substack{i\in Z_k\\ j\in Z_{r_i}}} \beta^{(r_i,j)})$$

where

$$\beta_1 = (u_1u_2\ldots u_{m_1})^s$$

for some $s \in Z_{m_1}$ and

$$\beta^{(r_i,j)} = (u_1^{(r_i,j)} u_2^{(r_i,j)} \ldots u_{n_i}^{(r_i,j)})^{m_1t_i+s}$$

REFERENCES

1. J.N. Kagno, Linear graphs of degree ≤ 6 and their groups, Amer. Jour. Math., 68(1946), 505-520.

2. C.Y. Chao, On a theorem of Sabidussi, Proc. Amer. Math. Soc., 15(1964), 291-292.

3. S.P. Mohanty, M.R. Sridharan and S.K. Shukla, On cyclic permutation groups and graphs, Jour. Math. Phy. Sci., 12(1978), 409-416.

4. F. Harary, Graph Theory, Addison-Wesley, Reading, Mass, 1969.

5. H. Wielandt, Finite Permutation Groups, Academic Press Inc., New York, 1964.

ORTHOGONAL MAIN EFFECT PLANS WITH VARIABLE NUMBER OF LEVELS FOR FACTORS

A. C. MUKHOPADHYAY
Indian Statistical Institute
203 B. T. Road, Calcutta 700 035

1. INTRODUCTION

Rao (1971) defined orthogonal arrays with variable number of levels for different factors. He also constructed a few such orthogonal arrays of strength two. Since then, some attempts have been made to construct these arrays of strength two, which can serve as saturated orthogonal main effect plans with variable number of levels for different factors (Pal(1976), Mukhopadhyay(1978)). When s is a prime number or a prime power, the one to one correspondence between the factors of saturated main effect plans of the type s^n and the points of a projective geometry constructed over a $GF(s)$ has been known for a long time. In such a projective geometry of n dimensions, the identification of a subspace of m dimensions $(0 \leq m < n)$ with a factor having s^{m+1} levels has been pointed out in Mukhopadhyay (1978). Now, to construct saturated orthogonal arrays of strength two i.e., main effect plans, it is essential that these subspaces of the $PG(n,s)$, representing distinct factors should be disjoint (two disjoint subspaces are also said to be parallel or skew). Hence the importance of enumerating parallel subspaces of m_i dimensions, $i = 1,2,...,k$, say, with each $m_i \geq 0$ contained in a $PG(n,s)$ (assuming there are factors with k distinct levels in the main effect plan, the levels being s^{m_i+1}, $i = 1,2,...,k$). The particular case when all m_i's are equal to, say, t is well known as the problem of t spreads. The problem of t-spreads of a projective geometry has been solved by Bruck and Bose (1964). The present paper lends heavily on the available results on t spreads and particularly on the affine representation of spreads given by Bruck and Bose (1964).

2. DEFINITIONS AND NOTATION

The definition of an orthogonal array (OA) of strenth d (≥ 2), given in Rao (1973) may be stated as follows :

Let A be a rectangular array with r rows and N columns. The ith row of the array representing the ith factor is written with the elements of the set M_i, $|M_i| = s_i$, $i = 1,2,\ldots,r$. For any d (≥ 2)-rowed submatrix of A, say, containing the rows $i_1,i_2,\ldots,i_d$, $\{i_1,i_2,\ldots,i_d\} \subset \{1,2,\ldots,r\}$. Let $n(j_{i_1},j_{i_2},\ldots,j_{i_d})$ denote the number of times a given d-tuple $(j_{i_1},j_{i_2},\ldots,j_{i_d})'$ occurs in the submatrix, $j_{i_k} \varepsilon M_{i_k}$, $k = 1,2,\ldots,d$. A is an <u>orthogonal array</u> of strength $d(\geq 2)$, if $n(j_{i_1},j_{i_2},\ldots,j_{i_d})$ is a constant, say, $n(s_{i_1},s_{i_2},\ldots,s_{i_k})$ for all $j_{i_k} \varepsilon M_{i_k}$, $k = 1,2,\ldots,d$ and for all choice of d rows of A. In case $s_1 = s_2 \ldots = s_r = s$, the array can be denoted as $OA[N,r,s,d]$.

The definition of parallel subspaces and t-spreads of a projective space may be stated as follows :

For any set of points $e_1,e_2,\ldots,e_t$ of a projective geometry, the symbol $< e_1,e_2,\ldots,e_t >$ represents the subspace spanned by the points $e_1,e_2,\ldots,e_t$ of the geometry. Two subspaces $< e_1,e_2,\ldots,e_t >$ and $< e_1',e_2',\ldots,e_{t'}' >$ are said to be <u>parallel</u> if and only if $< e_1,e_2,\ldots,e_t > \cap < e_1',e_2',\ldots,e_{t'}' > = \phi$.

Let Σ be a projective space. Let S be a collection of parallel $(t-1)$-dimensional subspaces of Σ. S is called a <u>t-spread</u> of Σ, provided each point of Σ is contained in one member of S. It is known that t-spreads exist if and only if the dimension of Σ, say, n is such that $n + 1$ is divisible by t.

In the present paper, the idea of spreads is generalised. We divide a projective space Σ into a collection S of parallel subspaces, say, denoted by Σ_i, $i = 1,2,\ldots,r$, the dimension of Σ_i being t_i, $i = 1,2,\ldots,r$ (all t_i's need not be same) such that each point of Σ is contained in one member of S.

3. DIVISION OF A PROJECTIVE SPACE INTO PARALLEL SUBSPACES

Bose's result (1947) on orthogonal main effect plans in the symmetrical factorial set up can be easily extended to the case where the number of levels of the factors may be different for the different factors, but the number of levels of each factor is some positive integral power of s, where s is a prime number or a prime power. The extension can be stated in the form of a theorem as follows :

<u>Theorem 3.1</u>. Let S be a collection of parallel subspaces Σ_i, $i = 1,2,\ldots,r$ of a PG(n,s), the dimension of Σ_i being t_i, $i = 1,2,\ldots,r$, such that each point of PG(n,s) is contained in one member of S. Then, there exists an orthogonal array of strength 2 with number of rows = r, number of columns = s^{n+1} and the number of levels of ith factor is s^{t_i+1}, $i = 1,2,\ldots,r$.

<u>Proof</u>. Construction of an $OA[s^{n+1}, (s^{n+1}-1)/s-1,s,2]$, identifying each point of PG(n,s) with a factor having s levels is well known and $(s^{t_i+1}-1)/s-1$ points of PG(n,s) comprising the subspace Σ_i of t_i dimensions can be identified, $i = 1,2,\ldots,r$. Any set of (t_i+1) linearly independent points of Σ_i can be utilised to construct the ith row of the orthogonal array with number of levels = s^{t_i+1}, $i = 1,2,\ldots,r$.

In the light of Theorem 3.1, division of a projective space into an exhaustive collection of parallel subspaces is of fundamental importance in the construction of orthogonal main effect plans, where the level of each factor can be written as a positive integral power of s, s being a prime number or a prime power.

In the affine representation of t-spreads of a projective space of dimension 2t - 1, what Bruck and Bose (1964) consider is as follows :

Let $J(\infty)$ denote a subspace of (t-1) dimensions, represented by an arbitrary basis $e_1,e_2,\ldots,e_t$. We can extend this to a basis of the projective space of (2t-1) dimensions, by adjoining t additional basis elements $e'_1,e'_2,\ldots,e'_t$. Let

$$J(\infty) = < e_1,e_2,\ldots,e_t >$$

and

$$J(X) = < e'_1 + x_1, e'_2 + x_2,\ldots,e'_t + x_t >,$$

where, $x_i = \sum_{j=1}^{t} x_{ij} e_j$, $i = 1,2,\ldots,t$.

Then a t-spread can be constructed which contains $J(\infty)$. Each member of S, other than $J(\infty)$ is of the form $J(X)$, where X belongs to a collection $\underline{\underline{C}}$ of matrices of t rows and columns with elements in the field over which the projective space is constructed, subject to the following conditions :

(i) $\underline{\underline{C}}$ contains the null matrix 0 and the identity matrix, I.

(ii) If X,Y are distinct matrices in $\underline{\underline{C}}$, then the matrix $X-Y$ is non-singular.

(iii) To each ordered pair of elements a,b of $J(\infty)$ with $a \neq 0$ there corresponds a (unique) matrix X in $\underline{\underline{C}}$ such that $a^X = b$.

Obviously, $|S| = s^t + 1$ and as we are dealing with finite geometrics the conditions (i) and (ii) for the class with $|S| = s^t + 1$ will suffice for a t-spread and condition (iii) is automatically satisfied there. Bruck and Bose (1964) ensure that such a collection of matrices, $\underline{\underline{C}}$, always exists.

In the procedure above, what is being done is essentially as follows :

We start with two parallel subspaces $J(\infty)$ and $J(0)$, both of (t-1) dimensions. Then, we construct the other subspaces $J(X)$, one for each $X \varepsilon \underline{\underline{C}}$. $\underline{\underline{C}}$ should satisfy the properties enumerated above and such a collection $\underline{\underline{C}}$ of $t \times t$ matrices, always exists. Now, if the dimension of $J(0)$ is $t' - 1$, $t' < t$, still we can follow the same procedure successfully. We can define a new class $\underline{\underline{C}}_1$ of $t' \times t$ matrices as follows :

For each matrix $X \varepsilon \underline{\underline{C}}$, the $t \times t$ submatrix containing the first t' rows of X, say X_1 is contained in $\underline{\underline{C}}_1$. All the matrices of $\underline{\underline{C}}_1$ are obtained this way only. Then

$$J^*(\infty) = < e_1, e_2, \ldots, e_t >$$

$$J^*(X_1) = < e_1' + \sum_{j=1}^{t} x_j e_j, \ldots, e_{t'}' + \sum_{j=1}^{t} x_{t'j} e_j > \quad \text{for each } X_1 \varepsilon \underline{\underline{C}}_1.$$

These subspaces can be easily proved to be parallel and exhaust all the points of the projective space of (t+t'-1)-dimensions if the original $J(\infty)$ and $J(X)$'s could exhaust the whole projective space of dimension $2t - 1$. The dimension of $J(\infty)$ is $t - 1$ and each $J^*(X_1)$ has dimension $t' - 1$. So, what we have proved can be stated in the following theorem.

<u>Theorem 3.2</u>. A PG(t+t'-1, s), $1 \leq t' \leq t$, can always be divided into a collection S of $s^t + 1$ parallel subspaces, one of these subspaces being of dimension $t - 1$ and the remaining s^t subspaces all of dimension $t' - 1$, such that every point of the PG(t+t'-1,s) is contained in one subspace.

Suppose Σ represents a PG(n,s), divided into a collection S of, say, r parallel subspaces, ith member of S being denoted by Σ_i, $i = 1,2,\ldots,r$, such that each point of Σ is contained in one member of S. Let the dimension of Σ_i be t_i, $i = 1,2,\ldots,r$. All the t_i's need not be necessarily equal. Let Σ^* be a subspace of dimension $t \leq \min(t_1,t_2,\ldots,t_r)$, parallel to Σ. Then $< \Sigma \cup \Sigma^* >$ is a projective space of dimension $n + t + 1$. Following the method of Theorem 3.2 and writing

$$J_i^*(\infty) = \Sigma_i = < e_{i1}, e_{i2}, \ldots, e_{it_{i+1}} >$$

and

$$J^*(0) = \Sigma^* = < e_1', e_2', \ldots, e_{t+1}' >,$$

we can construct a collection, say, S_i of $(s^{t_i+1}+1)$ parallel subspaces; one of them, namely $J^*(\infty)$ is of dimension t_i and each of the remaining s^{t_i+1} subspaces is of dimension t. The same procedure can be followed with each one of the subspaces Σ_i, $i = 1,2,\ldots,r$. Now it follows from the elementary properties of a projective space that if

$$\Sigma_{ik} \in S_i \text{ and } \Sigma_{i'k'} \in S_{i'}', \quad i \neq i',$$

$$\Sigma_{ik}, \Sigma_{i'k'} \neq J^*(0),$$

then

$$\Sigma_{ik} \cap \Sigma_{i'k'} = \phi$$

i.e., The subspaces are parallel.

Thus, we have been able to divide a $PG(n+t+1,s)$ into a collection of $(\sum_{i=1}^{r} s^{t_i+1}+1)$ parallel subspaces, r of these subspaces are the original Σ_i's of dimensions t_i' respectively, $i = 1,2,\ldots,r$ and each of the remaining $(\sum_{i=1}^{r} s^{t_i+1}-r+1)$ subspaces is of dimension t. Moreover, each point of $PG(n+t+1)$ is contained in one of the subspaces of this resulting collection. This result is stated in the following theorem.

<u>Theorem 3.3</u>. Let the points of $PG(n,s)$ be divided into a collection S of r parallel subspaces, the ith subspace Σ_i being of dimension t_i (≥ 0), $i = 1,2,\ldots,r$, so that each point of $PG(n,s)$ lies in one subspace of the collection. Then, the points of a $PG(n+t+1,s)$, with $0 \leq t \leq \mathrm{Min}(t_1,t_2,\ldots,t_r)$, can be divided into a collection of $(\sum_{i=1}^{r} s^{t_i+1}+1)$ parallel subspaces; $(\sum_{i=1}^{r} s^{t_i+1}-r+1)$ of these subspaces are of dimension t and the remaining r subspaces are of dimensions $t_1,t_2,\ldots,t_r$ respectively. Moreover each point of $PG(n+t+1,s)$ is contained in one subspace of the collection.

The use of Theorems 3.1 through 3.3 is illustrated in the following very simple examples :

Example 1. A PG(4,s) can be divided into a collection of (s^3+1) parallel subspaces, one of the subspaces being of dimension 2 and the remaining subspaces of dimension 1. The collection of subspaces exhaust all the points of PG(4,s). By a repeated application of Theorem 3.3, a PG(4+2t,s), $t \geq 0$, can be divided into a collection of $(s^{2t+3} + s^{2t+1} + \dots + s^3 + 1)$ parallel subspaces, one of these subspaces being a plane and the remaining all lines. The parallel subspaces together exhaust all the points of PG(4+2t,s), $t \geq 0$.

Example 2. A 3-spread of PG(5,s) can be constructed easily. Hence, by Theorem 3.3, a PG(7,s) can be divided into a collection of (s^6+s^3+1) parallel subspaces, (s^3+1) of these subspaces being planes and the remaining s^6 subspaces all lines. The collection of subspaces exhaust all the points of PG(7,s).

4. CONSTRUCTION OF PARALLEL SUBSPACES AND ORTHOGONAL MAIN EFFECT PLANS

Construction procedure of Section 3 remains valid in the light of the results on t-spreads and affine representation of spreads considered by Bruck and Bose (1964). But the actual construction hinges on the availability of X matrices satisfying the required properties. The existence is theoretically ensured and obtaining them by following the procedure indicated in Section 2 of Bruck and Bose (1964) is not difficult but may prove cumbersome, if not followed systematically. So, there should exist a simple practical method of obtaining them, so that the results can be fruitfully applied in the construction of orthogonal arrays of strength two. In the present section a few simple methods of construction of t-spreads and exhaustive parallel subspaces are taken up.

A. 2-spread of a PG(3,s). Let us write

$$J(\infty) = < e_1, e_2 >$$

$$J(0) = < e_1', e_2' >,$$

where e_1, e_2, e_1', e_2' forms a basis of the PG(3,s). Then, we can write

$$J(X) = < e_1' + (\beta - \alpha\gamma)e_1 + \alpha e_2,\ e_2' - \alpha\delta e_1 + \beta e_2 >, \text{ for all } \alpha, \beta \in GF(s),$$

where $x^2 + \gamma x + \delta$ is an irreducible polynomial over GF(s).

B. The division of the points of PG(4,s) into a collection of (s^3+1) parallel subspaces of which one is a plane and all the remaining are lines, such that the subspaces together exhaust all the points of PG(4,s).

Let $e_1, e_2, e_3, e'_1, e'_2$ form a basis of the PG(4,s). We can write

$J(\infty) = < e_1, e_2, e_3 >$

$J(X) = < e'_1 + \alpha_i e_1 + \alpha_j e_2 + \alpha_k e_3,\ e'_2 + \alpha_j e_1 + \alpha_k e_2 + (\beta_1\alpha_i + \beta_2\alpha_j + \beta_3\alpha_k) e_3 >$

$i,j,k = 0,1,\ldots,s-1$ and $\alpha_o, \alpha_1, \ldots, \alpha_{s-1}$ are the s distinct elements of GF(s). $\beta_1, \beta_2, \beta_3$ are any given elements of GF(s), each non-null, such that there exists an irreducible polynomial of degree 3 defined over GF(s), given by

$$x^3 + \beta_3 x^2 - \beta_2 x + \beta_1.$$

Such irreducible polynomials usually exists for $s > 2$ e.g., in case of

(i) GF(3), we can take the polynomial

$$x^3 + x^2 - x + 1.$$

(ii) GF(4), the polynomial

$$x^3 + (1+\alpha)x^2 - \alpha x + \alpha,$$

where the 4 elements of GF(4) are denoted by $0, 1, \alpha, 1+\alpha$ with $\alpha^2 + \alpha + 1 = 0$.

(iii) GF(5), the polynomial

$$x^3 + x^2 - 4x + 3.$$

Over GF(2), an irreducible polynomial of the type indicated here can never exist. But the following collection of exhaustive parallel subspaces were obtained by enumeration in case $s = 2$.

$J(\infty) = < e_1, e_2, e_3 >$

$J(0) = < e'_1, e'_2 >$

other J(X)'s are as follows.

$J(1) = < e_1' + e_1, e_2' + e_1 + e_2 >$

$J(2) = < e_1' + e_2, e_2' + e_3 >$

$J(3) = < e_1' + e_1 + e_2, e_2' + e_1 + e_2 + e_3 >$

$J(4) = < e_1' + e_3, e_2' + e_1 >$

$J(5) = < e_1' + e_1 + e_3, e_2' + e_2 >$

$J(6) = < e_1' + e_2 + e_3, e_2' + e_1 + e_3 >$

$J(7) = < e_1' + e_1 + e_2 + e_3, e_2' + e_2 + e_3 >$

C. t-spread of a $PG(2t-1,s)$, $t \geq 1$.

We can construct an $OA[(s^t)^2, s^t+1, s^t, 2]$ with the elements of $GF(s^t)$ as follows :

The first two rows of the OA constitute the all possible ordered pairs of elements of $GF(s^t)$. Denoting the first row by p_1 and the second row by p_1', any other row of the array can be written as a linear combination $p_1' + \alpha p_1$, where α is an element of $GF(s^t)$, $\alpha \neq 0$.

Now, the elements of $GF(s^t)$ upto isomorphism can be written as polynomials of degree $t-1$ with coefficients lying in its subfield $GF(s)$, with the help of an irreducible polynomial of degree t defined over $GF(s)$. In other words, each element of $GF(s^t)$ can be written as a t-tuple with the elements of the t-tuple lying in $GF(s)$. Let us replace each element of $GF(s^t)$ occuring in each row of the $OA[(s^t)^2,s^t+1,s^t,2]$ by its corresponding t-tuple written vertically downwards.

Now we have $(s^t+1)t$ rows divided into (s^t+1) sets, each set containing vertically all the s^t t-tuples formed by the elements of $GF(s)$, each s^t times in some order. Let $e_{i1},e_{i2},\ldots,e_{it}$ be such t rows of the ith set giving all the s^t t-tuples formed by the elements of $GF(s)$, each occurring s^t times, $i = 1,2,\ldots,s^t+1$. Then, we can form all the linear combinations $\beta_1 e_{i1} + \beta_2 e_{i2} \cdots + \beta_t e_{it}$ where β_i's are elements of $GF(s)$, $(\beta_1,\beta_2,\ldots,\beta_t) \neq (0,0,\ldots,0)$ and the t-tuples $(\beta_1,\beta_2,\ldots,\beta_t)$ and $(\beta_1',\beta_2',\ldots,\beta_t')$ are to be treated as identical if there exists a $\rho \in GF(s)$, $\rho \neq 0$ such that

$$(\beta_1,\beta_2,\ldots,\beta_t) = \rho(\beta_1',\beta_2',\ldots,\beta_t').$$

Let us replace the ith set of rows by all such distinct linear combinations of the rows $e_{i1}, e_{i2}, \ldots, e_{it}$. Now we have exactly $(s^t-1)/s-1$ rows in the set with all the elements lying in GF(s). After doing this for each set, i.e., for $i = 1,2,\ldots, s^t+1$, of the array we obtain a resulting array with $(s^{2t}-1)/s-1$ rows all written with the elements of GF(s). The resulting array so constructed is easily seen to be an OA of strength 2 with number of levels s for each row.

Let us consider the following 2t rows from the first two sets of rows of the array, viz.,

$$e_{11}, e_{12}, \ldots, e_{1t}, e_{21}, e_{22}, \ldots, e_{2t}.$$

These 2t rows give all s^{2t}, 2t-tuples formed by the elements of GF(s). Hence, identifying these 2t rows of the array as the basis, we can develop a projective geometry of (2t-1) dimensions, each of whose points can be identified with a unique row of the final orthogonal array containing $(s^{2t}-1)/s-1$ rows.

Each point has a unique representation as a linear combination of the 2t points in the basis. The $(s^t-1)/s-1$ points corresponding to the rows of the ith set constitute the ith subspace of dimension t-1. These subspaces are parallel and they exhaust all the points of the PG(2t-1,s) constructed. The basis of the ith subspace is given by $e_{i1}, e_{i2}, \ldots, e_{it}$, where $e_{ij} = \sum_{k=1}^{t} \beta_{1ijk} e_{1k} + \sum_{k=1}^{t} \beta_{2ijk} e_{2k}$ and β_{1ijk}'s and β_{2ijk}'s are elements of GF(s), $k = 1,2,\ldots,t$, $j = 1,2,\ldots,t$, $i = 1,2,\ldots,s^t+1$. The coefficients β_{1ijk}'s and β_{2ijk}'s are exactly known for each e_{ij} from the method of construction followed for obtaining the final orthogonal array. Writing

$$B_{1i} = \begin{bmatrix} \beta_{1i11} & \beta_{1i12} & \cdots & \beta_{1i1t} \\ & \vdots & & \\ \beta_{1it1} & \beta_{1it2} & \cdots & \beta_{1itt} \end{bmatrix}$$

and

$$B_{2i} = \begin{bmatrix} \beta_{2i11} & \beta_{2i12} & \cdots & \beta_{2i1t} \\ & \vdots & & \\ \beta_{2it1} & \beta_{2it2} & \cdots & \beta_{2itt} \end{bmatrix}$$

$i = 1,2,\ldots,s^t+1$, from the method of construction followed, we obviously get

$$B_{11} = I, \quad B_{21} = 0$$

and

$$B_{21} = 0, \quad B_{22} = I$$

and, both B_{1i} and B_{2i} are easily seen to be non-singular for all $i \neq 1,2$.

So, ith subspace, $i \geq 3$ can be equivalently written as

$$< e_{21} + x_1, e_{22} + x_2, \ldots, e_{2t} + x_t >$$

where

$$x_i = \sum_{j=1}^{t} x_{ij} e_{ij}, \quad i = 3,4,\ldots,t$$

and the matrix $X = B_{21}^{-1} B_{22}$.

So, we have an affine representation of spreads and the X matrices are easily known.

It should be noted that we have followed in this construction the same method as indicated by Bruck and Bose (1964). Only the introduction of orthogonal arrays help us obtain the X matrices easily and in a very straight forward manner.

Remark 1. The construction methods of this section coupled with Theorem 3.2 can be applied repeatedly to obtain the spreads or in general exhaustive collections of parallel subspaces in higher dimensional projective spaces.

Remark 2. It can be proved easily in method A of this section that the set of parallel lines constructed for PG(3,s) is closed in the sense, that if we start with any two parallel lines of the set and apply the stated method A of construction of an exhaustive set of parallel lines, we will always get the same set. Now, the number of lines parallel to a given line in PG(3,s) can be calculated to be s^4. So, let us take a line L and any of the s^4 lines parallel to it. Then apply method A to construct a set, say, S_1 of (s^2+1) parallel lines which contains L. Then we can take another line parallel to L, not contained in S_1. This line along with L will give rise to a set, say, S_2 of s^2+1 parallel lines including the line L. Now, $|S_1 \cap S_2| = 1$, because of the closedness property of S_1 and S_2. In this manner we can go on constructing such sets of parallel lines, each set containing the fixed line L, until we have exhausted all the s^4 lines parallel to L. So, we can construct by this method s^2 sets of 1-spreads of PG(3,s), each set containing a fixed line.

REFERENCES

1. R.H. Bruck and R.C. Bose, The construction of translation planes from projective planes, Journal of Algebra, 1(1964), 85-102.
2. R.C. Bose, Mathematical theory of symmetrical factorial experiments, Sankhyā,8 (1947), 107-166.
3. A.C. Mukhopadhyay, On the property P_t and balanced asymmetrical factorial plans, (1978), submitted to ARS Combinatoria.
4. S. Pal, Balanced optimal plans for factorial experiments, CSA Bull., 25(1976), 41-54.
5. C.R. Rao, Some combinatorial problems of arrays and applications to design of experiments, a survey of combinatorial theory, North Holland (edited by Srivastava, J.N. et al.), (1973).

ON MOLECULAR AND ATOMIC MATROIDS

H. NARAYANAN
Department of Electrical Engineering
Indian Institute of Technology
Bombay 400 076

M. N. VARTAK
Department of Mathematics
Indian Institute of Technology
Bombay 400 076

INTRODUCTION

In this paper we make a brief study of the properties of classes of matroids which we shall call molecular and atomic matroids. It has been shown, independently, in [1] and [2] that the set of definition of any matroid can be uniquely partitioned into subsets on which suitable minors can be defined. These minors turn out to be atomic matroids. This partition, called principal partition, is achieved by first breaking up the original matroid into molecular matroids and the latter are further partitioned into atomic matroids. A certain partial order is naturally associated with this partition. Both the partition as well as the partial order are invariants of the matroids under its group of automorphisms. It can be shown that there is an intimate relationship between the matroid union theorem and this partition. It seems appropriate therefore to study the indivisible units of this partition through the matroid union theorem. The results reported in this paper are taken essentially from [1]. They may be regarded as refinements of the results contained in [3]. Welsh's book on matroid theory [4] gives a very thorough discussion of the basic concepts.

PRELIMINARIES

A <u>matroid</u> on S is a pair (S,I) where S is a set and I is a class of subsets of S called independent which have the following properties :

(a) Every subset of an independent set is independent.

(b) Maximal independent sets contained in any subset of S have the same cardinality.

Maximal independent sets of a matroid are called <u>bases</u>. Complements of bases are called <u>cobases</u>. The cobases of a matroid $M = (S,I)$ form the bases of another matroid (S,I^*). This matroid is called the <u>dual</u> of M and is denoted by M^*. Minimal dependent sets of a matroid are called its <u>circuits</u>. A circuit of M^* is called a <u>cocircuit</u> of M. Let M be a matroid on S and let $T \subseteq S$. Then (T,I_T), where I_T is the class of independent sets of M contained in T, is a matroid called the <u>contraction</u> of M to T, and is denoted by $M \times T$. We denote $(M^* \times T)^*$ by $M.T$ and refer to it as the <u>reduction</u> of M to T. Matroids of the kind $(M \times T).Q$, $Q \subseteq T \subseteq S$, are referred to as <u>minors</u> of M.

Let $M_1 = (S,I_1)$ and $M_2 = (S,I_2)$ be two matroids on S. Then $M_1 \vee M_2 = (S,I_1 \vee I_2)$, where $I_1 \vee I_2$ is the class of subsets which are unions of members of I_1 and I_2, is a matroid referred to as the <u>union</u> of M_1 and M_2. We associate two functions on 2^S with a matroid M on S. The <u>rank function</u> ρ is defined by taking $\rho(T)$, $T \subseteq S$ to be the cardinality of a base of $M \times T$. By an abuse of notation denote $\rho(S)$ by $\rho(M)$. The <u>density function</u> d is defined by taking $d(T)$, $T \subseteq S$, to be the ratio $\frac{|T|}{\rho(T)}$. We may denote $d(S)$ by $d(M)$. We say M on S is <u>molecular</u> (<u>atomic</u>) iff $d(S) \geq d(T)$, for all $T \subseteq S$ ($d(S) > d(T)$, for all $T \subseteq S$). The following simple results on the matroid union are stated without proof.

<u>Theorem 1</u>. Let M_1,M_2 be matroids on S. Let $T \subseteq S$. Then

$$(M_1 \vee M_2) \times T = (M_1 \times T) \vee (M_2 \times T).$$

<u>Theorem 2</u>. Let M_1,M_2 be matroids on S. Then if $M_1 \vee M_2$ has no coloops,

$$\rho(M_1 \vee M_2) = \rho(M_1) + \rho(M_2).$$

MOLECULAR MATROIDS AND MATROID UNIONS

We now present our main result

<u>Theorem 3</u>. Let M_1,M_2 be two matroids on S. Let d_1,d_2,d_{12} denote the density functions of $M_1,M_2,M_1 \vee M_2$ respectively. Let

$$d_i(Q) = \max_{T \subseteq S} d_i(T), \quad i = 1,2.$$

Then

$$d_{12}(Q) = \max_{T \subseteq S} d_{12}(T).$$

Proof. Let $\rho_1, \rho_2, \rho_{12}$ denote the rank functions of $M_1, M_2, M_1 \vee M_2$ respectively. If $M_1 \vee M_2$ has no circuits, the result is trivial, by Theorem 1. We will suppose, therefore, that $M_1 \vee M_2$ has at least one circuit.

Let $R \subseteq S$ be such that $d_{12}(R) = \max_{T \subseteq S} d_{12}(T)$. Let R_C denote the set of coloops of $(M_1 \vee M_2) \times R$. Clearly, $R_C \neq R$. Suppose $R_C \neq \phi$. Then it is easy to see that $d((M_1 \vee M_2) \times R \times (R-R_C)) = d((M_1 \vee M_2) \times (R-R_C)) > d(R)$, which is a contradiction. We therefore conclude that $R_C = \phi$. So, by Theorem 2, it follows that

$$\rho_{12}(R) = \rho_1(R) + \rho_2(R).$$

Hence

$$(d_{12}(R))^{-1} = (d_1(R))^{-1} + (d_2(R))^{-1} \geq (d_1(Q))^{-1} + (d_2(Q))^{-1}.$$

But

$$\rho_{12}(Q) \leq \rho_1(Q) + \rho_2(Q)$$

so

$$(d_{12}(Q))^{-1} \leq (d_1(Q))^{-1} + (d_2(Q))^{-1}$$

and therefore

$$(d_{12}(R))^{-1} \geq (d_{12}(Q))^{-1}.$$

This proves the theorem.

Corollary 1. If M_1, M_2 on S are molecular matroids, then $M_1 \vee M_2$ is a molecular matroid.

By examining the steps of the proof of Theorem 3, it is easy to show the following.

Corollary 2. If M_1, M_2 on S are atomic matroids and $M_1 \vee M_2$ has a circuit, then $M_1 \vee M_2$ is atomic.

Corollary 3. If M_1, M_2 on S are molecular matroids and $M_1 \vee M_2$ has a circuit then

$$\rho(M_1 \vee M_2) = \rho(M_1) + \rho(M_2)$$

and

$$(d(M_1 \vee M_2))^{-1} = (d(M_1))^{-1} + (d(M_2))^{-1}.$$

For n, a positive integer, let M^n denote $M \vee \ldots \vee M$ (n times). Then from Corollary 3 we have (cf.[3]).

Corollary 4. Let M be a molecular matroid and n be the largest integer such that M^n has a circuit. Then (a) S can be partitioned into $(n+1)$ independent sets of M of which n are bases of M and (b) n is the integral part of $d(M)$.

Theorem 4. Let M be a molecular (atomic) matriod on S. Then M^* is molecular (atomic).

Proof. For any matroid M on S, it can be shown that

$$\rho(M) = \rho(M \times T) + \rho(M.(S-T)).$$

Suppose $T \subseteq S$ is such that $d(M^* \times T) \geq d(M^*)$. Then, since $M^* \times T = (M.T)^*$, it follows that

$$d((M.T)^*) \geq d(M^*)$$

that is,

$$\frac{|T|}{|T| - \rho(M.T)} \geq \frac{|S|}{|S| - \rho(M)} .$$

Hence

$$\frac{|S|}{|S - T|} \leq \frac{\rho(M)}{\rho(M) - \rho(M.T)},$$

so $d(M) \leq d(M \times (S-T))$. The theorem is then immediate.

Theorem 5. Let M_1, M_2 be molecular matroids on S. Then if $\rho(M_1) \geq \rho(M_2)$, there exists a base of M_2 that is independent in M_1.

Proof. Since M_1 is molecular, so is M_1^*. Now $M_1^* \vee M_2$ is molecular by Corollary 1.

Suppose $M_1^* \vee M_2$ has no circuits. Then clearly S is a base of $M_1^* \vee M_2$ and hence there exists a base of M_1 contained in a base of M_2. Since $\rho(M_1) \geq \rho(M_2)$ it follows that M_1 and M_2 have a common base.

Next, suppose $M_1^* \vee M_2$ has a circuit. Then there exist disjoint bases of M_1^* and M_2. Hence there exists a base of M_1 containing a base of M_2. This proves the theorem.

The next theorems lead to a different characterization of a molecular matroid.

Theorem 6. Let M be a matroid on S. Let there exist a family of pairwise disjoint bases $b_1, \ldots, b_n$ of M which covers S. Then M is molecular.

Proof. If $T \subseteq S$, then clearly T can be covered by n independent sets of $M \times T$. The theorem follows.

We say that two elements e_1, e_2 in a matroid M are parallel iff they form a circuit. The family of bases of M can be partitioned into three subfamilies, namely, bases which contain neither e_1 nor e_2, bases which contain e_1 and bases which contain e_2. There is a one to one correspondence between the members of the latter two families since for any base b, $(b-e_1) \cup e_2((b-e_2) \cup e_1)$ is a base iff $e_1 \in b(e_2 \in b)$.

This motivates the following construction.

Let e be an element of a matroid M_1 on S. We can then construct a matroid M_2 on S_2, where $S_2 = S \cup \{e_2 \ldots e_k\}$ $(S \cap \{e_2 \ldots e_k\} = \phi)$ by 'replacing e by k parallel elements $e, e_2, \ldots, e_k$' i.e., by defining M_2 as the matroid that has as bases all bases of M_1 and in addition has as bases the sets $(b-e) \cup e_i$ whenever b is a base of M_1 and $e \in b$.

This leads to the following theorem whose proof we omit.

<u>Theorem 7</u>. Let M be a matroid on S. Then the matroid $M(k)$, obtained by replacing each element of M by k parallel elements is molecular iff M is molecular.

The next theorem characterizes molecular matroids.

<u>Theorem 8</u>. Let M be a matroid on S. Then M is molecular iff there exist bases $b_1, \ldots, b_n$ of S such that each element of S belongs to precisely q of these bases.

<u>Proof</u>. Let the density of M be $n + p/q$, where n, q are positive integers and p is non-negative. Suppose M is molecular and $p = 0$. By Corollary 4, M has n pairwise disjoint bases. Suppose $p \neq 0$. Replace each element in M by q parallel elements. The resulting matroid $M(q)$ has density $qn + p$ and has $qn + p$ pairwise disjoint bases which cover the set on which $M(q)$ is defined. Since no two parallel elements can belong to the same base the $qn + p$ disjoint bases of $M(q)$ will correspond to $qn + p$ bases of M in which each element is repeated precisely q times. Suppose M is a matroid on S such that there exist bases $b_1, \ldots, b_n$ of M in which each element of S occurs precisely q times. $M(q)$ clearly has density n and has n pairwise disjoint bases. From Theorem 6 it follows that $M(q)$ is molecular and hence M is molecular. This proves the theorem.

We associate with a matroid M on S a group of automorphisms on S which preserve independence. The next theorem stated without proof, refers to this group.

Theorem 9. Let M be a matroid on S. Let $\max_{T \subseteq S} d(T) = q$. Then there exists a unique maximal set R such that $d(R) = q$, i.e., R is invariant under automorphisms of M.

Corollary 1. Let M be a non-molecular matroid on S. Then the group of automorphisms of M is not transitive on S.

IMPLICATIONS FOR GRAPHS

Let G be a graph with edge set S. The edge sets of spanning forests of G form the bases for a matroid which we shall refer to as P(G). Circuits of P(G) are the edge sets of circuits of G. If G_T is the subgraph of G on the set of edges $T \subseteq S$, then $P(G_T) = P(G \times T)$. The notions of density, molecularity etc., carry over to graphs naturally. Density of graph G on $S = \frac{|S|}{\text{rank of } G}$.

We would say a graph is molecular iff it has no subgraph of higher density.

We now translate some of our matroidal results to the class of graphs.

Let us, for brevity, refer to edge sets of spanning forests of G as forests of G.

Theorem 10. A graph G on S is molecular iff it has a family of pairwise disjoint forests $b_1, b_2, \ldots, b_n$ covering S, in which each edge is repeated the same number of times.

This follows from Theorem 8.

Theorem 11. Let G be a graph on S whose group of automorphisms is transitive on S. Then G is molecular.

Proof. This follows from Corollary 1 of Theorem 9.

In particular the complete graphs and the complete bipartite graphs are molecular. (As a matter of fact, they are atomic but we do not need this here). We therefore have

Corollary 1. If G is a complete graph or a complete bipartite graph on S, then S can be partitioned into k subforests of which (k-1) are forests.

<u>Theorem 12</u>. Let G_1 and G_2 be molecular graphs on S. Let rank of $G_1 \leq$ rank of G_2. Then there exists a forest of G_1 that is a subforest of G_2.

<u>Corollary 1</u>. Let G_1 and G_2 be graphs on S such that S can be partitioned into k_i disjoint forests (i = 1,2). Then if $k_1 \geq k_2$ there exists a forest of G_1 that is a subforest of G_2.

REFERENCES

1. H. Narayanan, Theory of matroids and network analysis, Ph.D. Thesis submitted to Department of Electrical Engineering, I.I.T., Bombay, February 1974.

2. N. Tomizawa, Strongly irreducible matroids and principal partition of a matroid into strongly irreducible matroids, Trans. I.E.C.E.(Japan), J 59-A, No.2(1976), 1-10. (This paper is in Japanese but the journal is translated into English).

3. J. Edmonds, Lehman's switching game and a theorem of Tutte and Nash-Williams, J. Res. N.B.S., Vol.69 B(1965), 73-77.

4. D. J. Welsh, Matroid Theory, Academic Press, 1976.

NEW INEQUALITIES FOR THE PARAMETERS OF AN ASSOCIATION SCHEME

A. NEUMAIER
Fachbereich Mathematik
Technische Universität Berlin
D-1000 Berlin 12

Let X be a finite set with v objects. An s-class <u>scheme</u> on X is a partition of the set of all 2-subsets of X into $s \geq 2$ nonempty classes. Two objects x and y are <u>i-th</u> <u>associates</u> if $x \neq y$ and x,y is in the i-th class of the partition. We define $k_i(x)$ as the number of i-th associates of x, and $p_{ij}(x,y)$ as the number of $z \in X$ which are i-th associates of x, and j-th associates of y. If we write $D_i = (d^i_{xy})$ with $d^i_{xy} = 1$ or 0 according as x and y are i-th associates or not then

$$D_i D_j = (p_{ij}(x,y)), \qquad \ldots \quad (1)$$

$$p_{ij}(x,x) = k_i(x)\delta_{ij}, \qquad \ldots \quad (1')$$

where δ_{ij} (= 1 or 0 according as $i = j$ or not) is the Kronecker symbol. An <u>association scheme</u> is a scheme with $k_i(x) = k_i$ for all $x \in X$, and $p_{ij}(x,y) = p^{\ell}_{ij}$ whenever x and y are ℓ-th associates.

For the following well-known results see e.g. Cameron, et. al [1]. The <u>Bose-Mesner-algebra</u> of an association scheme is the algebra V generated by the matrices I and $D_1,\ldots,D_s$. There is a basis $E_0,\ldots,E_s$ of V satisfying for $i,j = 0,\ldots,s$

$$E_i E_j = \delta_{ij} E_j. \qquad \ldots \quad (2)$$

Also, the Bose-Mesner-algebra is closed under pointwise multiplication $(a_{xy}) \circ (b_{xy}) = (a_{xy} b_{xy})$, whence

$$E_i \circ E_j = \sum_{\ell=0}^{s} q^{\ell}_{ij} E_{\ell} \qquad \ldots \quad (3)$$

for appropriate numbers q^{ℓ}_{ij}, which can be calculated from the parameters. The <u>Krein condition</u>

$$q^{\ell}_{ij} \geq 0 \quad \text{for } i,j,\ell = 0,\ldots,s \qquad \ldots \quad (4)$$

gives a restriction on the parameters.

The ranks f_i of E_i can also be calculated from the parameters; they satisfy $f_0 + \dots + f_s = v$, and the fact that all f_i must be integers places a severe restriction on the possible parameter sets. We prove here a new inequality for the ranks :

<u>Theorem 1</u>. The following inequalities hold :

$$\sum_{\ell:q_{ij}^{\ell}>0} f_\ell \le \begin{cases} f_i f_j & \text{for } i \ne j, \\ \frac{1}{2}f_i(f_i+1) & \text{for } i = j \end{cases} \qquad \dots \quad (5)$$

<u>Proof</u>. By the following lemma, the rank of $E_i \circ E_j$ is at most $f_i f_j$ if $i \ne j$, and $\frac{1}{2}f_i(f_i+1)$ if $i = j$. On the other hand, since the E_i are mutually orthogonal, (3) implies that the rank of $E_i \circ E_j$ is given by the left hand side of (5).

<u>Lemma</u>. (i) Let A be a matrix of rank f. Then AA has rank $\le \frac{1}{2}f(f+1)$.

(ii) Let A and B be matrices of the same size and of rank f, resp., g. Then AB has rank $\le fg$.

<u>Proof</u>. Write $A = (a_{xy})$, and let $x_1,\dots,x_f$ be the labels of f independent rows. Then each a_{xy} is a linear combination of $a_{x_1y},\dots,a_{x_fy}$. Hence each entry a_{xy}^2 of $A \circ A$ is a linear combination of $a_{x_1y}^2,\dots,a_{x_fy}^2$, $a_{x_1y}a_{x_2y},\dots,a_{x_{f-1}y}a_{x_fy}$, of which there are $f + \binom{f}{2} = \frac{1}{2}f(f+1)$ terms. This proves (i), and the proof of (ii) is completely analogous.

A 2-class association scheme is essentially the same as a <u>strongly regular graph</u> (see e.g. Seidel [2] for a definition). A strongly regular graph with $q_{22}^2 = 0$ is called a <u>Smith graph</u>. Cameron, et. al [1] show that q_{22}^0 and q_{22}^1 are nonzero. Hence Theorem 1 gives

<u>Theorem 2</u>. (i) The parameters of a strongly regular graph which is not a Smith graph satisfy

$$v \le \tfrac{1}{2}f_2(f_2+1). \qquad \dots \quad (6)$$

(ii) The parameters of a Smith graph satisfy

$$v \le \tfrac{1}{2}f_2(f_2+3). \qquad \dots \quad (7)$$

<u>Proof</u>. Apply Theorem 1 with $i = j = 2$, and observe that $f_0 + f_1 + f_2 = v$.

Example. The following parameter set for a strongly regular graph satisfies all previously known conditions for strongly regular graphs (as stated e.g. in Seidel [2]) but fails (6) :

$$v = 841,\ k = 200,\ \lambda = 87,\ \mu = 35,\ f_2 = 40.$$

Problem. Characterize those graphs for which (6) is satisfied with equality.

Remarks. 1. More inequalities can be obtained similarly by looking at $E_i \circ E_j \circ E_k$, etc., but it is not known whether they are really more restrictive than those of Theorems 1 and 2.

2. The special case of Theorem 2, where the graph has a rank 3 automorphism group, has been proved already in Cameron et. al [1].

3. Theorem 2 improves the absolute bound (see e.g. Seidel [2]) for strongly regular graphs; it is not known how Theorem 1 relates to the more general absolute bound mentioned in [1], Proposition 6.1.

REFERENCES

1. P.J. Cameron, J.M. Goethals, and J.J. Seidel, The Krein condition, spherical designs, Norton algebras and permutation groups, Proc. Kon. Ned. Acad. Wet. A81(1978), 196-206.

2. J.J. Seidel, Strongly regular graphs, an introduction, Proc. 7th Brit. Comb. Conf., Cambridge (LMS Lecture Notes Series 38) 1979, 157-180.

ON THE (4-3)-REGULAR SUBGRAPH CONJECTURE

K. R. PARTHASARATHY
S. SRIDHARAN
Department of Mathematics
Indian Institute of Technology
Madras 600 036

ABSTRACT

In this paper we establish the validity of Berge's Conjecture that every 4-regular simple graph has a 3-regular subgraph, for graphs with diameter 2.

1. INTRODUCTION

We consider ordinary (simple) graphs on p points with q lines and follow Harary [3] for general notation and terminology. The following Conjecture due to Berge is listed as an unsolved problem in Bondy and Murty ([1], page 246, problem 3) : Every 4-regular simple graph contains a 3-regular subgraph. We call this the (4,3)-regular subgraph problem. Clearly, it is enough to prove the statement for connected graphs. So, in what follows, we assume the <u>graph G to be connected</u>. In a recent paper [2], Chvátal, Fleishner, Seehan and Thomassen have established the validity of the Conjecture for all graphs with cyclic connectivity number $\lambda_c \geq 10$ and have shown that the smallest possible counter-example to the Conjecture should have $\lambda_c(G)$ either 6 or 8. Following a different approach we prove here that the Conjecture is valid for all graphs with diameter 2.

We first dispose off the case of graphs on even number of points :

<u>Lemma 1</u>. If G is a 4-regular graph of diameter 2 then $\lambda_c(G) \geq 4$.

<u>Proof</u>. Since G is Eulerian, $\lambda_c(G)$ is even, so that we have only to rule out the case $\lambda_c(G) = 2$. If possible let e_1, e_2 be the two edges constituting a cyclic disconnecting edge set of cardinality 2. Two cases arise :

(i) e_1 and e_2 are adjacent at a vertex v. Let the other two edges incident at v be e_3, e_4 and the other ends of these four edges be x, y, z and w, respectively. Then, in $G - e_1 - e_2$ we should have x, y belonging to one cyclic component C_1 and z,w belonging to another cyclic component C_2. Clearly there is another vertex t in C_2 and $d(x,t) > 2$, a contradiction.

(ii) The lines e_1 and e_2 are non-adjacent. Let $e_1 = (x,z)$ and $e_2 = (y,w)$, with x,y and z,w belonging to the two different cyclic components C_1 and C_2 of $G-e_1-e_2$. Then there exist vertices $u \in C_1$ and $t \in C_2$ with $d(u,t) > 2$, a contradiction.

Corollary 1. If a 4-regular graph G is of diameter 2 and has an even number of points, then G has a 1-factor.

Proof. From Lemma 1, $\lambda_c(G) \geq 4$ and as observed by Chvátal, Fleishner, Seehan and Thomassen [2], G has a 1-factor by Tutte's theorem.

To prove the Conjecture for graphs on an odd number of points we state the following modified

Conjecture (2) : If G is a 4-regular graph on $p = 2n+1$ points it has a vertex v such that G-v has a 3-regular spanning subgraph.

Chvâtal, Fleishner, Seehan and Thomassen proved this conjecture (in a slightly modified form) for graphs with $\lambda_c(G) \geq 10$. That the Conjecture is not true in general has been demonstrated by S.A. Choudum by the construction of a counterexample (a graph with 105 points). We prove the Conjecture for graphs with diameter 2, by the following procedure.

For a graph $G = (V,E)$ and $v \in V$, let $N(v) = \{u \in V | (u,v) \in E\}$ and $N^*(v) = V-v-N(v)$. For any point $v \in V$ of a graph G of diameter ≥ 2, $N^*(v) \neq \phi$. To prove Conjecture (2) we have to verify that there is a $v \in V$ such that $\langle N^*(v)\rangle$ has a 1-factor. Our method is to determine the possible degree sequences of $\langle N^*(v)\rangle$ and to verify that these are forcibly 1-factorable.

Here we make a note on terminology. A 1-factor of a graph is a spanning regular subgraph of degree 1. A graph G is said to be 1-factorable if it has a 1-factor and is said to be 1-factorizable if it can be expressed as the superposition of k line-disjoint 1-factors. A graphical degree sequence is said to be forcibly 1-factorable if every graph realizing it is 1-factorable. It is forcibly 1-factorizable if every graph realizing it is 1-factorizable. It is to be pointed out that this is at variance with some of the usual terminology.

For a given $v \in V$, let $B(v)$ be the bipartite graph on $N(v) \cup N^*(v)$ induced by the edges of G between $N(v)$ and $N^*(v)$. It is easily seen that if $N(v)$ has q_1 lines, $B(v)$ has $2(6-q_1)$ lines and that the degree sequence at $N^*(v)$ in $B(v)$ is just a partition of $12-2q_1$. We call this the deficiency sequence at $N^*(v)$. The degree sequence of $\langle N^*(v) \rangle$ is simply the complement of this deficiency sequence in the degree sequence 4^m where $m = p-5$. The following lemma can also be easily verified.

Lemma 2. If G is a 4-regular graph on $2n+1$ points there exists a $v \in V$ such that $\delta(\langle N^*(v) \rangle) > 0$.

Proof. Observe that $\delta(\langle N^*(v) \rangle) = 0$ iff there is a $u \in N^*(v)$ such that $N(u) = N(v)$. The rest of the proof is just a routine verification that such a u does not exist in $N^*(v)$ for every v, by considering the different possible configurations for $\langle N(v) \rangle$. We omit the details.

In Table 1 we list all the possible deficiency sequences and the corresponding degree sequences of $\langle N^*(v) \rangle$ for which $\delta > 0$. The deficiency sequences are listed in non-increasing mode and the degree sequences of $\langle N^*(v) \rangle$ in non-decreasing mode. The degree sequences of $\langle N^*(v) \rangle$ are of the form $1^a 2^b 3^c 4^d$. A general plan to verify Conjecture (2) is to prove that all such degree sequences are forcibly 1-factorable. As forcibly 1-factorable degree sequences have not yet been characterized in general, we limit the problem to graphs of diameter 2. Then the number of points (for odd graphs) is at most 15 (see e.g., Bondy and Murty [1], page 15, Problem 1.7.5) and only those partitions in the table with 2,4,6,8 or 10 parts and with no part equal to 4 have to be scanned to verify whether they are forcibly 1-factorable. Of course, some of these graphs may have diameter > 2. The discussion of the above five cases will be given in the next section and this establishes

Theorem 1. If G is a 4-regular graph with diameter 2, then G has a 3-regular subgraph.

We observe that if $p \geq 9$ and $q_1 \geq 2$ then $\lambda_c \leq 8$ so that we have proved the Berge Conjecture for a class of graphs not covered by [2].

2. THE FIVE CASES

Case 1. $p = 7$, $|N^*(v)| = 2$.

From the last column of the table it is seen that the only 2-part partitions to be considered are those with serial numbers 2,3,4 and 7 of which 2,3 and 4 are not graphical. The partition corresponding to 7 is 1^2 which forcibly realizes K_2, which obviously has a 1-factor.

Case 2. $p = 9$, $|N^*(v)| = 4$.

The four-part graphical sequences are at serial numbers 6,11,16,19,24,25 and 38 and these uniquely realize K_4, K_4-x, K_4-P_3, C_4, $K_{1,3}$, P_4 and $2P_2$. All of these except $K_{1,3}$ obviously have 1-factors. The first four of these correspond to $q_1 \geq 2$ and we will show that the case $K_{1,3}$ is reducible to one of these by a different choice of the vertex v.

From Figure 1 we see that if u_1 is joined to x and one of y,z or t then

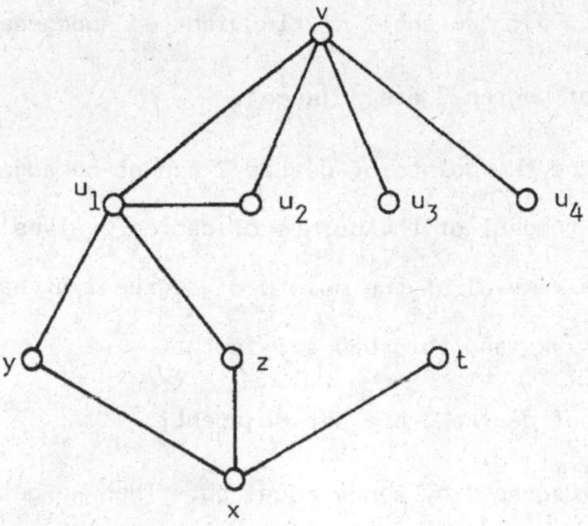

FIGURE 1

$\langle N(u_1)\rangle$ has two lines. A similar argument holds for the vertex u_2. In the remaining case, suppose u_1 is joined to y and z. Since u_2 should receive two more degrees, it should be adjacent to one of y and z. Again $\langle N(u_1)\rangle$ has two lines. Thus replacing v by u_1 we are in one of the cases ($q_1 \geq 2$) already discussed.

<u>Case 3</u>. $p = 11$, $|N^*(v)| = 6$.

As seen from the table the only feasible 6-part graphical sequences are in serial numbers 13,18,21,27,29,33,40,42,45, and 50. Of these the first six correspond to $q_1 \geq 1$ which implies G has girth 3. We consider these cases first, in terms of q_1, denoting the degree sequences of $\langle N^*(v)\rangle$ by π.

3.1 $q_1 = 3$, $\pi = 3^6$.

We know that the smallest number of points in a cubic graph with a bridge is ten. So $\langle N^*(v)\rangle$ is a bridgeless cubic graph and by Petersen's theorem, it has a 1-factor.

3.2 $q_1 = 2$.

First let $\pi = \pi_1 \equiv 13^5$. Any graph with degree sequence π_1 has a pendant edge. Removal of the points incident with the pendant edge gives the degree sequence $\theta : 2^2 3^2$. There is only one graph with this degree sequence, namely K_4-x and this has a 1-factor. This 1-factor together with the pendant edge gives a 1-factor of $\langle N^*(v)\rangle$.

Next let $\pi = \pi_2 \equiv 2^2 3^4$. We shall distinguish two subcases.

<u>Subcase (i)</u> : The points of degree 2 are adjacent.

Here we note that the two points of degree 2 cannot be adjacent to a point of degree 3. For if so, the removal of the points of degree 2 gives the sequence 13^3 which is not graphical. Now the removal of the points of degree 2 gives the sequence $\theta = 2^2 3^2$ which realises uniquely K_4-x and this has a 1-factor.

<u>Subcase (ii)</u> : The points of degree 2 are not adjacent.

Join the points of degree 2 by a new edge ab. Then we get a cubic graph on six points. As seen in Case 3.1 this graph has a 1-factor. If this 1-factor does not contain the new edge ab, then the original graph has a 1-factor. Suppose this 1-factor contains the new edge ab. The removal of this 1-factor gives a 2-factor. This 2-factor can be either C_6 or $C_3 \cup C_3$. If it is C_6, it has a 1-factor not containing ab. So let the 2-factor be $C_3 \cup C_3$. Now the addition of the 1-factor removed, gives the graph of Figure 2. The removal of the 1-factor shown in the figure gives C_6, which in turn has

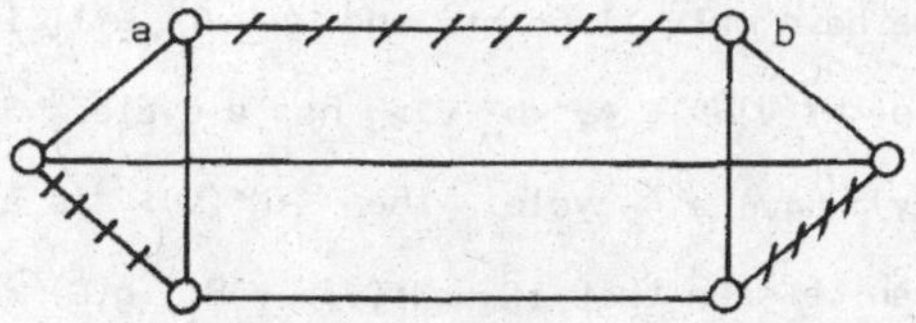

FIGURE 2

a 1-factor which is in the original graph.

3.3 $q_1 = 1$.

First let $\pi = \pi_1 \equiv 1^2 3^4$. Every graph with degree sequence π_1 has 2 points of degree 1. If the vertices of degree 1 are adjacent then the graph is $K_4 \cup K_2$ which has a 1-factor. So let an end point be adjacent to a point of degree 3. Removal of this end point and its adjacent vertex gives the sequence $\theta = 12^2 3$ which uniquely realises the graph $K_4 - P_3$ and this has a 1-factor.

Next let $\pi = \pi_2 \equiv 12^2 3^3$. We distinguish two subcases.

Subcase (i). Let the pendant edge be incident with a point t of degree 3.

It can be seen that this point t cannot be adjacent to two points of degree 2. Removal of the points incident with the pendant edge gives two sequences $\theta_1 = 2^4$, $\theta_2 = 12^2 3$. But θ_1 gives uniquely C_4 and θ_2 gives uniquely $K_4 - P_3$ and each of these has a 1-factor.

Subcase (ii). Let the end point be adjacent to a point s of degree 2.

Then s is not adjacent to a point of degree 2. The removal of the points incident with the pendant edge gives $\theta_3 = 2^2 3^2$ which uniquely realises $K_4 - x$ and has a 1-factor.

Finally let $\pi = \pi_3 \equiv 2^4 3^2$. Consider the points, say a and b, of degree 2. It is easily seen that these are always adjacent. Removal of the points a and b gives the sequences $\theta_1 = 12^2 3$, $\theta_2 = 2^4$, $\theta_3 = 12^2 3$ and these are forcibly 1-factorable.

3.4 $q_1 = 0$ (The partitions at serial numbers 40,42,45,50 have to be considered here).

If $g = 3$ there exists v such that $q_1 \geq 1$ and this case has already been considered. So let $g \neq 3$. Then $g = 4$ because a 4-regular graph of $g = 5$ has

for $t_i \in \{0,1,\ldots,\frac{n_i}{m_1}\}$. Therefore β is always in $\langle\sigma\rangle$. Since every permutation of σ preserves adjacency in G, we have $\Gamma(G) = \langle\sigma\rangle$. Hence $\langle\sigma\rangle$ is graphical. This completes the proof of Theorem 2.

We will use Theorem 2 to characterise graphical cyclic permutation groups of order p^n where p is a prime greater than five.

Theorem 3. Let σ be a permutation of order p^n where p is a prime greater than five and $n \neq 0$. $\langle\sigma\rangle$ is graphical if and only if cyclic form of σ has at least two cycles of length more than one.

Proof. Since the order of σ is p^n, its cyclic form has only cycles of length p^m where $0 \leq m \leq n$.

We know that the group generated by a single cycle of length m where $m \neq 1$, it not graphical (see [2]). Even if we add cycles of length one to such a cycle the group generated by the new permutation is not graphical. Suppose there is a graph G such that $\Gamma(G)$ is generated by a permutation $(u_1u_2\ldots u_m)(u_{m+1})(u_{m+2})\ldots(u_n)$ where $m > 1$. It is clear that if $[u_i,u_j]$ is an edge in G for some $i \in Z_m$ and $j = m+1, m+2,\ldots,n$, then $[u_i,u_j]$ is an edge for all $i \in Z_m$. Hence the automorphism group of $\langle u_1,u_2,\ldots,u_n\rangle$ is $\langle(u_1u_2\ldots u_m)\rangle$, which is false. Therefore $\langle\sigma\rangle$ is not graphical, if its cyclic form has only one cycle of length not equal to one.

Let the cyclic form of σ consist of r non-trivial cycles of lengths p^{m_1}, $p^{m_2},\ldots,p^{m_r}$ where $r \geq 2$ and $0 \leq m_1 \leq m_2 \leq \ldots \leq m_r$. Since p^{m_1} divides p^{m_i} for $i \in Z_r$, it follows from Theorem 2 that the group generated by these cycles is graphical. Even if we add cycles of length one, we conclude that $\langle\sigma\rangle$ is graphical. This completes the proof of the theorem.

Corollary 1. A permutation group H of prime order $p > 5$ is graphical if and only if the cyclic form of its generator has at least two cycles each of length p.

Proof. Since a permutation group of prime order is cyclic, the generator of H has order p. So the result follows from Theorem 4.

at least 17 points and we have only 11 points and $g \geq 6$ will imply diameter $d \geq 3$. Since $q(\langle N^*(v)\rangle) = 6 \geq p(\langle N^*(v)\rangle) = 6$, $\langle N^*(v)\rangle$ has a cycle. Since a 3-cycle is not possible first let $\langle N^*(v)\rangle$ have a 4-cycle. Then $\langle N^*(v)\rangle$ is either H_1 or H_2 of Figure 3. However, it can be seen that if $\langle N^*(v)\rangle = H_1$, $g(G) = 3$. Therefore H_2

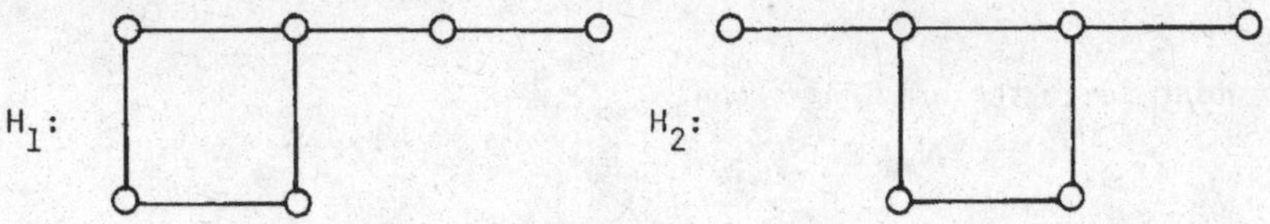

FIGURE 3

is the only graph possible and this has a 1-factor. Next let $\langle N^*(v)\rangle$ have a 5-cycle. Then H_3 of Figure 4 is the only graph possible for $\langle N^*(v)\rangle$ and this has a 1-factor. If $\langle N^*(v)\rangle$ has a 6-cycle then

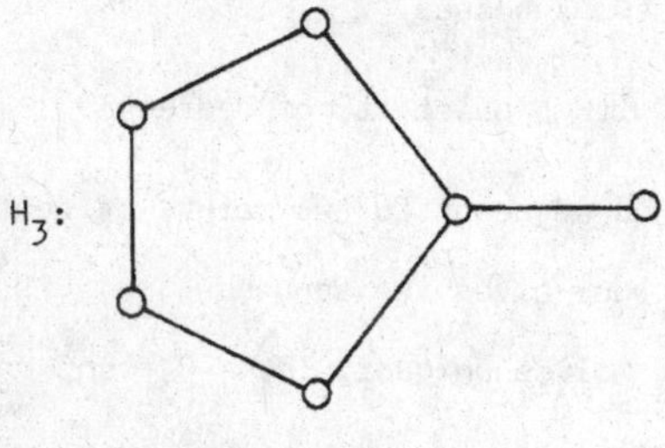

FIGURE 4

$\langle N^*(v)\rangle = C_6$ which has a 1-factor.

<u>Case 4</u>. $p = 13$; $|N^*(v)| = 8$.

Let $q_1 \geq 1$. From the table the only feasible 8-part graphical partitions are in serial numbers 23,31,35. We consider these in terms of q_1, denoting the degree sequence of $\langle N^*(v)\rangle$ by π.

4.1 $q_1 = 2$, $\pi = 3^8$.

As seen in 3.1, $\langle N^*(v)\rangle$ is a bridgeless cubic graph and has a 1-factor.

4.2 $q_1 = 1$.

First let $\pi = \pi_1 \equiv 13^7$. Removal of the points incident with the pendant edge gives the sequence $\theta_1 = 2^2 3^4$ and this is forcibly 1-factorable as seen in 3.2, hence π_1 is forcibly 1-factorable.

Next let $\pi = \pi_2 \equiv 2^2 3^6$. Here we distinguish two subcases.

Subcase (i) : The points of degree 2 are adjacent.

Removal of the points of degree 2 gives the sequences $\theta_2 = 2^2 3^4$ and $\theta_3 = 13^5$ which are forcibly 1-factorable as seen in 3.2. This 1-factor with the edge joining the points of degree 2 shows that π_2 is forcibly 1-factorable.

Subcase (ii) : The points of degree 2 are not adjacent.

Join the points of degree 2 by a new edge $e = ab$. Then we proceed as in subcase (ii) of 3.2 to get a 1-factor of any realisation of $\pi_2 = 2^2 3^6$ without containing the new edge ab.

4.3 $q_1 = 0$.

We have to consider the partitions in serial numbers 44,47,52.

First let $\pi = \pi_1 \equiv 1^2 3^6$. If the points of degree 1 are adjacent then $\langle N^*(v) \rangle = H \cup K_2$ where H is a cubic graph on 6 points. Therefore $\langle N^*(v) \rangle$ has a 1-factor.

If an end point is adjacent to a point of degree 3 the removal of the points incident with the pendant edge gives the sequence $\theta_1 = 12^2 3^3$ and this is forcibly 1-factorable as seen in 3.3. Hence π_1 is forcibly 1-factorable.

Next let $\pi = \pi_2 \equiv 12^2 3^5$. Now Figure 5 gives possible structures near the pendant edge. Removal of the points of the pendant edge give the sequences $\theta_2 = 2^2 3^4$, $\theta_3 = 13^5$,

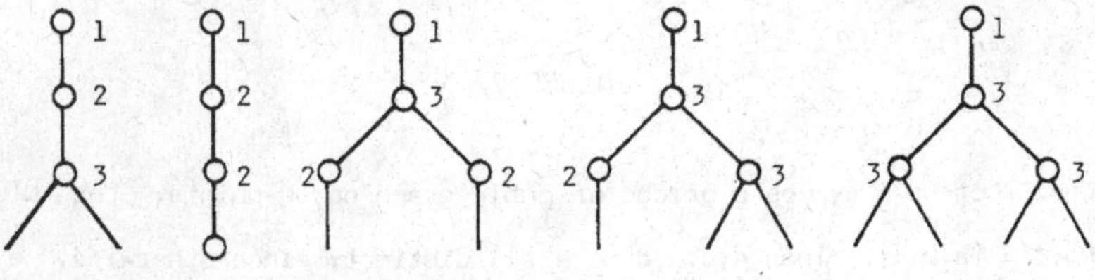

FIGURE 5

$\theta_4 = 1^2 3^4$, $\theta_5 = 12^2 3^3$ and $\theta_6 = 2^4 3^2$ and these sequences have been verified to be forcibly 1-factorable. Thus π_2 is forcibly 1-factorable.

Finally let $\pi = \pi_3 \equiv 2^4 3^4$. Consider any realisation of π. If the points of degree 2 are not adjacent, join them by two new edges. Then we get a cubic graph on eight points and we proceed as in Subcase (ii) of 3.2 to get a 1-factor without containing the new edges. So let there be an edge ab between two points of degree 2 (see Figure 6).

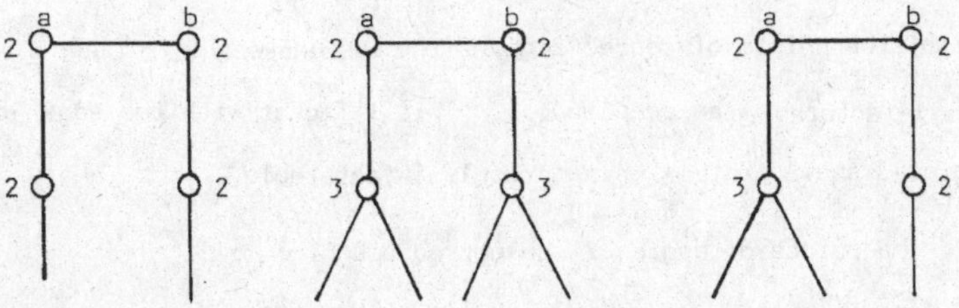

FIGURE 6

The removal of a,b gives the sequences $\theta_7 = 1^2 3^4$, $\theta_8 = 2^4 3^2$ and $\theta_9 = 12^2 3^3$. They have already been verified to be forcibly 1-factorable. Thus π_3 is forcibly 1-factorable.

<u>Case 5</u> : $p = 15$; $|N^*(v)| = 10$.

From the table the only 10-part graphical sequences are in serial numbers 37,49,54.

5.1 $q_1 = 1$, $\pi = 3^{10}$.

We shall show that the graph of Figure 7 is the only cubic graph H on 10 points

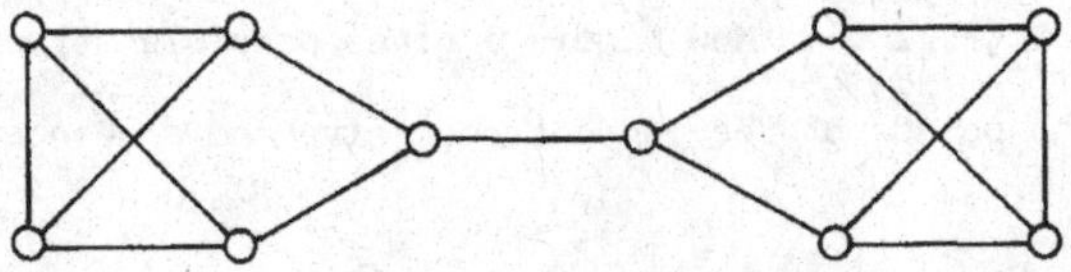

FIGURE 7

with a bridge. Let $e = uv$ be a bridge of cubic graph on 10-points. Let $N(u) = \{c,d,v\}$, $N(v) = \{a,b,u\}$. Then a,b,c,d are all distinct, since otherwise, e will lie on a cycle and hence e will not be bridge. The points v,a,b and u,c,d should lie in different blocks B_1 and B_2 respectively. Since the graph is 3-regular, each of these blocks has two more points. Let m,n be in B_1. Now ab $\notin E(B_1)$, for if ab $\in E(B_1)$ then either $d(m) \neq 3$ or $d(n) \neq 3$. Therefore am, an $\in E(B_1)$. Similarly bm, bn $\in E(B_1)$ and mn $\in E(B_1)$. Similar arguments hold for B_2. Hence the graph in Figure 7 is the only cubic graph with a bridge on 10 points, and this has a 1-factor. All other cubic graphs on 10 points are bridgeless and so have a 1-factor by Petersen's theorem.

5.2 $q_1 = 0$.

We can assume $g \neq 3$. Then $g = 4$. The possible sequences are in serial numbers 49 and 54.

First let $\pi = \pi_1 \equiv 13^9$. The removal of the points of the pendant edge gives the sequence $2^2 3^6$ and this has been verified to be forcibly 1-factorable in 4.2. Hence π_1 is forcibly 1-factorable.

Next let $\pi = \pi_2 \equiv 2^2 3^8$. We distinguish two subcases.

Subcase (i) : The points a,b of degree 2 are adjacent.

The removal of the points a,b gives the sequence $\theta = 1^2 3^6$ and this was shown to be forcibly 1-factorable in 4.3. Hence π_2 is forcibly 1-factorable.

Subcase (ii) : The points a,b of degree 2 are not adjacent.

Introduce a new edge ab. Then we get a cubic graph on 10 points. This graph cannot be the graph of Figure 7 since the graph of Figure 7 has a triangle after the removal of any line whereas in our case $g = 4$. Therefore $\langle N^*(v) \rangle$ is a bridgeless cubic graph and hence has a 1-factor. If the 1-factor does not contain ab, it is in the original graph. Otherwise we proceed as in Subcase (ii) of 3.2 to get a 1-factor of realisation of π_2 without containing the edge ab.

REFERENCES

1. J.A. Bondy and U.S.R. Murty, Graph Theory with Applications, The Macmillan Press Ltd., 1976.

2. V. Chvátal, H. Fleishner, J. Seehan and C. Thomassen, Three-regular subgraphs of four-regular graphs, Journal of Graph Theory, Vol.3, 4(1979), 371-386.

3. F. Harary, Graph Theory, Addison-Wesley, Reading, 1969.

TABLE 1

POSSIBLE PARTITIONS OF $\langle N^*(v)\rangle$

$q_1=q(\langle N(v)\rangle)$ (1)	$\langle N(v)\rangle$ (2)	Deficiency sequence (3)	Possible sequences of $\langle N^*(v)\rangle$ (4)	Serial Number (5)
5	K_4-x	2	24^d	1
		1^2	3^24^d	2
4	C_4			
	K_4-P_3	31	134^d	3
		2^2	2^24^d	4
		21^2	23^24^d	5
		1^4	3^44^d	6
3	$K_{1,3},K_3\cup K_1$			
	P_4	3^2	1^24^d	7
		321	1234^d	8
		31^3	13^34^d	9
		2^3	2^34^d	10
		2^21^2	$2^23^24^d$	11
		21^4	23^44^d	12
		1^6	3^64^d	13
2	$2K_2,P_3\cup K_1$	3^22	1^224^d	14
		3^21^2	$1^23^24^d$	15
		32^21	12^234^d	16
		321^3	123^34^d	17
		31^5	13^54^d	18
		2^4	2^44^d	19
		2^31^2	$2^33^24^d$	20
		2^21^4	$2^23^44^d$	21
		21^6	23^64^d	22
		1^8	3^84^d	23

Table 1 (Contd.)

(1)	(2)	(3)	(4)	(5)
1	$\overline{K_4-x}$	$3^3 1$	$1^3 3 4^d$	24
		$3^2 2^2$	$1^2 2^2 4^d$	25
		$3^2 2 1^2$	$1^2 2 3^2 4^d$	26
		$3^2 1^4$	$1^2 3^4 4^d$	27
		$3 2^3 1^3$	$1 2^3 3 4^d$	29
		$3 2 1^5$	$1 2 3^5 4^d$	30
		$3 1^7$	$1 3^7 4^d$	31
		2^5	$2^5 4^d$	32
		$2^4 1^2$	$2^4 3^2 4^d$	33
		$2^3 1^4$	$2^3 3^4 4^d$	34
		$2^6 1^6$	$2^2 3^6 4^d$	35
		$2 1^8$	$2 3^8 4^d$	36
		1^{10}	$3^{10} 4^d$	37
0	$\overline{K}_4$	3^4	$1^4 4^d$	38
		$3^3 2 1$	$1^3 2 3 4^d$	39
		$3^3 1^3$	$1^3 3^3 4^d$	40
		$3^2 2^3$	$1^2 2^3 4^d$	41
		$3^2 2^2 1^2$	$1^2 2^2 3^2 4^d$	42
		$3^2 2 1^4$	$1^2 2 3^4 4^d$	43
		$3^2 1^6$	$1^2 3^6 4^d$	44
		$3 2^4 1$	$1 2^4 3 4^d$	45
		$3 2^3 1^3$	$1 2^3 3^3 4^d$	46
		$3 2^2 1^5$	$1 2^2 3^5 4^d$	47
		$3 2 1^7$	$1 2 3^7 4^d$	48
		$3 1^9$	$1 3^9 4^d$	49
		2^6	$2^6 4^d$	50
		$2^5 1^2$	$2^5 3^2 4^d$	51
		$2^4 1^4$	$2^4 3^4 4^d$	52
		$2^3 1^6$	$2^3 3^6 4^d$	53
		$2^2 1^8$	$2^2 3^8 4^d$	54
		$2 1^{10}$	$2 3^{10} 4^d$	55
		1^{12}	$3^{12} 4^d$	56

ENUMERATION OF LATIN RECTANGLES VIA SDR'S

C. R. PRANESACHAR
Department of Applied Mathematics
Indian Institute of Science
Bangalore 560 012

ABSTRACT

Presented here is a summary of the results on the theory of enumeration of Latin rectangles and their connection with the theory of SDR's (Systems of Distinct Representatives). We shall touch only the salient features in the chronological development. In the last section we indicate how the problem of finding the number of Latin rectangles can be tackled, in general, with the help of SDR's.

1. INTRODUCTION

Counting finite patterns or configurations having certain specified properties has been an ever fascinating subject of Combinatorics and Graph Theory. In this paper, we are interested in the enumeration of Latin rectangles, a problem whose origin can be traced back to the eighteenth century.

Though the problem of finding the number of all Latin rectangles of given order in general remains unsolved, simpler cases have been successfully disposed. This problem seems to have drawn much attention in the forties and the early fifties of the current century. There have been occasional papers during the rest of the period. Some of the pioneers are Jacob, Kerawala, Yamamoto, Riordan, Erdös, Kaplansky and Hall. Touchard and Moser are somewhat indirectly connected with problem. A recent paper on this problem is by Stein. (Specific references will be given later).

Before we embark upon the development of the subject proper we introduce some preliminary definitions and notation in Section 2. Section 3 is a survey of the literature on the enumeration problem; as a rule the results have been stated chronologically. Section 4 describes the relation between Latin rectangles on one hand and SDR's on the other and also gives a general method to tackle the problem.

2. DEFINITIONS AND NOTATION

If X is a finite set, we denote its cardinality by $|X|$. The set $\{i : 1 \leq i \leq n$ of positive integers is denoted by J_n. If X is a (finite) set, the set of all its partitions will be written as $\underline{\underline{P}}(X)$. The members of a partition will be called blocks. Thus any two blocks of a partition are disjoint subsets of X and if $P \varepsilon \underline{\underline{P}}(X)$, then $\Sigma_{B\varepsilon P} |B| = |X|$.

If $(A_i : i \varepsilon J_n)$ is a family of finite sets, then $\underline{\underline{D}}(A_1 \times \ldots \times A_n)$ or $\underline{\underline{D}}(\prod_{i\varepsilon J_n} A_i)$ will denote the set of ordered n-tuples of the cartesian product $A_1 \times A_2 \times \ldots \times A_n$ with distinct coordinates. Such ordered n-tuples are called SDRs. If P is a partition of X and P contains k blocks with cardinalities $n_1, \ldots, n_k$, then we shall denote by $m(P)$ the number

$$(-1)^{(n_1-1)+\ldots+(n_k-1)} (n_1-1)! \ldots (n_k-1)!$$

$$= (-1)^{|X|-|P|} (n_1-1)! \ldots (n_k-1)! \qquad \ldots \qquad (1)$$

If X is an n-set, then a Latin rectangle of order r by s based on X is a rectangular array (a_{ij}), $1 \leq i \leq r$, $1 \leq j \leq s$ with each row and each column containing distinct elements. Generally we take $X = J_n$. (Euler used Latin letters to construct such square arrays and hence the term 'Latin Square'). The number of $r \times s$ Latin rectangles with entries from an n-set, say, J_n is denoted by $L(r,s,n)$. Here $r \leq n$, $s \leq n$ and $L(r,s,n) = L(s,r,n)$. The number $L(r,n,n)$ will be written $L(r,n)$; in this case each row of the Latin rectangle is permutation of all the n elements of the basic n-set. When $r = s = n$, we have a Latin square of order n and $L(n,n,n)$ will be written $L(n)$.

In the case $s = n$, if the first row is normalized (i.e., $1,2,\ldots,n$ occur in their natural order), then the number of Latin rectangles obtained thus is written $K(r,n)$. We have

$$L(r,n) = n!\, K(r,n). \qquad \ldots \qquad (2)$$

Again in the case $r = s = n$, if the first row as well as the first column is normalized, then the number of Latin rectangles so obtained is written $R(n)$. We have

$$L(n) = n!\,(n-1)!\,R(n) \qquad \ldots \quad (3)$$

Two permutations σ and τ of $1,2,\ldots,n$ are said to be discordant if they agree at no place, i.e., $\sigma(i) \neq \tau(i)$, $1 \leq i \leq n$.

The number of permutations discordant with $(1,2,\ldots,n)$ is denoted by D_n. Such permutations are called derangements.

The number of permutations discordant with each of

$(1,2,3, \ldots, n)$, $(2,3,4, \ldots, 1)$

is denoted by U_n.

D_n and U_n are respectively called rencontres and ménages numbers.

Finally, the number of permutations discordant with each of

$(1,2,3, \ldots, n-1,n)$, $(2,3,4, \ldots, n,1)$, $(3,4,5, \ldots, 1,2)$

is denoted by V_n.

The numbers D_n, U_n, V_n are respectively given by

$$D_n = n!\,[1 - \frac{1}{1!} + \frac{1}{2!} - \ldots + \frac{(-1)^n}{n!}] \qquad \ldots \quad (4)$$

$$U_n = n! - \frac{2n}{2n-1}\binom{2n-1}{1}(n-1)! + \frac{2n}{2n-2}\binom{2n-2}{2}(n-2)! \ldots + (-1)^n \frac{2n}{n}\binom{n}{n} 0!, (n > 1) \qquad \ldots \quad (5)$$

and

$$V_n = g_o n! - g_1.(n-1)! + g_2.(n-2)! - \ldots + (-1)^i g_i.(n-i)! + \ldots + (-1)^n g_n.0!,$$

where

$$g_i = \begin{cases} \sum\limits_{2\alpha+\beta+\gamma=i} \frac{n}{(n-i)} \binom{n+\alpha-i-1}{\alpha}\binom{n-\alpha-1}{\gamma}\binom{n-i}{\beta} 2^\beta, & i < n \\ 3 + \sum\limits_{\alpha=1}^{[n/2]} \frac{n}{\alpha}\binom{n-\alpha-1}{\alpha-1}, & i = n \end{cases} \qquad \ldots \quad (6)$$

D_n was known to Euler, U_n was given by Touchard [17] in 1933 and V_n by Moser [10] in 1967. Some of these proofs are also given in Ryser [15].

3. A BRIEF SURVEY OF RESULTS

The story begins with Euler. He seems to have been the first to have defined a Latin square (perhaps entirely for a different purpose). It is not difficult to evaluate $L(n)$, the number of Latin squares of order n, for small values of n, say, $n \leq 5$.

The numbers L(6), L(7), L(8), L(9) were respectiyely found by Frolov (1890), Sade (1948), Wells (1967), Bammel and Rothstein (1975). See [1] for complete references. Computers have helped in finding L(n) for some values of $n > 9$.

The first few of the known values of R(n) are given below :

R(1) = R(2) = R(3) = 1; R(4) = 4: R(5) = 56; R(6) = 9408; R(7) = 16,942,080;

R(8) = 535,281,401,856; R(9) = 377,597,570,964,258,816.

The corresponding L(n) can now be found by using equation (2.3).

For lower and ypper bounds of R(n) one is referred to [1]. They are given below:

$$1!\,2!\,\ldots\,(n-2)! \le R_n \le (n-1)!\,[(n-2).(n-2)!][(n-3)^2.(n-2)!] \ldots [(n-k)^{k-1}.(n-k)!] \ldots [2^{n-3}\,2!][1^{n-2}\,1!] \quad \ldots \quad (1)$$

The lower bound is by Marshall Hall [5] and the 'crude' upper bound can be derived by the reader easily.

We now come to Latin rectangles.

Clearly,

$$L(1,s,n) = \frac{n!}{(n-s)!} \quad \ldots \quad (2)$$

It is also easy to see that $K(2,n) = D_n$ and hence by (2.2) and (2.4) we have

$$L(2,n) = (n!)^2\,[1 - \frac{1}{1!} + \frac{1}{2!} - \ldots + \frac{(-1)^n}{n!}] \quad \ldots \quad (3)$$

The number L(2,s,n) is found without much difficulty. In fact,

$$L(2,s,n) = \frac{n!}{((n-s)!)^2} \sum_{k=0}^{s} (-1)^k \binom{s}{k} (n-k)! \quad \ldots \quad (4)$$

The number K(3,n) offers some difficulty. The first attempt seems to have been made by Jacob [6] in 1930. He found a recurrence relation for K(3,n). But it was erroneous and was later corrected by Kerawala [8] in 1941. It runs as follows:

$$(n+3)\,\mu_{n+5} = (n+4)\,(n^2+8n+17)\,\mu_{n+4} + (n+3)(n+4)(n^2+8n+17)\,\mu_{n+3} + (n+3)(n+4)(n^2+8n+13)\,\mu_{n+2} + 2(n+2)(n+3)(n+4)(n^2+5n+3)\,\mu_{n+1} - 4(n+1)(n+2)(n+3)(n+4)^2\,\mu_n \quad \ldots \quad (5)$$

where we have written μ_n for K(3,n) for convenience.

Meanwhile K. Yamamoto found out a simple explicit formula for $K(3,n)$ in late thirties. It is given by

$$K(3,n) = n! \sum_{\alpha+\beta+\gamma=n} (-1)^{\beta}\, 2^{\gamma}\frac{\alpha!}{\gamma!}\binom{3\alpha+\beta+2}{\beta}. \quad \ldots \quad (6)$$

Using a result of Irving Kaplansky [7], Riordan [11] found $L(3.n)$ by obtaining a symbolic expression for it in terms of U_n (the ménage polynomials) in 1944. Two years later, in 1946, he [12] condensed it into a remarkably crisp formula in terms of D_n and U_n as given by (2.4) and (2.5).

$$K(3,n) = \sum_{k=o}^{[n/2]} \binom{n}{k} D_k D_{n-k} U_{n-2k}, (U_o = 1) \quad \ldots \quad (7)$$

Again in 1952, Riordan [13] found an 'impure' recurrence relation for $K(3,n)$ simpler than Kerawala's using Yamamoto's formula given by (6) above. The method is very elegant. This recurrence relation is

$$\mu_n = n^2 \mu_{n-1} + n(n-1)\mu_{n-1} + 2n(n-1)(n-2)\mu_{n-3} + \lambda_n, \ (\mu_o = 1, \mu_1 = 0, \mu_2 = 0) \quad \ldots \quad (8)$$

where $\mu_n = K(3,n)$, as usual, and

$$\lambda_n + n\lambda_{n-1} = -(n-1)2^n, \ (\lambda_o = 1).$$

In the meantime asymptotic expressions were found for $L(r,n)$. It is easy to see that $L(2,n) \sim (n!)^2 e^{-1}$ and it is not very difficult to show that $L(3,n) \sim (n!)^3 e^{-3}$ using Riordan's expressions. So it was conjectured that, in general,

$$L(r,n) \sim (n!)^r e^{-\binom{r}{2}} \quad \ldots \quad (9)$$

as long as r was much less than n. Erdös and Kaplansky [3] showed in 1946 that this was true for $r < (\log n)^{3/2}$ and conjectured that it was true for $r < n^{1/3}$. This was later proved by Yamamoto [18] in 1951. C.M. Stein [16] improved it in 1978.

Not much was done in the case of four-line Latin rectangles except that Moser found the numbers V_n as already noted. (See (2.6)). Yamamoto seems to have obtained some results in connection with $L(4,n)$. Just as it is possible to express $K(3,n)$ in terms of D_n and U_n (vide (7)), it is hoped that $K(4,n)$ is expressible in terms of D_n, U_n and V_n. But no such expression is so far found. It should also be mentioned here that recurrence relations exist for D_n and U_n, namely,

$$D_n = n\,D_{n-1} + (-1)^n$$

$$(n-2)U_n = n(n-2)\,U_{n-1} + n\,U_{n-2} + 4(-1)^{n+1} \qquad \ldots \qquad (10)$$

and that no such recurrence relation is known for V_n.

It may also be observed that the methods used to find $K(r,n)$ for different values of r did not have any unifying approach, in the sense that, they could not work for higher orders.

This is where Möbius functions come into picture. Almost a quarter of a century ago, Rota [14] had been convinced that Möbius functions could prove powerful in solving many enumeration problems. It will be seen in the next section how using the technique of Möbius inversion formula one can give a unified approach to the problem of enumeration of Latin rectangles. This method yields $L(2,s,n)$ and $L(3,s,n)$ easily and $L(4,s,n)$ without much trouble.

4. A GENERAL APPROACH

We convert the problem of enumeration of Latin rectangles into what we shall call a 'One-one problem' on a finite family of finite sets.

<u>One-one problem on r sets</u> : Let $F = \{A_1, \ldots, A_r\}$ be a family of r finite sets (not necessarily disjoint) and $X = \bigcup_{i \in J_r} A_i$. The problem is to determine the number of all maps $f : X \to J_n$ such that f/A_i (f restricted to A_i) is one-one for each $i \in J_r$.

We may denote this number by $M(F,n)$. It is clear that this number depends on the cardinalities of the $2^r - 1$ Boolean atoms for the family F, namely

$$\left(\bigcap_{i \in T} A_i\right) - \left(\bigcup_{i \notin T} A_i\right)$$

where $T \subset J_r$ and also on n.

In fact the expression for $M(F,n)$ will be a polynomial in n of degree $|X|$.

We consider the cases $r = 2$ and $r = 3$.

(a) If $F = \{A_1, A_2\}$, $|A_1 - A_2| = p$, $|A_2 - A_1| = q$, $|A_1 \cap A_2| = r$, then

$$M(F,n) = \begin{bmatrix} n \\ r \end{bmatrix} \begin{bmatrix} n-r \\ p \end{bmatrix} \begin{bmatrix} n-r \\ q \end{bmatrix} \qquad \ldots \qquad (1)$$

where $\begin{bmatrix} \alpha \\ \beta \end{bmatrix}$ means $\frac{\alpha!}{(\alpha-\beta)!}$ if $\alpha \geq \beta$ and zero otherwise

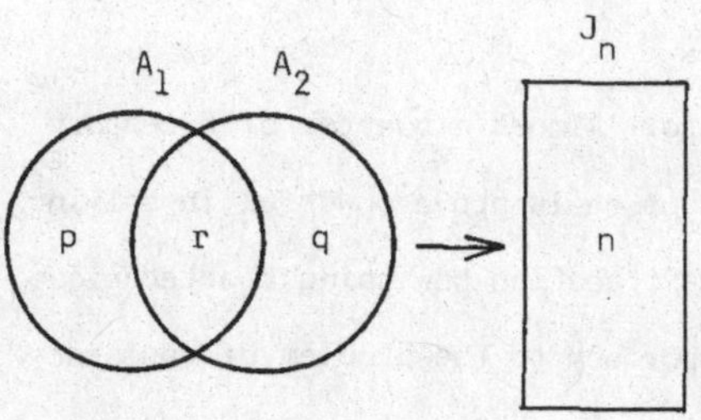

(a)

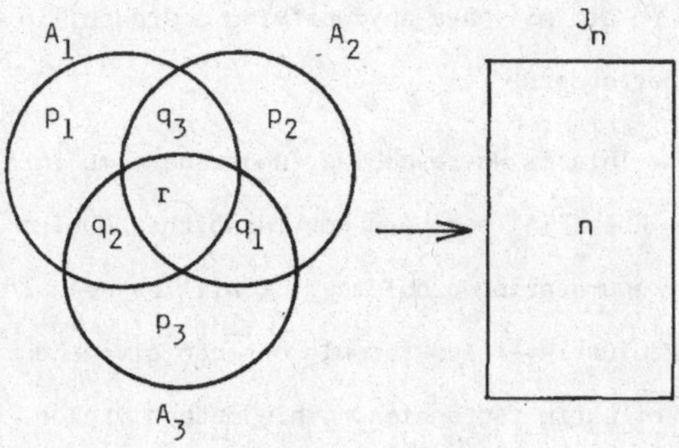

(b)

(b) If $F = \{A_1, A_2, A_3\}$ and the cardinalities of the atoms are as indicated in the diagram (b) then

$$M(F,n) = \begin{bmatrix} n \\ q_1+q_2+q_3+r \end{bmatrix} \begin{bmatrix} n-q_2-q_3-r \\ p_1 \end{bmatrix} \begin{bmatrix} n-q_3-q_1-r \\ p_2 \end{bmatrix} \begin{bmatrix} n-q_1-q_2-r \\ p_3 \end{bmatrix} \qquad \ldots \qquad (2)$$

When $r = 4$, $M(F,n)$ is a little more complicated with more terms occuring in the formula.

Now the problem of finding $L(r,s,n)$ can be posed equivalently as a one-one problem on a certain family of $(r+s)$ sets. This is done as follows : let

$$X = J_r \times J_s = \{(i,j) : 1 \leq i \leq r,\ 1 \leq j \leq s\}.$$

Define subsets $R_i (1 \leq i \leq r)$ and $C_j (1 \leq j \leq s)$ of X by

$$R_i = \{(i,j) : 1 \leq j \leq s\},$$

$$C_j = \{(i,j) : 1 \leq i \leq r\}.$$

Let $F = \{R_1, \ldots, R_r, C_1, \ldots, C_s\}$. Then one can see that $L(r,s,n) = M(F,n)$.

What the authors in [2] have proved is that the problem of finding $L(r,s,n)$ can be reduced to a one-one problem on $\min\{r,s\}$ sets. Hence 2-line, 3-line and 4-line Latin rectangles can be enumerated since we know the expressions for $M(F,n)$ when F

has 2,3 or 4 sets.

It is here that the theory of SDRs is used. If $(A_i : i \in J_n)$ is a finite family of n finite sets, as already defined, a member $(x_1,\ldots,x_n)$ in the cartesian product $A_1\times\ldots\times A_n$ is called an SDR (or a matching) if the x_i are all different. A well known criterion for the existence of such an SDR was given by Philip Hall in 1935. A not so well known formula is that for the number of SDRs. It is given by

$$|\underline{\underline{D}}(A_1\times\ldots\times A_n)| = \sum_{P\in\underline{\underline{P}}(J_n)} m(P) \prod_{C\in P} |\bigcap_{i\in C} A_i| \qquad \ldots \qquad (3)$$

where all the symbols involved are described in Section 2. The reader is referred to [4], Chapter XI, Section D, Proposition D24 for a proof by using Möbius inversion. It is this formula we are actually interested in. We write down explicitly for the cases $n = 2,3,4$.

$$|\underline{\underline{D}}(A\times B)| = |A|.|B| - |AB|$$

$$|\underline{\underline{D}}(A\times B\times C)| = |A|.|B|.|C| - |A|.|BC| - |B|.|CA| - |C|.|AB| + 2|ABC|$$

and

$$|\underline{\underline{D}}(A\times B\times C\times D)| = |A|.|B|.|C|.|D| - \sum_6 |A|.|BCD| + \sum_3 |AB|.|CD| + 2\sum_4 |A|.|BCD| - 6|ABCD|,$$

where the number below each Σ indicates the number of terms governed by it. (Here intersection is denoted by juxtaposition.)

We shall, by an example, indicate how this formula helps to find $L(r,s,n)$. We consider the case $r = 3$. Let A,B,C be any three finite sets. Dropping the cardinality symbols (i.e., writing A for $|A|$), we have

$$\underline{\underline{D}}(A\times B\times C) = A.B.C - A.BC - B.CA - C.AB + 2ABC.$$

Then

$$L(3,s,n) = (A.B.C - A.BC - B.CA - C.AB + 2ABC)^s$$

where we shall presently explain the meaning of the right side. Expand the right side using the multinomial theorem in the usual manner treating A,B,C,BC,CA,AB,ABC as different symbols. A typical term in the expansion would be

$$\frac{s!}{\alpha!\beta_1!\beta_2!\beta_3!\gamma!}(-1)^{\beta_1+\beta_2+\beta_3}2^{\gamma}[A^{\alpha+\beta_1} * B^{\alpha+\beta_2} * C^{\alpha+\beta_3} * (BC)^{\beta_1} * (CA)^{\beta_2} * (AB)^{\beta_3} * (ABC)^{\gamma}],$$

with $\alpha+\beta_1+\beta_2+\beta_3+\gamma = s$.

Here we have used $*$ as the product symbol. The entire expression in the square brackets is to be interpreted as follows :

Consider three sets P,Q,R such that $|P - Q\cup R| = \alpha + \beta_1$, $|Q - R\cup P| = \alpha + \beta_2$, $|R - P\cup Q| = \alpha + \beta_3$, $|QR - P| = \beta_1$, $|RP - Q| = \beta_2$, $|PQ - R| = \beta_3$, $|PQR| = \gamma$, as shown in the figure.

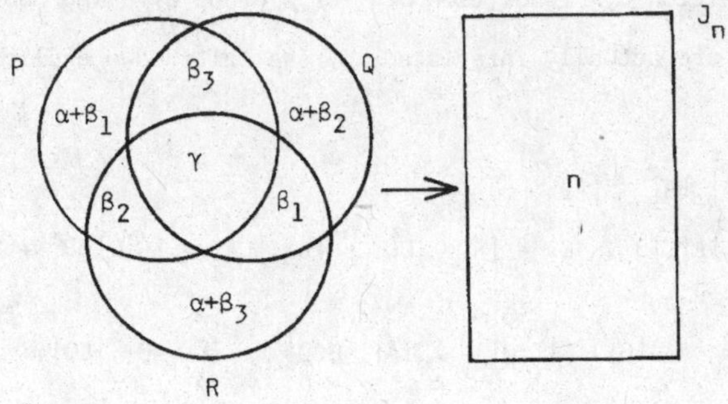

Consider the one-one problem of finding $M(F,n)$ where $F = \{P,Q,R\}$, i.e., the number of maps $f : P\cup Q\cup R \to J_n$ such that, f/P, f/Q, f/R are one-one. Then this number is taken as the definition of the expression in the square brackets. As already noted this is equal to

$$\begin{bmatrix} n \\ \beta_1+\beta_2+\beta_3+\gamma \end{bmatrix}\begin{bmatrix} n-\beta_2-\beta_3-\gamma \\ \alpha+\beta_1 \end{bmatrix}\begin{bmatrix} n-\beta_3-\beta_1-\gamma \\ \alpha+\beta_2 \end{bmatrix}\begin{bmatrix} n-\beta_1-\beta_2-\gamma \\ \alpha+\beta_3 \end{bmatrix}.$$

Hence

$$L(3,s,n) = \sum_{\alpha+\beta_1+\beta_2+\beta_3+\gamma=s} (-1)^{\beta_1+\beta_2+\beta_3} 2^{\gamma} \frac{s!}{\alpha!\beta_1!\beta_2!\beta_3!\gamma!}\frac{n!}{(n-\beta_1-\beta_2-\beta_3-\gamma)!} \times$$

$$\frac{(n-\beta_2-\beta_3-\gamma)!(n-\beta_3-\beta_1-\gamma)!(n-\beta_1-\beta_2-\gamma)!}{((n-s)!)^3}.$$

Simplifying we find that

$$L(3,s,n) = \frac{n!s!}{((n-s)!)^3} \sum_{\alpha+\beta+\gamma=s} (-1)^{\beta} 2^{\gamma} \frac{((n-s+\alpha)!)^2}{\alpha!\gamma!} \binom{3n-3s+3\alpha+\beta+2}{\beta} \quad \ldots \quad (4)$$

By putting $s = n$, we get $L(3,n)$ which is same the expression given in (3.6).

An exactly similar method yields the number of 4-line rectangles :

$$L(4,s,n) = \frac{n!s!}{((n-s)!)^4} \sum \frac{(-1)^{\Sigma\beta_i+\varepsilon}\, 2^{\Sigma\delta_i}\, 6^{\varepsilon}}{\alpha!\Pi(\beta_i!)\Pi(\gamma_i!)\Pi(\delta_i!)\varepsilon!} \times T \times S \quad \ldots \quad (5)$$

where the sum is over all

$$\alpha + \sum_{i=1}^{6} \beta_i + \sum_{i=1}^{3} \gamma_i + \sum_{i=1}^{4} \delta_i + \varepsilon = s.$$

Further

$$\Pi(\beta_i!) = \prod_{i=1}^{6} \beta_i! \quad \text{etc.,}$$

$$T = \sum_{\theta_1,\theta_2,\theta_3 \geq 0} \binom{\beta_1+\gamma_1}{\theta_1}\binom{\beta_6+\gamma_1}{\theta_1} \theta_1! \binom{\beta_2+\gamma_2}{\theta_2}\binom{\beta_5+\gamma_2}{\theta_2} \theta_2! \binom{\beta_3+\gamma_3}{\theta_3} \times$$

$$\binom{\beta_4+\gamma_3}{\theta_3} \theta_3! \; (n-(\Sigma\beta_i+2\Sigma\gamma_i+\Sigma\delta_i+\varepsilon)+\theta_1+\theta_2+\theta_3)!$$

and

$$S = (n-k+\alpha+\beta_4+\beta_5+\beta_6+\delta_1)! \; (n-k+\alpha+\beta_2+\beta_3+\beta_6+\delta_2)! \times$$

$$(n-k+\alpha+\beta_1+\beta_3+\beta_5+\delta_3)! \; (n-k+\alpha+\beta_1+\beta_2+\beta_4+\delta_4)!$$

A proof of the validity of this procedure involves a simple application of Möbius inversion formula [2].

Using this procedure one can also find the number of Latin rectangles in a more generalized sense. For example, we can find

$$L \begin{bmatrix} A & A & \ldots & A \\ B & B & \ldots & B \\ C & C & \ldots & C \end{bmatrix}_{3\times s}$$

which stands for the number of $3\times s$ Latin rectangles with additional constraint that the elements of the first row should come from a finite set A, those of the second row from a finite set B and those of the third from a finite set C. (A,B,C need not be disjoint nor distinct). Finally it should be noted here that K. Lebensold [9] has given a necessary and sufficient condition for the existence of such a Latin rectangle (of order $r \times s$) with the above constraint, which itself is a generalization of Hall's marriage criterion.

REFERENCES

1. R. Alter, How many Latin squares are there?, Amer. Math. Monthly, 82 (1975), 632-634.
2. K.B. Athreya, C.R. Pranesachar and M.M. Singhi, On the number of Latin rectangles and chromatic polynomial of $L(K_{r,s})$, European Journal of Combinatorics (to appear).
3. P. Erdös and I. Kaplansky, The asymptotic number of Latin rectangles, Amer. Jour. Math., 68 (1946), 230-236.
4. J.E. Graver and M.E. Watkins, Combinatorics with Emphasis on the Theory of Graphs, Springer-Verlag, 1977.
5. M. Hall, Jr., An existence theorem for Latin squares, Bull. Amer. Math. Soc., 51(1945), 387-388.
6. S.M. Jacob, The enumeration of Latin rectangle of depth three, Proc. Lond. Math. Soc. (2), Vol.31 (1930), 329-336.
7. I. Kaplansky, On a generalisation of the "Probleme des rencontres", Amer. Math., Monthly, 46 (1939), 159-161.
8. S.M. Kerawala, The enumeration of the Latin rectangle of depth three by means of of difference equation, Bull. Calcutta Math. Soc., 33(1941), 119-127.
9. K. Lebensold, Disjoint matchings of graphs, J. Comb. Theory (B), 22 (1977), 207-210.
10. W.O.J. Moser, The number of very reduced $4 \times n$ Latin rectangles, Canad. J. Maths., 19(1967), 1011-1017.
11. J. Riordan, Three-line Latin rectangles, Amer. Math. Monthly, 51(1944), 450-452.
12. J. Riordan, A recurrence relation for three-line Latin rectangles, Amer. Math. Monthly, 59(1952), 159-162.
13. J. Riordan, Three-line Latin rectangles, Amer. Math. Monthly, 53(1946), 18-20.
14. G.G. Rota, On the foundations of combinatorial theory I. Theory of Möbius functions, Z. Wahrscheinlichkeitstheorie, 2 (1964), 340-368.
15. H.J. Ryser, Combinatorial Mathematics, The Mathematical Association of America, 1963.
16. C.M. Stein, Asymptotic evaluation of the number of Latin rectangles, J. Comb. Theory (A), 25 (1978), 38-49.
17. J. Touchard, Sur un probleme de permutations, C.R.Acad.Sci. Paris, 198(1934), 631-633.
18. K. Yamamoto, On the asymptotic number of Latin rectangles, Japanese Journal of Math., 21 (1951), 113-119.

NEARLY LINE REGULAR GRAPHS AND THEIR RECONSTRUCTION

S. RAMACHANDRAN*
Aditanar College
Tiruchendur, Tamil Nadu, 628 216

ABSTRACT

Some new invariants of a point (line) of a graph, mostly based on the neighborhood degree sequence are defined. Imposing conditions on these invariants, some classes of graphs are reconstructed (line reconstructed). A characterization theorem for line regilar graphs and another for regular graphs are given.

1. INTRODUCTION

The conjecture that every graph with three or more points is reconstructible from its collection of point-deleted subgraphs seems according to all available data, to be true. Several classes of graphs have already been reconstructed. In this paper, we reconstruct some classes of graphs including line regular ones. To this end, we define a few concepts and describe the classes in terms of these concepts. Incidentally we also characterize line regular graphs in terms of two of these concepts. Terms not defined here can be found in Harary [3].

<u>Definition 1</u>. Let G be a graph and $v \varepsilon V(G)$. The <u>neighborhood degree sum</u> of v (NDS(v)) is defined to be $\Sigma \deg(v')$ where the summation is taken over all vertices v' adjacent to v.

<u>Definition 2</u>. If x is a line of G, <u>degree of x</u> (deg(x)) is defined to be the number of lines adjacent to x. A graph is called <u>line regular</u> if all lines have the same degree.

<u>Definition 3</u>. Let x be a line of a graph G. <u>The line neighborhood degree sum</u> of x (LNDS (x)) is defined to be $\Sigma \deg(x')$ where the summation is taken over all lines x' adjacent to x.

---

* This research was done when the author was at Madurai Kamaraj University with UGC Fellowship.

Definition 4. For any point v of a graph G, the incident line degree sum of v(ILDS(v)) is defined to be Σ deg(x) where the summation is taken over all lines x incident with v. Note that

$$\mathrm{ILDS}(v) = \mathrm{NDS}(v) + r(r-2)$$

where r is the degree of v.

Definition 5. A graph is called an S-graph if its degree sequence can be found out from the set of its point-deleted subgraphs.

The following theorem on S-graphs is proved by Manvel [4].

Theorem 1. (Manvel [4]). A graph G is an S-graph if at least one of the following hold in G :

(a) No point of minimum degree lies on a triangle.

(b) Every point of maximum degree is adjacent to at least one of any two nonadjacent points of G.

(c) The minimum degree in G is at most three.

(d) The maximum degree in G is at least $|V(G)| - 4$.

Definition 6. The adjacent degree pair (ADP) of a line $x = uv$ is the unordered pair (d_1, d_2) where d_1 = deg u and d_2 = deg v.

Definition 7. A class G^* of graphs is called recognisable if, for each graph G in G^*, every graph having the same collection of point-deleted subgraphs as G is in G^*.

2. REGULARITY CONCEPTS

Theorem 2. Let G be a connected graph. The following statements are equivalent.

(1) G is line regular.

(2) All points of G have the same NDS.

(3) $G \neq P_4$ and all lines of G have the same LNDS.

Proof. (1) → (2) If G is regular it is obvious. If G is not regular, let points u and v, having degrees a and b ($a \neq b$) respectively be adjacent. Now every line should have degree $a + b - 2$. Since u has degree a, the ADP of any line through

u must be (a,b) so that this line has degree $a + b - 2$. Similarly the other end of any line through v has degree a. Since the graph is connected, proceeding like this we get that any point has degree a or b and each line joins a point of degree a and a point of degree b. Hence $a.b$ is the NDS of any point of G.

(2) → (3) Let r be the NDS of a point of G and δ be the minimum degree in G. If v is a point of G having degree greater than $\frac{r}{\delta}$, then $NDS(v) > r$ which contradicts the hypothesis. Hence the maximum degree of a point in G is $\leq \frac{r}{\delta}$. Now for any point u of degree δ, since $NDS(u) = r$, every point adjacent to u must have degree $\frac{r}{\delta}$. Hence $\frac{r}{\delta}$ is an integer. Now take a point w of degree $\frac{r}{\delta}$. Since the minimum degree in G is δ and $NDS(w) = r$, every point adjacent to w has degree δ. Since G is connected and has only finite number of points, the degree of any point of G is either δ or $\frac{r}{\delta}$ and every line joins a point of degree δ with a point of degree $\frac{r}{\delta}$. Hence every line of G has $LNDS(\delta + \frac{r}{\delta} - 2)^2$ and evidently $G \neq P_4$.

(3) → (1) All lines of G have the same LNDS implies all points of $L(G)$ (line graph of G) have the same NDS. Hence by the same arguments as above, it follows that $L(G)$ is either a regular graph (when $\delta = \frac{r}{\delta}$) or a bidegree one and if it is a bidegree one, no two points of the same degree are adjacent. When $L(G)$ is a bidegree graph, if at least one of the degrees is greater than two, then $K_{1,3}$ is an induced subgraph of $L(G)$ giving a contradiction (since a line graph cannot have $K_{1,3}$ as an induced subgraph). Hence when $L(G)$ is a bidegree graph it is $K_{1,2}$ which implies G is P_4 contradicting (3). Hence $L(G)$ is regular and this implies G is line regular.

<u>Note 2.1</u>. From the above proof it is clear that "a connected graph G is line regular iff all lines have the same ADP".

<u>Theorem 3</u>. A graph is regular iff all points have the same ILDS.

<u>Proof</u>. It is obvious that in a regular graph, all points have the same ILDS.

Conversely, let G be a graph in which all points have the same ILDS. If possible, let G be not regular. If v is a point of degree r, then $ILDS(v) = r(r-2) + NDS(v)$. Since all points have the same ILDS, as r increases, NDS decreases.

Therefore, as G is not regular, all points of G do not have the same NDS. Consider a point u of maximum degree Δ. Its NDS is the minimum NDS. Hence at least one point adjacent to u has degree $\leq \frac{1}{\Delta}$ (minimum NDS). Hence the minimum degree $\delta \leq \frac{1}{\Delta}$ (minimum NDS). Hence $\delta.\Delta \leq$ minimum NDS. Now consider a point w of degree δ. NDS(w) must be maximum NDS. Since maximum degree of a point in G is Δ, we have

$$\text{maximum NDS} = \text{NDS}(w) \leq \delta.\Delta \leq \text{minimum NDS}.$$

This is a contradiction since all points of G do not have the same NDS. Hence G must be regular.

3. RECONSTRUCTION

Since set reconstruction conjecture is true for 3 point graphs, we consider only graphs with at least four points in this section.

<u>Theorem 4</u>. Line regular graphs are set-reconstructible.

<u>Proof</u>. Let A be the set of point deleted subgraphs of a connected graph G. If all the members of A have the same number of lines, then G is regular and hence any member of A can be completed to G. Otherwise, if G is line regular, G has no triangles and hence the degree sequence of G can be found from A (Manvel [4]). Hence the set of neighborhood degree sequences of points of G can be found from A and hence the set of NDS of points of G. By Theorem 2 line regularity of G can be recognised from A. Now a subgraph of G obtained by deleting a point of maximum degree can be completed to G uniquely as the points of minimum degree in this subgraph are precisely the points that were adjacent to the deleted point of G.

In the following theorem, we prove that a graph is reconstructible if the difference between any two distinct NDS is sufficiently large.

<u>Theorem 5</u>. Let G be a graph such that the difference between any two distinct NDS is at least 2m and let G have a point v of degree d, $1 < d \leq m$ satisfying at least one of the following :

(1) v lies on a triangle.

(2) v is adjacent to a point of degree t and no point of G other than v and its neighbors have degree $t - 1$ in G.

(3) v is adjacent to a point of degree at most d.

Then G is reconstructible.

Proof. From the collection F of point-deleted subgraphs, the collection of neighborhood degree sequences of the points of G can be found out. Hence whether G satisfies the hypothesis of the theorem or not can be decided from F. Consider a subgraph H of G obtained by deleting a point v, where v satisfies the conditions stated. We will prove that the points that were adjacent to v in G can be located in H and hence H can be augmented to G uniquely.

When a point of degree d is deleted, the loss in the NDS of any other point is at most $2d - 1$ which is less than $2m$. Hence the NDS loss of each point in H can be found out. Points of H that have lost more than d in NDS must be adjacent to v in G. Points that have lost less than d in NDS cannot be adjacent to v in G. A point that has lost d in NDS is either "adjacent to v in G and adjacent to no other neighbour of v" or "not adjacent to v, but adjacent to every neighbour of v". So if we are able to locate one point w adjacent to v, then a point of NDS loss d is adjacent to v iff it is not adjacent to w. Thus in this case, adjacency of each point of H with v can be determined and hence G reconstructed.

However, we can always locate at least one point in H that was adjacent to v in G as follows. If (1) holds, then H has a point which has lost more than d in NDS and this point is adjacent to v in G. If (2) holds, then any point in H with degree $t - 1$ is adjacent to v in G and there is at least one such point in H. If (3) holds, then H has a point u with degree at most $d - 1$ which has lost at least d in NDS. Hence u must be adjacent to v in G (because if u is not adjacent to v in G, it will lose at most $d - 1$ in its NDS in H).

Corollary 5.1. If G is an S-graph satisfying the conditions of Theorem 5, then G is set-reconstructible.

Corollary 5.2. If G is a graph (S-graph) such that the difference between any two distinct NDS is at least 2Δ, then G is reconstructible (set-reconstructible).

We give below some examples of graphs reconstructible by Corollary 5.2.

Example 1. Let the graphs G and H be as in Figure 1.

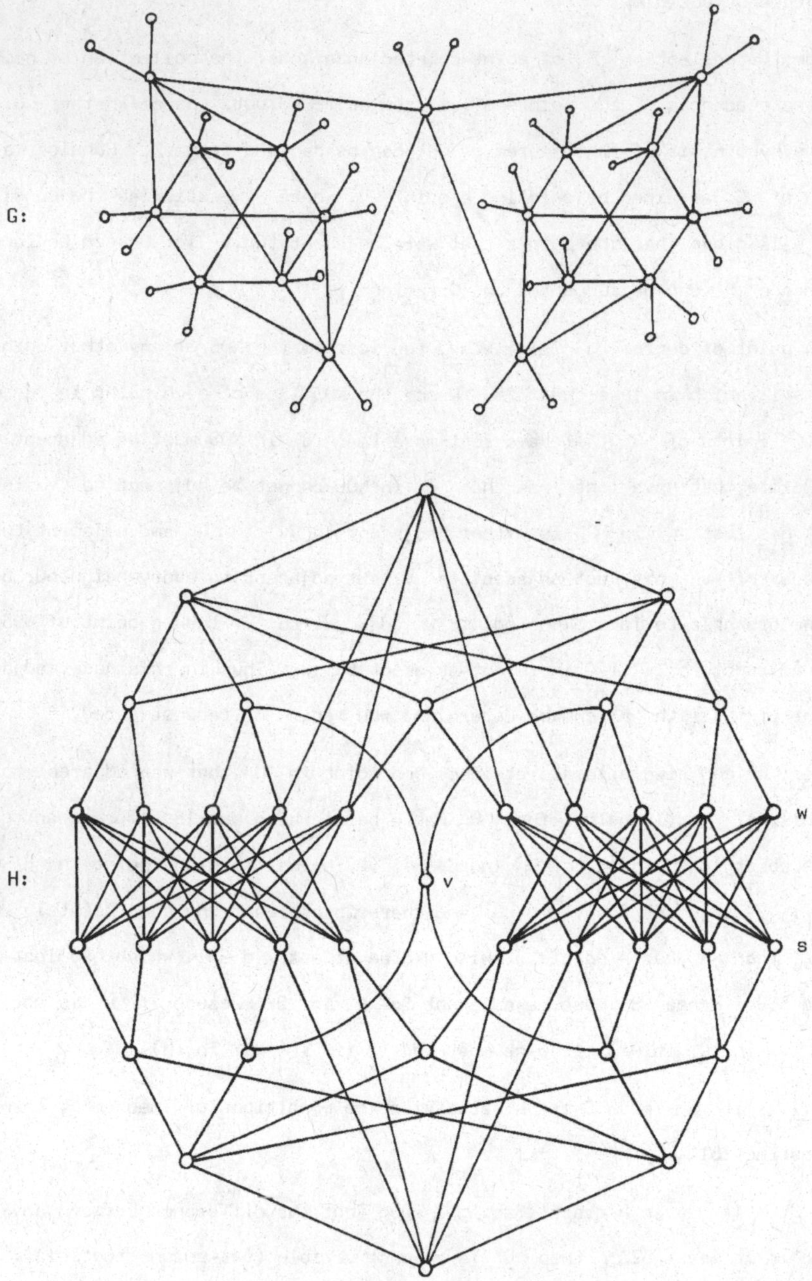

FIGURE 1

Consider the graph $L = H_1 \cup H_2 \cup \ldots \cup H_{34}$ where $H_i \simeq H$ for each i. Each H_i has precisely two points with neighborhood degree sequence (6,6,6,6,4) and call them w_i and s_i. Let L^* be the graph obtained from L by adding the lines $w_i\, s_{i+1}$ for $i = 1,2,\ldots,33$ and the line $w_{34}\, s_1$. Now consider the graph $G \cup L^*$. Call the endpoints in this graph as $u_1, u_2, \ldots, u_{34}$ and the points with neighborhood degree sequence (4,4,4,4) as $v_1, v_2, \ldots, v_{34}$. Let E be the graph obtained from $G \cup L^*$ by identifying the point u_i with v_i ($i = 1,2,\ldots,34$). The graph E is connected, has 1275 (= 17+ 34+34 × 36) points and has no cutpoints. Any point of E has degree 6,5 or 4, neighborhood degree sequence (6,6,6,6,6,4), (6,6,6,6,5,5), (6,4,4,4,4) or (6,6,5,5) and NDS 34 or 22. Hence E is reconstructible by Corollary 5.2.

Example 2. Let n be any integer ≥ 2. Consider the graph $E^* = F_1 \cup F_2 \cup \ldots \cup F_n$ where $F_i \simeq E$ for each i. Also for each i, let $a_i b_i$ be a line of F_i with ADP (6,6). Let E_n be the graph obtained from E^* by deleting the lines $a_i b_i$, $i = 1,2, \ldots, n$ and adding the lines $b_i\, a_{i+1}$ for $i = 1,2,\ldots,n-1$ and the line $b_n a_1$. E_n is connected, has $1275n$ points and no cutpoints. Also any point of E_n has degree 6,5 or 4 and NDS 34 or 22. Hence E_n for each $n \geq 2$ satisfies the hypothesis of Corollary 5.2, and hence is reconstructible.

Theorem 6. If G has an endpoint and if the difference between any two distinct NDS is at least two, then G is set-reconstructible.

Proof.(Outline) By Theorem 1, G is an S-graph. Consider $G - v$, where v is an endpoint in G. A point w of $G - v$ is called a qualified point if the NDS loss of w is 1 and for every other point u in $G - v$, "u is adjacent to w iff the NDS loss of u is 1". G is the graph obtained from $G - v$ by introducing an endpoint adjacent to a qualified point. (Note that if w_1 and w_2 are two qualified points and u is any other point, then u adjacent to w_1 iff u adjacent to w_2).

The graph given in Figure 2 satisfies the hyothesis of Theorem 6 and hence is reconstructible.

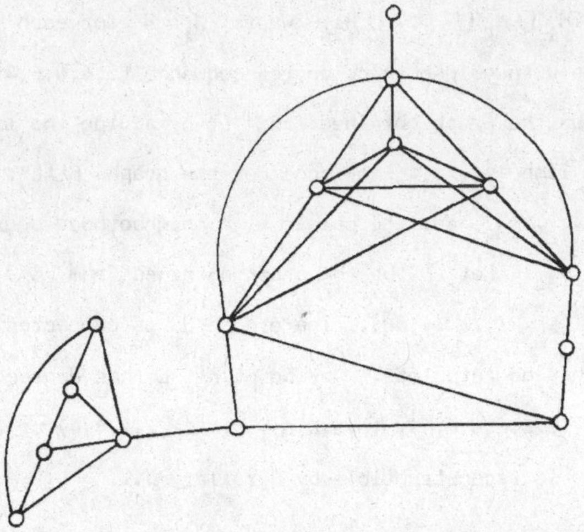

FIGURE 2

When a point of degree x is removed, a point of degree y adjacent to it loses at least $x + y - 2 + y - 1$ and at most $(x+y-2) + (y-1) + \min\{x-1, y-1\} = x + 2y - 4 + \min\{x, y\}$ in its ILDS, whereas a point of degree z not adjacent to it loses at most $\min\{x, z\}$ in its ILDS. We use this fact to reconstruct a class of graphs in the following theorem.

Definition 8. For a point v of a graph G, we define $f(v) = \max\{\deg v_i | v_i \text{ adj } v\}$ and $r(v) = a + 2.f(v) - 4 + \min\{\deg v, f(v)\}$. We define $r(G) = \min\{r(v) | v \in v(G)\}$.

Theorem 7. If in graph (S-graph) G, the difference between any two distinct ILDS is at least $r(G) + 1$, then G is reconstructible (set-reconstructible).

Proof. From the collection (set, if G is an S-graph) of point-deleted subgraphs of G, the degree sequence of G can be determined. Hence for each point-deleted subgraph G_u given, we can determine neighborhood degree sequemce of u in G and $r(u)$. Hence $r(G)$ can also be determined from the given point-deleted subgraphs. From the set of neighborhood degree sequences, we can calculate the set of ILDS and hence verify whether the difference between any two distinct ILDS is at least $r(G) + 1$ or not. Thus the situation is recognizable from the subgraphs given.

If G has a point w such that $f(w) = 1$, then G is either a star or is disconnected and hence is set-reconstructible by [4]. Hence let $f(w) > 1$ for every point w of G. Consider a point-deleted subgraph G_v such that $r(v) = r(G)$. Since $f(v) \geq 2$, if $\deg_G v = x$, then $r(G) \geq x + 2.2 - 4 + 1 = x + 1$. Also the ILDS loss of any point in G_v is at most $r(G)$. Since the difference between any two distinct ILDS is at least $r(G) + 1$, the ILDS loss of each point of G_v can be calculated. A point of degree $z - 1 (> 0)$ in G_v is adjacent to v in G iff its loss in ILDS is at least $x + 2z - 3$ (if it were not adjacent to v, its loss will be at most $z - 1$ and $z - 1 < x + 2z - 3$ since $z - 1 > 0$). Since G is connected, the isolated points of G_v are also adjacent to v in G. Thus for each point u of G_v we can determine whether u is adjacent to v or not. Hence G is determined uniquely from G_v.

The graphs E and E_n defined in Examples 1 and 2 are examples of graphs satisfying the hypothesis of Theorem 7.

4. LINE RECONSTRUCTION

Since it is known from [2] and [1] that "G is line reconstructible $\Leftrightarrow$ $L(G)$ is reconstructible" and "G is reconstructible $\Rightarrow$ G is line reconstructible", we see that the theorems in the previous section also establish the line reconstructiblity of some classes of graphs. Here we give some more results on line reconstruction.

Suppose we know the degree sequence of G and a line deleted subgraph H of G. Write the degrees of the points of G in non-decreasing order and similarly for H. The entries in the first sequence in positions where the two sequences differ give the degrees of the ends of the line whose removal gives H. This fact is used in the proof of some of the theorems that follow.

When a line x with ADP$(m(x), n(x))$, $m(x) \geq n(x)$ is deleted, the loss in the LNDS of a line is at most $L(x)$ where $L(x) = 2m(x) + n(x) - 3$ if $n(x) > 1$ and $L(x) = 2m(x) + n(x) - 4$ if $n(x) = 1$. In this process, the loss in LNDS for a line adjacent to x is at least the degree of x in G and that for a line not adjacent to x is at most four and in no case the loss for a line not adjacent to x is greater than the loss for a line adjacent to x. We use these facts in the following theorem on line reconstruction.

Theorem 8. If G is a graph such that the difference between any two distinct LNDS is at least r + 1 where $r = \min_{x \in X(G)} L(x)$, then G is line reconstructible.

Proof. Since reconstruction conjecture and hence line reconstruction conjecture is true for disconnected graphs and graphs having at most 7 points, we will prove the theorem only for connected graphs having at least 8 points. From the line deleted subgraphs, the degree sequence of G and the ADP of each line of G can be found. So the degrees of the lines of G are known. When a line x is deleted, the degree of only the neighbors of x gets reduced by one and hence the degrees of the neighbors of x can be found as follows : Write the degrees of the lines of G except the deleted line in non-decreasing order. Write the degrees of the lines of G - x similarly. The entries in the first sequence in positions where the two sequences differ give the degrees of the neighbors of the line x. Hence the LNDS of each line can be found. As we know the ADP of each line, r can be calculated and hence the graphs we are considering are recognisable from the collection of line deleted subgraphs.

Now consider a subgraph G - x, where the ends of x in G have degrees m and n, $m \geq n$, and L(x) = r. Calculate the LNDS of each line of G - x. Since a line loses at most r in its LNDS and the difference between any two distinct LNDS is at least r + 1, the LNDS loss of each line of G - x can be found out. Using this loss in LNDS, we will find out the lines that were adjacent to x in G. These lines will be called qualified lines (qualified to be adjacent to x). A point of G - x is called a qualified point if every line through this point is a qualified line. The ends of x in G - x are qualified points and hence each qualified line passes through a qualified point and there exists a pair of non-adjacent qualified points in G - x such that every qualified line passes through one of them.

Case 1. n = 1. Since G is connected and has at least 8 points, m has to be at least two. Also G - x has only one isolated point u, and x must pass through u.

If m = 2, then the maximum loss in LNDS for any line is one and all the lines that have lost one in LNDS are concurrent and precisely one of them passes through a point v of degree one in G - x. (If more than one of them pass through endpoints, then there are points whose LNDS differ by just one, contradicting the hypothesis). This v is the other qualified point. Joining u and v in $G - x$ we get G.

If $m \geq 3$, then there will be precisely $m - 1$ lines that have lost at least m in LNDS. These $m - 1$ (≥ 2) lines are the qualified lines and they will all be concurrent at a qualified point v. Joining u and v in $G - x$, we get G.

Case 2. $n = 2$.

Case 2(a). $m=2$. If x lies on a triangle in G, then precisely two lines have lost three in LNDS and they meet at a point. The other end of each of these lines has degree just one in $G - x$ and joining these two endpoints in $G - x$ we get G. If x does not lie on a triangle but lies on a C_4, then no line has lost more than two in LNDS and precisely three lines have lost two in LNDS. These three lines form a P_4 and the ends of this P_4 have degree one in $G - x$. Joining these endpoints in $G - x$, we get G.

If x does not lie on a cycle of length less than five, then precisely two lines y and z have lost 2 in LNDS. These two are the qualified lines and each is incident with at least one point of degree 1. Joining a point of degree 1 on y and a point of degree 1 on z in $G - x$, we get G. If one of the qualified lines, (say y) has ADP(1,1) in $G - x$, then any end of y can be chosen as u and both choices for u give the same graph G.

Case 2(b). $m = 3$. Lines that have lost at least 4 in LNDS are qualified lines (there may be other qualified lines also). If there are three of them then these are precisely the qualified lines. These three lines must form a P_4 and at least one of the four points involved is not qualified (since G is connected and has at least 8 points) and hence we will be able to select a unique pair of nonadjacent qualified points such that each qualified line passes through one of them. Clearly one end of x is a point of degree 1 and the other is the point of degree 2 not adjacent to this endpoint on this P_4. G is got from $G - x$ by adding the line x.

If there are precisely two lines that have lost at least 4 in LNDS, then these two meet at a qualified point u. In this case there is only one line that has lost

3 in LNDS. This line is qualified and at least one of its ends v is a point of degree one in $G - x$. Joining u and v in $G - x$, we get G.

Case 2(c). $m > 3$. In this case, a line adjacent with x will lose at least 4 in LNDS and a line not adjacent with x will lose at most 3 in LNDS. So in $G - x$, qualified lines can be determined. All but one of the qualified lines (at least three) will be concurrent at a qualified point u. The other qualified line will be incident with at least one point v of degree one. Joining u and v in $G - x$, we get G.

Case 3. $n \geq 3$. In this case, any line adjacent to x will lose at least 5 in LNDS whereas a line not adjacent to x will lose at most 4 in LNDS. So from the loss in LNDS, the qualified lines can be found out. We can choose two non-adjacent qualified points u and v such that $m - 1$ of the qualified lines are concurrent at u and the remaining $n - 1$ qualified lines are concurrent at v. It can be easily proved that the unordered pair (u,v) is unique. (When $m = n = 3$, it is unique because G is connected and has more than 4 points). G is the graph obtained from $G - x$ by adding the line uv.

The graphs G and H in Figure 3 and Figure 4 are graphs satisfying conditions of Theorem 8. Also for each integer $n(\geq 2)$, we construct a graph D_n satisfying the hypothesis of Theorem 8 as follows : Let $D = H_1 \cup H_2 \cup \ldots \cup H_n$, where each H_i is isomorphic to the graph H in Figure 4. In each H_i, label the points u and v

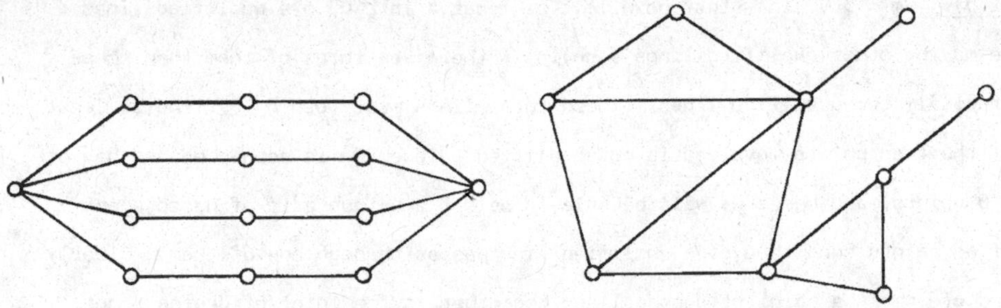

G H

FIGURE 3

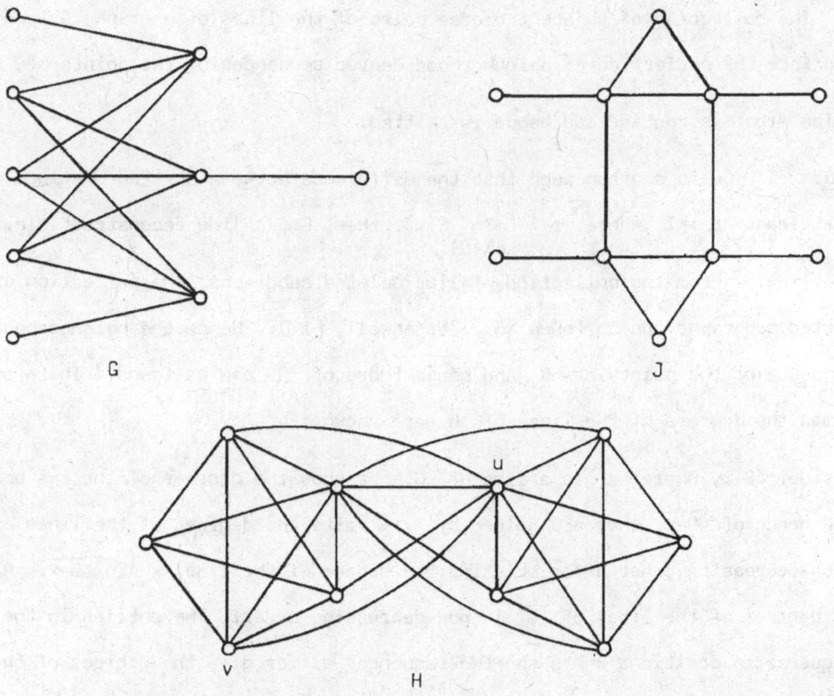

FIGURE 4

of H as u_i and v_i respectively. Let D_n be the graph obtained from D by deleting the lines $u_i v_i$, $i = 1,2,\ldots,n$ and adding the lines $v_i\, u_{i+1}$, $i = 1,2,\ldots, n-1$ and the line $v_n u_1$. Now D_n is clearly a graph satisfying the hypothesis of Theorem 8.

<u>Definition 9</u>. For any line uv of G, the sum of the ILDS of u and v in $\overline{G}$ is called the <u>complement line neighborhood degree sum</u> (CLNDS) of the line uv.

When a line x with ADP $(a(x), b(x))$, $a(x) \leq b(x)$ is deleted, the gain in CLNDS of a line is at most $C(x)$, where $C(x) = 3p - 2a(x) - b(x) - 2$ if $b(x) \leq p - 2$ and $c(x) = 3p - 2a(x) - b(x) - 3$ if $b(x) = p - 1$ where p is the number of points in the graph considered. A line adjacent to x gains at least $3p - a(x) - 2b(x) - 3$ and a line not adjacent to x gains at most 2,3 or 4 according as $b = p - 1$, $b = p - 2$ or $b \leq p-3$ and in no case the gain of a line not adjacent to x is more than that for a line adjacent to x. We use these facts to line reconstruct a class of graphs in Theorem 10.

Lemma 9. The collection of adjacent degree pairs of the lines of a graph G can be calculated from the collection of neighborhood degree sequences of the points of G.

Proof. The proof is routine and hence is omitted.

Theorem 10. If G is a graph such that the difference between any two distinct CLNDS is at least $r + 1$ where $r = \min_{x \in X(G)} C(x)$, then G is line reconstructible.

Proof. (Outline) From the collection of line deleted subgraphs, the collection of point deleted subgraphs can be found (D.L. Greenwell, [1]). Hence the neighborhood degree sequence of the points of G and hence those of $\overline{G}$ can be found. Therefore by the lemma the degrees of the lines of $\overline{G}$ are known.

Consider $G-x$, where x is a line of G. We know the degrees of the two points of G and hence of $\overline{G-x}$ that are joined by x. Write the degrees of the lines of $\overline{G-x}$ in non-decreasing order after deleting the degree of the line x in $\overline{G-x}$. Also write the degrees of the lines of $\overline{G}$ in non-decreasing order. The entries in the second sequence in positions where the two sequences differ give the degrees of the lines of $\overline{G}$ that share a point of incidence with the line x of G. (Only the degrees of such lines increase by one each in $G-x$). Thus for each line x of G, we can calculate the CLNDS and hence the situation is recognisable from the collection of line deleted subgraphs of G.

Let x be a line of G with ADP(a,b), $a \leq b$ and $C(x) = r$. In $G-x$, find the CLNDS of each line. Since any line gains a maximum of r in CLNDS and the difference between any two distinct CLNDS is at least $r + 1$, the CLNDS gain of each line of $G-x$ can be found out. Using this gain in CLNDS, we will find out the lines of $G-x$ that were adjacent to x in G. The position of x in $G-x$ can be determined from these lines by a procedure similar to that in the proof of Theorem 9 above and hence G is known upto isomorphism.

The graphs G,H of Figure 4, the graph G of Figure 3 and the complement of the graph J in Figure 4 satisfy the hypothesis of Theorem 10. But all these graphs satisfy the hypothesis of Theorem 8 also. The graph H of Figure 3 satisfies the hypothesis of Corollary 5.2. But neither this graph nor its complement satisfies the hypothesis of Theorem 10.

4. CONCLUSION

We have seen that some classes of graphs are reconstructible using the concepts of NDS, LNDS, ILDS and CLNDS thereby widening the class of reconstructible graphs. We suggest the following problems which are similar to the results proved above.

(1) Characterize graphs in which all lines have the same CLNDS. (Line regular graphs have this property).

(2) Reconstruct graphs that have only two distinct NDS.

(3) Reconstruct graphs such that the difference between any two distinct NDS is at least 2δ, where δ is the minimum degree.

ACKNOWLEDGEMENT

We are thankful to Dr. T.V.S. Jagannathan and Dr. S.A. Choudum for helpful discussions and the referee for valuable comments.

REFERENCES

1. D.L. Greenwell, Reconstructing graphs, Proc. Amer. Math. Soc., 30(1971), 431-433.
2. D.L. Greenwell and R.L. Hemminger, Reconstructing Graphs, The Many Facets of Graph Theory, Springer-Verlag (1969), 91-114.
3. F. Harary, Graph Theory, Addison-Wesley, Mass., 1969.
4. B. Manvel, On reconstructing graphs from their sets of sub-graphs, J. Comb. Theory, B 21(1976), 156-165.

ON RECONSTRUCTING SEPARABLE DIGRAPHS

S. RAMACHANDRAN*
Mathematics Department
Aditanar College, Tiruchendur 628 216

ABSTRACT

Separable digraphs without endpoints are reconstructed from point-deleted subdigraphs for each of which the degree pair of the deleted point is also known.

RESULTS

Unless otherwise stated, the digraphs dealt with in this note will be finite and may have loops and multiple arcs. The <u>degree pair</u> of a point v in a digraph G is the ordered pair (od (v), id (v)) where od (v) and id (v) are respectively the outdegree of v in G. (A loop gives indegree one and outdegree one to the base point). A digraph is said to be <u>connected</u> if every two points are joined by a path (not necessarily directed). A digraph is <u>disconnected</u> if it is not connected. A cutpoint of a connected digraph is a point whose removal disconnects the digraph. A connected digraph is <u>separable</u> if it contains a cutpoint. A <u>block</u> is a maximal connected subdigraph that is not separable. Each arc of the digraph lies in exactly one block. However the loops at any cutpoint lie on all the blocks having that cutpoint. An <u>endpoint</u> is a point joined to just one other point. Note that an endpoint can have any indegree and outdegree. It can also have loops at it. Let G^1 be a subset of the points, arcs and loops of G. (The endpoints/basepoints of the arcs and loops in G^1 need not themselves be in G^1). Now $G-G^1$ denotes that subdigraph of G which is obtained by deleting G^1 and all arcs and loops of G which are joined to points of G^1. For any set S of points of G, the <u>induced subdigraph</u> <S> is the maximal subdigraph of G with point set S.

---

* Presently at Mathematics Department, Madurai Kamaraj University, Madurai-625 021, under UGC TF Programme.

In [4], the following conjecture is proposed.

CONJECTURE A. If G and H are digraphs with points u_i, $i = 1,2,\ldots,n$ and v_i, $i = 1,2,\ldots,n$ $(n \geq 2)$ respectively such that for each i, u_i and v_i have the same degree pair and $G-u_i \simeq H-v_i$, then $G \simeq H$.

This conjecture is weaker than the Digraph Reconstruction Conjecture (DRC) and is true for digraphs with at most four points that have no multiple arcs. Hence any class of digraphs satisfying the DRC must satisfy Conjecture A and any counterexample to Conjecture A must be a counterexample to DRC. But none of the known counterexamples to the DRC is a counterexample to Conjecture A. (Stockmeyer [7]). In [5], some classes of digraphs including one infinite family of counterexamples to the DRC given in [6] are shown to satisfy Conjecture A. Here we prove Conjecture A for separable digraphs without endpoints. The proof closely resembles that given in Bondy [1] for separable graphs without endpoints.

The following lemma of Harary and Palmer [2] is a generalization of a result of Kelly [3].

Lemma 1. Let G and H be digraphs with points u_i, $i = 1,2,\ldots,n$ and v_i, $i = 1,2,\ldots,n$ $(n \geq 2)$ respectively such that for each i, u_i and v_i have the same degree pair and $G-u_i \simeq H-v_i$. Let Y be any digraph of order less than n. Suppose there are α distinct subdigraphs of G isomorphic to Y and that point u_i of G is in α_i of these subdigraphs; that there are β distinct subdigraphs of H isomorphic to Y and that point v_i is in β_i of these subdigraphs. Then $\alpha = \beta$ and $\alpha_i = \beta_i$ $(1 \leq i \leq n)$.

Lemma 1 remains true if 'subdigraph' is replaced by 'induced subdigraph' throughout. We will refer to this version as Lemma 1(a).

Note. It is an easy consequence of Lemma 1 that u_i and v_i have the same number of loops.

Lemma 2. Let G and H be digraphs as in Lemma 1. Suppose G has blocks $B_1,B_2,\ldots,B_m$ $(m > 1)$ and H has blocks $C_1,C_2,\ldots,C_n$. Then $m = n$ and the blocks can be relabelled so that $B_i \simeq C_i (1 \leq i \leq n)$.

Proof. The proof is same as that of Lemma 1.2 in Bondy [1].

Theorem 3. Let G and H be as in Lemma 1. If G and H are separable digraphs without endpoints, then $G \simeq H$.

Proof. Let B_1 be an 'end' block of G (that is, a block containing just one cutpoint of G) such that no endblock of G or H has order less than b_1, the order of B_1. (The assumption that B_1 is in G results in no loss of generality). Let u be the cutpoint joining B_1 to the rest of G. Let $G_1 = G - (B_1 - u)$ and $G_1^1 = G - (B_1 - \{u,w\})$ where w is a point in B_1 joined to u by an arc and $w \neq u$. Then G_1^1 is a proper induced subdigraph of G and hence by Lemma 1(a) there is an induced subdigraph H_1^1 of H isomorphic to G_1^1, say, $\varphi(G_1^1) = H_1^1$, $\varphi(u) = v$. (Note : u and v are not necessarily corresponding vertices). Let H_1 be the digraph obtained from H_1^1 by deleting its endpoint. Then it is clear that $H_1 = \varphi(G_1)$ and by Lemma 2, H_1 has one block fewer than H, that block, C_1 say, being isomorphic to B_1. Now H is obtained from H_1^1 by adding $b_1 - 2$ points, some arcs and possibly some loops. Since no end block of H has order less than that of B_1, and since H has no endpoints, it is easy to see that those arcs and loops can only be incident with v,p (the endpoint of H_1^1) and the $b_1 - 2$ new points. Thus v is a cutpoint of H and it follows that the subdigraph of H on v,p and these $b_1 - 2$ new points is isomorphic to C_1. It now remains to show that there is an isomorphism of B_1 and C_1 mapping u onto v. Denote by $B_1^{s,t}(G_1^{s,t})$ the digraph obtained from B_1 (G_1) by adding s + t isolated points and joining each of tem to u by an arc, s of the arcs directed away from u and the remaining t arcs directed towards u. Define $C_1^{s,t}$ and $H_1^{s,t}$ analogously. It will suffice to prove $B_1^{s,t} \simeq C_1^{s,t}$ for some s and t, not both zero.

Now u_i is a cutpoint in G iff $G - u_i$ is disconnected. Hence, by hypothesis, G and H have the same number of cutpoints of the same degree pair and loop structure. Hence u and v have the same degree pair and loop structue (as the degree pair and loop structure of all other cutpoints are identical in G and H). Let the common degree pair of u and v be (r_1+r_2+t, s_1+s_2+t) where t is the common number of loops at u and v. Suppose that apart from loops, $r_1 + s_1$ arcs (r_1 directed away from u and s_1 directed towards u) of G_1 and $r_2 + s_2$ arcs (r_2 directed away from u and s_2 directed towards u) of B_1 are joined to u. Then it is clear that $r_1 + s_1$ arcs (r_1 directed away from v and s_1 directed towards v) of H_1 and $r_2 + s_2$ arcs (r_2 directed away from v and s_2 directed towards v) of C_1 are joined to v. Evidently at least one among r_1 and s_1 is nonzero (say $r_1 > 0$, the argument when $s_1 > 0$ being similar). If $B_1^{1,0}$ occurs α times in $G_1^{r_2,s_2}$, it occurs $\alpha + r_1$ times in G. But then $B_1^{1,0}$ occurs α times in $H_1^{r_2,s_2}$ (since $G_1^{r_2,s_2} = H_1^{r_2,s_2}$) and $\alpha + r_1$ times in H (by Lemma 1). This implies that $B_1^{1,0} = C_1^{1,0}$ and hence $G \simeq H$.

Finally we observe that the discussion on the reconstructibility of graphs with endpoints given in Section 3 of Bondy [1] holds as it is in the case of digraph reconstruction conjecture also.

REFERENCES

1. J.A. Bondy, On Ulam's conjecture for separable graphs, Pacific J. Math.,31(1969). 281-288.
2. F. Harary and E. Palmer, On the problem of reconstructing a tournament from subtournaments, Monatsh. Math., 71(1967), 14-23.
3. P.J. Kelly, A congruence theorem for trees, Pacific J. Math., 7(1957), 961-968.
4. S. Ramachandran, A digraph reconstruction conjecture, Graph Theory Newsletter, 8(1979) No.4, Abstract 11.
5. S. Ramachandran, On a new digraph reconstruction conjecture, (1979), preprint.
6. P.K. Stockmeyer, The falsity of the reconstruction conjecture for tournaments, J. Graph Theory, 1(1977), 19-25.
7. P.K. Stockmeyer, Personal Communication (1979).

DEGREE SEQUENCES OF CACTI

A. RAMACHANDRA RAO
Indian Statistical Institute
203, B. T. Road, Calcutta 700 035

In this paper, we consider only undirected graphs with no loops and no multiple edges.

Following Berge [1], we define a <u>cactus</u> or a Husimi tree to be a connected graph in which each block is an edge or a cycle. Also a graph is said to have <u>property</u> Q if it is connected and each block is an edge or a triangle. A graph is said to have <u>property</u> H_k if it is connected and every block is a cycle on k vertices.

If $\pi = (d_1, d_2, \ldots, d_n)$ and P is a property, then we say that π is <u>potentially</u> P if there is a realisation of π (i.e., a graph with degree sequence π) with property P. Also π is said to be <u>forcibly</u> P if π is graphic and every realisation of π has property P.

If P is the property that the graph is a cactus, we refer to potentially P (respectively, forcibly P) sequences as potentially cactaceous (respectively, forcibly cactaceous) sequences.

In this paper we characterise potentially P and forcibly P sequences for each of the three properties: being a cactus, Q, and H_k.

Throughout this paper we let $\pi = (d_1, d_2, \ldots, d_n)$, where $d_1 \geq d_2 \geq \ldots \geq d_n$ are non-negative integers.

We first give a characterisation of potentially Q sequences.

<u>Theorem 1</u>. Let $n \geq 2$. Then π is potentially Q if and only if the following five conditions are satisfied, where $b_2 = 3n - 3 - \sum_{i=1}^{n} d_i$:

(i) $d_n \geq 1$,

(ii) $\sum_{i=1}^{n} d_i$ is even and $\geq 2(n-1)$,

(iii) $b_2 \geq \frac{\beta}{2}$ where β is the number of odd d_i's in π,

(iv) if $n \geq 3$ then $b_2 \geq \alpha$ where α is the number of 1's in π,

(v) if $b_2 > 0$ then $\beta > 0$.

Proof. Suppose first that π is potentially Q and G is a realisation of π with property Q. Then it is easy to see that the number of triangle-blocks in G is the cyclomatic number $m - n + 1$ where m is the number of edges in G, and so the number of cut edges in G is

$$m - 3(m-n+1) = 3n - 3 - \sum_{i=1}^{n} d_i.$$

Thus b_2 is the number of blocks of order 2. Now (i) and (ii) follow from the fact that G is connected. To prove (iii), we observe that every vertex with odd degree is incident with at least one cut edge. To prove (iv), we note that if $n \geq 3$, then the pendant vertices of G are incident with distinct cut edges, and (v) follows from the fact that if G has a cut edge then G has (at least two) vertices with odd degree.

Conversely let π satisfy the conditions (i) - (v) of the theorem. Then we prove by induction on n that there exists a realisation of π with property Q. We first make some simple observations. Firstly β is an even integer. Also by (iii), $\sum d_i \leq 3n - 3$. Now if $n = 2$ then $\pi = 1^2$ is potentially Q. If $n = 3$, then $\pi = 21^2$ or $\pi = 2^3$, and both these are potentially Q. Thus we assume the theorem for sequences of length at most $n - 1$ and let π be a sequence of length $n \geq 4$ satisfying (i) - (v). Note that $d_1 = 1$ leads to a contradiction by (i) and (ii).

If $d_1 = 2$, then $\pi = 2^n$ or $2^{n-2}1^2$. The latter is potentially Q and the former gives a contradiction to (v). So we take $d_1 \geq 3$. We consider two cases.

Case (1). Assume that $\sum d_i \geq 2n$ and π contains at least two 2's. Then let π' be the sequence obtained from π by deleting two 2's and subtracting 2 from d_1. We will show that π' (after being rearranged in non-increasing order) satisfies conditions (i) - (v) with $n - 2$ replacing n.

That π' satisfies (i) and (ii) is trivial. Let b_2', β', α' denote the values of b_2, β, α, respectively, for the sequence π'. Then it is easy to see that

$$b_2' = b_2, \ \beta' = \beta, \ \alpha' = \alpha \text{ or } \alpha' = \alpha + 1.$$

Hence π' satisfies (iii) and (v). If π' violates (iv), then $n \geq 5$, $b_2 = \alpha$ and $\alpha' = \alpha + 1$. Hence $d_1 = 3$ and $\Sigma d_i = 3n - 3 - \alpha$. Now if γ is the number of 2's in π, then $\Sigma d_i = 3n - \gamma - 2\alpha$, hence $\gamma + \alpha = 3$. Now by (iii), $\alpha \geq \frac{\beta}{2} > 0$ (note $d_1 = 3$) and by the case under consideration, $\gamma \geq 2$. Hence $\alpha = 1$, $\gamma = 2$ and $\beta = 2$, a contradiction since $n \geq 5$. Thus π' satisfies (iv) also. Hence by induction hypothesis, there exists a realisation H of π' with property Q. Attaching a triangle at the vertex with degree $d_1 - 2$, we get a realisation of π with property Q.

<u>Case (2)</u>. Assume that $\Sigma d_i = 2(n-1)$ or π contains at most one 2. Since $\Sigma d_i \leq 3n-3$, it easily follows that $d_n = 1$. Now if there is an odd $d_j \geq 3$ in π, choose and fix one such d_j. Otherwise let $j = 1$. Define π' to be the sequence obtained from π by deleting the term d_n and subtracting 1 from d_j. We will show that π' (after being rearranged in non-increasing order) satisfies conditions (i) - (v) with $n - 1$ replacing n.

That π' satisfies (i) and (ii) is trivial. Denoting by b_2', β' and α' the values of b_2, β and α for π', we can easily see that

$$b_2' = b_2 - 1, \ \alpha' = \alpha - 1 \text{ and } \beta' = \beta \text{ or } \beta' = \beta - 2.$$

Thus π' satisfies (iv). If π' violates (iii), then $b_2 = \frac{\beta}{2}$ and $\beta' = \beta$. Hence $\frac{\beta}{2} \geq \alpha$ by (iv) and d_j is even. Thus $\frac{\beta}{2} \geq \alpha = \beta > 0$, a contradiction which proves that π' satisfies (iii). If π' violates (v), then $\beta = 2$ and d_j is odd. Clearly then $\alpha = 1$, $\Sigma d_i \geq 2n$ and so π contains at most one 2. Hence $\Sigma d_i \geq 3n - 3$, a contradiction to (iv). Thus π' satisfies (v) also. Hence by induction hypothesis, there exists a realisation H of π' with property Q. Attaching a pendant edge at the vertex with degree $d_j - 1$, we get a realisation of π with property Q. This completes the proof of the theorem.

We now show that none of the five conditions in Theorem 1 is redundant. The sequences 32^210, 1^4, 3^22^3, $4^22^21^2$ and 2^4 show respectively that the conditions (i), (ii), (iii), (iv) and (v) cannot be dropped.

We also mention that the proof of Theorem 1 in fact gives an algorithm to construct a realisation of π with property Q whenever one exists.

Corollary. Let $n \geq 3$. Then there exists a connected realisation of π in which every block is a triangle if and only if the following three conditions are satisfied.

(i) $d_n \geq 2$,

(ii) d_i is even for all i,

(iii) $\sum_{i=1}^{n} d_i = 3n - 3$.

The corollary follows easily noting that we are looking for graphs with no cut edges (b_2=0). We mention that some authors define cacti as the graphs considered in this corollary. Sometimes Husimi trees are defined as graphs with property H_k and we characterise their degree sequences in the following theorem.

Theorem 2. Let $k \geq 3$ be a fixed integer and let $n \geq k$. Then π is potentially H_k if and only if the following three conditions are satisfied.

(i) $d_n \geq 2$,

(ii) d_i is even for all i,

(iii) $\sum_{i=1}^{n} d_i = \frac{2k}{k-1}(n-1)$.

The proof of this theorem is similar to that of Theorem 1 and is omitted.

Lemma. Let $n \geq 2$. Then π is potentially cactaceous if and only if either π is potentially Q or $d_n \geq 2$, all d_i's are even and $\sum_{i=1}^{n} d_i < 3n - 3$.

Proof. If π is potentially Q then π is potentially cactaceous. If $d_n \geq 2$, all d_i's are even and $\Sigma d_i < 3n - 3$, then clearly π has at least $b_2 + 3$ members equal to 2 where b_2 is defined in Theorem 1. Now the sequence π' obtained from π by deleting b_2 2's satisfies the conditions of the corollary to Theorem 1 with $n - b_2$ replacing n. If H is a realisation of π' with property Q, then a cactus with degree sequence π is obtained by enlarging some terminal triangle of H to a cycle of length $b_2 + 3$.

Conversely let π be potentially cactaceous and G a realisation which is a cactus. If G has a cut edge, we will prove that there exists a realisation of π with property Q. So let G have a cut edge and let C be a cycle of length ≥ 4. If C is a terminal block, then it can be replaced by a triangle and the cut edge replaced by a path of suitable length. If C contains at least two cut vertices x and y of G, we may take that there is a cut edge uv in G such that every path from x to u uses some edge of C. Now join the two vertices adjacent to x on C, delete the two edges of C at x, create a vertex of degree 2 by splitting the cut edge uv and identify it with x. Repeating these steps we finally obtain a realisation of π with property Q and so π is potentially Q. Next suppose G has no cut edge. Then we show that after removing some 2's, π becomes potentially Q. If G has a terminal block which is not a triangle, it may be replaced by a triangle, thus dropping some 2's from π. Let C be a cycle of length ≥ 4 containing at least two cut vertices x and y of G. Then join the two vertices adjacent to x on C, delete the two edges of C at x and identify x with some vertex of degree 2 in a terminal triangle which is connected to C through the vertex y. In this process one 2 is lost from the degree sequence. Repeating these steps we see that π becomes potentially Q after dropping some 2's, hence $d_n \geq 2$, all d_i's are even and $\Sigma d_i \leq 3n - 3$. Equality in the last inequality implies π is potentially Q and the lemma is proved.

<u>Theorem 3</u>. Let $n \geq 2$. Then π is potentially cactaceous if and only if the conditions (i) - (iv) of Theorem 1 are satisfied.

This theorem follows easily from the Lemma and Theorem 1. Also our proof gives a method of obtaining a realisation of π which is a cactus whenever one exists.

Before going on to forcibly Q sequences, we mention that a graph with property Q on n vertices with m edges exists if and only if $n - 1 \leq m \leq [\frac{3n-3}{2}]$; the same result holds if property Q is replaced by: cactaceous.

<u>Theorem 4</u>. The sequence π is forcibly Q if and only if it is one of the following.

(i) $\pi_1 = (n-1,\ 2^{2k},\ 1^{n-2k-1}) : 4 \leq 2k \leq n-1,$

(ii) $\pi_2 = (r+2,\ s+2,\ t+2,\ 1^{n-3}) : r \geq s \geq t \geq 0,\ r + s + t = n - 3,$

(iii) $\pi_3 = (r+1,\ s+1,\ 1^{n-2}) : r \geq s \geq 0,\ r + s = n - 2.$

Proof. It is easy to check that each of π_1, π_2 and π_3 has a unique realisation and it has property Q, proving the if part of the theorem.

To prove the converse, let π be forcibly Q and G a realisation of π. If G has a triangle T then we show that every other block of G intersects T. Otherwise there exists a path xuv where $T = \{x,y,z\}$, $u, v \notin T$ and x,v are not adjacent. Now removing the edges xz and uv and adding the edges zu and xv, we get a realisation of π with a 4-cycle, a contradiction. It now follows easily that if G has at least two triangles then $\pi = \pi_1$. If G has exactly one triangle then $\pi = \pi_2$. If G has no triangle then G is a tree. If the diameter of G is at least 4, then there exiot pendant vertices x_1 and x_2 and edges x_1y_1 and x_2y_2 such that y_1 and y_2 are distinct and not adjacent. Now deleting the edges x_1y_1 and x_2y_2 and adding the edges x_1x_2 and y_1y_2 we get a disconnected realisation of π, a contradiction. Thus the diameter of G is at most 3 and $\pi = \pi_3$. This completes the proof of the theorem.

Theorem 5. The sequence π is forcibly cactaceous if and only if π is one of $\pi_1, \pi_2, \ldots, \pi_6$ where π_1, π_2 and π_3 were defined in Theorem 4 and

$\pi_4 = (n-2,\ 2^{2k+1},\ 1^{n-2k-2}) : 3 \leq 2k + 1 \leq n - 1,$

$\pi_5 = (n-3,\ 2^{n-1}) : n$ is odd and $n \geq 5,$

$\pi_6 = (n-3,\ 2^{n-2},\ 1) : n$ is even and $n \geq 6.$

Proof is omitted.

The proof of the following theorem is simple and is omitted.

Theorem 6. The sequence π is forcibly H_3 if and only if n is odd and $\pi = (n-1,\ 2^{n-1})$. The only forcibly H_k sequence is 2^k for $k = 4$ and 5 and there is no forcibly H_k sequence for $k > 5$.

One can study potentially P and forcibly P sequences for various properties P related to those considered in this paper. We mention a few here:

1. connected and each block is complete (equivalently, block graph of a connected graph),
2. line graph of a tree,
3. connected and each block is minimal.

After writing this paper, the author came to know that Beineke and Schmeichel [2] had, by a different method, characterised potentially cactaceous degree sequences with the number $\frac{1}{2}(\sum_{i=1}^{n} d_i - 2n+2)$ of cycles small. They also proved our Theorem 5.

REFERENCES

1. C. Berge, Graphs and Hypergraphs, North-Holland Publishing Company, Amsterdam, 1973.

2. L. W. Beineke and E. F. Schmeichel, Degrees and cycles in graphs, Second International Conf. on Comb. Math., (Ed. A. Gewirtz and L.V. Quintas), Ann. N.Y. Acad. Sci., 319(1979), 64-70.

A SURVEY OF THE THEORY OF POTENTIALLY P-GRAPHIC AND FORCIBLY P-GRAPHIC DEGREE SEQUENCES

S. B. RAO
INDIAN STATISTICAL INSTITUTE
203 B. T. Road, Calcutta 700 035

ABSTRACT

A sequence $\pi = (d_1,\ldots,d_p)$ of length p is said to be graphic if there exists a graph G with $V(G) = \{u_1,\ldots,u_p\}$ such that the degree of u_i in G is equal to d_i, for every i, $1 \le i \le p$; and G is referred to as a realization of π. Let P be an invariant graph theoretic property. A sequence π is said to be potentially P-graphic if there exists a realization of π with the property P; π is said to be forcibly P-graphic if π is graphic and every realization of π has the property P. Indicating the different unified approaches in the theory of potentially P-graphic and forcibly P-graphic sequences, we survey the known results on potentially P-graphic and forcibly P-graphic sequences for various properties P. We provide an extensive bibliography on these topics and mention several unsolved problems and conjectures.

1. POTENTIALLY P-GRAPHIC AND FORCIBLY P-GRAPHIC SEQUENCES

For a graph G with vertex set $V(G) = \{u_1,\ldots,u_p\}$, the sequence $\pi(G) = (d_1,\ldots,d_p)$ where $d_i = d_G(u_i)$, the degree of u_i in G, that is, the number of edges in G incident at u_i, is called the <u>degree sequence</u> of G. The degree sequence of a graph is an invariant of G, it is rather a weak invariant, for, there may be several non-isomorphic graphs with the same degree sequence. For example, $\pi = (3,2^2,1^3)$, where $b_i^{n_i}$ means that the degree b_i occurs exactly n_i times in π; has the following nonisomorphic realizations G_1 and G_2 of Figure 1. Another example of this nature is

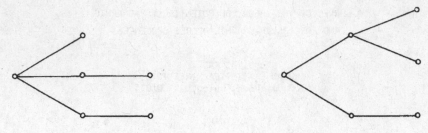

Figure 1. The graphs G_1 and G_2, respectively.

$\pi = (4^2,3^6,2,1^6,0^3)$. The nonisomorphic graphs G_3 and G_4 of Figure 2 have this as the degree sequence. All sequences in this section are in non-increasing order and

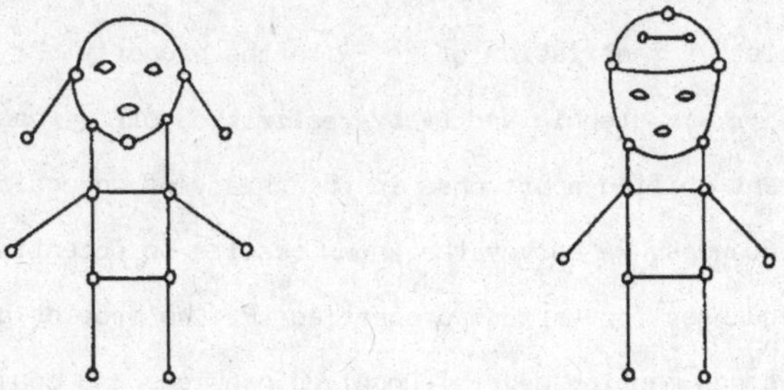

Figure 2. The graphs G_3 and G_4, respectively.

have integral terms. A sequence $\pi = (d_1,\ldots,d_p)$ is said to be graphic if there exists a graph G with $V(G) = \{u_1,\ldots,u_p\}$ such that $\pi(G) = \pi$; such a graph G is referred to as a realization of π; and p the length of π. Define $\underline{\underline{G}}(\pi)$ to be the set of all nonisomorphic graphs G with $V(G) = \{u_1,\ldots,u_p\}$ and $\pi(G) = \pi$. Let P be an invariant property of graphs. A graphic sequece π is said to be potentially P-graphic if there exists at least one $G \in \underline{\underline{G}}(\pi)$ such that G has the property P; the graphic sequence π is said to be forcibly P-graphic if every $G \in \underline{\underline{G}}(\pi)$ has the property P. More generally, if P,Q are two sets of invariant properties of graphs, then a graphic sequence π is said to be forcibly PQ-graphic if there exists a $G \in \underline{\underline{G}}(\pi)$ having the property Q and every $H \in \underline{\underline{G}}(\pi)$ having the property Q has also the property P. In case Q is the empty set, every forcibly PQ-graphic sequence is a forcibly P-graphic sequence. In this section, mentioning the various unified approaches developed by several authors to solve potentially P-graphic and forcibly PQ-graphic sequences, we review the solved and some unsolved problems for several properties P and Q. The corresponding results for digraphic sequences will be dealt with in Section 2. Section 3 deals with some related sequences. We only review those properties P for which complete characterizations exist. The rest are given as unsolved problems with references.

We start with necessary and sufficient conditions for a sequence $\pi = (d_1,\ldots,d_p)$ to be graphic. The first one, due to Havel [106], Hakimi [96], Senior [200], Kleitman and Wang [124] is algorithmic in nature. The second one is due to Erdos-Gallai [76], derivable from one by Fulkerson, Hoffman and MacAndrew [84] which is somewhat existential in nature, but nonetheless uses the same construction as in Havel-Hakimi's theorem. To state the former one we need a definition : If $\pi = (d_1,\ldots,d_p)$, then the residual sequence obtained after laying off d_j from π, is the sequence π' of length $p-1$ defined as :

$$\pi' = \begin{cases} d_1-1,\ldots,d_{d_j}-1,\ d_{d_{j+1}},\ldots,d_{j-1},\ d_{j+1},\ldots,d_p, & \text{if } d_j < j; \\ d_1-1,\ldots,d_{j-1}-1,\ d_{j+1}-1,\ldots,d_{d_j+1}-1,\ d_{d_j+2},\ldots,d_p, & \text{if } d_j \geq j. \end{cases}$$

Theorem 1.1.(Kleitman-Wang [124]) A sequence π is graphic if and only if π' is graphic; and a realization G of π then may be obtained from a realization H of π' by adjoining a new vertex labelled u_j to H and joining it to the vertices whose degrees are reduced by one in going from π to π'. It may be remarked that Havel-Hakimi result is the case when $j = 1$.

Next we state a theorem of [84] mentioned above. To this end we say a multigraph G has the odd cycle property if whenever C_1 and C_2 are two vertex disjoint odd cycles of G, then some vertex of C_1 is adjacent in G to a vertex of C_2.

Theorem 1.2.(FHM [84]) Let G be a multigraph of order p with the odd cycle property, with $V(G) = \{1,2,\ldots,p\}$. Let $\pi = (d_1,\ldots,d_p)$ be a sequence of non-negative integers with even sum. Then G has a sub-graph H with degree sequence π if and only if for every S,T,U partition of $\{1,2,\ldots,p\}$(empty sets not excluded)

$$\sum_{i\in S} d_i \leq \sum_{i\in T} d_i + \sum_{\substack{i\in S\\ j\in S\cup U}} c_{ij},$$

where

$$c_{ij} = \begin{cases} 1 & \text{if } (i,j) \in E(G) \\ 0 & \text{if } (i,j) \notin E(G). \end{cases}$$

From the above theorem we have the following theorem of Erdös-Gallai [76] mentioned earlier.

Theorem 1.3. If π is a sequence with $\sum_{i=1}^{p} d_i$ even, then π is graphic if and only if for every integer r with $1 \leq r < p$, the following inequality is satisfied :

$$\sum_{i=1}^{r} d_i \leq r(r-1) + \sum_{i=r+1}^{p} \min(d_i, r). \qquad \ldots(1.1)$$

This theorem was extended to r-multigraphs by Chungphaisan [52] using Theorem 1.2. For a proof of these results using the method of flows in networks the reader is referred to Chapter 6 of Berge [8] and Chapter 6 of Chen [47]. The unimultigraphic sequences were determined by Hakimi [96], Senior [200] and only recently a characterization of unigraphic sequences was given by Koren [128,129]. One of the main results of Koren [129] may be stated as

Theorem 1.4. If $\pi = (d_1, \ldots, d_p)$ is a unigraphic sequence with $d_2 > d_{p-1}$ $(p \geq 4)$, then for some r, $1 \leq r < p$, we have equality in (1.1).

For some earlier results on unigraphic sequences the reader is referred to Johnson [113, 114]; and Suho-yen R. Li [202].

The first problem on potentially P-graphic sequences was solved by Edmonds [70] where p is k-edge connectedness and may be stated as a

Theorem 1.5. A graphic sequence $\pi = (d_1, \ldots, d_p)$ is potentially k-edge connected - graphic if and only if

1) $d_i \geq k$ for every i, $1 \leq i \leq p$,

and

2) $\sum_{i=1}^{p} d_i \geq 2(p-1)$ if $k = 1$.

The technique of Edmonds [70] uses the following facts : If G is a realization of π having vertices u,v,x,y such that $uv, xy \in E(G)$ and $ux, vy \notin E(G)$, then the graph H obtained from G by deleting the edges uv, xy and adding the edges ux, vy, (this process is called a simple interchange) is a realization of π (possibly nonisomorphic to G). In fact if C is a closed alternating trail in G and $\overline{G}$, then the graph obtained from G by interchanging the edges of G and $\overline{G}$ on this closed trail is a realization of π, (this process may be called a generalised interchange).

The method of Edmonds [70] may be generalized as follows to the first unified approach to the problems of characterization of potentially P-graphic sequences and may be described as : Choose a measure $\mu(P)$ of nearness to the property P. Out of all realizations of π choose one realization for which $\mu(P)$ is the smallest possible; and under the given conditions on π show that this realization itself has the property P, using some generalized interchanges. For example, if P is the property that a graph is connected, $\mu(P)$ may be chosen as the number of components; and if P is the property that a graph is hamiltonian, $\mu(P)$ may be chosen as the number of components in a 2-factor provided π has a realization with a 2-factor. The method was also used in Rao and Rao [179], Das and Rao [191], Bankfalvi and Bankfalvi [3], Schmeichel and Hakimi [101], Lovász [151]. We refer to this as method of interchanges.

The second unified approach for characterizing potentially P-graphic sequences is that of Kleitman and Wang [124] : Construct an appropriate realization of π suitably choosing the degrees and laying them off successively. Using this method Wang, Kleitman [211] characterized potentially k-vertex connected-graphic sequences.

Theorem 1.6. A graphic sequence $\pi = (d_1,\ldots,d_p)$ is potentially k-vertex connected ($k \geq 2$), if and only if,

1) $d_i \geq k$ for every i, $1 \leq i \leq p$,

2) $\sum_{i=1}^{k-1} d_i \leq \frac{1}{2} \sum_{i=1}^{p} d_i + \binom{k-1}{2} - (p-k)$.

Moreover if (2) is not satisfied but (1) is satisfied, then π is potentially (k-1)-vertex connected.

Rao and Rao [179] characterized potentially k-factorable sequences which are potentially connected k-factorable sequences using the method of interchanges along alternating 4-cycles.

Theorem 1.7. A potentially k-factorable sequence $\pi = (d_1,\ldots,d_p)$ is potentially connected k-factorable if and only if for every $s < p/2$ the following condition is satisfied :

$$\sum_{i=1}^{s} d_i < s(p-s-1) + \sum_{j=1}^{s} d_{p+1-j} . \qquad \ldots \quad (1.2)$$

Note that (1.2) is necessary and sufficient for a potentially 2-factorable sequence to be potentially hamiltonian-graphic. A constructive proof of this result was given by Chungphaisan [56]. Potentially hamiltonian bipartite sequences were characterized in Chungphaisan [54].

Rao and Rao [179] conjectured that π is potentially k-factorable if and only if π, as well as $\pi - k = (d_1-k,\ldots,d_p-k)$ are graphic. This conjecture, in a slightly general form, was proved by Kundu [134,135,136] using generalized interchanges along alternating cycles, and by the laying off technique by Kleitman and Wang [1]. Lovász [151] gave an elegant proof of the conjecture by using the interchange technique. Kundu's theorem [134] may be stated as :

Theorem 1.8. If $\pi = (d_1,\ldots,d_p)$ and $\pi' = (k_1,\ldots,k_p)$ are graphic sequences (not necessarily in the nonincreasing order) with $k_i \le d_i$ and $|k_i - k_j| \le 1$ for all i and j, then π has a realization G containing a realization H of π' as a spanning subgraph.

Using the method of Rao and Rao [179], potentially traceable-graphic sequences and some general results were obtained by Schmeichel and Hakimi [101]. Using the laying off technique, an elegant characterization of potentially self-complementary sequences was given by Clapham and Kleitman [62].

Theorem 1.9. A graphic sequence $\pi = (d_1,\ldots,d_p)$ is potentially self-complementary if and only if $p = 4n$ for some n and

(i) $d_i + d_{4n+1-i} = 4n - 1$ for $i = 1,\ldots,2n$,

(ii) $d_{2j} = d_{2j-1}$ for $j = 1,\ldots,n$;

or $p = 4n + 1$ for some n and

(i) $d_i + d_{4n+2-i} = 4n$, for $i = 1,\ldots,2n+1$,

(ii) $d_{2j} = d_{2j-1}$, for $j = 1,\ldots,n$.

In fact, in these cases, there is a self-complementary graph G with degree sequence π such that there is an isomorphism σ of G onto $\overline{G}$, the complement of G such that the permutation σ is the product of disjoint cycles of length 4 together with a fixed point in the case when $p = 4n + 1$.

A simple set of conditions to determine whether or not a sequence satisfying (i) and (ii) in Theorem 1.9 is graphic is given by Clapham [63].

A characterization of forcibly self-complementary sequences was given by Rao [189] by using the interchange technique and certain new invariants of self-complementary graphs. In fact this is the only nontrivial property P for which both potentially P-graphic and forcibly P-graphic sequences have been characterized.

<u>Theorem 1.10</u>. Let $\pi = (b_1^{n_1},\ldots,b_m^{n_m})$ where $n_i > 0$, $\sum_{i=1}^{m} x_i = b = 4N$ and $b_1 > \ldots > b_m$, be a sequence where $b_i^{n_i}$ means that b_i occurs exactly n_i times in π. Then π is forcibly self-complementary-graphic if and only if

(i) m is even, $m = 2k$, say,

and for every i, $1 \leq i \leq k$, the following conditions are satisfied

(ii) $n_i = 2$ or 4

(iii) $n_i = n_{m+1-i}$,

(iv) $b_i = 4N - 1 - b_{m+1-i}$; and

(v) $b_i = 4N - 1 - \frac{n_i}{2} - \sum_{j=1}^{i-1} n_i$

with the convention that if $i = 1$, then the summation on the right hand side is zero.

The corresponding result in the case $p = 4N + 1$, is also given in Rao [187]. The characterizations of potentially self-complementary bipartite sequences and forcibly self-complementary bipartite sequences were given in Gangopadhyay [86,87].

By using the laying off technique and the fact that line-graphicness and planarity are hereditary, characterizations of forcibly line-graphic (resp. planar) sequences were given in Rao [185,188] and that of forcibly outer-planar graphic sequences was given in Choudum [51]. Recently the present author Rao [193] developed a theory of forcibly P-graphic sequences, where P is any hereditary invariant property of graphs.

Theorem 1.11. (Rao [185]) Let $\pi = (d_1,\ldots,d_p)$.

(a) If $d_p \geq 3$, then π is forcibly line-graphic if and only if
$$\pi = 4,3^4;\ \pi = 4^6 \quad \text{or} \quad \pi = (p-1)^p.$$

(b) If π has even sum and $d_p \leq 2$; n_i is the number of terms in π equal to i, i = 1,2; and $k = p - n_1 - n_2$; and if $k \geq 4$, then π is forcibly line-graphic if and only if

(1) Equality holds in the Erdös-Gallai condition (1.1) for $r = k$,

(2) $d_1 = k$ and $2n_2 + n_1 \leq k$.

(c) If π is as in (b) with $k \leq 3$, then π is forcibly line-graphic if and only if π is one of the following 8 sequences $(4,3^2,2^2)$; $(4^3,2^3)$; $(3^3,2,1)$; $(3^3,1^3)$, $(3^2,2^2)$; $(3^2,2,1^2)$; $(3,2^2,1)$ and $(4,2^4)$. A graphic sequence $(d_1,\ldots,d_p)$ is said to be a Type 1 sequence if $d_5 \geq 4$ and a Type 2 sequence if it is not a Type 1 sequence but $d_6 \geq 3$.

Theorem 1.12. (Rao [188]) If π is a graphic sequence, $\pi = (d_1,\ldots,d_p)$ with $p \geq 6$, then π is forcibly planar if and only if π satisfies one of the following conditions:

(1) π is neither a Type 1 nor a Type 2 sequence.

(2) $\pi = 4^6$ or $(5^4,3^2,2)$ or $(6,5^3,3^3)$ or $(6^4,3^4)$.

(3) $\pi = (p-1,5^3,3^3,1^{p-7})$.

(4) $d_6 = d_5 = 3$, $d_3 = 4$ and $d_1 + d_2 + n_1 = 2p - 2$
where n_1 is the number of terms of π which are equal to 1.

(5) $d_6 = d_3 = 3$, $d_2 \geq 4$; and $d_1 = p - 1$ or $d_1 + d_2 + n_1 = 2p - 2$ or $2p - 3$.

(6) $d_6 = d_2 = 3$; and $d_1 = p - 2$ or $p - 1$.

Using the method of construction and the fact that the induced subgraph on the maximum degree vertices of a total graph is again a total graph, the forcibly total graphic sequences were determined in Rao [190].

Theorem 1.13. A sequence π of length p is forcibly total-graphic if and only if $\pi = 2^3$ or 4^6 or $\pi = 2n, (n+1)^n, 2^n$ where $p = 2n + 1$.

In one of the interesting papers on forcibly P-graphic sequences, Bondy and Chvátal [20] presented a unified approach to a wide variety of problems concerning forcibly P-graphic sequences for several invariant properties P of graphs by the method of k-closure of a simple graph G of order n. Let k be a nonnegative integer. Then P is said to be k-stable if whenever $G - uv$ has property P and $d_G(u) + d_G(v) \geq k$, then G itself has the property P. For example the property of containing a Hamiltonian cycle is n-stable and G is Hamiltonian if $C_n(G)$ is. (The k-closure $C_k(G)$ of a simple graph is the graph obtained from G by recursively joining pairs of nonadjacent vertices with degree sum at least k.) This condition for a graph to be Hamiltonian is shown to imply well known conditions of Chvátal [58] and Las Vergnas [139] which generalize earlier results due to Dirac [67], Pósa [178] and Bondy [13].

<u>Theorem 1.14</u>.(Chvátal [58]) If a graphic $\pi = (d_1,\ldots,d_p)$ with $p \geq 3$, $d_1 \leq \ldots \leq d_p$ satisfies the condition, for every k, such that $d_k \leq k < p/2$ implies that $d_{p-k} \geq p - k$, then π is forcibly Hamiltonian.

The above theorem does not characterize the set of all forcibly Hamiltonian-graphic sequences and as shown by Nash-Williams [164] (see also Jackson [108]) the sequence $\pi = (2n)^{4n+1}$ is forcibly Hamiltonian but fails to satisfy the condition in the above theorem. However Theorem 1.14 is the best possible result in the following sense. A sequence $\pi' = (d_1',\ldots,d_p')$ is said to majorize $\pi = (d_1,\ldots,d_p)$ if $d_i' \geq d_i$ for all i, $1 \leq i \leq p$. If a graphic sequence fails to satisfy the condition of Theorem 1.14 for some $k < p/2$, then π is majorized by the sequence $(K^k, (p-k-1)^{p-2k}, (p-1)^k)$ which has, in fact, a unique realization and this realization is non-Hamiltonian. Schmeichel and Hakimi [99] proved the following interesting theorem.

<u>Theorem 1.15</u>. If π is a graphic sequence as in Theorem 1.14, and if G is a realization of π, then G is bipartite and has all even cycles or G is pan-cyclic.

For several interesting results which are in the same vein as Theorem 1.14, the reader is referred to Bondy and Chvátal [20] and the references contained therein and also to the expository article of Hakimi and Schmeichel [102] on degree sequences.

It is shown by the present author in [193] that, in principle, for any hereditary property P the set of all forcibly P-graphic sequences can be characterized. To this end we need a definition : If π_1, π_2 are two graphic sequences, then $\pi_1 \leq \pi_2$, in words π_2 captures π_1, if π_2 has a realization having a realization of π_1 as an induced subgraph. By using Theorem 1.2 a characterization of π_2 capturing π_1 on a given set of vertices is presented. Further, if $\underline{\underline{A}}(P)$ is the set of all degree sequences of minimal forbidden graphs for the hereditary property P (which exist since P is hereditary) and $\underline{\underline{A}}_o(P)$ is the minimal elements under the partial order $\leq$ on $\underline{\underline{A}}(P)$ then a graphic sequence π is forcibly P-graphic if and only if for no π_1 in $\underline{\underline{A}}_o(P)$, π captures π_1. It is conjectured that for any hereditary property P, the set $\underline{\underline{A}}_o(P)$ of minimal elements under the partial order $\leq$ on $\underline{\underline{A}}(P)$, is finite. In Rao [193] it is shown that $\underline{\underline{A}}_o(P)$ is finite, verifying the above conjecture, for several hereditary properties which include 1) P = perfect graphic, 2) P = planarity, 3) P = chromatic number $\leq$ k, where k is a fixed positive integer. Further, it is shown, that if $\underline{\underline{B}}$ is any set of graphic sequences such that for any two distinct sequences π_1, π_2 in $\underline{\underline{B}}$ neither $\pi_1 \leq \pi_2$ nor $\pi_2 \leq \pi_1$; such a $\underline{\underline{B}}$ is called an independent set, then there exists an invariant property P such that $\underline{\underline{A}}_o(P) = \underline{\underline{B}}$. Thus the above conjecture may be restated as : If $\underline{\underline{B}}$ is any independent set of graphic sequences, then $\underline{\underline{B}}$ is finite. The details of all these results are presented in Rao [193].

2. POTENTIALLY P-DIGRAPHIC AND FORCIBLY P-DIGRAPHIC DEGREE-PAIR SEQUENCES

Let G be a directed graph with $V(G) = \{u_1,\ldots,u_p\}$ and $d_i^+ = d_G^+(u_i)$ be the number of arcs of G with initial vertex u_i called the out-degree of u_i, and $d_i^- = d_G^-(u_i)$ be the number of arcs of G with terminal vertex u_i, called the in-degree of u_i. The sequence of ordered pairs

$$\pi(G) = ((d_1^+,d_1^-),\ldots,(d_p^+,d_p^-))$$

is called the <u>degree-pair</u> sequence of G. A sequence π of ordered pairs is said to be digraphic if there is a digraph G with $V(G) = \{u_1,\ldots,u_p\}$ such that $\pi(G) = \pi$, and G is referred to as a realization of π. Let P be an invariant property of digraphs then potentially P-digraphic and forcibly P-digraphic sequences are defined analogous to the ones in Section 1. Excellent references for the available results on potentially P-digraphic sequences are Berge [8], Chen [47] and Thomassen [203]. The existence of a digraph with given degree-pair sequence $\pi = ((d_1^+,d_1^-),\ldots,(d_p^+,d_p^-))$

is equivalent to the existence of a $p \times p$ matrix with entries zeros and ones and trace 0 and the ith row sum is d_i^+ and i^{th} column sum is d_i^-; whereas the existence of a multidigraph with degree-pair sequence π is equivalent to the existence of a $p \times p$ matrix with nonnegative integers satisfying the above conditions. Using the method of flows in networks [82], supply demand theorem, the characterization of digraphic degree-pair sequence are presented in Berge [8] and Chen [44,45,46,47]. The characterization of unimultidigraphic sequences is given only recently in Rao [192], which actually determines all such sequences; whereas Johnson [115] determines the uni-di-tree sequences. Potentially strongly connected digraphic sequences were characterized Beineke-Harary [6]. Potentially self-complementary and potentially self-converse digraphic sequences were characterized by Das [67]. For a characterization of potentially k-factorable digraphic sequences the reader is referred to Kundu [134] and Kleitman and Wang [124]. All the known results on forcibly hamiltonian-digraphic sequences can be had from Thomassen [204]. In this connection we mention the following interesting result of Meyniel [156] (Bondy and Thomassen [21]) which generalizes a theorem of Ghouila-Houri [88]).

Theorem 2.1. If G is a strongly connected digraph of order p and for any two non-adjacent vertices u,v of G we have $d_G(u) + d_G(v) \geq 2p - 1$, where $d_G(u) = d_G^+(u) + d_G^-(u)$; then G has an hamiltonian di-circuit.

Nash-Williams [160] asks the following interesting question : If G is a directed graph such that its out-degree sequence and in-degree sequence (arranged in nondecreasing way) both satisfy Pósa's condition [178] (Herary [103], Berge [8]), must G contain an hamiltonian di-circuit? For all the references on these and other topics on potentially and forcibly P-digraphic sequences, the reader is referred to the solved and unsolved problems sections of this paper.

3. RELATED SEQUENCES

It seems reasonable to associate a 'degree' to each edge of a graph and develop a new theory of degree sequences. Such a theory was initiated in Patrinos and Hakimi [176], who associated an unordered pair of integers with each edge of G of order p

and size q, representing the degrees of its end vertices. This q-sequence is referred to as the integer-pair sequence of the graph G, and such graphic sequences were characterized in [176]. A characterization of unimultigraphic integer-pair sequences was given in Rao and Taneja [190a]. Potentially connected graphic integer-pair sequences were characterized in Achuthan [165] (see 16 of the solved problem section and 15 of the unsolved problem section). It may be remarked that for graphs G without isolated vertices, the integer-pair sequence determines its degree sequence.

The second related sequence is the frequency partition of a graph G defined as follows : If π is the degree sequence of G then write $\pi = (b_1^{n_1},\dots,b_m^{n_m})$ where $b_i \neq b_j$ whenever $1 \leq i \neq j \leq m$, n_i is a positive integer $\sum_{i=1}^{m} n_i = p$ and the symbol $b_i^{n_i}$ means that the degree b_i occurs exactly n_i times in π. The sequence $(n_1,\dots,n_m)$ is called the frequency partition of G. A sequence of integers $(n_1,\dots,n_m)$ is <u>graphic</u> if there exists a graphic sequence π with frequency partition $(n_1,\dots,n_m)$. Graphic frequency partitions were determined by Chinn [48]. Rao, Vhat-Nayak and Naik [11,194] characterized potentially hamiltonian and potentially Eulerian frequency partitions; whereas M. Rao [152] and Reid [196] characterized potentially tree-graphic frequency partitions. For related results on undirected and directed graphs the reader is referred to the references in 15 of the solved problems section and 16 of the unsolved problems section.

4. SOLVED PROBLEMS

1. GRAPHIC SEQUENCES (DIRECTED, UNDIRECTED, MULTIGRAPHS AND MULTIDIGRAPHS) : Ayoub and Frisch [2], Brownlee [34], Chen [44,45,46,47], Chungahaisan [52], Dewdney [68], Eggleton [71], Eggleton and Holton [72,73], Erdős and Gallai [76], Ford and Fulkerson [82], Fulkerson [83], Fulkerson, Hoffman and McAndrew [84], Hakimi [96,97], Havel[106], Johnson [115,116], Kleitman [122], Kleitman and Wang [124], Koren [130], Ore [166,167], Owen and Trent [171], Reid [195], Ryser [197,198], Senior [200], Sridharan and Parthasarathy [201].

2. UNIGRAPHIC AND UNIDIGRAPHIC SEQUENCES : Hakimi [96], Johnson [113,114,115], Koren [126,127,128], Parthasarathy[175], Rao [192],Suho-yen R. Li [202].

3. POTENTIALLY k-EDGE-CONNECTED SEQUENCES : Edmons [70], Wang [213], Wang and Kleitman [214].

4. POTENTIALLY k-VERTEX-CONNECTED SEQUENCES : Hakimi [98], Rao and Rao [183], Wang and Kleitman [211].

5. POTENTIALLY k-FACTORABLE SEQUENCES (DIRECTED, UNDIRECTED) : Kleitman and Wang [124], Kundu [134,135,136], Lovász [151], Rao [186] (2-factorable self-complementary sequences).

6. POTENTIALLY HAMILTONIAN, POTENTIALLY CONNECTED k-FACTORABLE SEQUENCES : Rao [184], Rao [187] (hamiltonian self-complementary sequences).

7. POTENTIALLY STRONGLY-CONNECTED DIGRAPHIC SEQUENCES : Beineke and Harary [6].

8. POTENTIALLY SELF-COMPLEMENTARY, SELF-CONVERSE SEQUENCES (DIRECTED, UNDIRECTED) : Clapham and Kleitman [62], Clapham [63], Das [67], Gangopadhyay [86] (self-complementary bipartite sequences).

9. FORCIBLY SELF-COMPLEMENTARY SEQUENCES : Rao [189], Gangopadhyay [87] (self-complementary bipartite sequences).

10. FORCIBLY LINE-GRAPHIC SEQUENCES : Rao [185,193].

11. FORCIBLY PLANAR-GRAPHIC SEQUENCES : Farell [80], Rao [188,193].

12. FORCIBLY TOTAL-GRAPHIC SEQUENCES : Rao [190].

13. FORCIBLY HAMILTONIAN (2-FACTORABLE) SELF-COMPLEMENTARY SEQUENCES : Rao [186,187].

14. POTENTIALLY TOURNAMENT-SEQUENCES : Chang and Sharp [38], Eplett[74], Harary and Moser [104], Kleitman [122], Lendau [138], Moon and Moser [158].

15. FREQUENCY PARTITIONS AND RELATED SEQUENCES : Alspach and Reid [1], Behzad and Chartrand [4], Bhat-Nayak [10], Bhat-Nayak, Naik and Rao [11], Johnson [117], Kapoor, Polimeni and Wall [121], Lesniak, Polimeni and Roberts [146], Rao, Bhat-Nayak and Naik [194], M. Rao [152], Rein [196].

16. INTEGER PAIR-SEQUENCES OF GRAPHS AND DIGRAPHS : Achuthan [165], Patrinos and Hakimi [176], Rao and Taneja [190a].

5. UNSOLVED PROBLEMS

1. POTENTIALLY BIPARTITE SEQUENCES.

2. POTENTIALLY PAN-CYCLIC SEQUENCES.

3. POTENTIALLY HAMILTONIAN-DIGRAPHIC SEQUENCES.

4. POTENTIALLY LINE-GRAPHIC SEQUENCES.

5. POTENTIALLY TOTAL GRAPHIC SEQUENCES.

6. POTENTIALLY k-CONNECTED, k-VERTEX-CONNECTED DIGRAPHIC SEQUENCES.

7. POTENTIALLY PLANAR-GRAPHIC SEQUENCES AND RELATED PROBLEMS : For partial results refer to : Bottger and Harders [24], Bowen [21], Choudum [49,50], Chvátal [57], Cook [65,66], Etourneau [79], Fisher [81], Grunbaum [91,94], Grünbaum [93], Hakimi-Schmeichel [100], Hawkins, Hill, Reeve and Tyrel [107], Tendrol [110,111,111a,112], Jucovic[118, 119], Malkevitch [154,155], Owens [173,174], Pelikan [177], Schmeichel [199],Zaks [217].

8. FORCIBLY HAMILTONIAN SEQUENCES AND GENERALIZATIONS (GRAPHS AND DIGRAPHS) : For partial results refer to : Bondy [13,14,15,16,18,19,22,23], Bondy-Chvátal [20], Bondy-Thomassen [21], Bermond [30,31,32], Bollobás and Hobbs [33], Chartrand, Kapoor and Lick [43], Chvâtal [58,59,60,61], Dirac [69], Erdös [77], Erdös and Hobbs [78], Ghouila-Houri [88], Grotschel [90], Haggavist and Thomassen [95], Hakimi and Schmeichel [99,102], Jackson [108,109], Jung [120], Kronk [132,133], Las Vergnas [140,141], Lowin [143], Lesniak [145], Lick [147], Meyniel [156], Moon and Moser [157], Nash-Williams [159, 160,161,162,163,163a,164], Ore [168,169], Overbeck-Lavisch [170], Owens [174], Pósa [178], Thomassen [203,204], Woodall [215,216].

9. FORCIBLY k-CONNECTED SEQUENCES (k-EDGE CONNECTED, k-VERTEX CONNECTED OF GRAPHS, DIGRAPHS) (Partial results) : Bondy and Chvátal [20], Bollobás [35], Chartrand and Harary [40], Boesch [25], Chartrand, Kapoor, Kronk [41,42], Goldsmith and White [89], Grünbaum [94].

10. FORCIBLY k-FACTORABLE SEQUENCES (GRAPHS, DIGRAPHS).

11. UNIDIGRAPHIC SEQUENCES.

12. POTENTIALLY ORIENTED DIGRAPHIC SEQUENCES.

13. FORCIBLY SELF-COMPLEMENTARY DIGRAPHIC SEQUENCES.

14. FORCIBLY k_3-SEQUENCES.

15. POTENTIALLY P-GRAPHIC INTEGER-PAIR SEQUENCES, AND FORCIBLY P-GRAPHIC INTEGER-PAIR SEQUENCE FOR PROPERTIES P NOT COVERED IN 17 OF SOLVED PROBLEMS.

16. POTENTIALLY P-GRAPHIC FREQUENCY PARTITIONS AND FORCIBLY P-GRAPHIC FREQUENCY PARTITIONS FOR PROPERTIES P NOT COVERED IN 15 OF SOLVED PROBLEMS.

ACKNOWLEDGEMENT

Part of this work was done while the author was visiting the School of Mathematical Sciences, Madurai-Kamaraj University, Madurai during November - December 1978. I thank Professor K. Nagarajan for providing me this opportunity.

REFERENCES

1. B. Alspach and K.B. Reid, Degree frequencies in digraphs and tournaments, J.Graph Theory, 2(1978), 241-249.

2. J.N. Ayoub and I.T. Frisch, Degree realization of undirected graphs in reduced form, J. Franklin Inst., 289(1970), 303-312.

3. M. Bankfalvi and Zs. Bankfalvi, Alternating hamiltonian circuits in two-coloured complete graphs, Theory of graphs (P. Erdös et al. Eds.), Tihany Conference, 1968, 11-18.

4. M. Behzad and G. Chartrand, No graph is perfect, Amer. Math. Monthly, 74(1967), 962-963.

5. M. Behzad and J.E. Simpson, Eccentric sequences and eccentric sets in graphs, Discrete Mathematics, 16(1976), 187-194.

6. L.W. Beineke and F. Harary, Local restrictions for various classes of directed graphs, J. London Math. Soc., 40(1965), 87-95.

7. L.W. Beineke and E.F. Schmeichel, On degrees and cycles in graphs (to appear).

8. C. Berge, Graphs and Hypergraphs, North-Holland Publishing Co., Amsterdam,1973.

9. C. Berge and M.L. Vergnas, On the existence of subgraphs with degree constraints, Proc. Kon. Aka. Amersta., 81(2), 1978.

10. V.N. Bhat-Nayak, Characterization of 3-perfect graphic sequences, Proc. Indian National Academy, 41(1975), 228-244.

11. V.N. Bhat-Nayak, R.N. Naik and S.B. Rao, Frequency partitions. Forcibly pancyclic and Forcibly non-hamiltonian degree sequences, Discrete Maths., 20(1977), 93-102.

12. V.N. Bhat-Nayak and R.N. Naik, Forcibly 2-variegated sequences (to appear).

13. J.A. Bondy, Properties of graphs with constraints on degrees, Studia Sci. Math. Hungariea, 4(1969), 473-475.

14. J.A. Bondy, Cycles in graphs, Combinatorial structures and their applications (R.K. Guy et al Eds.), Gordon and Breach, New York, 1970, 15-18.

15. J.A. Bondy, Variations on the hamiltonian theme, Canad. Math. Bull., 15(1972), 57-62.

16. J.A. Bondy, Pancyclic graphs, J. Combinatorial Theoru, 11(1971), 80-84.

17. J.A. Bondy, Pancyclic graphs-II, Proc. Louisiana Conference, Utilitas Math. Publ. Inc., Winnipeq, 1971, 167-172.

18. J.A. Bondy, Large cycles in graphs, Proc. Louisiana Conference, Utilitas Math. Publ. Inc., Winnipeg, 1972, 583-590.

19. J.A. Bondy, Large cycles in graphs, Discrete Math., 1(1971), 121-132.

20. J.A. Bondy and V. Chvátal, A method in graph theory, Discrete Math., 15(1976), 111-135.

21. J.A. Bondy and C. Thomassen, A short proof of Meyneil's theorem, Discrete Math., 19(1977), 195-197.

22. J.A. Bondy, A remark on two sufficient conditions for hamiltonian cycles, Discrete Math., 22(1978), 191-193.

23. J.A. Bondy, Cycles in digraphs, Proc. South-Eastern Conference on Combinatorics, Graph Theory and Computing, Utilitas Math. Publ. Inc., 1977, 91-98.

24. G. Bottger and H. Harders, Note on a problem of S.L. Hakimi concerning planar graphs without parallel elements, SIAM J. Applied Maths., 12(1964), 838-839.

25. F.T. Boesch, The strongest monotone degree condition for n-connectedness of a graph, J. Combinatorial Theory, 168(1974), 162-165.

26. F.T. Boesch and F. Harary, Unicyclic realizations of a degree list, Networks, 8(1978), 93-96.

27. R. Bowen, On sums of valencies in planar graphs, Canad. Math. Bull., 9(1966), 111-114.

28. E.A. Bouder and E.R. Canfield, The asymptotic number of labelled graphs with given degree sequences, J. Combinatorial Theory, 24A(1978), 296-307.

29. D.H. Bent and T.V. Narayana, Computation of the number of score sequences in round-robin tournaments, Canad. Math. Bull.

30. J.C. Bermond, On hamiltonian walks, Proc. 5th British Comb. Conference (Nash-Williams et al Eds.), Utilitas Math., Publ. Inc., Winnipeg, 1975, 41-51.

31. J.C. Bermond, Thesis, University of Paris, XI, Orsay, 1975.

32. J.C. Bermond, Hamiltonian graphs, Selected topics in Graph Theory (L.W. Beineke et al Eds.), Academic Press, New York, (1978), 127-167.

33. B. Bollobás and A.M. Hobbs, Hamiltonian cycles in regular graphs, Annals of Discrete Maths., 3(1978), 43-48.

34. A. Brownlee, Directed graph realization of degree-pairs, Amer. Math. Monthly, 75(1968), 36-38.

35. B. Bollobás, On graphs with equal edge-connectivity and minimum degree, Discrete Maths., 28(1979), 321-323.

36. R.A. Brauldi, A note on degree sequences of graphs. Canad. Math. Bull., (1980) (to appear).

37. P.A. Catlin, Graph decompositions satisfying extremal degree constraints, J. Graph Theory, 2(1978), 165-176.

38. Chang M. Bang and Henry Sharp Jr., Score vectors of tournaments, J. Combinatorial Theory, 26B(1979), 81-84.

39. P.D. Chwathe, 2-2 perfect graphical degree sequences, (this proceedings).

40. G. Chartrand and F. Harary Graphs with prescribed connectivities, Theory of graphs, The Tihany Conference (P. Erdös et al, Eds.), Akad. Kiado., Budapest, (1968), 61-63.

41. G. Chartrand, S.F. Kapoor and H.V. Kronk, A sufficient condition for n-connectedness of graphs, Mathematika, 15(1968), 51-52.

42. G. Chartrand, S.F. Kapoor and H.V. Kronk, A generalization of hamiltonian connected graphs, J. Math. Pures et Appl., 48(1969), 109-116.

43. G. Chartrand, S.F. Kapoor and D.R. Lick, n-Hamiltonian graphs, J. Combinatorial Theory, 9(1970), 308-312.

44. W.K. Chen, On the realization of a (p,s) digraph with prescribed degrees, J. Franklin Institute, 281(1966), 406-422.

45. W.K. Chen, On d-invariant transformations of (p,s)-digraphs, J. Franklin Institute, 291(1971), 89-100.

46. W.K. Chen, On equivalence of realizability conditions of a degree sequence, IEEE Trans-Circuit Theory, CT-20(1973), 260-262.

47. W.K. Chen, Applied Graph Theory, North-Holland, Amsterdam, 13(1976).

48. P.Z. Chinn, The frequency partition of a graph, Recent trends in Graph Theory, Lecture Notes in Maths., Spring-Verlag, Berlin, 186(1971), 69-70.

49. G.A. Choudum, Some 4-valent, 3-connected, planar almost pancyclic graphs, Discrete Maths., 18(1977), 125-129.

50. S.A. Choudum, Existence of a family of planar almost pancyclic graphs, Proc. Symp. on Graph Theory (A.R. Rao, Ed.), ISI Lecture Notes series 4, Mac-Millan & Co. India Ltd., 1979, 151-161.

51. S.A. Choudum, Characterization of forcibly outer-planar degree sequences (this proceedings).

52. V. Chungphaisan, Conditions for sequences to be r-graphic, Discrete Maths., 7(1974), 31-38.

53. V. Chungphaisan, Sequences realisable by graphs with hamiltonian squares, Canad. Math. Bull., 17(1975), 629-631.

54. V. Chungphaisan, Construction of hamiltonian graphs and bigraphs with prescribed degrees, J. Combinatorial Theory, 1978.

55. V. Chungphaisan, Factors of graphs and degree sequences, Nanta. Math.,

56. V. Chungphaisan, Construction of hamiltonian graphs with prescribed degrees, Proc. 6 South-Eastern Conference on Combinatorics, Graph Theory and Computing, 1975, 161-167.

57. V. Chvátal, Planarity of graphs with given degrees of vertices, Nieuw. Arct. Wisk., XVII(1969), 47-60.

58. V. Chvátal, On hamiltonian ideals, J. Combinatorial Theory, 12B(1972), 163-168.

59. V. Chvátal, Tough graphs and hamiltonian circuits, Discrete Maths., 5(1973), 215-228.

60. V. Chvátal, New directions in hamiltonian graph theory, New directions in the theory of graphs (F. Harary Ed.), Academic Press, New York, 1973.

61. V. Chvátal and P. Erdös, A note on hamiltonian circuits, Discrete Maths., 2(1972), 111-113.

62. C.R.J. Clapham and D.J. Kleitman, The degree sequences of self-complementary graphs, J. Combinatorial Theory, 20B(1976), 67-74.

63. C.R.J. Clapham, Potentially self-complementary sequences, J. Combinatorial Theory, 20B(1976), 75-79.

64. C.R.J. Clapham, Hamiltonian arcs in self-complementary graphs, Discrete Maths., 8(1974), 251-255.

65. R.J. Cook, Vertex degrees of graphs on orientable surfaces, Utilitas Mathematica, 15(1979), 281-290.

66. R.J. Cook, Vertex degrees of planar graphs, J. Combinatorial Theory, 26B(1979), 337-345.

67. P. Das, Characterization of potentially self-complementary, self-converse degree-pair sequences for digraphs (this proceedings).

68. A.K. Dewdney, Degree sequences in complexes and hypergraphs, Proc. Amer. Math., Soc., 53(1975), 535-540.

69. G.A. Dirac, Some theorems on abstract graphs, Proc. Lond. Math. Soc., 2(1952), 69-81.

70. J. Edmonds, Existence of k-edge connected ordinary graphs with prescribed degrees, J. Res. Nat. Bur. Stand., 68B(1964), 73-74.

71. R. Eqqleton, Graphic sequences and graphic polynomials a report : Infinite and finite sets, Coll. Math. Soc. J. Bolyai, 10(1975), 385-392, North Holland Publ. Inc., Amsterdam.

72. R.B. Eggleton and D.A. Holton, Graphic sequences, Proc. Australian Combinatorial Mathematics, VI, Springer-Verlag, Lecture Notes in Mathematics, 748(1979),1-10.

73. R.B. Eggleton and D.A. Holton, The graph of type (o,∞) realization of a graphic sequence, Ibid., 41-54.

74. W.J.R. Eplett, Self-converse tournaments, Canad. Math. Bull., 22(1979), 23-27.

75. P. Erdös and T. Gallai, On maximal paths and circuits of graphs, Acta. Math. Sci. Hung., 10(1959), 337-356.

76. P. Erdös and T. Gallai, Graphs with given degrees of vertices (Hungarian), Mat. Lapok, 11(1960), 264-274.

77. P. Erdös, Remarks on a paper of Pósa, Publ. Math. Inst. Hung. Aca. Sci.,7(1962), 227-228.

78. P. Erdös and A.M. Hobbs, A class of hamiltonian regular graphs (to appear).

79. E. Etourneau, Existence and connectivity of planar graphs having 12 vertices of degree 5 and n-12 vertices of degree 6, Infinite and finite sets, (A. Hajnal et al, Ed.), North-Holland Publ. Inc., Amsterdam, 10(1975).

80. E.J. Farell, On Graphical partitions and planarity, Discrete Maths., 18(1977), 149-153.

81. J.C. Fisher, An existence theorem for simple convex polyhedron, Discrete Maths., 7(1974), 75-97.

82. L.R. Ford and D.R. Fulkerson, Flows in Networks, Princeton Univ. Press, Princeton 1962.

83. D.R. Fulkerson, Zero-one matrices with zero trace, Pacific J. Math., 10(1960), 831-836.

84. D.R. Fulkerson, A.J. Hoffman and M.H. McAndrew, Some properties of graphs with multiple edges, Canad. J. Math., 17(1965), 166-177.

85. D.A. Gale, A theorem on flows in networks, Pacific J. Maths., 7(1957), 1073-1082.

86. T. Gangopadhyay, Characterization of potentially bipartite self-complementary pair-sequences, Discrete Maths., (submitted for publication).

87. T. Gangopadhyay, Characterization of forcibly bipartite self-complementary pair-sequences (this proceedings).

88. A. Ghouila-Houri, Une condition suffisante existence d'un circuit hamiltonien, C. R. Acad. Sci., 251(1960), 494-497.

89. D.L. Goldsmith and A.T. White, On graphs with equal edge connectivity and minimum degree, Discrete Math., 23(1978), 31-36.

90. A.M. Grotschel, Graphs with cycles containing given paths, Annals of Discrete Maths., Studies in integer-programming (P.L. Hammer et al, Eds.), 1(1977), 233-245.

91. B. Grünbaum, Convex Polytopes, Wiley, New York, 1967.

92. B. Grünbaum and T. Motzkin, The number of hexagons and the simplicity of geodesics on certain polyhedra, Canad. J. Maths., 15(1963), 744-751.

93. B. Grünbaum, Problem session : International Conference on Comb. Structures and their Applications (R.K. Guy et al, Eds.), Gordon and Breach, New York, (1970), 491-492, Problems 1.1, 2.1 and 2.2.

94. B. Grünbaum, Polytopal graphs; Studies in Graph Theory, Part II, Studies in Math., 12, Math. Assoc. of America, Washington,(1975), 201-224.

95. R. Haggavist and C. Thomassen, On pan-cyclic digraphs, J. Comb. Theory, 20B(1976), 20-40.

96. S.L. Hakimi, On realizability of a set of integers and the degrees of the vertices of a linear graph - I,II, SIAM J. Appl. Math., 10(1962), 496-506 and 11(1963), 135-147.

97. S.L. Hakimi, On the degrees of the vertices of a directed graph, J. Franklin Inst., 279(1965), 290-308.

98. S.L. Hakimi, On the existence of graphs with prescribed degrees and connectivity, SIAM J. Appl. Math., 26(1974), 154-164.

99. S.L. Hakimi and E.F. Schmeichel, Pancyclic graphs and a conjecture of Bondy and Chvâtal, J. Combinatorial Theory, 17B(1974), 22-34.

100. S.L. Hakimi and E.F. Schmeichel, On planar graphical degree sequences, SIAM J. Appl. Math., 32(1977), 598-609.

101. S.L. Hakimi and E.F. Schmeichel, On the existence of a traceable graph with prescribed vertex degrees, Ars. Combinatoria, 4(1977), 69-80.

102. S.L. Hakimi and E.F. Schmeichel, Graphs and their degree sequences : A Survey. Theory and applications of graph theory (Y. Alavi et al, Eds.), Springer-Verlag, Lecture Notes in Mathematics, 642(1978), 225-235.

103. F. Harary, Graph Theory, Addison Weley, Reading, 1969.

104. F. Harary and L. Moser, The theory of round robin tournaments, Amer. Math. Monthly, 73(1966), 231-246.

105. F. Harary and G. Prins, Enumeration of locally restricted digraphs, Canad. J. Maths., 18(1966), 853-860.

106. V. Havel, A remark on the existence of finite graphs (Hungarian), Casopis P. Mat., 80(1955), 477-480.

107. A. Hawkins, A. Hill, J. Reeve and J. Tyre, On certain polyhedra, Math. Gaz., 50(1966), 140-144.

108. W.B. Jackson, Hamiltonian cycles in regular 2-connected graphs, J. Graph Theory.

109. W. B. Jackson, Edge disjoint hamiltonian cycles in regular graphs of large order, J. Lond. Math. Soc., GBR 19(1979), 13-16.

110. S. Jendrol, On the face-vectors and vertex vectors of maps, Coll. Math. Soc. Jan. Bol. Hungary 18-Combinatorics (P. Erdös et al, Eds.), 1976, 629-633.

111. S. Jendrol, On the toroidal analogue of Eberhards theorem, Proc. Lond. Math. Soc., 25(1972).

112. S. Jendrol, On the face vector of a simple map, Proc. Graph Theory Symp., Prague 1974 (Academia 1975), 311-314.

113. R.H. Johnson, Simple separable graphs, Pacific J. Math., 56(1975), 143-158.

114. R.H. Johnson, The diameter and radius of simple graphs, J. Combinatorial Theory, 17B(1974), 188-198.

115. R.H. Johnson, Simple directed trees, Discrete Maths., 14(1976), 257-264.

116. R.H. Johnson, A note on applying a theorem of Tutte to graphical sequences, J. Combinatorial Theory, 18B(1975), 42-45.

117. R.H. Johnson, Frequency partitions of trees, Proc. 8th South-Eastern Conference on Combinatorics, Graph Theory and Computing, Utilitas Maths. Publ., Winnipeg (1977), 419-422.

118. E. Jucovic, On polyhedrol realizability of certain sequences, Canad. Math. Bull., 12(1969), 31-39.

119. E. Jucovic, On the number of hexagons in a map, J. Combinatorial Theory, 10(1971), 232-236.

120. H.A. Jung, On maximal circuits in finite graphs, Annals of Discrete Maths., 3(1978) (B. Bollobás Ed.), 129-144.

121. S.F. Kapoor, A.D.Polimeni and C.E. Wall, Degree sets for graphs, Fund. Math., 95(1977), 189-194.

122. D.J. Kleitman, The number of tournament score sequences for a large number of players; Comb. Structure and their applications (R.K. Guy Ed.), Gordon and Breach, New York, 1967.

123. D.J. Kleitman, Minimal number of multiple edges in realization of an incidence sequence without loops, SIAM J. Appl. Maths., 18(1970), 25-28.

124. D.J. Kleitman and D.L. Wang, Algorithms for constructing graphs and digraphs with given valences and factors, Discrete Maths., 6(1973), 79-88.

125. D.J. Kleitman, M. Koren and S.Y.R.Li, On the existence of simultaneous edge disjoint realizations of degree sequences with few edges, SIAM J. Appl. Maths., 32(1977), 619-626.

126. M. Koren, Extreme degree sequences of simple graphs, J. Combinatorial Theory, 15B(1973), 213-224.

127. M. Koren, Realizations of a sum of sequences by a sum graph, Israel J. Math., 15(1973), 396-403.

128. M. Koren, Pairs of sequences with a unique realization, J. Combinatorial Theory, 21B(1976), 224-234.

129. M. Koren, Sequences with a unique realization by simple graphs, J. Combinatorial Theory, 21B(1976), 235-244.

130. M. Koren, Graphs with degrees from prescribed intervals, Discrete Maths., 15(1976), 253-261.

131. M. Koren, Edge disjoint realizations of two forest realizable sequences, SIAM J. Appl. Math.,

132. H.V. Kronk, Variations of a theorem Pósa; The many facets of graph theory (G. Chartrand et. al, Eds.) Lecture Notes in Mathematics, Springer-Verlag, 110(1969), 193-197.

133. H.V. Kronk, A note on k-path hamiltonian graphs, J. Combinatorial Theory, 7(1969), 104-106.

134. S. Kundu, The k-factor conjecture is true, Discrete Maths., 6(1973), 367-376.

135. S. Kundu, Factorization of graphs, Discrete Maths., 8(1974), 41-48.

136. S. Kundu, Generalizations of the k-factor theorem, Discrete Maths., 9(1974), 173-177.

137. S. Kundu, Disjoint representation of tree-realizable sequences, SIAM J. Appl. Math., 26(1974), 103-107.

138. H.G. Landau, On dominance relations and the structure of animal societies III, the condition for a score structure, Bull. Math. Biophys., 15(1955), 143-148.

139. M. Las Vergnas, Sur l'existence des cycles hamiltoniens dans un graphe, C. R. Acad. Sci. Paris, 270A(1970), 1361-1364.

140. M. Las Vergnas, Sur une properiete des arbres maximaux dans un graphe, C. R. Acad. Sci. Paris, 272(1971), 1297-1300.

141. M. Las Vergnas, Sur les arborescences dans un graphe oriente, Discrete Math., 15(1976), 27-39.

142. M. Las Vergnas, Degree constrained subgraphs and matroids, Infinite and finite sets, Coll. Math. Bolya Janos, 10(1975), 1473-1502.

143. M. Lewin, On maximal circuits in directed graphs, J. Combinatorial Theory, 18B(1975), 125-129.

144. L. Lesniak, Eccentric sequences in graphs, Periodica Math. Hungarica.

145. L. Lesniak, On n-hamiltonian graphs, Discrete Maths., 14(1976), 165-169.

146. L. Lesniak, A.D. Polimeni and J. Roberts, Asymmetric digraphs and degree sets, 7th South Eastern Conference on Combinatorics, Graph Theory and Computing, Utilitas Math. Publ. Inc., Winnipeg ( ), 421-432.

147. D.R. Lick, A sufficient condition for hamiltonian connectedness, J. Combinatorial Theory, 8(1970), 444-445.

148. L. Lovász, On decomposition of graphs, Studia Sci. Math. Hungarica, 1(1966), 237-238.

149. L. Lovász, On factorization of graphs, Calgary Conference (R. K. Guy Ed.), Gordon & Breach (1970), 243-246.

150. L. Lovász, Subgraphs with prescribed valencies, J. Combinatorial Theory, 8(1970), 391-416.

151. L. Lovász, Valencies of graphs with 1-factors, Periodica Math. Hungarica, 5(1974), 149-151.

152. T. Mahadeva Rao, Frequency sequences in graphs, J. Combinatorial Theory, 17B(1974), 19-21.

153. V.V. Menon, On the existence of trees with given degrees, Sankhyā, 26(1964), 63-68.

154. J. Malkevitch, Properties of planar graphs with uniform face structure, Mem. Amer. Math. Soc., 99(1970).

155. J. Malkevitch, On the lengths of cycles in planar graphs; Recent Trends in Graph Theory (M. Capabinaco et al, Eds.), Springer-Verlag, Lecture Notes in Mathematics, 186(1971).

156. M. Meyniel, Une condition suffisante d'existence d'un circuit hamiltonien dans un graph oriente, J. Combinatorial Theory, 14B(1973), 137-147.

157. J.W. Moon and L. Moser, On hamiltonian bipartite graphs, Israel, J. Math., 1(1963), 163-165.

158. J.W. Moon and L. Moser, An extension of Landau's theorem on tournaments, Pacific J. Maths., 13(1963), 1343-1345.

159. C.St.J.A. Nash-Williams, On hamiltonian circuits in finite graphs, Proc. Amer. Math. Soc., 17(1966), 466-467.

160. C.St.J.A. Nash-Williams, Hamiltonian circuits in graphs and digraphs, The many facets of graph theory (G. Chartrand et. al, Eds.), Lecture Notes in Mathematics, Springer-Verlag, 110(1969), 237-243.

161. C.St.J.A. Nash-Williams, Hamiltonian lines in graphs whose vertices have sufficiently large valencies, Proc. Int. Symp. on Combinatorics, Hungary 1969, North-Holland Publ. Company, Amsterdam, 1970, 813-819.

162. C.St.J.A. Nash-Williams, Edge disjoint hamiltonian graphs with vertices of large valencies, Studies in Pure Mathematics (L. Mirsky Ed.), Academic Press ,1971.

163. C.St.J.A. Nash-Williams, Hamiltonian arcs and circuits, Recent Trends in Graph Theory, Lecture Notes in Mathematics, Springer-Verlag, 186(1971), 197-209.

163a. C.St.J.A. Nash-Williams, Unexplored and semi-explored territories in graph theory, Proc. of the 1971 Ann Arbor Graph Theory Conference, Academic Press 1973.

164. C.St.J.A. Nash-Williams, Valency sequences which force graphs to have hamiltonian circuits - Interim report, University of Waterloo, Waterloo, Canada.

165. Nirmala Achuthan, Characterization of potentially connected integer-pair sequences (this proceedings).

166. O. Ore, Studies on directed graphs-I, Annals of Maths., 63(1956), 383-406.

167. O. Ore, Graphs and subgraohs, Trans. Amer. Math. Soc., 84(1957), 109-137.

168. O. Ore, A note on hamiltonian circuits, Amer. Math. Monthly, 67(1960), 55.

169. O. Ore, Hamiltonian connected graphs, J. De. Math. Pures et Appl., 42(1963), 21-27.

170. M. Overbeck-Larisch, Hamiltonian paths in oriented graphs, J. Combinatorial Theory, 21B(1976), 76-80.

171. A.B. Owens and H.M. Trent, On determining minimal singularities for the realization of an incidence matrix, SIAM J. Appl. Math., 15(1967), 406-418.

172. A.B. Owens, On determining the minimum number of multiple edges for an incidence sequences, SIAM J. Appl. Math., 18(1970), 238-240.

173. A.B. Owens, On the planarity of regular incidence sequences, J. Combinatorial Theory, 11B(1971), 201-212.

174. P.J. Owens, On regular graphs and hamiltonian circuits including answers to some questions of Joseph Zaks, Rocky J. Maths.

175. K.R. Parthasarathy, Enumeration of ordinary graphs with given partition, Canad. J. Math., 20(1968), 40-47.

176. A.N. Patrinos and S.L. Hakimi, Relations between graphs and integer-pair sequences, Discrete Math., 15(1976), 347-358.

177. J. Pelikan, Valency conditions for the existence of certain subgraphs, Tihany Conference (P. Erdös et.al, Eds.), (1968), 251-258.

178. L. Pósa, A theorem concerning hamiltonian lines, Magyar Tud. Akad. Mat. Kutato. Inst. Köln, 7(1962), 225-226.

179. A. Ramachandra Rao and S.B. Rao, On factorable degree sequences, J. Combinatorial Theory, 13B(1972), 185-191.

180. A. Ramachandra Rao, The clique number of a graph with a given degree sequence, Proc. Symposium on Graph Theory (A.R. Rao Ed.), MacMillan & Co. India Ltd., I.S.I. Lecture Notes Series 4(1979), 251-267.

181. A. Ramachandra Rao, Degree sequences of Cacti (this proceedings).

182. A. Ramachandra Rao, An Erdős-Gallai type result on the clique number of a realisation of a degree sequence (unpublished).

183. S.B. Rao and A. Ramachandra Rao, Existence of 3-connected graphs with prescribed degrees, Pacific J. Math., 33(1970), 203-207.

184. S.B. Rao, Contribution to the theory of directed and undirected graphs, Doctoral Thesis, Indian Statistical Institute, Calcutta, (1970).

185. S.B. Rao, Characterization of forcibly line-graphic degree sequences, Utilitas Mathematica, 11(1977), 357-366.

186. S.B. Rao, Characterization of self-complementary graphs with 2-factors, Discrete Mathematics, 17(1977), 225-233.

187. S.B. Rao, Solution of the Hamiltonian problem for self-complementary graphs, J. Combinatorial Theory, 28B(1979), 13-41.

188. S.B. Rao, Characterization of forcibly planar degree sequences, Ars Combinatoria (to appear).

189. S.B. Rao, Characterization of forcibly self-complementary degree sequences, Discrete Mathematics (submitted for publication).

190. S.B. Rao, Characterization of forcibly total graphic sequences, Jour. of Combinatorics, Information and Systems Sciences (submitted for publication).

190a. S.B. Rao and A. Taneja, Characterization of unipseudographic and unimultigraphic integer-pairs sequences, Discrete Mathematics (submitted for publication).

191. S.B. Rao and P. Das, Alternating Eulerian trails with prescribed degrees in two-edge colored complete graphs, Discrete Mathematics (submitted for publication).

192. S.B. Rao, Characterization of unimultidigraphic sequences, (under preparation).

193. S.B. Rao, Towards a theory of forcibly hereditary P-graphic degree sequences (this proceedings).

194. S.B. Rao, V.N. Bhat-Nayak and R.N. Naik, Characterization of frequency partitions of Eulerian graphs, Proc. Symposium on Graph Theory, (A. Ramachandra Rao Ed.), ISI Lecture Notes series 4, The MacMillan Company of India Ltd., 1979, 124-138.

195. K.B. Reid, Extension of graphical sequences, Utilitas Mathematica, 12(1977), 255-261.

196. K.B. Reid, Score sets for tournaments, Proc. 9th South-Eastern Conference, Utilitas Mathematica Publ. Inc., Winnipeg, (to appear).

197. H.J. Ryser, Matrices of zeros and ones, Bull. Amer. Math. Soc., 66(1960), 442-464.

198. H.J. Ryser, Combinatorial properties of matrices of zeros and ones, Canad. Math. Bull., 9(1957), 371-377.

199. E. Schmeichel, The cycle structure and planarity of graphs under degree constraints, Ph.D. Thesis, Northwestern Illinois Univ., Evanston, (1974).

200. J.K. Senior, Partitions and their representating graphs, Amer. J. Math., 73(1951), 663-689.

201. M.R. Sridharan and K.R. Parthasarathy, Enumeration of graphs and digraphs with local restrictions, J. Math. Phy. Sci., 11(1977), 483-490.

202. Suho-yen R. Li, Graphic sequences with unique realization, J. Combinatorial Theory, 19B(1975), 42-68.

203. C. Thomassen, An Ore-type condition implying a digraph to be pancyclic, Discrete Math., 19(1977), 85-92.

204. C. Thomassen, Long cycles with constraints on the degrees, surveys in combinatorics (B. Bollobás Ed.), Proc. 7th British Combinatorial Conference, London Math., Society Lecture Notes Series, 38(1979), 211-228.

205. W.T. Tutte, A short proof of the factor theorem for finite graphs, Canad. J. Math., 6(1954), 347-352.

206. W.T. Tutte, The factorization of linear graphs, J. London Math. Soc., 22(1947), 107-111.

207. W.T. Tutte, The factors of graphs, Canad. J. Maths., 4(1952), 314-328.

208. W.T. Tutte, The 1-factors of oriented graphs, Proc. Amer. Math. Soc., 4(1953), 922-937.

209. W.T. Tutte, Spanning subgraphs with specified valencies, Discrete Maths., 9(1974), 97-108.

210. W.T. Tutte, The subgraph problem, Annals of Discrete Maths., 3(1978), 289-295.

211. D.L. Wang and D.J. Kleitman, On the existence of n-connected graphs with prescribed degrees ($n \geq 2$), Net Works, 3(1973), 225-239.

212. D.L. Wang, On degree sequences of digraphs without 2-cycles, Proc. 7th South-Eastern Conference, Utilitas Math. Publ. Inc., (1976), 509-515.

213. D.L. Wang, Construction of maximally edge-connected graph with prescribed degrees, Stud. Appl. Math., 55(1976), 87-92.

214. D.L. Wang and D.J. Kleitman, A note on n-edge connectivity, SIAM J. Appl. Math., (1974).

215. D.R. Woodall, Sufficient conditions for circuits in graphs, Proc. Lond. Math. Soc., 24(1972), 739-755.

216. D.R. Woodall, A sufficient condition for Hamiltonian circuits, J. Combinatorial Theory, 25B(1978), 184-186.

217. J. Zaks, Recent results in graph theory, Proc. 7th South-Eastern Conf., Utilitas Math. Publ. Inc., (1976), 527-532.

TOWARDS A THEORY OF FORCIBLY HEREDITARY P-GRAPHIC SEQUENCES

S. B. RAO
Indian Statistical Institute
203 B. T. Road, Calcutta 700 035

ABSTRACT

Let P be a hereditary invariant property of graphs. In this paper it is shown that, in principle, the set of all forcibly P-graphic sequences can be characterized using a theorem of Fulkerson, Hoffman and McAndrew and a partial order $<<$ on any given set of graphic sequences A defined as: for two elements π_1, π_2 say that $\pi_1 << \pi_2$, if there exists a graph G with degree sequence π_2 having a graph H with degree sequence π_1 as an induced subgraph. Let P be a hereditary invariant property of graphs and $\underline{\underline{A}}(P)$ be the set of all degree sequence of the minimal forbidden graphs for the property P, and $M(P)$ be the set of all minimal elements in $\underline{\underline{A}}(P)$ under the partial order $<<$ defined above. Conjecture: For any hereditary property P, the set $M(P)$ is finite. We verify this conjecture for several hereditary properties P which include (1) P = Perfect-graphic, (2) P = Chromatic number of the graph $\leq k$, where k is a fixed positive integer; and (3) P = Planarity.

1. INTRODUCTION AND DEFINITIONS

All graphs considered in this paper are finite and have neither loops nor multiple edges. The degree sequence of a graph G will be denoted by $\pi(G)$. A sequence $\pi = (d_1,\ldots,d_p)$ of length p is said to be <u>graphic</u> if there exists a graph G with $V(G) = \{u_1,\ldots,u_p\}$ such that $d_G(u_i)$, the degree of u_i in G, is equal to d_i, for every i, $1 \leq i \leq p$; and G is referred to as a <u>realization</u> of π and $V(\pi) = \{u_1,\ldots,u_p\}$. Let $\underline{\underline{G}}(\pi)$ be the set of all nonisomorphic realizations of π and P an invariant property of graphs. A sequence π is said to be <u>forcibly P-graphic</u> if π is graphic and every realization of π has the property P. A property P of graphs is said to be <u>hereditary</u> if whenever a graph G has the property P and H is an induced subgraph of G, then H has also the property P. The aim of this paper is to show in Section 2 that, in principle, for any hereditary property P, the set of all forcibly P-graphic sequences can be characterized, using a theorem of Fulkerson, Hoffman and McAndrew [2] and a partial order $<<$ defined on any set $\underline{\underline{A}}$ of graphic sequences as follows : For two sequences π_1, π_2 in $\underline{\underline{A}}$, say that $\pi_1 << \pi_2$, in words π_2 <u>captures</u> π_1, if π_2 has a realization having a realization of π_1 as an induced subgraph. Several conjectures on sets of graphic sequences are mentioned and the equivalence of some of them is established in Section 3. In Section 4 we verify these conjectures for several hereditary properties P which include (1) P = perfect-graphic, (2) P = chromatic number is $\leq k$, where k is a fixed positive integer and (3) P = planarity.

Let P be a hereditary property of graphs and $\underline{\underline{G}}(P)$ be the set of all graphs with property P; $\underline{\underline{F}}(P)$ be the set of all graphs not having the property P and $\underline{\underline{F}}_o(P)$ be the subset of $\underline{\underline{F}}(P)$ consisting of those G such that for every $v \in V(G)$, the graph G-v has the property. Elements of $\underline{\underline{F}}_o(P)$ are called "<u>the minimal forbidden subgraphs</u>" for the property P. Other pertinent definitions are given at the appropriate junctures. For definitions not given here and notation not explained, the reader is referred to Harary [12].

2. CHARACTERIZATION OF FORCIBLY HEREDITARY P-GRAPHIC SEQUENCES

We start this section with a simple lemma in folklore, whose proof is easy and is omitted.

<u>Lemma 2.1</u>. If P is a hereditary property, then a graph $G \in \underline{\underline{G}}(P)$ if and only if G has no induced subgraph isomorphic to a member of $\underline{\underline{F}}_o(P)$.

<u>Lemma 2.2.</u> If π_1, π_2 are graphic sequences such that π_2 captures π_1, then for any realization H of π_1, H occurs as an induced subgraph of some realization of π_2.

<u>Proof</u>. Since $\pi_1 << \pi_2$, there exists a $G \in \underline{\underline{G}}(\pi_2)$ having a $H_1 \in \underline{\underline{G}}(\pi_1)$ as an induced subgraph. Now if $H \in \underline{\underline{G}}(\pi_1)$, then replacing an induced subgraph of G isomorphic to H_1 by the graph H so that degrees of the corresponding vertices are same in both H_1 and H, we obtain a $G^* \in \underline{\underline{G}}(\pi_2)$ such that H occurs as an induced subgraph.

Next we prove that the relation $\ll$ is a partial order on any set of graphic sequences. More precisely we have

Lemma 2.3. If $\underline{\underline{A}}$ is any nonempty set of graphic sequences, then $(\underline{\underline{A}}, \ll)$ is a partially ordered set with minimal elements. Further, given any $\pi \in \underline{\underline{A}}$, there exists a minimal element π_1 such that $\pi_1 \ll \pi$.

Proof. Note that if $\pi_1 \ll \pi_2$, then the length of π_1 is less than or equal to that of π_2. This and definition of $\ll$ imply that $\ll$ is a reflexive and antisymmetric relation. That $\ll$ is transitive follows from Lemma 2.2. Therefore $(\underline{\underline{A}}, \ll)$ is a partially ordered set. Further for any $\pi' \in \underline{\underline{A}}$, the number of $\pi_1 \in \underline{\underline{A}}$ with $\pi_1 \ll \pi'$ is finite. Now it is not difficult to see that the lemma holds.

Now we present a preliminary characertization of forcibly hereditary P-graphic sequences which will be refined in a series of lemmas to follow.

Theorem 2.4. If P is a hereditary property of graphs, then a graphic sequence π is forcibly P-graphic if and only if for no $\pi_1 \in M(P)$, the relation $\pi_1 \ll \pi$ holds, where $M(P)$ is the set of all minimal elements of the set $(\underline{\underline{A}}_o(P), \ll)$ and $\underline{\underline{A}}_o(P)$ is the set of all degree sequences of the minimal forbidden subgraphs in $\underline{\underline{F}}_o(P)$ for the property P.

Proof. Suppose π is forcibly P-graphic and for some $\pi_1 \in M(P)$, $\pi_1 \ll \pi$; and a $H \in \underline{\underline{G}}(\pi_1)$ with $H \in \underline{\underline{F}}_o(P)$. Then by Lemma 2.2, there exists a $G \in \underline{\underline{G}}(\pi)$ having H as an induced subgraph. This implies, by Lemma 2.1, that $G \in \underline{\underline{F}}(P)$. Thus for no $\pi_1 \in M(P)$, the relation $\pi_1 \ll \pi$ holds.

Conversely, if $G \in \underline{\underline{G}}(\pi)$ and $G \in \underline{\underline{F}}(P)$, then by Lemma 2.1, there exists $H \in \underline{\underline{F}}_o(P)$ such that H is an induced subgraph of G. Let $\pi_1 = \pi(H) \in \underline{\underline{A}}_o(P)$. Then by Lemma 2.3, there exists $\pi_2 \in M(P)$ such that $\pi_2 \ll \pi_1$; this implies that $\pi_2 \ll \pi$. Thus π is forcibly P-graphic.

To improve on the characterization of forcibly P-graphic sequences of Theprem 2.4 we first prove the following

Lemma 2.5. If $\pi = (d_1, \ldots, d_p)$, $d_1 \geq \ldots \geq d_p$ and $\pi' = (e_1, \ldots, e_n)$, $e_1 \geq \ldots \geq e_n$ with $\pi' \ll \pi$ and $G \in \underline{\underline{G}}(\pi)$ with $V(G) = \{u_1, \ldots, u_p\}$ having a $H \in \underline{\underline{G}}(\pi')$ with

$V(H) = \{u_{i_1},\ldots,u_{i_n}\}$ as an induced subgraph, then there exists $G^* \in \underline{\underline{G}}(\pi)$ having a $H^* \in \underline{\underline{G}}(\pi')$ with $V(H^*) = \{u_{j_1},\ldots,u_{j_n}\}$ and $d_{H^*}(u_{j_i}) = e_i$, $1 \leq i \leq n$, where $(j_1,\ldots,j_n)$ is a permutation of $(i_1,\ldots,i_n)$ with $j_1 < \ldots < j_n$.

<u>Proof</u>. Let $d_{i_s} \leq d_{i_k}$ but $e_s > e_k$, for some s,k, $1 \leq s,\ k \leq n$. Let $t = e_{i_s} - e_{i_k}$. By a simple counting argument, it can be easily shown that there exist vertices $Z_1,\ldots,Z_t$ outside $V(H)$ such that $(u_{i_k},z_j) \in E(G)$, $(u_{i_s},z_j) \notin E(G)$ for $1 \leq j \leq t$. Also there are vertices $w_1,\ldots,w_t$ in $V(H) - \{u_{i_s},u_{i_k}\}$ such that $(u_{i_s},w_j) \in E(G)$ but $(u_{i_k},w_j) \notin E(G)$, $1 \leq j \leq t$. Let G^* be the graph obtained from G as follows :

$$V(G^*) = V(G),$$

$$E(G^*) = E(G) \cup \{(u_{i_k},w_j),(u_{i_s},Z_j) : 1 \leq j \leq t\} - \{(u_{i_s},w_j),(u_{i_k},Z_j) : 1 \leq j \leq t\}.$$

Note that $G^* \in \underline{\underline{G}}(\pi)$ and in the induced subgraph H^* in G^*, $d_{H^*}(u_{i_j}) = e_j$ if $j \neq s,k$, and $d_{H^*}(u_{i_s}) = e_k$, $d_{H^*}(u_{i_k}) = e_s$ and $e_k < e_s$. Repeating this procedure we finally get a required realization of π.

We shall further improve upon Theorem 2.4 by proving a series of lemmas, which culminate in giving a characterization of π_1 capturing π_2, using a theorem of Fulkerson, Hoffman and McAndrew. To this end we need a definition. A graph G is said to have the <u>odd cycle property</u> if whenever two odd cycles of G are vertex disjoint, then there is an edge of G joining these two odd cycles.

<u>Theorem 2.6</u>. (FHM [2]) If G is a graph of order p, $V(G) = \{1,\ldots,p\}$, with the odd cycle property, then G has a subgraph (not necessarily induced) with degree sequence $(d_1,\ldots,d_p)$ if and only if the following two conditions (C.1) and (C.2) are satisfied :

(C.1) $\sum_{i=1}^{p} d_i$ is even,

(C.2) for any three subsets S,T,U (empty sets not excluded) which partition $V(G)$ we have

$$\sum_{i\in S} d_i \leq \sum_{i\in T} d_i + \sum_{i\in S,\, j\in S\cup U} c_{ij},$$

where
$$c_{ij} = \begin{cases} 1, & \text{if } (i,j) \in E(G), \\ 0, & \text{otherwise.} \end{cases}$$

We now present necessary and sufficient conditions for a sequence π to capture π^* on a given set of vertices. First we treat the case when $\pi^* = (0^n)$.

Lemma 2.7. If $\pi = (d_1,\dots,d_p)$ is a sequence of nonnegative integers and $V(\pi) = \{1,2,\dots,p\}$ where $d_1 \geq \dots \geq d_n$ and $d_{n+1} \geq \dots \geq d_p$, then π captures $\pi^* = (0^n)$ on $A = \{1,\dots,n\}$ if and only if the conditions (C.1), and (C.3) given below are satisfied :

$$\text{(C.3)} \quad \sum_{i=1}^{s_1} d_i + \sum_{i=1}^{s_2} d_{n+i} \leq \sum_{i=1}^{t_1} d_{n+1-i} + \sum_{i=1}^{t_2} d_{p+1-i} + s_1(p-n-t_2) + s_2(p-t_1-t_2-1),$$

for all nonnegative integers s_1,s_2,t_1,t_2 satisfying the inequalities $s_1 + t_1 \leq n$, $s_2 + t_2 \leq p - n$ where if one of s_1,s_2,t_1,t_2 is zero, then the corresponding sum should be taken, with the usual convention, as zero.

Proof. The necessity of the conditions follow by a simple counting argument. The sufficiency follows by applying Theorem 2.6 to the graph G with $V(G) = \{1,\dots,p\}$ and $(u_i,u_j) \in E(G)$ $(i \neq j)$ if and only if at least one of i,j is greater than n, which clearly has the odd cycle property.

The following lemma, which is a slight improvement on Lemma 2.7, will be used in Section 4.

Lemma 2.8. If $\pi = (d_1,\dots,d_p)$ captures $\pi^* = (0^n)$ on $A = \{1,\dots,n\}$, then π captures π^* on a set B, with $|B| = n$, of the minimum possible degrees in π.

Proof. Let $\underline{\underline{G}}_o(\pi)$ be the set of all graphs $G \in \underline{\underline{G}}(\pi)$ such that G has an independent set I with $|I| = n$. Out of all $\underline{\underline{G}}_o(\pi)$ and independent sets I of G with $|I| = n$, choose one graph G_o and an independent set I_o of G_o with $|I_o| = n$ such that $\sum_{u \in I_o} d_{G_o}(u)$ is the minimum possible. We shall prove that in G_o a set, of cardinality n, of the minimum possible degrees is independent. Assume then that there exists a $y \in V(G_o) - I_o$ such that $d_{G_o}(y) < d_{G_o}(x)$ for some x in I_o, which in particular implies that $n \geq 2$. Let $y_1, \dots, y_k$ be the vertices in $I_o - \{x\}$, adjacent to y in G_o. Since (x,y_i) is not in $E(G_o)$, $1 \leq i \leq k$, and $d_{G_o}(x) > d_{G_o}(y)$, it follows that there are vertices $x_1,\dots,x_k$ such that $(x,x_i) \in E(G_o)$, $1 \leq i \leq k$, but none of (y,x_i), $1 \leq i \leq k$, belongs to $E(G_o)$. Let G^* be the graph obtained from G_o as follows :

$$E(G^*) = E(G) \cup \{(y,x_i),(x,y_i) : 1 \le i \le k\} - \{(x,x_i),(y,y_i) : 1 \le i \le k\}.$$

Then $\pi(G^*) = \pi(G) = \pi$ and $I' = (I \cup \{y\}) - \{x\}$ is an independent set of cardinality n such that

$$\sum_{Z\varepsilon I'} d_{G_o}(Z) < \sum_{Z\varepsilon I_o} d_{G_o}(Z),$$

contradicting the selection of I_o.

Now we present a characterization of π capturing a general π^* on a given set of vertices.

<u>Theorem 2.9</u>. A sequence $\pi = (d_1,\dots,d_p)$ with $d_1 \ge \dots \ge d_n$, $d_{n+1} \ge \dots \ge d_p$ and $V(\pi) = \{1,\dots,p\}$, captures a graphic sequence $\pi^* = (e_1,\dots,e_n)$ with $d_i \ge e_i$, $1 \le i \le n$, on the set $A = \{1,\dots,n\}$ if and only if (C.1), and (C.4) given below are satisfied :

(C.4) $\pi' = (d'_1,\dots,d'_n, d_{n+1},\dots,d_p)$ satisfies the condition (C.3) where $(d'_1,\dots,d'_n)$ is a rearrangement of $(d_1-e_1,\dots,d_n-e_n)$ in the nonincreasing order.

<u>Proof</u>. This follows from Lemma 2.7 since π captures π^* if and only if π' captures the graphic sequence (0^n) on A.

From the above Theorem we shall deduce, by taking $\pi^* = ((n-1)^n)$, the following theorem, announced in Rao [6], which presents necessary and sufficient conditions, for π with $V(\pi) = \{1,\dots,p\}$ to have a realization in which $\{1,\dots,n\}$ is complete. These conditions are similar to the Erdös-Gallai [12] criterion for the graphicness of a sequence, their proof is lengthy and, as mentioned in Rao [7], imitates the Erdös-Gallai proof of the graphicness of a sequence (Harary [12]).

<u>Lemma 2.10</u>. The sequence $\pi = (d_1,\dots,d_p)$ with $V(\pi) = \{1,\dots,p\}$, $d_1 \ge \dots \ge d_n$, $d_{n+1} \ge \dots \ge d_p \ge 0$ captures $\pi^* = ((n-1)^n)$, $n \ge 1$ on $\{1,\dots,n\}$ if and only if (C.1), and (C.5) below are satisfied :

(C.5) $$\sum_{i=1}^{s_1} d_i + \sum_{i=1}^{s_2} d_{n+i} \le (s_1+s_2)(s_1+s_2-1) + \sum_{i=s_1+1}^{n} \min(s_1+s_2, d_i-n+s_1+1) + \sum_{i=n+s_2+1}^{p} \min(s_1+s_2, d_i)$$

for all integers s_1, s_2 such that $0 \le s_1 \le n$ and $0 \le s_2 \le p-n$.

Proof. The necessity is easy to prove. To prove the sufficiency we show that condition (C.4) of Theorem 2.9 is satisfied with $\pi^* = ((n-1)^n)$. Let s_1, s_2, t_1, t_2 be nonnegative integers such that $0 \le s_1 + t_1 \le n$, $0 \le s_2 + t_2 \le p-n$. To prove (C.4) it is enough to show that the right hand side of (C.5) is less than or equal to the right hand side of (C.3) + $(n-1)(s_1-t_1)$. Note that

$$\sum_{i=s_1+1}^{n} \min(s_1+s_2,\, d_i-n+s_1+1) \le s_1(n-s_1) + (n-s_1-t_1)s_2 + \sum_{i=1}^{t_1} (d_{n+1-i}-n+1)$$

and

$$\sum_{i=n+s_2+1}^{p} \min(s_1+s_2,\, d_i) \le (s_1+s_2)(p-n-s_2-t_2) + \sum_{i=1}^{t_2} d_{p+1-i}.$$

Using these inequalities it is easy to see by (C.5) that the required inequality mentioned above holds.

We remark that Theorems 2.4 and 2.9 give a characterization, at least in principle, of forcibly P-graphic sequences for any hereditary property P of graphs. But simple criteria for π to be forcibly line-graphic, forcibly planar-graphic were presented in Rao [8,9] by the method of laying off technique developed by Kleitman and Wang (see Rao [10] for a description) to construct a canonical realization of a graphic sequence.

In an attempt to further improve on Theorems 2.4 and 2.9 we pose, in Section 3, several interesting conjectures on sets of degree sequences under the partial order <<, and prove the equivalence of some of them. In Section 4 we verify these conjectures for several hereditary properties.

3. SOME CONJECTURES AND RESULTS ON INDEPENDENT SETS OF GRAPHIC SEQUENCES

A set $\underline{\underline{A}}$ of graphic sequences is said to be independent if $\pi_1, \pi_2 \in \underline{\underline{A}}$ and $\pi_1 \ne \pi_2$, then neither $\pi_1 << \pi_2$ nor $\pi_2 << \pi_1$ holds, that is, no two distinct members of $\underline{\underline{A}}$ are comparable under <<. Clearly every subset of an independent set of graphic sequences is independent. It is not difficult to check by Zorn's Lemma of set theory that any independent set of graphic sequences is contained in a maximal independent set.

Two hereditary properties P_1, P_2 are said to be <u>equivalent</u> if $\underline{\underline{G}}(P_1) = \underline{\underline{G}}(P_2)$, that is, P_2 is a characterization of P_1. A hereditary property P is said to be <u>reducible</u> if there exist two hereditary properties P_1 and P_2, neither equivalent to P, such that $\underline{\underline{G}}(P) = \underline{\underline{G}}(P_1) \cup \underline{\underline{G}}(P_2)$ in which case P is denoted by $P_1 \vee P_2$; P is <u>irreducible</u> otherwise.

Now we are ready to state several interesting conjectures.

<u>Conjecture 3.1</u>. For any irreducible hereditary property P, the set $M(P)$, defined in Theorem 2.4 is finite.

<u>Meta Conjecture 3.2</u>. For any hereditary property P, the set $M(P)$ is finite.

<u>Conjecture 3.3</u>. If $\underline{\underline{A}}$ is an independent set of graphic sequences, then $\underline{\underline{A}}$ is finite.

A hereditary property P is said to be <u>finitary</u> if $M(P)$ is finite.

<u>Conjecture 3.4</u>. If P_1 and P_2 are finitary hereditary properties, then so is $P = P_1 \vee P_2$.

The following Lemma 3.5 gives several examples of irreducible properties as demonstrated in Corollary 3.6.

<u>Lemma 3.5</u>. If P is a hereditary property such that whenever $G, H \in \underline{\underline{G}}(P)$, the disjoint union of G and H, denoted by $G \circ H$, also belongs to $\underline{\underline{G}}(P)$, then P is irreducible.

<u>Proof</u>. Suppose that P is reducible with $\underline{\underline{G}}(P) = \underline{\underline{G}}(P_1) \cup \underline{\underline{G}}(P_2)$. Then there is a $G_1 \in \underline{\underline{G}}(P)$ such that $G_1 \notin \underline{\underline{G}}(P_1)$. This implies that there exists $H_1 \in \underline{\underline{F}}_o(P_1)$ such that H_1 is an induced subgraph of $G_1 \in \underline{\underline{G}}(P)$. Since $H_1 \in \underline{\underline{G}}(P)$ and $H_1 \notin \underline{\underline{G}}(P_1)$, we have that $H_1 \in \underline{\underline{G}}(P_2)$. Similarly there exists a $H_2 \in \underline{\underline{G}}(P)$ such that $H_2 \in \underline{\underline{G}}(P_2)$ but $H_2 \notin \underline{\underline{G}}(P_1)$. Both $H_1, H_2 \in \underline{\underline{G}}(P)$ and therefore by hypothesis $H = H_1 \circ H_2 \in \underline{\underline{G}}(P)$. But H is neither in $\underline{\underline{G}}(P_1)$ nor in $\underline{\underline{G}}(P_2)$, a contradiction.

<u>Corollary 3.6</u>. Let P be any one of the properties (1) through (8) given below. Then P is irreducible.

(1) planarity, (2) chromatic number $\leq k$, where k is a fixed positive integer, (3) perfect graphic in the sence of Berge [1], (4) interval graphic, (5) comparability

graphic, (6) line-graphic, (7) the clique number $\leq k$, where k is a fixed positive integer and (8) any hereditary property P for which each member of $\underline{\underline{F}}_o(P)$ is connected.

Problem 3.7. Characterize irreducible hereditary properties (refer to Lemma 3.11).

Conjecture 3.8. If P is a hereditary property, then there exist finitely many irreducible properties $P_1, \ldots, P_n$ such that $P = P_1 \vee \ldots \vee P_n$.

Now we prove the equivalence of conjectures 3.2 and 3.3. More precisely, we prove the following :

Theorem 3.9. A set $\underline{\underline{A}}$ of graphic sequences is independent if and only if there exists a hereditary property P such that $M(P) = \underline{\underline{A}}$. In particular, Conjectures 3.2 and 3.3 are equivalent.

Proof. To prove the sufficiency, observe that no two distinct elements of $M(P)$ are comparable by the minimality, under <<, therefore $M(P) = \underline{\underline{A}}$ is an independent set.

To prove the necessity, define an invariant property P of graphs as follows : A graph G is said to be have the property P if G has no induced subgraph H with $\pi(H) \in \underline{\underline{A}}$. Clearly P is a hereditary property, further if $\pi \in \underline{\underline{A}}$ and $G \in \underline{\underline{G}}(\pi)$, then $G \in \underline{\underline{F}}(P)$. Now if $G-v$ does not have the property P for some $v \in V(G)$, then $G-v$ has an induced subgraph H_1 with $\pi(H_1) = \pi_1 \in \underline{\underline{A}}$, which implies, in particular that $\pi_1 << \pi$, this is not possible since length of π_1 is less than π and both π, $\pi_1 \in \underline{\underline{A}}$ which is independent. Thus $G \in \underline{\underline{F}}_o(P)$, and $\pi \in \underline{\underline{A}}_o(P)$ of Theorem 2.4, and $\underline{\underline{A}} \subseteq \underline{\underline{A}}_o(P)$.

We are now ready to show that $M(P) = \underline{\underline{A}}$. Suppose $\pi \in \underline{\underline{A}} \subset \underline{\underline{A}}_o(P)$. Then by Lemma 2.3, there exists a $\pi_1 \in M(P)$ such that $\pi_1 << \pi$. By definition there exists a realization $H \in \underline{\underline{F}}_o(P)$ of π_1 which is an induced subgraph of some $G \in \underline{\underline{G}}(\pi)$. As $G \in \underline{\underline{F}}_o(P)$, it follows that $G = H$ and $\pi = \pi_1 \in M(P)$ implying that $\underline{\underline{A}} \subseteq M(P)$. To prove the reverse inequality, let $\pi \in M(P)$. Since there exists a $G \in \underline{\underline{F}}_o(P)$ with $\pi(G) = \pi$, it follows by definition of P that there exists an induced subgraph H of G with $\pi(H) \in \underline{\underline{A}}$. Let $\pi(H) = \pi_1$. Then note that $\pi_1 << \pi$. Since $\underline{\underline{A}} \subseteq M(P)$, we have that $\pi_1 \in M(P)$. Since π_1, $\pi \in M(P)$, an independent set and $\pi_1 << \pi$, it follows that $\pi = \pi_1$. Therefore $\pi \in \underline{\underline{A}}$ and $M(P) = \underline{\underline{A}}$.

Lemma 3.10. For every positive integer n, there exists an irreducible hereditary property P with $|M(P)| = n$.

Proof. In view of Theorem 3.9 and Lemma 3.5 it is enough to exhibit an independent set of graphic sequences of cardinality n each of which has only connected realizations. To this end we observe that any set of graphic sequences of the same length p is independent; and if p is sufficiently large there is such a set of graphic sequences. For example if $p \geq n + 2$, the following set of independent set of graphic sequences may be chosen. For every i, $1 \leq i \leq n$, let

$$\pi_i = ((p-1)^i, i^{p-1}) \text{ and } \underline{\underline{A}} = \{\pi_1, \ldots, \pi_n\}.$$

Now we give an example of an irreducible hereditary property not covered by Lemma 3.5.

Lemma 3.11. Let P be the property that for a graph G the independence number $\alpha(G) \leq s - 1$, that is G is $\overline{K}_s$-free. Then P is irreducible.

Proof. Suppose that there are properties P_1 and P_2 such that $P = P_1 \vee P_2$, that is $\underline{\underline{G}}(P) = \underline{\underline{G}}(P_1) \cup \underline{\underline{G}}(P_2)$. Since $\overline{K}_s \notin \underline{\underline{G}}(P)$, there exist $\overline{K}_{s_1} \in \underline{\underline{F}}_o(P_1)$ and $\overline{K}_{s_2} \in \underline{\underline{F}}_o(P_2)$. Without loss of generality assume that $s_1 \leq s_2 \leq s$. If $s_2 < s$, then $\overline{K}_{s-1} \in \underline{\underline{G}}(P)$ but belongs to neither $\underline{\underline{G}}(P_1)$ nor $\underline{\underline{G}}(P_2)$. Therefore $s_2 = s$. If $s_1 < s$, then since $\underline{\underline{F}}_o(P_2) \neq \{\overline{K}_s\}$, it follows that there exists a $H \in \underline{\underline{F}}_o(P_2)$ such that $H \neq \overline{K}_s$ and by the minimality of H it follows that $\alpha(H) \leq s - 1$. But then the graph G obtained from $\overline{K}_{s_1} \circ H$ by adding all edges joining a vertex of $\overline{K}_{s_1}$ to a vertex of H, $\overline{K}_{s_1} + H$ in the notation of Harary [12], belongs to $\underline{\underline{G}}(P)$ but to neither $\underline{\underline{G}}(P_1)$ nor $\underline{\underline{G}}(P_2)$. Thus we proved that $s_1 = s$ also. Now since $\underline{\underline{F}}_o(P_1) \neq \{\overline{K}_s\} \neq \underline{\underline{F}}_o(P_2)$, there are graphs H_i, in $\underline{\underline{F}}_o(P_1)$, $i = 1,2$, such that both are different from $\overline{K}_s$ in which case $\alpha(H_i) \leq s - 1$ by the minimality of H_i and $H_1 + H_2$ is in $\underline{\underline{G}}(P)$, but neither in $\underline{\underline{G}}(P_1)$ nor in $\underline{\underline{G}}(P_2)$, a contradiction.

Lemma 3.12. Given any positive integer n, there exists a reducible property $P = P_1 \vee P_2$ such that P_1, P_2 are irreducible with $|M(P_i)| = 1$, for $i = 1,2$ but $|M(P)| \geq n$.

Proof. Let P_1 be the property that a graph is K_{n+1}-free and P_2 be the property that a graph is $\overline{K}_{n+1}$-free. By Lemmas 3.5 and 3.11 both P_1 and P_2 are irreducible properties. Further $\underline{\underline{F}}_o(P_1) = \{K_n\}$ and $\underline{\underline{F}}_o(P_2) = \{\overline{K}_n\}$. Let $P = P_1 \vee P_2$, then P is hereditary and if $H \in \underline{\underline{F}}(P)$ then $|V(H)| \geq 2n + 1$. By Lemma 2.3 the following graphic sequences π_i, $1 \leq i \leq n$, belong to $M(P)$:

$$\pi_1 = ((n+1)^n, n, 1^n) \text{ and}$$

$$\pi_i = (n+i, (n+1)^{n-1}, n, 2^{i-1}, 1^{n-i}), \; 2 \leq i \leq n.$$

Therefore $|M(P)| \geq n$.

4. VERIFICATION OF CONJECTURE 3.1 FOR SEVERAL HEREDITARY PROPERTIES

In this section we verify Conjecture 3.1 for several hereditary properties P and determine $M(P)$ in some cases. First we remark that if $\underline{\underline{F}}_o(P)$ is finite then obviously $M(P)$ is finite. Such properties P include the property of line-graphicness in which case $|\underline{\underline{F}}_o(P)| = 9$ (as shown by Beineke, refer to Harary [12]) and $M(P)$ can easily be seen to be equal to $\{(3,1^3); (3^4,2); (4^3,3^2); (5,3^4); (3^4,1^2); (3^4,3^2)\}$. Note $(3^4,1) << (5^2,3^4)$; $(3,1^3) << (4^2,3^3,1)$; and $(3^4,2) << (4^2,3^2,2^2)$.

If P is the property that the chromatic number ≤ 2, then $M(P) = \{2^3; 2^5\}$; note here that $\underline{\underline{F}}_o(P)$ is infinite. Next we prove the following

<u>Theorem 4.1</u>. If P is the hereditary property of being perfect-graphic in the sense of Berge [1], then $M(P) = \{(2^5); (2^7); (4^7)\}$.

To prove this theorem it is enough to prove the following

<u>Theorem 4.2</u>. If π has a non-perfect graphic realization, then π captures one of $\pi_1 = (2^5)$, $\pi_2 = (2^7)$ and $\pi_3 = (4^7)$.

<u>Proof</u>. Suppose that this theorem is false and that π^* is a sequence of the minimum possible length for which the theorem is false. Let G be a non-perfect graph in $\underline{\underline{G}}(\pi^*)$. If $G-v$ is a non-perfect graph for some $v \in V(G)$, then by the minimality of π^*, the sequences $\pi(G-v)$ captures one of π_1, π_2 and π_3 and since $\pi(G-v) << \pi$, so is π. Thus we have

<u>Assertion 1</u>. If $G \in \underline{\underline{G}}(\pi^*)$ and G is not perfect-graphic, then G is vertex critical.

Out of all $G \in \underline{\underline{G}}(\pi^*)$ with G not perfect, choose one graph G^*, say, having an odd cycle $C^*_{2t+1} = (x_1,\ldots,x_{2t+1})$, $t \geq 2$, with the minimum possible number $m(\pi^*) = m$ of chords. If $m(\pi^*) \geq 2$, then by a theorem of Meyniel [5], G^* is perfect-graphic. Thus we may assume that $m = 0$ or 1. If $m = 0$, then since π^* does not capture π_1 or π_2, we have that $t \geq 4$. Then the subgraph of G^* induced on $V(C^*_{2t+1})$ may be replaced by the two cycles $(x_1,\ldots,x_5)$; $(x_6,\ldots,x_{2t+1})$ to obtain a $H \in \underline{\underline{G}}(\pi^*)$ such that the 5-cycle is an induced subgraph of H implying that π^* captures π_1. Thus it follows that $m = 1$. If now $t \geq 3$, then C^*_{2t+1} has exactly one chord (x_1,x_{i_o}), $i_o \neq 2, 2t+1$. By the minimality of m and the assumption, it follows that $i_o = 3$ or $2t$. Without loss of generality assume that $i_o = 3$. Replacing the subgraph of G^* induced on $V(C^*_{2t+1})$ by the cycle $(x_1,x_2,x_5,x_4,x_3,x_6,x_7,\ldots,x_{2t+1})$ together with chord (x_1,x_3) we get a realization of π^*, which implies that π^* captures π_1. Thus we have proved

Assertion 2. $m = 1$ and $t = 2$.

Let without loss of generality (x_1,x_4) be the chord of C^*_5. Let

$$A = \{y/(y,x_1) \notin E(G^*),\ y \neq x_1,x_3\}.$$

Assertion 3. $|A| \geq 2$.

Proof. If $|A| \leq 1$, then $d_{G^*}(x_1) \leq 2$. Since G^* is a vertex-critical non-perfect-graph, so is $\overline{G}^*$ by the weak-perfect graph theorem of Lovász [4]. By a theorem of Sachs [11], we have

$$2w(\overline{G}^*) - 2 \leq d_{\overline{G}^*}(x_1) \leq 2,$$

where $w(H)$ is the maximum size of a complete subgraph in H. Since $(x_1,x_3) \in E(\overline{G}^*)$, we have that $w(\overline{G}^*) = 2$ and since the strong-perfect graph conjecture is known to be true for graphs H with $w(H) = 2$ it follows that $\overline{G}^* = C_{2s+1}$ for some $s \geq 2$. This implies as in the proof of Assertion 2 that $\pi(\overline{G}^*)$ captures π_1 or π_2 which in turn implies that $\pi(G^*)$ captures π_1 or π_3, contradicting the selection of π^*.

If A is independent in G^*, then so is $A \cup \{x_1\}$ and therefore $\alpha(G^*) \geq |A| + 1$. This implies by a theorem of Sachs [11], mentioned above, that $p - |A| - 2 \leq d_{G^*}(x_1) \leq p - 2\alpha + 1 \leq p - 2|A| - 1$ and therefore by Assertion 3 we have

<u>Assertion 4</u>. G^* has at least one edge in A.

Now we prove the following assertion by the method of interchanges described in Rao [10].

<u>Assertion 5</u>. If $(y,x_1) \in E(G^*)$ with $y \neq x_2, x_4$, then $(y,x_4) \in E(G^*)$.

<u>Proof</u>. If $(y,x_1) \in E(G^*)$ with y specified as above, then $y \neq x_5$ and

$$\mu = (x_1,x_4,y,x_1,Z_1,Z_2)$$

where Z_1,Z_2 are vertices in A such that $(Z_1,Z_2) \in E(G^*)$, is an alternating $(G^*,\overline{G}^*)$ trail; and the graph H obtained by interchanging the edges in μ is a realization of π^* having a C_5 induced subgraph on $(x_1,\dots,x_5)$ implying that π^* captures π_1.

Since $\alpha(G^*) \geq 2$, we have by a theorem of Sachs [11] that $d_{G^*}(x_4) \leq p - 3$ and by Assertion 5, there exists a $x_6 \in A$ such that $(x_4,x_6) \in E(\overline{G}^*)$.

<u>Assertion 6</u>. If $Z \in A$ and $Z \neq x_6$, then $(Z,x_6) \in E(\overline{G}^*)$.

<u>Proof</u>. If $(Z,x_6) \in E(G^*)$, then $\mu = (x_1,x_4,x_6,Z)$ is an alternating $(G^*,\overline{G}^*)$ cycle and as in the proof of Assertion 5, we obtain a contradiction.

Let $B = \{y/(y,x_6) \in E(G^*)$ with $y \notin \{x_2,x_3,x_5\}\}$. Since $w(G^*) \geq 3$, we have that

$$d_{G^*}(x_6) \geq 2w(G^*) - 2 \geq 4$$

and this implies that $B \neq \phi$.

<u>Assertion 7</u>. B is complete in G^*.

<u>Proof</u>. If $y_1,y_2 \in B$ with $y_1 \neq y_2$ and $(y_1,y_2) \in E(\overline{G}^*)$, then

$$\mu = (x_6,y_1,y_2,x_6,x_4,x_1)$$

is an alternating $(G^*,\overline{G}^*)$trail and as in the proof of Assertion 5 we obtain a contradiction.

We now complete the proof of the theorem as follows. By Assertions 2 and 7 it follows that $w(G^*) \geq |B| + 2 = s + 2$. Also since $(x_6,x_4) \in E(\overline{G}^*)$, Assertion 6 and theorem of Sachs [11] imply that

$$2w(G^*) - 2 \leq d_{G^*}(x_6) \leq s + 3,$$

implying that $s = 1$ and $w(G^*) = 3$. This in particular implies that $(x_5,y) \varepsilon E(\overline{G}^*)$ where y is the unique vertex in B; and that (x_6,x_2), (x_6,x_3), $(x_6,x_5) \varepsilon E(G^*)$. Then

$$\mu = (x_6,x_5,y,x_6,x_1,x_4)$$

is an alternating $(G^*,\overline{G}^*)$ trail and as in the proof of Assertion 5 we have that π^* captures π_1: this is a contradiction.

Next we prove that the property the chromatic number of a graph being $\leq k$, where k is a fixed positive integer, is finitary. To this end we need a Lemma of Rao [7] whose original proof is quite lengthy and uses a generalization of Theorem 3.9. We present an elementary proof, more for its method which will be used repeatedly in Theorem 4.3.

<u>Lemma 4.2</u>. If $\pi = (d_1,\ldots,d_p)$ is graphic with $d_1 \geq \ldots \geq d_p$, $V(\pi) = \{u_1,\ldots,u_p\}$, and w is a fixed positive integer such that $d_w \geq 2w-3$, then π captures $\pi^* = ((w-1)^w)$ on $\{u_1,\ldots,u_w\}$.

<u>Proof</u>. It is enough to prove, by Lemma 3, that π captures π^* on some set A. The proof of this fact is by induction on w. If $w = 2$, then the result follows by a theorem of Kleitman and Wang [3] by laying off u_2 and constructing a realization of π by the algorithm mentioned therein. Assume then that the results hold for $w - 1 \geq 2$ and let π be a graphic sequence satisfying the conditions of the lemma. By the inductive hypothesis π captures $\pi^{**} = ((w-2)^{w-1})$. Out of all $G \varepsilon \underline{\underline{G}}(\pi)$ choose one G_o, say, such that $A = \{u_1,\ldots,u_{w-1}\}$ is complete in G_o and u_w is joined to the maximum number of vertices in A (this is possible by Lemma 3). We shall prove that u_w is joined to all vertices in A. Suppose $(u_w,u_{i_o}) \notin E(G_o)$ for some i_o, $1 \leq i_o \leq w - 1$. Let A_1 (respectively, A_2) be the set of all vertices not in A adjacent to u_{i_o} (respectively, u_w) but not to u_w (respectively, u_{i_o}). Let A_3 be the set of vertices not in A which are adjacent to both u_{i_o} and u_w. Since $d_w \geq 2w - 3$ it follows that $|A_i \cup A_3| \geq w - 1$, for i, 1,2.

<u>Claim 1</u>. If $A_3 \neq \phi$, then A_3 is complete in G_o.

Proof. If $u,v \in A_3$ and $(u,v) \notin E(G_o)$, then the graph G^* with $E(G^*) = E(G) - (u_w,v) - (u_{v_o},u) + (u_w,u_{i_o}) + (u,v)$ has $\pi(G^*) = \pi(G)$ and contradicts the selection of G_o.

If now A_1 or A_2 is empty, then $|A_3| \geq w - 1$ and $A_3 \cup \{u_w\}$ is complete in G_o implying that $w(G_o) \geq w$. Therefore $A_i \neq \phi$, $i = 1,2$.

Claim 2. $B = A_2 \cup A_3$ is complete in G_o.

Proof. Let u,v be distinct vertices in B with $(u,v) \notin E(G_o)$. If u,v are in distinct A_i, $i = 2,3$, then as in the proof of Claim 1, we obtain a contradiction. If $u,v \in A_2$, then

$\mu = (u_{i_o},u,v,u_{i_o},u_w,x)$ where $x \in A_1$ is an alternating $(G_o,\overline{G}_o)$ trail, and interchanging the edges in μ we get a $H \in \underline{\underline{G}}(\pi)$ contradicting the property of G_o.

Now $B \cup \{u_w\}$ is complete in G_o and therefore $w(G_o) \geq w$, a contradiction.

Now we prove the following

Theorem 4.3. Let P be the property that a graph G has chromatic number $\gamma(G) \leq k$, where k is a fixed positive integer. Then P is finitary (note however that $\underline{\underline{F}}_o(P)$ is infinite if $k \geq 2$).

In fact we prove a bit more

Theorem 4.4. If $\pi = (d_1,\ldots,d_p)$, $d_1 \geq \ldots \geq d_p$, is a graphic sequence with $V(\pi) = \{1,\ldots,p\}$ and the number $N(w)$ of degrees $d_i \geq w - 1$ is at least $f(w) = (16w^3 + 8w^2+6w+1)$ for some fixed positive integer w, then π captures $\pi^* = ((w-1)^w)$ on $A = \{1,2,\ldots,w\}$.

Proof. The proof is by induction on w. If $w = 2$, then the result is clearly true. Assume that the result is true for $w - 1$ and let π be a graphic sequence satisfying the hypothesis of the theorem. If $d_w \geq 2w - 3$, then by Lemma 4.2, π captures π^* and therefore we have

$$d_w \leq 2w - 4. \qquad \ldots \qquad (1)$$

Out of all $G \in \underline{\underline{G}}(\pi)$ which induce complete subgraph on $B = \{1,\ldots,w-1\}$ choose one G_o such that w is joined to the maximum number of vertices in B. We claim that

<u>Proof</u>. Suppose that $(x,y) \varepsilon E(G_o)$ with both x,y in B_4^* or x in B_3 and y in B_4^*. If $Z \varepsilon A_3$, let $\mu = (Z,w,i_o,Z,x,y)$; and if $A_3 = \phi$, then let $Z_i \varepsilon A_i$, $i = 1,2$ and $\mu = (Z_2,w,i_o,Z_1,y,x)$. Interchanging the edges on this μ we get a realization $G' \varepsilon \underline{\underline{G}}(\pi)$ which contradicts the selection of G_o.

Now we are ready to complete the proof of the theorem. Clearly

$$|B_4^*| \geq p - w(16w^2+8w+5).$$

By hypothesis $N(w) \geq f(w)$, and this implies, by the pigeon-hole principle, that there are at least w points in B_4^* whose degree in G_o is at least $w - 1$; all of which by Claim 3, are adjacent to all points in A and in particular to the vertex w which implies that $d_{G_o}(w) \geq 2w - 2$, a contradiction.

Now we collect various properties P (without proof) for which $\underline{\underline{F}}_o(P)$ is infinite but $M(P)$ is finite. The proofs of these use the well known forbidden graphs characterizations of these properties.

	Property P	M(P)
1.	Planarity	the length of any $\pi \varepsilon M(P)$ is at most 25; therefore $M(P)$ is finite.
2.	Chromatic number is at most 3	$\{(3^4),(5,3^5),(4,3^6),(4^7)\}$
3.	Interval-graphic	$\{(2^4),(2^5),(2^6),(3^3,1^3),(4,3^3,1^3),$ $(5,4,3^2,2^2,1),(5^2,4^2,2^3)\}$
4.	Outer-planarity	$\{(3^4),(5,3^5),(3^2,2^3),(3^2,2^4),(3^2,2^5),(4^2,2^3)\}$
5.	Comparability graphic	$\{(2^5),(2^7),(4,3^4,1^2),(5^2,4^3,2^2),(3^6),(4^7),$ $(4^3,2^3),(5^3,3^3,2),(3,2^3,1^3),(4^3,2^3,1^2),$ $(3^2,2^2,1),(5^3,4^3,3),(5,3^3,4^2,2,1),(5^2,4,3^4),$ $(5,4^3,3,2^2),(4^3,2^4),(4^3,3,2^2,1),(4^3,3^2,2)\}$

REFERENCES

1. C. Berge, Graphs and Hypergraphs, North-Holland Publishing Company, Amsterdam, 1976.

2. D.R. Fulkerson, A.J. Hoffman and M.H. McAndrew, Some properties of graphs with multiple edges, Canad. J. Math., 17 (1965), 166-177.

3. D.J. Kleitman and D.L.Wang, Algorithms for constructing graphs and digraphs with given valences and factors, Discrete Maths., 6 (1973), 79-88.

4. L. Lovász, Normal hypergraphs and the perfect graph conjecture, Discrete Math., 2 (1972), 253-267.

5. H. Meyniel, On the perfect graph conjecture, Discrete Maths., 16 (1976), 339-342.

6. A. Ramachandra Rao, The clique number of a graph with a given degree sequence, Proc. Symposium on Graph Theory (A.R. Rao, Ed.), ISI Lecture Notes Series 4, MacMillan and Co. India Ltd., 4 (1979), 251-267.

7. A. Ramachandra Rao, An Erdös-Gâllai type result on the clique number of a realization of a degree sequence, (unpublished).

8. S.B. Rao, Characterization of forcibly line-graphic degree sequences, Utilitas Mathematica, 11 (1977), 357-366.

9. S.B. Rao, Characterization of forcibly planar degree sequences, Ars Combinatoria, (to appear).

10. S.B. Rao, A survey of the theory of potentially P-graphic and forcibly P-graphic degree sequences, (this proceedings).

11. H. Sachs, On the Berge conjecture concerning perfect graphs, Comb. Structures and their applications (R. Guy et.al Eds.), Calgary International Conference, Gordon and Breach, (1970), 377-384.

12. F. Harary, Graph Theory, Addison Wesley, New York, 1972.

THE MINIMAL FORBIDDEN SUBGRAPHS FOR GENERALIZED LINE-GRAPHS*

S. B. RAO
Indian Statistical Institute
203 B. T. Road, Calcutta 700 035

N. M. SINGHI
School of Mathematics
Tata Institute of Fundamental Research
Colaba, Bombay 400 005

K. S. VIJAYAN
Indian Statistical Institute
203 B. T. Road, Calcutta 700 035

ABSTRACT

It is shown, by exhibiting the list, that there are exactly 31 nonisomorphic minimal forbidden subgraphs for generalized line graphs, that is, graphs which are representable by a subset of the root system D_n for some n. This solves a problem of Hoffman. Further, it is proved that the family of minimal forbidden subgraphs for graphs whose adjacency matrix has the least eigen-value at least -2, is finite.

1. INTRODUCTION AND STATEMENT OF THE MAIN THEOREM

All graphs considered in this paper are finite and have neither loops nor multiple edges. For a graph G, the symbols $V(G)$, $E(G)$ denote the vertex set, edge set, respectively, of G. Our graphs include the graph with $V(G) = \phi = E(G)$ called the null graph. The line-graph $L(G)$ of a graph G is defined to be the graph whose vertex set is $E(G)$ with two vertices in $L(G)$ adjacent if and only if the corresponding edges of G have a common vertex. Next, a cocktail party graph $CP(n)$ is the result of removing a 1-factor from the complete graph of order $2n$; we let $CP(0)$ to be the null graph. Let G be a graph with $V(G) = \{v_1,\ldots,v_p\}$ where $p \geq 0$ and $a = (a_1,\ldots,a_p)$ be a sequence of nonnegative integers. Then the generalized line-graph $L(G; a_1,\ldots,a_p) = L(G(\underline{\underline{a}}))$ is the graph consisting of $L(G)$ together with $CP(a_1),\ldots,CP(a_p)$ with the

* This work was done while the second author was visiting Math-Stat. Division of I.S.I. during November-December 1979.

following additional adjacencies : vertex (v_i,v_j) of $L(G)$ is adjacent to every vertex of $CP(a_i)$ and to every vertex of $CP(a_j)$. We illustrate this definition by considering the example G in Figure 1 where G has vertices v_1,v_2,v_3,v_4 and $a = (2,1,0,3)$.

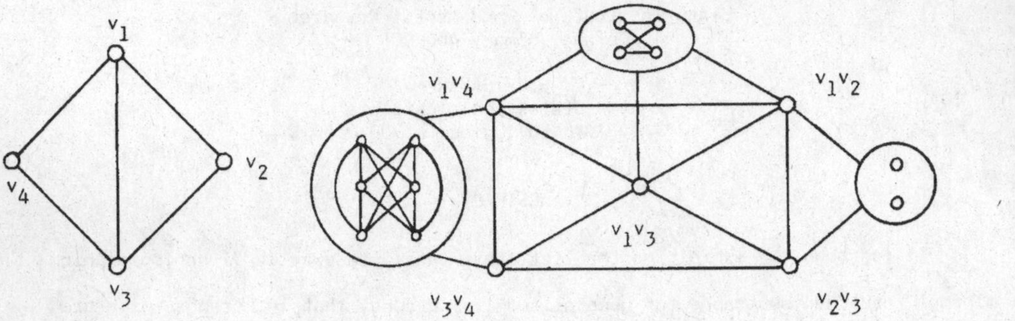

FIGURE 1: The graph $L(G(\underline{a}))$.

A graph H is said to be a generalized line-graph if $H \simeq L(G(\underline{\underline{a}}))$ for some graph G and some sequence a where the symbol $\simeq$ denotes isomorphism as graphs.

Generalized line-graphs were first defined by Hoffman [4] who also proved the following interesting result

<u>Theorem 1.1</u> (Hoffman [4]). A graph G has the least eigen-value -2 if and only if (i) G is a generalized line-graph, or (ii) G is represented by a subset of the root system E_8.

For more details about this see a recent papar of Cameron, Goethals, Seidel and Shult [2], which subsumes, improves and generalizes all earlier work on graphs G with the least eigen-value -2. One of the results proved in [2] gives an interesting relationship between generalized line-graphs and root systems.

<u>Theorem 1.2</u> ([2]). A graph G is a generalized line-graph if and only if G is represented by a root system D_n, for some n.

The following is an immediate consequence of the above theorem, and can be easily verified directly also.

<u>Theorem A</u>. If G is a generalized line-graph then any induced subgraph of G is also a generalized line-graph.

The aim of this paper is to study the family of nonisomorphic minimal graphs which are not generalized line-graphs, called "<u>The minimal forbidden graphs</u>" for the generalized line-graphs. These minimal graphs exist by Theorem A. We in fact show, by exhibiting the list of such graphs, that there are exactly 31 such graphs and each graph has at most six vertices. This gives a characterization of generalized line-graphs in the spirit of the characterizations of Van Rooij and Wilf [6] and Beineke [1] of line-graphs. This also solves a problem raised by Hoffman [5]. We can now state the main theorem of this paper.

<u>Main Theorem B</u>. A graph H is a generalized line-graph if and only if none of the 31 graphs G_i, $1 \leq i \leq 31$ of Figure 2 occurs as an induced subgraph of H.

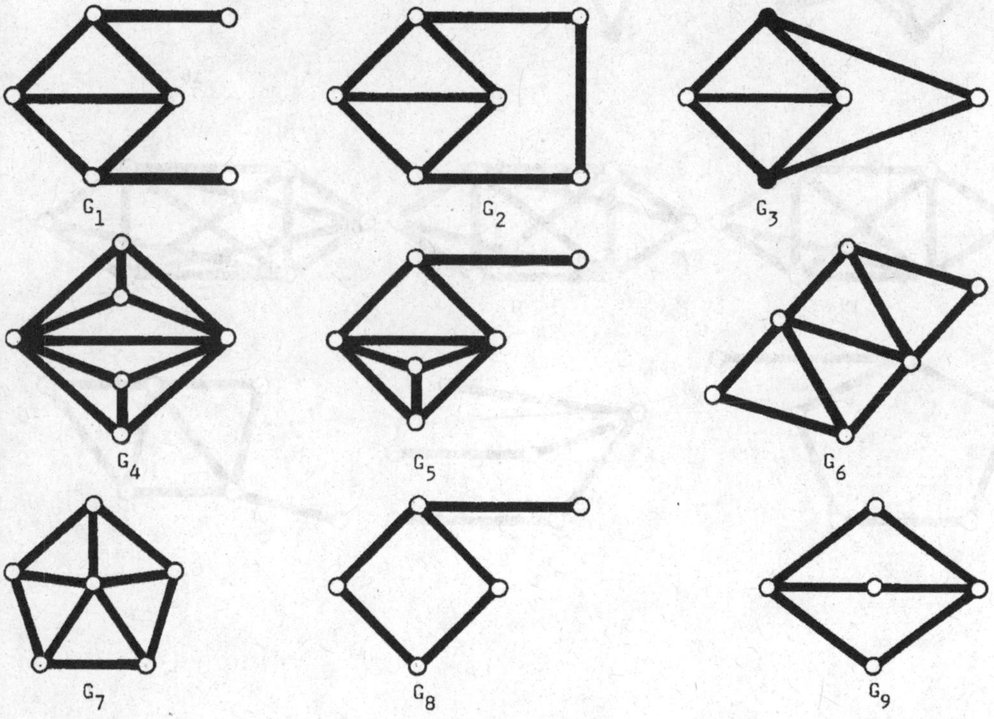

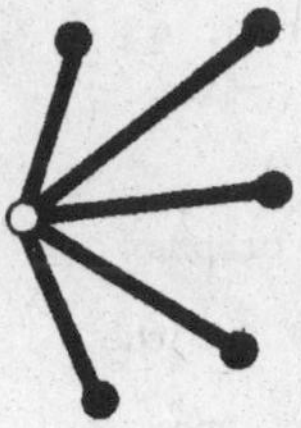

G_{10}

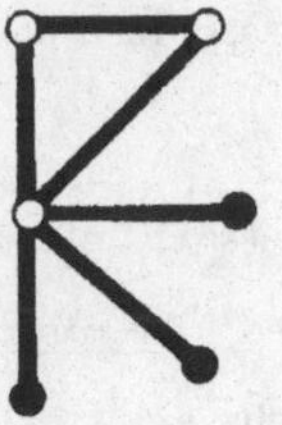

G_{11}

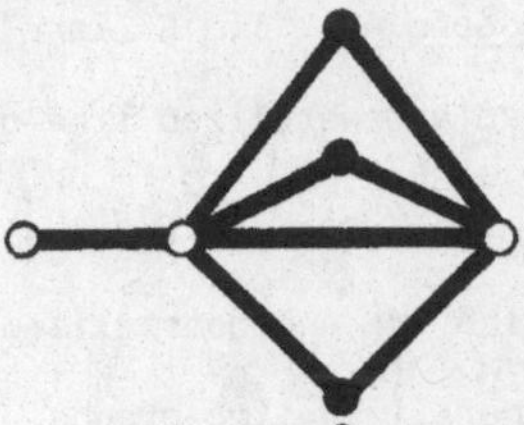

G_{12}

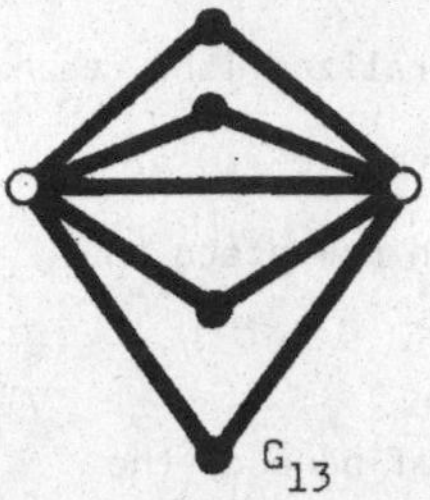

G_{13}

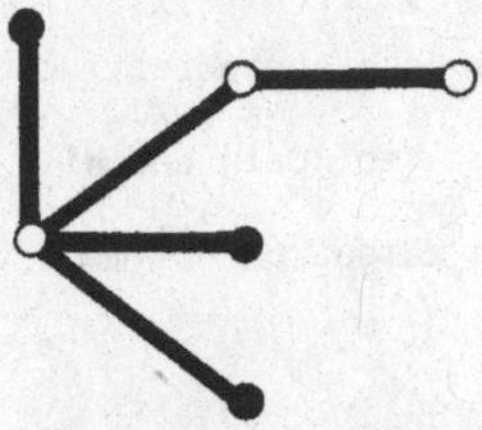

G_{14}

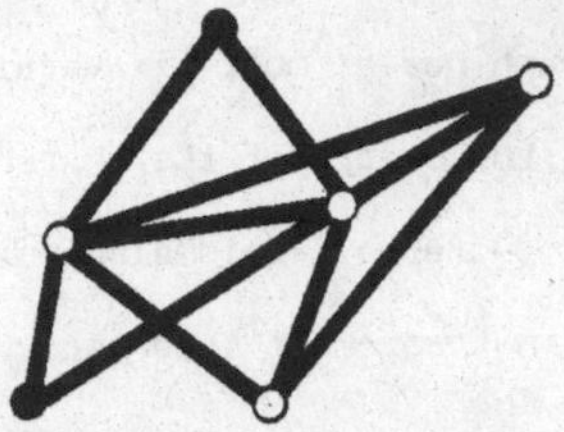

G_{15}

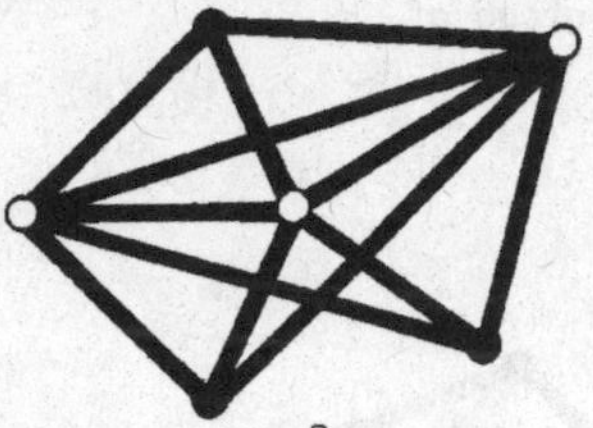

G_{16}

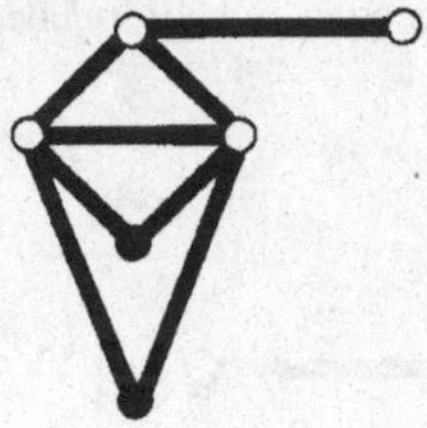

G_{17}

G_{18}

G_{19}

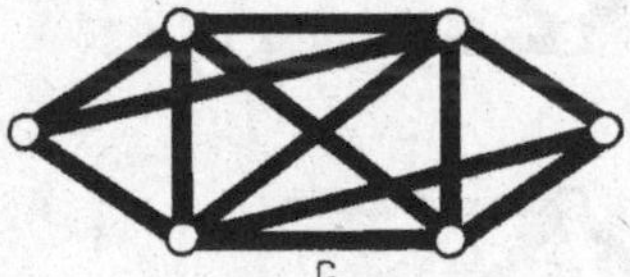

G_{20}

G_{21}

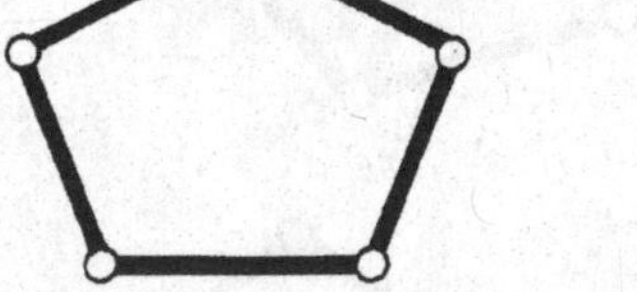

G_{22}

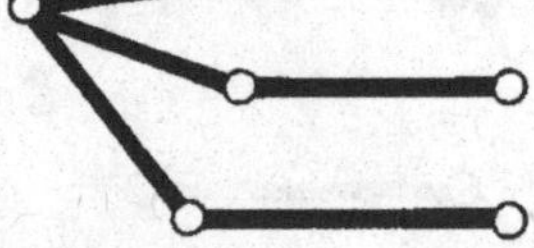

G_{23}

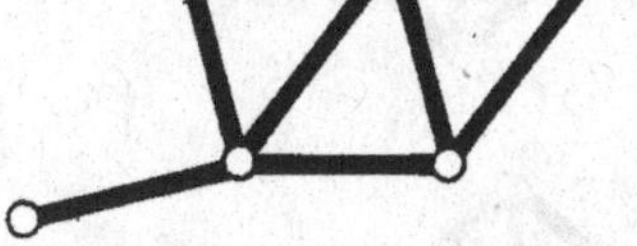

G_{24}

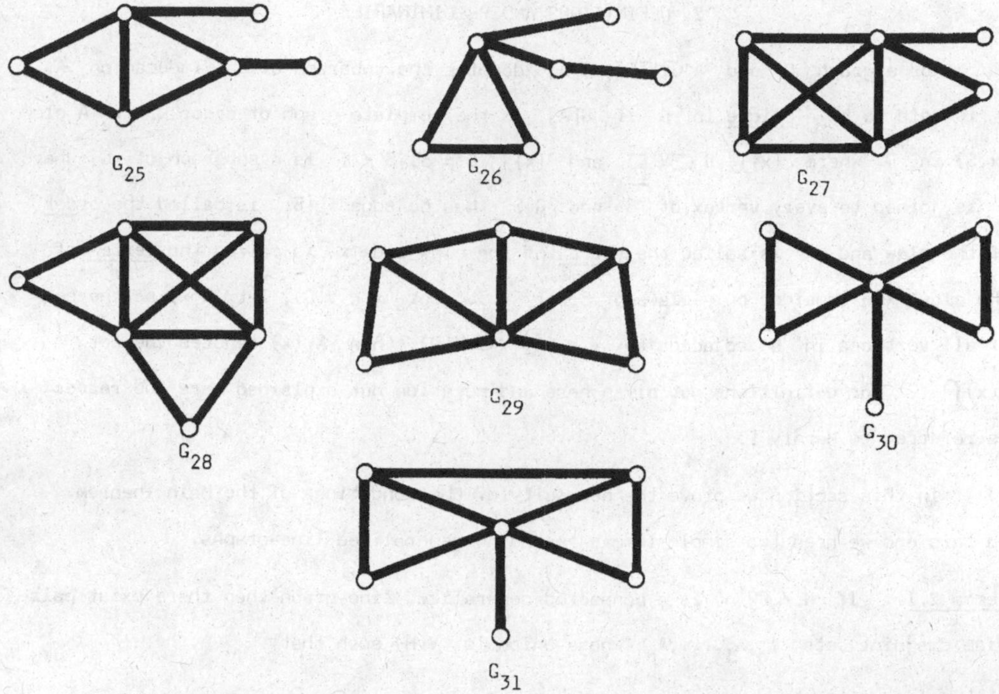

FIGURE 2 : The 31 forbidden graphs G_i, $1 \leq i \leq 31$, for generalized line-graphs.

As an application we also study the minimal forbidden graphs for the family of all graphs whose least eigen-value is at least -2, and we prove

Theorem C. The set of minimal forbidden graphs for the family of all graphs whose least eigen-value is at least -2 is finite.

A complete list of these minimal forbidden graphs will be given in a subsequent communication.

In Section 2 we give various definitions and prove the necessity part of the Main Theorem B, the sufficiency will be proved in Section 3. The Section 4 is devoted to prove Theorem C.

2. DEFINITIONS AND PRELIMINARIES

For a graph G and $A \subset V(G)$, $G[A]$ denotes the subgraph of G induced on A. A is said to be a <u>clique</u> in G if $G[A]$ is the complete graph of order $|A|$. A claw (x,S) in G where $\{x\} \cup S \subset V(G)$ and $\{x\} \cap S = \phi$, $S \neq \phi$ is a subgraph of G where x is joined to every vertex of S and $G[S]$ has no edges; $|S|$ is called the <u>order</u> of the claw and x is called the <u>vertex</u> of the claw and x is called the <u>vertex</u> of the claw. We consider only claws of order ≥ 3. For $x \varepsilon V(G)$, let $N(x)$ be the set of all vertices in G adjacent to x. If $Y \subset V(G)$, then $N_Y(x)$ denotes the set $N(x) \cap Y$. For definitions not given here and notation not explained here the reader is referred to Harary [3].

In this section we prove the necessity of the conditions of the Main Theorem. To this end we need two simple lemmas regarding generalized line-graphs.

<u>Lemma 2.1</u>. If $H \neq CP(n)$ is a connected generalized line-graph then there exist pair-wise disjoint sets $A_o, A_1, \ldots, A_p$ whose union is $V(H)$ such that

(1) $H[A_o]$ is a connected line-graph of order p,

(2) $H[A_i]$ is a Cocktail party graph, $1 \leq i \leq p$,

(3) any two vertices in the same A_i have the same degree in H and have the same neighborhood in A_o and further this neighborhood is a clique in H,

(4) if $u \varepsilon A_i$, $v \varepsilon A_j$ with $1 \leq i \neq j \leq p$, then $uv \varepsilon E(H)$.

<u>Proof</u>. Let $H = L(G_i; a_1, \ldots, a_p)$ where p is the order of G, let A_i be the vertices of the cocktail party graph associated with the i-th vertex and let $A_o = V(G)$. Then conditions (1),(2),(3) and (4) are satisfied.

<u>Lemma 2.2</u>. The seven graphs H_i of Figure 3 are generalized line graphs.

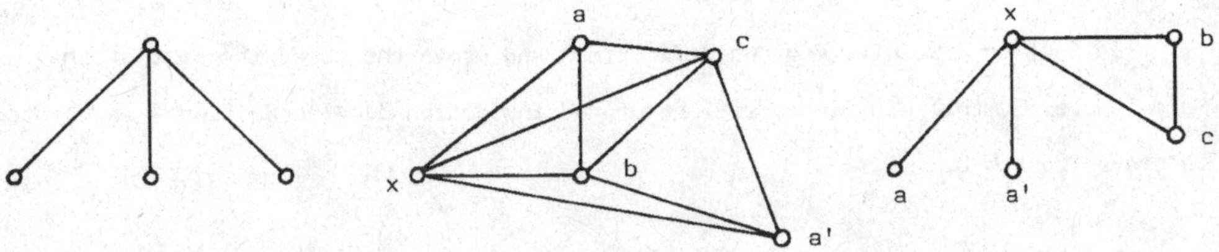

H_1 H_2 H_3

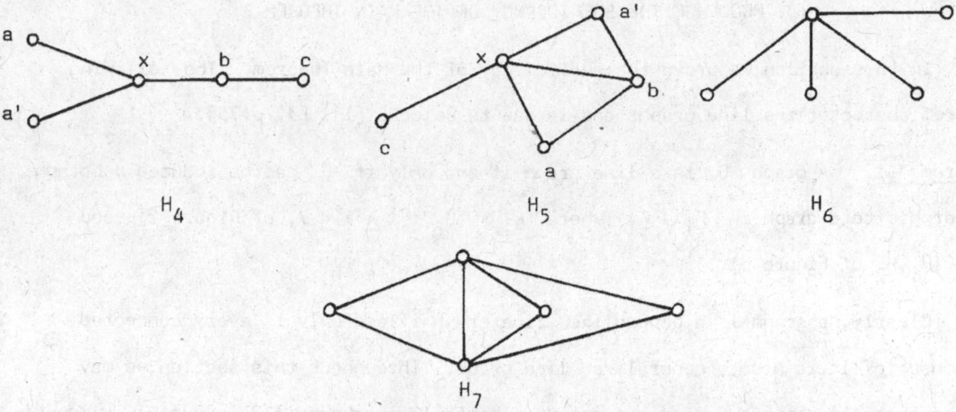

FIGURE 3 : Some generalized line-graphs which are not line graphs.

Proof. Note that these 7 graphs are not line-graps, for, all except H_2 contain a 3-claw and H_2 is forbidden for line-graphs. However,

H_1 = L( ; 1,0,0); H_2 = L( ; 0,1,0,0);

H_3 = L( ; 1,0,0,0); H_4 = L( ; 1,0,0,0);

H_5 = L( ; 0,1,0,0); H_6 = L(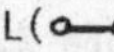 ; 1,1); H_7 = L( ; 1,0,0).

Now we are ready to give the

Proof of the necessity of the Main Theorem : It can be easily checked that for $1 \leq i \neq j \leq 31$, the degree sequence of G_i is equal to the degree sequence of G_j if and only if $\{i,j\} = \{17,24\}$ and further G_{17} is not isomorphic to G_{24}. Therefore these 31 graphs are mutually nonisomorphic. Using Lemma 3.1 it can be proved that G_i, $1 \leq i \leq 31$, is not a generalized line-graph, where the only vertices possible in cocktail party graphs are denoted by thick circles. Further it can be seen that for any vertex v of G_i, $1 \leq i \leq 31$, any given connected component of $G_i - v$ is either a line graph or is one of the graphs H_i, $1 \leq i \leq 7$ of Figure 3; and therefore by Lemma 3.2, it follows that $G_i - v$ is a generalized line graph. Then by Theorem A the necessity of the conditions of the Main Theorem B follows.

3. PROOF OF THE SUFFICIENCY OF THE MAIN THEOREM B

In this section we prove the sufficiency of the Main Theorem. The following theorem characterizes line graphs and is due to Beineke [1], [3, p.75].

Theorem 3.1. A graph G is a line graph if and only if G has no induced subgraph isomorphic to a graph in $\underline{\underline{F}}_1 \cup \underline{\underline{E}}_1$ where $\underline{\underline{F}}_1 = \{G_i : 1 \leq i \leq 7$, of Figure 2$\}$ and $\underline{\underline{E}}_1 = \{H_1, H_2$ of Figure 3$\}$.

Clearly, a graph is a generalized line graph if and only if every connected component of it is also a generalized line graph. Throughout this section we may assume that all graphs G are connected. The following Lemma 3.2 is useful in identifying, under certain conditions, the cocktail party graphs of G as will be demonstrated in Lemma 3.3.

Lemma 3.2. If G has no induced subgraph isomorphic to a graph in $\underline{\underline{F}}_2 = \{G_i : 1 \leq i \leq 17$, of Figure 2$\}$, and (x,S) is a 3-claw in G, then exactly one of the following is true :

(1) $G \simeq K_{1,3}$.

(2) $G \simeq K_{1,4} = H_6$ or $G \simeq H_7$ of Figure 3.

(3) There exists a set $A \subseteq V(G)$ with $x \in A$ and vertices $a, a' \in A \cap S$ such that $G[A]$ is identical with H_i, $i = 3, 4$ and 5.

Proof. Assume that (1) and (2) are not satisfied. Since G is connected, there exists a vertex $y \notin \{x\} \cup S = S_1$, say, such that $N(y) \cap S_1$ is nonempty. Now using the hypothesis that G has no induced subgraph isomorphic to G_8 or G_9 it can be checked, by considering the cases depending on the number of vertices of S_1 to which y is joined, that $G[S_2]$ where $S_2 = S_1 \cup \{y\}$, is isomorphic to H_i, $3 \leq i \leq 7$. If $G[S_2]$ is isomorphic to H_6 or H_7, again as condition (2) is not satisfied there is a vertex $z \notin S_2$ such that $N(z) \cap S_2$ is nonempty. Again as above, now using the fact that G has no induced subgraph isomorphic to G_i, $8 \leq i \leq 17$, it can be checked that $G[S_2 \cup z]$ is isomorphic to one of the three graphs of Figure 4 in which case

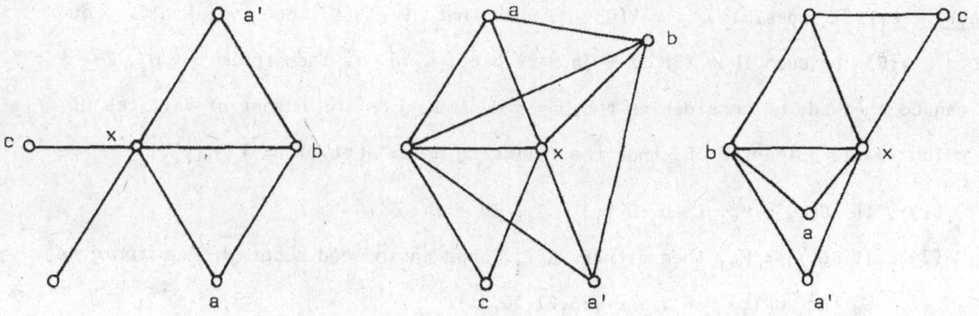

FIGURE 4

the labeled subgraphs shown in Figure 4 are identical with H_5.

HYPOTHESIS 3.3. From now onwards we shall assume that G is not isomorphic to the generalized line-graphs H_1, H_6, H_7, CP(n) and L(o—o ; a_1, a_2); and also that the order of G is at least 4.

Now we define a subset X_1 of V(G) which later on in Lemma 3.8, under certain conditions, will be identified with union of vertex sets of possible Cocktail party subgraphs of G. For any graph G let X_1 be the set of all vertices a of V(G) with the property that there are 4 more vertices a', x,b,c such that the induced subgraph G[Y] where Y = {a,a',x,b,c} is identical with the labeled graph H_i, $2 \leq i \leq 5$, of Figure 3. The vertex a' is referred to as a mate of a. In Corollary 3.7 we shall prove under certain conditions, that a mate, if it exists, is unique. Note that if $a \in X_1$, then its mate a' also belongs to X_1 and $a a' \notin E(G)$. Let $X_2 = V(G) - X_1$. Note in general that X_2 may be empty. In Lemma 3.8 we show that $X_2 \neq \phi$.

The following Lemma 3.4 follows from Theorem 4.1 and Lemma 3.2.

Lemma 3.4. If G has no induced subgraph isomorphic to a graph in $\underline{\underline{F}}_2$, then $G[X_2]$ is a line-graph (the case $X_2 = \phi$ is not excluded). Further, if G not a line-graph, then $X_1 \neq \phi$.

In the next lemma we shall prove that for any $a \in X_1$, a and its mate have the same neighborhood.

Lemma 3.5. If G has no induced subgraph isomorphic to a graph in $\underline{\underline{F}} = \{G_i : 1 \leq i \leq 31$ of Figure 2}, then for any $a \in X_1$, N(a) = N(a') where a' is a mate of a.

<u>Proof</u>. Let, if possible, $v \in V(G)$ be such that $v \in N(a)$ but $v \notin N(a')$. Let $x,b,c \in V(G)$ be such that for $Y = \{a,a',x,b,c\}$, $G[Y]$ is isomorphic to H_i, $2 \leq i \leq 5$. It can be checked, by considering the cases depending on the number of vertices of Y to which v is joined in G, that the following hold. Let $Y_1 = Y \cup \{v\}$.

(1) If $G[Y] = H_2$, then $G[Y_1] \simeq G_i$, $18 \leq i \leq 31$.

(2) If $G[Y] = H_3$, then either $G[Y_1]$ has an induced subgraph isomorphic to G_8 or $G[Y_1] \simeq G_i$, $i \in \{26,27,30,31\}$.

(3) If $G[Y] = H_4$, then $G[Y_1]$ has an induced subgraph isomorphic to G_8 or $G[Y_1] \simeq G_i$, $i \in \{22,23,24,25,26\}$.

(4) If $G[Y] = H_5$, then $G[Y_1]$ has an induced subgraph isomorphic to G_8 or $G[Y_1] \simeq G_i$, $i \in \{24,25,27,28,29,31\}$.

Therefore the lemma follows from the hypothesis.

<u>Lemma 3.6</u>. If G has no induced subgraph isomorphic to a member of $\underline{\underline{F}}$ and $v \in N(a)$, where $a \in X_1$ and a' is a mate of a, then the following hold :

(1) There is at most one vertex $v' \neq v$ with $v' \in N(a) = N(a')$ such that $v' \notin N(v)$.

(2) If there is such a vertex v', then $v \in X_1$ and conversely.

<u>Proof</u>. By Lemma 3.5 we have that $N(a) = N(a')$. Now (1) follows from the fact that G has no induced subgraph isomorphic to G_8 or G_9. Suppose that there exists such a vertex v'. Then since G_8, G_9 and G_6 are forbidden for G, it follows that $N(v) - \{a,a'\} = N(a) - \{v,v'\} = N(v') - \{a,a'\}$. Let now x,b,c be vertices of G such that for $Y = \{a,a',x,b,c\}$, the graph $G[Y] \equiv H_i$, $2 \leq i \leq 5$. If v,v' are different from x in case $G[Y] \equiv H_3$ or H_4; x,b in case $G[Y] \equiv H_5$ and x,b,c in case $G[Y] \equiv H_2$, then for $Y_1 = \{v,v',x,b,c\}$, by hypothesis it can be checked that the graph $G[Y_1] \equiv H_i$, $2 \leq i \leq 5$. This implies that $v,v' \in X_1$. Even in the exceptional cases, it is not difficult to check that $v,v' \in X_1$. Conversely, if $v \in X_1$, then by Lemma 3.5 $N(v) = N(v')$ where v' is a mate of a and $vv' \notin E(G)$. Therefore $v' \in N(a) = N(a')$ and $v' \notin N(v)$.

From now onwards we assume that G has no induced subgraph isomorphic to a member of $\underline{\underline{F}}$. From Lemma 3.6 we deduce the following

Corollary 3.7. (1) If $a \in X_1$, then $N_{X_2}(a)$ is a clique in G provided $X_2 \neq \phi$.

(2) Each connected component of size at least 2 of $G[X_1]$ is a cocktail party graph $CP(t)$ for some $t(\geq 2)$. In particular, each $a \in X_1$ has a unique mate a'.

(3) If $a, a_1 \in X_1$ are in the same component of $G[X_1]$, then $N_{X_2}(a) = N_{X_2}(a_1)$.

Proof. (1) is immediate from Lemma 3.6. To prove (2) we show that if $a \in X_1$, then there is exactly one vertex a' such that $N(a) = N(a')$. First by Lemma 3.6 there is one such vertex a'. Suppose there is another vertex a^* such that $N(a) = N(a') = N(a^*)$. Let x,b,c be vertices with $G[Y] \cong H_i$ $2 \leq i \leq 5$, where $Y = \{a,a',x,b,c\}$. Now it can be checked that $G[Y \cup a^*]$ is one of the graphs G_{11}, G_{12}, G_{14} and G_{16}, a contradiction. To complete the proof of (2) it is enough to show that the connected component A containing a is $(X_1 \cap N(a)) \cup \{a,a'\}$ provided this set has cardinality at least 2. This is true, for in the other case there exists a $y \in A$ and not in the set above, joined to some $v \in N(a)$ with $v \neq a,a'$. Let v' be a mate of v. Then since $N(v) = N(v')$, we have $vy, v'y \in E(G)$ which imply that $\{v,v',a,a',y\}$ induces G_9. The uniqueness of the mate follows since $G \ncong L(\circ\!\!-\!\!\circ\,; a_1,a_2)$ and G_{10} is forbidden for G. The condition (3) also follows from the fact that $N(a) - \{a_1,a_1'\} = N(v) - \{a,a'\}$.

Using Lemma 3.6 and Corollary 3.7 we immediately have the following

Lemma 3.8. There is a unique partition $P = \{A_1,\dots,A_\ell\}$ of X_1 such that

(1) $G[A_i]$ is a cocktail party graph $CP(n_i)$ for some positive integer n_i.

(2) If $n_i > 1$, then $G[A_i]$ is a connected component of $G[X_1]$.

(3) If $n_i = 1$ and $A_i = \{a,a'\}$, then a and a' are isolated vertices of $G[X_1]$.

(4) $X_2 \neq \phi$, $N_{X_2}(a) = N_{X_2}(a_1)$ for $a,a_1 \in X_1$ if a,a_1 are in the same A_i; and $G[X_2]$ is a connected graph.

Definition 3.9. For $x \in A_i \in P$ of Lemma 3.8 define $K_i = N_{X_2}(x)$.

Note that the definition of K_i is independent of the choice of x by (4) of Lemma 3.8; and K_i, by Lemma 3.6, is a clique in G. Further, $K_i \neq K_j$ if $i \neq j$ as $G \nsupseteq L(\circ\!\!-\!\!-\!\!\circ;\ a_1, a_2)$.

To state the next lemma we need a definition. A triangle or a clique of size 3 in a graph G is said to be an odd triangle if there exists a $v \in V(G)$ joined to exactly an odd number of vertices of the triangle; otherwise the triangle is even.

Lemma 3.10. The set $S = \{K_1, \ldots, K_\ell\}$ of cliques defined in (3.9) above satisfies the following conditions :

(1) Each $x \in X_2 = V(G) - X_1$ is in at most two cliques K_i of S.

(2) Each edge of $G[X_2]$ is in at most one K_i of S.

(3) If $|K_i| \geq 3$, then K_i is a maximal clique of $G[X_2]$.

(4) If a vertex x is in two cliques, say K_{i_1} and K_{i_2} of S, then $N(x) = \{x\} \cup K_{i_1} \cup K_{i_2}$.

(5) If $|K_i| = 3$, and $G[X_2]$ is not isomorphic to one of the graphs of Figure 5, then K_i is an odd triangle of $G[X_2]$.

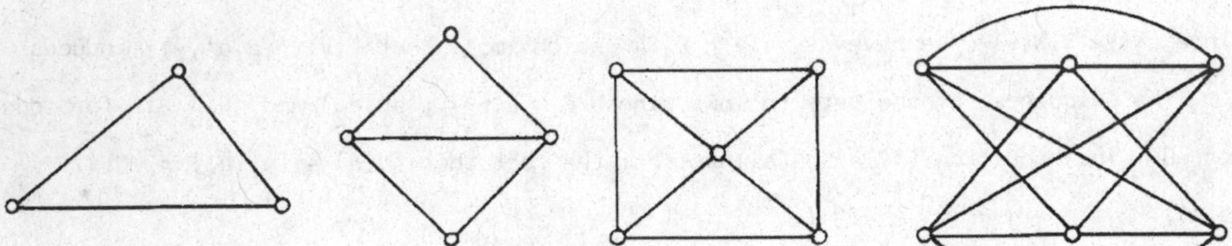

FIGURE 5

(6) If $|K_i| = 2$, then K_i is not contained in an odd triangle of $G[X_2]$.

(7) If $K_i = \{x\}$, then $L = N_{X_2}(x') \cup \{x\}$ is a maximal clique of $G[X_2]$; and either $|L \cap K_i| \leq 1$ for all i, $1 \leq i \leq \ell$ or $L = K_i$ for some i.

Proof. If one of (1) through (4) are not satisfied, then G has an induced subgraph isomorphic to $G_{10}; G_{13}; G_{16}; G_{10}$ respectively; if (5) is not satisfied one of G_{10}, G_{12},

G_{15} or G_{17} occurs as an induced subgraph of G; if (6) is not satisfied, then G_{12}, G_{15} or G_{16} occurs in G; and if (7) is not satisfied, since $G \not\simeq H_6$ and $L(\circ\!\!-\!\!-\!\!\circ; a_1,a_2)$, one of $\{G_{10},G_{14},G_{11},G_8,G_{17}\}$ occurs in G or $G[L]$ contains the second graph of Figure 5, with x as a degree 3 vertex, as an induced subgraph implying that the degree 2 vertices of the second graph belong to X_1. Therefore (1) through (7) are satisfied.

Now we are ready to present the

<u>Proof of the sufficiency of the Main Theorem</u> :

First by Lemma 3.4, $G[X_2]$ is a line-graph and, by Lemma 3.8, is a connected graph. We consider two cases; the proof in each case is similar to the proof of Theorem 8.4, p.76 of Harary [3].

<u>Case 1</u>. $G[X_2]$ has no edge belonging to only even triangles of $G[X_2]$.

Let S_1 be the set of all maximal cliques of $G[X_2]$ containing an odd triangle, together with edges not contained in any of these cliques; and vertices which belong to exactly one of these cliques and edges. By (1) through (6) of Lemma 3.10, S_1 contains the set $S = \{K_1,\ldots,K_\ell\}$ of Lemma 3.10. Let $S_1 = \{K_1,\ldots,K_\ell, K_{\ell+1},\ldots,K_p\}$ and H be the intersection graph of the family S_1 with v_i representing the clique K_i, $1 \leq i \leq p$. Then $L(H) \simeq G[X_2]$ and $K_i = \{(u_i,u_j) | (u_iu_j) \in E(H)\}$, $1 \leq i \leq p$. By (1) through (6) of Lemma 3.10, it is not difficult to check that $G \simeq L(H;a_1,\ldots,a_\ell,o,\ldots,o)$, where $a_i = |A_i|$, $1 \leq i \leq \ell$, and $\{A_1,\ldots,A_\ell\}$ is the partition P of X_1 of Lemma 3.8.

<u>Case 2</u>. $G[X_2]$ has an edge belonging to only even triangles of $G[X_2]$.

It is easy to check (see Theorem 8.4, p.76 of Harary [3]) that the connected graph $G[X_2]$ is one of the graphs of Figure 5 of Lemma 3.10. Then using condition (7) of Lemma 3.10 it is not difficult to check that G is a generalized line-graph.

4. AN APPLICATION

In this section we shall prove, using the Main Theorem of this paper, the following Theorem 4.1. For relevant definitions refer to [2].

Theorem 4.1. The family $\underline{\underline{F}}^{**}$ of minimal forbidden subgraphs for graphs whose adjacency matrix has the least eigen-value at least -2 is finite.

Proof. Let $\underline{\underline{F}}$ (respectively, $\underline{\underline{F}}^*$) be the family of minimal forbidden subgraphs for generalized line-graphs (respectively, graphs representable by a subset of the root system E_8). By Theorem 1.1 every connected component of graphs whose adjacency matrix has the least eigen value at least -2 is either a generalized line-graph or is represented by a subset of E_8. Also by Theorem 4.3 of [2], a graph represented by a subset of the root system E_8 has at most 36 vertices with maximum valency at most 28. Therefore the order of any graph in $\underline{\underline{F}}^*$ is at most 37. Any graph G in $\underline{\underline{F}}^{**}$ is connected and has an induced subgraph isomorphic to a graph H, say, of $\underline{\underline{F}}$. If $V(G) = V(H)$, then by the Main Theorem B, $|V(G)| \leq 6$. Let $x \in V(G) - V(H)$. If $G - x$ is connected, then since H occurs in $G - x$ it follows by the minimality of G in $\underline{\underline{F}}^{**}$ that $G - x$ is representable by a subset of E_8 and therefore $|V(G)| \leq 37$. If $G - x$ is disconnected, it is clear that there exists a vertex y, not belonging to the component of $G - x$ which contains H, such that $G - y$ is connected and as above $|V(G)| \leq 37$. Thus always $|V(G)| \leq 37$ and therefore $\underline{\underline{F}}^{**}$ is finite.

It can be checked that G_8 through G_{21} are in $\underline{\underline{F}}^{**}$. The complete list $\underline{\underline{F}}^{**}$ is deferred to a subsequent communication.

REFERENCES

1. L.W. Beineke, Derived Graphs and Digraphs, Beitrage zur Graphentheorie, (H. Sachs et.al, Eds.) Teub-ner, Leipzig, 1968, 17-33.
2. P.J. Cameron, J.M. Goethals, J.J. Seidel and E.E. Shult, Line graphs, root systems and Elliptic geometry, Journal of Algebra, 43(1976), 305-327.
3. F. Harary, Graph Theory, Addison Wesley Publ. Inc., Reading, 1972.
4. A.J. Hoffman, $-1 -\sqrt{2}$? in Combinatorial Structures and their Applications, (R. Guy, Ed.), Gordon and Breach, New York, 1970, 173-176.
5. A.J. Hoffman, On graphs whose least eigen value exceeds $-1 -\sqrt{2}$, J. Linear Algebra and its Applications, 16(1977), 153-166.
6. A.C.M. Van Rooij and H.S. Wilf, The interchange graph of a finite graph, Acta Math. Acad. Sci. Hungarica, XVI(1965), 263-270.

SPECTRAL CHARACTERIZATION OF THE LINE GRAPH OF K_ℓ^n

S. B. RAO
Indian Statistical Institute
203, B. T. Road, Calcutta 700 035

N. M. SINGHI*
School of Mathematics
Tata Institute of Fundamental Research, Bombay 400 005

K. S. VIJAYAN
Indian Statistical Institute
203, B. T. Road, Calcutta 700 035

ABSTRACT

The line graph of the regular complete multipartite graph K_ℓ^n is shown to be characterised by its spectrum except for $(\ell,n) = (1,8), (4,2), (2,4)$.

1. INTRODUCTION

All graphs considered in this paper are finite and have neither loops nor multiple edges. For definitions and notations we follow Biggs [1]. Let K_ℓ^n be the regular complete n-partite graph of degree $(n-1)\ell$ and of order $n\ell$. It is well known [4-8,10] that any graph cospectral with the line graph $L(K_1^n)$ (respectively, $L(K_\ell^2)$) is isomorphic to $L(K_1^n)$ (respectively, $L(K_\ell^2)$) except when $n = 8$ (respectively, $\ell = 4$) in which case there are exactly 3 (respectively, 1) exceptions and all these exceptions are Seidel-equivalent to $L(K_1^8)$ (respectively, $L(K_4^2)$). In this paper we prove, using a result of Hoffman and Ray-Chaudhuri [8] that any graph cospectral with $L(K_\ell^n)$, where $\ell \geq 2$ and $n \geq 3$, is isomorphic to $L(K_\ell^n)$ except when $(n,\ell) = (4,2)$ in which case there are exactly 8 exceptions all Seidel-equivalent to $L(K_2^4)$. This is achieved by first proving that any graph cospectral with $L(K_\ell^n)$ is regular unless $(n,\ell) = (3,1)$ and next showing that any regular graph cospectral to K_ℓ^n is isomorphic to K_ℓ^n. Some of the arguments of this paper are implicitly or explicitly contained in Bussemaker, Cvetkovic and Seidel [3]. For a geometric characterisation of $L(K_\ell^n)$ with

---

* This paper was written while the second author was visiting Indian Statistical Institute during November-December 1979.

$(n-1)\ell > 10$ the reader is referred to Ramachandra Rao [9].

The <u>adjacency matrix</u> of G of order p with $V(G) = \{v_1,\ldots,v_p\}$ is the $p \times p$ matrix $A = A(G)$ where entries a_{ij} are given by

$$a_{ij} = \begin{cases} 1 & \text{if } (v_i, v_j) \in E(G), \\ 0 & \text{otherwise.} \end{cases}$$

The spectrum of a graph G is the set of numbers which are eigen-values of $A(G)$, together with their multiplicities as eigen-values of $A(G)$.

2. MAIN RESULTS

<u>Proposition 1</u>. K_ℓ^n is a strongly regular graph with spectrum given by

$$\begin{pmatrix} -\ell & 0 & (n-1)\ell \\ n-1 & n(\ell-1) & 1 \end{pmatrix}.$$

<u>Proof</u>. Omitted. See [2] for definition and details of strongly regular graphs.

<u>Proposition 2</u>. If G is a regular graph, cospectral to K_ℓ^n, then G is isomorphic to K_ℓ^n.

<u>Proof</u>. If A is the adjacency matrix of G, then its eigen-values are $-\ell$, 0, $(n-1)\ell$, with multiplicities $n-1$, $n(\ell-1)$, 1 respectively. Let J be the matrix of size $n\ell \times n\ell$, with each entry equal to 1. Since G is regular and multiplicity of $(n-1)\ell$ is 1, it is connected. Also $AJ = JA = (n-1)\ell\, J$.

Put $B = A^2 + \ell A - \ell(n-1)J$. B can be reduced to the diagonal form, using an invertible matrix N, such that the eigen-value $(n-1)\ell$ of A corresponds to $n\ell$ of J. It is easily checked that this reduced matrix NBN^{-1} is zero. Hence $B = 0$. The Hoffman polynomial [7] $p(x)$ of G is given by

$$p(x) = \frac{1}{(n-1)\ell}[x^2 + \ell x]$$

Since $p(x)$ has degree 2, G is strongly regular. Let p_{11}^1, p_{11}^2 be the parameters of the graph. From the defining relations,

$$A^2 = (n-1)\ell\, I + p_{11}^1 A + p_{11}^2 (J - I - A)$$

$$= [(n-1)\ell - p_{11}^2]\, I + (p_{11}^1 - p_{11}^2)\, A + p_{11}^2 J.$$

Since I, J, A are linearly independent, comparing this relation with $B = 0$, one gets

$$(n-1)\ell = p_{11}^2.$$

This implies that for any two nonadjacent vertices x, y of G, the number of vertices z, adjacent to x, but nonadjacent to y is zero, and hence the relation $x \sim y$ iff x is not adjacent to y is an equivalence relation on the vertices of G. Comparing the degree of a vertex in each equivalence class, it is easy to see that each class has ℓ elements and that there are n classes.

The following is a well known result.

<u>Proposition 3</u>. For any matrix N, the nonzero eigen-values of NN^t and N^tN are the same with the same multiplicities.

<u>Proof</u>. (Omitted)

<u>Proposition 4</u>. The line graph of K_ℓ^n, (with $(n-1)\ell > 2$) denoted by $L(K_\ell^n)$ has spectrum given by

$$\begin{pmatrix} -2 & (n-2)\ell - 2 & (n-1)\ell - 2 & 2[(n-1)\ell - 1] \\ \binom{n}{2}\ell^2 - n\ell & (n-1) & n(\ell-1) & 1 \end{pmatrix}$$

<u>Proof</u>. Let N be the vertex edge incidence matrix of the graph K_ℓ^n, A its adjacency matrix, A_L the adjacency matrix of its line graph, then the following equations hold [1]

i) $N^tN = A_L + 2I$

ii) $NN^t = A + D$

where D is the diagonal matrix whose entries are degrees of vertices in K_ℓ^n. Using this, along with Propositions 1 and 3 gives the result.

Remark. It may be noted that the eigen-value $(n-1)\ell - 2$ will not occur if $\ell = 1$. Also, -2 will not be an eigen-value if $(n-1)\ell \leq 2$.

Proposition 5. If G and H are graphs with the same spectrum and if G is regular, so is H.

Proof. Let $\lambda_1 \geq \lambda_2 \geq \dots \geq \lambda_n$ be the eigen-values of G (also of H), having n vertices.

Let $d_1 \geq d_2 \geq \dots \geq d_n$ be the degrees of vertices of H. Since λ_1 is the degree of G, we have

$$\sum_{i=1}^{n} \lambda_i^2 = 2 \text{ (number of edges of } G) = n\lambda_1 \qquad \dots \qquad (1)$$

Let A be the adjacency matrix of H. Put $u = (1,1,\dots,1)$, the all one vector. Considering the Rayleigh quotient, and noting that the quotient lies between the maximum and minimum eigen-values, [1], one gets

$$\frac{u A^2 u^t}{u u^t} \leq \lambda_1^2$$

But

$$\frac{u A^2 u^t}{u u^t} = \frac{1}{n} \sum_{i,j} (A^2)_{ij} = \frac{1}{n} \sum_{i=1}^{n} d_i^2 .$$

Hence

$$\sum_{i=1}^{n} d_i^2 \leq n \lambda_1^2 .$$

But

$$\sum_{i=1}^{n} d_i = \text{trace } A^2$$

$$= \sum_{i=1}^{n} \lambda_i^2$$

$$= n \lambda_1 . \qquad \dots \qquad (2)$$

Using Cauchy-Schwarz inequality,

$$n\lambda_1^2 \geq \sum_{i=1}^{n} d_i^2 \geq (\sum_{i=1}^{n} d_i)^2/n = n\lambda_1^2$$

i.e.,
$$\sum_{i=1}^{n} d_i^2 = n\lambda_1^2 \qquad \ldots \quad (3)$$

Using (2) and (3), it follows that $d_i = \lambda_1$ for $i = 1,\ldots,n$, i.e., H is regular

<u>Proposition 6</u>. If $G = L(H)$, the line graph of the graph H, and G is regular with the same spectrum as that of $L(K_\ell^n)$, then H is regular or $H = K_{1,3}$. ($K_{m,n}$ denotes the complete bipartite graph, with a bipartition into m and n vertices).

<u>Proof</u>. Suppose H is not regular. Since G is regular, H is bipartite. Also, since the leading eigen-value of G has multiplicity 1, it is connected and hence so is H.

Let the vertices of H be partitioned into two parts R, and S with r, s elements respectively. Assume that the degree of each vertex in R is u, that of each vertex in S is v. Let M be the incidence matrix of H and $A = \begin{pmatrix} 0 & B \\ B^t & 0 \end{pmatrix}$ be the adjacency matrix of H, where B is an $r \times s$ matrix. $MM^t = \begin{pmatrix} uI & B \\ B^t & vI \end{pmatrix}$, $M^tM = A_L + 2I$, A_L being the adjacency matrix of G.

Using Proposition 3, and noting that $u + v$ is an eigen-value of MM^t, it may be seen that $u + v - 2$ is the corresponding eigen-value of A_L, this eigen-value being of multiplicity 1 (as G is connected).

If $[x_1, x_2,\ldots,x_r ; y_1,\ldots,y_s]^t$ is the eigen vector of MM^t corresponding to $u + v$, then

$$[-vx_1, -vx_2,\ldots, -vx_r ; uy_1, uy_2,\ldots,uy_s]^t$$

is an eigen vector of MM^t, corresponding to the eigen-value 0 ; and conversely . Hence the multiplicity of 0 as an eigen-value of MM^t is 1. Using Proposition 4, we have

$$r + s = 1 + n\ell \qquad \ldots \qquad (1)$$

$$ru = s.v \qquad \ldots \qquad (2)$$

$$u + v - 2 = 2[(n-1)\ell - 1] \qquad \ldots \qquad (3)$$

$$ru = \frac{n(n-1)}{2}\ell^2 \qquad \ldots \qquad (4)$$

The equations (2), (3), (4) are obtained by observing that $ru = sv$ is the number of edges in H and $u + v - 2$ is the degree of G.

Using (1) - (4), eliminating r, s, v, one gets

$$u^2(n\ell+1) - 2u(n-1)\ell(n\ell+1) + n(n-1)^2\ell^3 = 0 \qquad \ldots \qquad (5)$$

This equation in u gives the two possible degrees of vertices in H. Using $r \geq v$, one gets

$$\frac{n(n-1)}{2}\frac{\ell^2}{u} \geq v,$$

$$n(n-1)\ell^2 \geq 2.u.v = 2.\frac{n(n-1)^2\ell^3}{n\ell + 1},$$

using equation (5). Hence $(n-2)\ell \leq 1$.

Either $n = 3$, $\ell = 1$ or $n = 2$ and ℓ is arbitrary. When $n = 2$, from (5) the value of u is given by

$$u = \ell \pm \frac{\ell}{\sqrt{2\ell + 1}},$$

which is not an integer for $\ell \geq 1$. When $n = 3$, $\ell = 1$, (5) gives $u = 1$ or 3. In this case, H is the graph $K_{1,3}$.

Remark. If in the proposition, one assumes that $(n-1)\ell > 2$, then the case $H = K_{1,3}$ cannot occur.

Theorem. Let G be a graph with the same spectrum as that of $L(K_\ell^n)$. If $(n-1)\ell > 2$, then G is isomorphic to $L(K_\ell^n)$ except when

$$(\ell,n) = (1,8),\ (2,4),\ (4,2).$$

<u>Proof</u>. Let G be a graph with the same spectrum as that of $L(K_\ell^n)$. For $\ell = 1$, it is known [4,5,6], that $L(K_\ell^n)$ is unique except when $n = 8$. For $n = 2$, it is known [4, 10] that $L(K_\ell^2)$ is unique except when $\ell = 4$.

Since $L(K_\ell^n)$ is regular, by Proposition 5, G is regular and hence has degree $2[(n-1)\ell - 1]$, using Proposition 4. When $(n-1)\ell > 2$, the smallest eigen-value of G is -2. Using a theorem of Hoffman and Raychaudhuri [8] G is the line graph of a graph H, if degree of G is more than 16. By Proposition 6, H is regular and the eigen-values of H may be computed using (i), (ii) in the proof of Proposition 4 and the multiplicities are given by Proposition 3. This turns out to be the same spectrum as that of K_ℓ^n.

By Proposition 2, H is isomorphic to K_ℓ^n.

If the degree of G is ≤ 16, then

$$(n-1)\ell \leq 9.$$

The possible values of $\ell, n, \ell \geq 2, n \geq 3$ are

$$(\ell,n) = (2,5),\ (2,4),\ (2,3),\ (3,4),\ (3,3),\ (4,3).$$

Using the list in [3] all these parameters are eliminated except (2,4). Hence the theorem follows.

<u>Remark</u>. The number of exceptions of type (1,8) is 3 [5,4], that of type (4,2) is 1 [10,4], and that of type (2,4) is 8[3]. Graphs of the same type are Seidel equivalent.

REFERENCES

1. N. Biggs, Algebraic Graph Theory, Camb. Tracts in Math., 67, 1974.
2. R.C. Bose, Strongly regular graphs, partial geometries and partially balanced designs, Pacific J. Math., 1963.
3. F.C. Bussemaker, D.M. Cvetkovic, J.J. Seidel, Graphs related to exceptional root systems, Colloquia Math., Soc. Janos Bolyai 18 (1976), 185-191.
4. P.J. Cameron, J.M. Goethals, J.J. Seidel, E.E. Shult, Line graphs, root systems and elliptic Geometry, J. Alg., 43 (1976), 305-327.
5. S. Chang, The uniqueness and nonuniqueness of triangular association scheme, Sci. Record 3 (1959), 604-613.
6. A.J. Hoffman, On the exceptional case in the characterisation of arcs of a complete graph, IBM J. Res. Dev., 4 (1960), 487-496.

7. A.J. Hoffman, On the polynomial of a graph, Amer. Math. Monthly, 70 (1963), 30-36.

8. A.J. Hoffman, D.K. Raychaudhuri, On the spectral characterisation of regular line graphs. Unpublished.

9. A. Ramachandra Rao, A characterisation of a class of regular graphs J.C.T., 10 (1971), 264-274.

10. S.S. Shrikhande, The uniqueness of the L_2 association scheme, Ann. Math. Stat., 30 (1959), 781-798.

11. A.J. Hoffman, B.A. Jamil, On the line graph of a complete tripartite graph, Lin. Mult. Algebra, 5 (1977), 19-25.

BALANCED ARRAYS FROM ASSOCIATION SCHEMES AND SOME RELATED RESULTS

G. M. SAHA*
Instituto de Matemática
Universidade Federal da Bahia, Salvador-BA, Brasil

ABSTRACT

A method of construction of (m+1)-symbol balanced arrays of strength two from certain association schemes of m classes is described in this paper. Reduction of these arrays to the incidence matrices of balanced incomplete block designs is also studied.

1. INTRODUCTION

After the introduction of balanced arrays by Chakravarty (1957,1961) as fractional replicates of factorial arrangement, many research workers have contributed to the methods of construction of these arrays. But most of these arrays have a small number s of symbols, namely, $s = 2,3$ or 4. Little has been done on the construction of balanced arrays with larger number of symbols.

In this paper, we present a method of construction of balanced arrays of strength two in (m+1) symbols from a special class of m-class cyclic association schemes, introduced by Saha et. al [4]. The method is described with an illustration in Section 3, In Section 4, we consider the reduction of these arrays into the incidence matrices of balanced incomplete block designs.

2. DEFINITIONS AND PRELIMINARIES

Definition 1. A $v \times b$ matrix $B = (b_{ij})$, $b_{ij} \in \Sigma = \{\sigma_1,\sigma_2,\ldots,\sigma_s\}$, a set of s symbols, is called a balanced array (BA) of strength t with v constraints and b assemblies (or runs), if in every $t \times b$ submatrix of B, a column vector $(i_1\ i_2\ \ldots\ i_t)'$ with $i_j \in \Sigma$ $(j = 1,2,\ldots,t)$ occurs $\lambda_{i_1 i_2 \ldots i_t}$ times, where $\lambda_{i_1 i_2 \ldots i_t}$ remains the

---

* On leave from Indian Statistical Institute, Calcutta.

same for all permutations of $i_1, i_2, \ldots, i_t$. The constants v,b,s,t and $\lambda_{i_1 i_2 \ldots i_t}$ are called the parameters of the BA, and we denote it by $BA(v,b,s,t;\{\lambda_{i_1 i_2 \ldots i_t}\})$.

In the particular case of $s = 2$, we normally take 0 and 1 as our symbols, i.e., $\Sigma = \{0,1\}$, and the BA is denoted by $BA(v,b,2,t;\ \mu_0,\mu_1,\ldots,\mu_t)$ where $\mu_j = \lambda_{i_1 i_2 \ldots i_t}$, j being the number of unities in $(i_1,i_2,\ldots,i_t)$.

<u>Definition 2</u>. Given v symbols $(1,2,\ldots,v)$, a relation satisfying the following conditions is said to be an association scheme with m classes :

(a) Any two symbols are either 1st., or 2nd.,..., or m-th. associates, the relation of association being symmetric, i.e., if the symbol α is an i-th associate of the symbol β, then β is an i-th associate of α too.

(b) Each symbol has exactly n_i i-th associates, the number n_i being independent of the symbol taken.

(c) If any two symbols α and β are i-th associates, then the number of symbols which are j-th associates of α and k-th associates of β is p^i_{jk}, and is independent of the pair of i-th associates α and β. Clearly $p^i_{jk} = p^i_{kj}$.

The numbers v, n_i, p^i_{jk}, $1 \leq i,j,k \leq m$, are the parameters of the association scheme, denoted by $A(v,n_i,p^i_{jk};\ i,j,k = 1,2,\ldots,m)$.

Saha et. al [4] define a new class of cyclic association schemes, called NC_m association schemes, as follows.

Let v be a prime or power of a prime and x, a primitive element of the Galois Field GF(v). It is well known that $x^{v-1} = 1$, and that all the elements of GF(v) can be expressed as $0,x^0,x^1,x^2,\ldots,x^{v-2}$. Let C be a multiplicative group of GF(v), and let $m \geq 2$ divide $(v-1)$, say, $s = (v-1)/m$. Then x^m has order s. Let $C_1 = \{x^{qm} | 0 \leq q \leq s-1\}$ be the cyclic subgroup generated by x^m, and $C_1,C_2,\ldots,C_m$ be the cosets of the factor subgroup C/C_1. Then $C_j = \{x^{j-1+qm} | 0 \leq q \leq s-1\} = x^{j-1}.C_1$; $1 \leq j \leq m$.

<u>Definition 3</u>. For any two symbols α and β, $\alpha,\beta \in GF(v)$, β will be said to be an i-th associate of α if and only if $(\alpha-\beta)$ belongs to C_i, $1 \leq i \leq m$. Such an association <u>relation</u> may be called NC_m (m-class new cyclic) association relation.

It is shown in Saha et. al [4] that for v satisfying

$$(v-1)/2 \equiv 0 \mod m, \text{ if } v \text{ is odd;}$$

$$(v-1) \equiv 0 \mod m, \text{ if } v \text{ is even;}$$

an NC_m association relation is an association scheme of m classes with the parameters :

$$v, n_i = s, p^i_{jk} = |(1+C_{j-i+1}) \cap C_{k-i+1}|,$$

where (j-i) and (k-i) are to be reduced mod m. An interesting property of an NC_m association scheme is the following :

Since $p^i_{jk} = |(1+C_u) \cap C_{u'}|$ where $u-1 \equiv j-i$, mod m and $u'-1 \equiv k-i$, mod m, we have :

$$\sum_{u=0}^{m-1} p^i_{[j+u][k+u]} = \text{the same for all } i, \text{ say, } \gamma_{jk},$$

where $[j+u] = j+u$, mod m, if $(j+u)$ is not a multiple of m; $= m$ otherwise; and $[k+u]$ is similarly defined.

3. THE METHOD OF CONSTRUCTION

Let $A(v, n_i, p^i_{jk};\ i,j,k = 1,2,\dots,m)$ be an m-class association scheme defined on a set Σ of v symbols, the parameters of which satisfy

$$\sum_{u=0}^{m-1} p^i_{[j+u][k+u]} = \gamma_{jk}, \text{ (same for all } i)$$

where $[j+u]$ and $[k+u]$ have the same meaning as in Section 2.

Define the matrices M_u as

$$M_u = (m^u_{ij})_{v \times v}, \quad u = 0,1,2,\dots,m-1,$$

with

$$m^u_{ij} = 0, \text{ for } i = j \in \Sigma;$$

$$= [k+u], \text{ for } i \neq j \in \Sigma,$$

where the symbols i and j are k-th associates.

One can then prove the following

<u>Theorem 1</u>. The matrix $B = [M_o | M_1 | \dots | M_{m-1}]$ is a balanced array for (m+1) symbols $(0,1,2,\dots,m)$ in mv assemblies and v constraints, and is of strength two with the

following λ-parameters :

$$\lambda_{00} = 0,\ \lambda_{i0} = \lambda_{0i} = 1,\ \lambda_{ij} = \lambda_{ji} = \gamma_{ij};\ i,j \in \{1,2,\ldots,m\}.$$

Proof. It is easy to see that in any 2-rowed submatrix of B the vector $\binom{0}{0}$ does not appear at all, while the vectors $\binom{0}{i}$ and $\binom{i}{0}$ appear each exactly once. Hence $\lambda_{00} = 0$, $\lambda_{0i} = \lambda_{i0} = 1$. To show that $\lambda_{ij} = \lambda_{ji} = \gamma_{ij}$, we observe that in any 2-rowed submatrix of M_u, the vector $\binom{i}{j}$, $i,j \in \{1,2,\ldots,m\}$ appears in p^a_{ij} columns, where the two symbols corresponding to the two rows considered are a-th associates. Hence in B, $\binom{i}{j}$ appears in γ_{ij} columns.

Example 1. Let $v = 7$ and $m = 3$. $x = 3$ is a primitive element of GF(7). Then $C_1 = (1,6)$, $C_2 = (2,5)$ and $C_3 = (3,4)$. Hence, denoting by P^i the matrices (p^i_{jk}), $i = 1,2,3$, we have :

$$P^1 = \begin{bmatrix} 0 & 1 & 0 \\ 1 & 0 & 1 \\ 0 & 1 & 1 \end{bmatrix},\quad P^2 = \begin{bmatrix} 1 & 0 & 1 \\ 0 & 0 & 1 \\ 1 & 1 & 0 \end{bmatrix},\quad P^3 = \begin{bmatrix} 0 & 1 & 1 \\ 1 & 1 & 0 \\ 1 & 0 & 0 \end{bmatrix},$$

which clearly shows that this association scheme satisfies all the properties required for the construction of balanced arrays. The method of construction described above yields :

$$M_0 = \begin{bmatrix} 0 & 1 & 2 & 3 & 3 & 2 & 1 \\ 1 & 0 & 1 & 2 & 3 & 3 & 2 \\ 2 & 1 & 0 & 1 & 2 & 3 & 3 \\ 3 & 2 & 1 & 0 & 1 & 2 & 3 \\ 3 & 3 & 2 & 1 & 0 & 1 & 2 \\ 2 & 3 & 3 & 2 & 1 & 0 & 1 \\ 1 & 2 & 3 & 3 & 2 & 1 & 0 \end{bmatrix}$$

Therefore $B = [M_0 | M_1 | M_2]$, where the matrix M_1 is obtained from M_0 by changing the symbols 0,1,2,3 to 0,2,3,1 respectively, and the matrix M_2 is obtained from M_0 by changing the symbols 0,1,2,3 to 0,3,1,2 respectively, is a four-symbol balanced array of strength two in 7 constraints and 21 assemblies.

Remark 1. It can be observed that to construct a BA the way described in this section, one need not have to start always with an association scheme. In fact, an NC_m association relation, as in Definition 4, also leads to a balanced array, provided

w is joined to all vertices in B. Suppose that $(w,i_o) \notin E(G_o)$ for some i_o, $1 \leq i_o \leq w-1$. Define A_i, $1 \leq i \leq 3$, as in the proof of the Lemma 4.3; and it can be verified then that A_3 is complete in G_o if $A_3 \neq \phi$; and also if $A_2 \neq \phi$, $(A_1 \neq \phi)$ then every point of A_2 (A_1) is joined to all vertices of A_3 and if both A_1, A_2 are nonempty then every point of A_1 is joined to all points of A_2. Further, if $A_3 = \phi$, then by (1) A_1 and A_2 are both nonempty.

<u>Claim 1.</u> $d_{i_o} \leq 3w - 2$.

<u>Proof.</u> Suppose that the contrary holds.

If $A_1 = \phi$, then $|A_3| \geq 2w$ implying that $d_w \geq 3w-2$ (or $w(G_o) \geq 2w$). Thus we may assume that $A_1 \neq \phi$. If $A_3 = \phi$, $A_1 \cup A_2$ is complete in G_o, $|A_2| \geq 1$ and therefore $|A_1| \geq 2w$ which imply that $w(G_o) \geq w$ and by Lemma 4.3 we have a contradiction to the selection of G_o. Therefore $A_3 \neq \phi$. By what has been already proved before Claim 1, we have, for any $x \in A_3$, that $d_{G_o}(x) \geq 2w$, with $x > w$, contradicting (1), proving Claim 1.

Let $B_o = A$, $B_1 = A_1 \cup A_2 \cup A_3$. By (1) and Claim 1 we have that $1 \leq |B_1| \leq 4w$. Let B_2 be the set of all vertices in $V_1 = V - (B_o \cup B_1)$ which are joined to at least one vertex in B_1; by (1) we have $|B_2| \leq 8w^2$.

<u>Claim 2.</u> B_2 is nonempty.

<u>Proof.</u> Suppose B_2 is empty. Since $N(w) \geq f(w)$ by (1) we have that V_1 is not independent in G_o. Suppose $(x,y) \in E(G_o)$ with x,y in V_1. If $A_3 \neq \phi$ and $Z \in A_3$ let $\mu = (Z,w,i_o,Z,x,y)$ and if $A_3 \neq \phi$ let $Z_i \in A_i$, $i = 1,2$ and $\mu = (Z_2,i_o,w,Z_1,y,x)$. Interchanging edges along this μ we get $G' \in \underline{\underline{G}}(\pi)$ which contradicts the selection of G_o.

BY (1) and Claim 1 we have that $|B_2| \leq 8w^2$. Let B_3 be the set of all vertices in $V_2 = V_1 - B_2$ which are joined to at least one vertex in B_1. Clearly $|B_3| \leq 16w^3$ and no vertex of B_3 is joined to a vertex in B_1. Now as in the proof of Claim 2 it can be shown that B_3 is nonempty. Let $B_4^* = V_2 - B_3$.

<u>Claim 3.</u> B_4^* is an independent set in G_o and no vertex of B_4^* is joined to a vertex in $B_1 \cup B_2 \cup B_3$.

it satisfies some symmetry conditions. Saha et. al [4] showed that for an NC_m association relation, $q^i_{jk}(\alpha,\beta)$ remains a constant for all ordered pairs (α,β) where β is an i-th associate of α, and that

$$\sum_{u=0}^{m-1} q^i_{[j+u][k+u]}(\alpha,\beta) = \text{same for all } i$$
$$= \sigma_{jk} \text{ (say)},$$

where $q^i_{jk}(\alpha,\beta)$ denotes the number of symbols which are simultaneously j-th associates of α and k-th associates of β, where β is an i-th associate of α (α not necessarily being an i-th associate of β), and $[j+u]$, $[k+u]$ have the same meanings as explained earlier. Thus, if an NC_m association relation satisfies $\sigma_{jk} = \sigma_{kj}$, it will yield a balanced array by the method of construction described in this section. For example, for $v = 7$, $m = 2$, the NC_m association relation satisfies these conditions, and hence leads to a balanced array as shown in Example 2 below. Exactly these types of arrays were constructed by Saha and Gupta [3] from the incidence matrices of a series of balanced ternary designs, which are alternatively constructible from the NC_m association relations with $m = 2$ for any prime power v.

Example 2. $v = 7$, $m = 2$; $q^1_{11}(\alpha,\beta) = 1$; $q^1_{12}(\alpha,\beta) = 1$, $q^1_{21}(\alpha,\beta) = 2$; $q^1_{22}(\alpha,\beta) = 1$; $q^2_{11}(\alpha,\beta) = 1$, $q^2_{12}(\alpha,\beta) = 2$, $q^2_{21}(\alpha,\beta) = 1$, $q^2_{22}(\alpha,\beta) = 2$.

Thus the NC_m association relation satisfies the required symmetry conditions. Hence $B = [M_0 | M_1]$ is a $BA(7,14,3,2;\{\lambda_{i_1 i_2}\})$, where,

$$M_0 = \begin{bmatrix} 0 & 1 & 1 & 2 & 1 & 2 & 2 \\ 2 & 0 & 1 & 1 & 2 & 1 & 2 \\ 2 & 2 & 0 & 1 & 1 & 2 & 1 \\ 1 & 2 & 2 & 0 & 1 & 1 & 2 \\ 2 & 1 & 2 & 2 & 0 & 1 & 1 \\ 1 & 2 & 1 & 2 & 2 & 0 & 1 \\ 1 & 1 & 2 & 1 & 2 & 2 & 0 \end{bmatrix}$$

and M_1 is constructible from M_0 by permuting the symbols (0,1,2) to (0,2,1).

Remark 2. It is not necessary that only the NC_m association schemes satisfy the required conditions to construct a BA. As an example, we got the T_2-association scheme (i.e., the triangular association scheme of two classes) for $v = 10$ symbols. In fact, this scheme has

$$n_1 = 6,\ n_2 = 3;\quad P^1 = \begin{bmatrix} 3 & 2 \\ 2 & 1 \end{bmatrix},\quad P^2 = \begin{bmatrix} 4 & 2 \\ 2 & 0 \end{bmatrix},$$

which clearly satisfies

$$p_{11}^1 + p_{22}^1 = p_{11}^2 + p_{22}^2, \text{ and } p_{12}^1 + p_{21}^1 = p_{12}^2 + p_{21}^2,$$

and hence leads to a three-symbol balanced array of strength two for 10 constraints in 20 assemblies.

4. SOME BIB DESIGNS

We can construct incidence matrices of balanced incomplete block designs (BIBD) by replacing the $(m+1)$ symbols $(0,1,2,\ldots,m)$ of the balanced arrays of Section 3 by the symbols 0 and 1. It is observed that in each column of B the symbol 0 occurs once and each of the symbols $(1,2,\ldots,m)$ occurs $s = (v-1)/m$ times, while in each row of B, the symbol 0 occurs m times and each of the symbols $(1,2,\ldots,m)$ occurs $(v-1)$ times. Let us, therefore, define :

(i) N_1 = a (0,1) - matrix obtained from B by replacing any c $(1 \leq c \leq m)$ symbols of $(1,2,\ldots,m)$ by 1, and the remaining $(m-c)$ symbols by 0;

(ii) N_2 = a (0,1) - matrix obtained from B by replacing 0 by 1, any $c(1 \leq c \leq m)$ symbols of $(1,2,\ldots,m)$ by 1, and the remaining $(m-c)$ symbols by 0.

Since N_1 and N_2 are also two-symbol balanced arrays of strength two, it is clear that they are the incidence matrices of BIB designs with the following parameters :

(i) $N_1 : v_1 = v,\ b_1 = mv,\ r_1 = c(v-1),\ k_1 = cs,\ \lambda_1 = c(cs-1);$

(ii) $N_2 : v_2 = v,\ b_2 = mv,\ r_2 = c(v-1) + m,\ k_2 = cs + 1,\ \lambda_2 = c(cs+1);$

where $s = (v-1)/m$.

Thus, we arrive at the following

<u>Theorem 2</u>. BIB designs with the parameters (i) and (ii) given above <u>exist</u> for any prime power v, any integer m that divides $(v-1)$ and any integer c, $1 \leq c \leq m$, whenever we have :

(a) $(v-1)/2 \equiv 0 \mod m$, if v is odd;

(b) $(v-1) \equiv 0 \mod m$, if v is even.

Since the association schemes used are cyclic, it is easily realised that these BIB designs are directly constructible through the well-known method of differences from the following m "initial blocks" :

(i) Design N_1 : $I_j = (C_{[i_1-j]}, C_{[i_2-j]}, \ldots, C_{[i_c-j]})$;

(ii) Design N_2 : $I_j = (0, C_{[i_1-j]}, C_{[i_2-j]}, \ldots, C_{[i_c-j]})$;

where

$j = 0,1,2,\ldots,m-1$; $1 \leq c \leq m$; $(i_1,i_2,\ldots,i_c)$ is any c-subset of $(1,2,\ldots,m)$;

$[i_u-j] = (i_u-j) \mod m$, if this is non-zero;

$= m$, if $(i_u-j) \mod m$ is zero;

$u = 1,2,\ldots,c$;

C_i's being the cosets of the factor subgroup C/C_1 as described in Section 2.

Remark 3. The difference - set solutions of these BIB designs appear to be known in literature. But a study through the NC_m association schemes, as done in this section, offers an alternative and simpler proof that these initial blocks yield BIB designs. And it also reveals the fact that there is really a great variety of choices for the initial blocks of these BIB designs.

REFERENCES

1. I.M. Chakravarti, Fractional replication in asymmetrical factorial designs and partially balanced arrays, Sankhyā, 17(1957), 143-164.
2. I.M. Chakravarti, On some methods of construction of partially balanced arrays, Ann. Math. Statist., 32(1961), 1181-1185.
3. G.M. Saha and T.K. Gupta, On construction of three-symbol partially balanced arrays of strength two, Jour. Ind. Soc. Agril. Statist., 29(1977),
4. G.M. Saha and A.C. Kulshreshtha and A. Dey, On a new type of m-class cyclic association scheme and designs based on the scheme, Ann. Statist., 1(1973), 985-990.

SOME FURTHER COMBINATORIAL AND CONSTRUCTIONAL ASPECTS OF GENERALIZED YOUDEN DESIGNS

BIKAS KUMAR SINHA
Indian Statistical Institute
203 B. T. Road, Calcutta 700 035

ABSTRACT

The motivation for discussing the combinatorial and constructional aspects of the Generalized Youden Designs (GYDs) is largely based on optimality considerations (Kiefer [4]). In this article, we supplement the work of Kiefer by presenting additional new results and observations.

1. INTRODUCTION

Latin Square (LS) arrangements provide the simplest type of statistical designs eliminating heterogeneity in two directions (along the rows and columns). These were extended to Youden Square Designs (YSDs) in [9] to cover up the cases of incomplete blocks. However, the Generalized Youden Designs (GYDs) introduced by Kiefer [1] offer the experimenter maximum flexibility regarding the choice of the 'design parameters' in his experimental set-up. A subclass of the GYDs, known as regular GYDs, cover up the LSDs and the YSDs as its initial members. The complementary subclass consists of designs, known as nonregular GYDs. Over the last twenty years, Kiefer's classical work on 'optimal designs' has revealed many interesting optimality properties of the GYDs (both regular and nonregular) and it is only recently that some combinatorial and constructional aspects of such designs have been discussed, Ruiz and Seiden [6], Kiefer [4]. The regular GYDs do not pose any serious constructional problem; in particular, as is well-known, the LSDs always exist and the YSDs may be constructed by applying Smith-Hartley technique to Symmetrical Balanced Incomplete Block Designs (SBIBDs). In fact, the existence (and the actual construction as well) of a regular GYD is essentially dependent on the existence of some BIBD. As regards the non-regular GYDs, Ruiz and Seiden [6] discussed geometric methods of construction of some families of them. Subsequently, some basic techniques have been discussed in Kiefer [4].

Particularly, his patchwork methods appear to be very powerful. In order to give an adequate publicity to this method, we intend, in this article, to supplement his work by incorporating some observations on his methods and by presenting some new families of the GYDs with solutions attainable by the patchwork methods.

In Section 2, we present a formal definition of the GYDs and discuss some combinatorial results. In Section 3, we record our observations and new findings on the nonregular GYDs.

2. GENERALIZED YOUDEN DESIGNS (GYDs)

To start with, for the sake of completeness, we present the following definition of a GYD.

Definition 1. A $(k \times b)$ array formed of v symbols $(1,2,\ldots,v)$ is called a GYD with parameters v,b and k if the following conditions are satisfied:

(a) treating the rows as blocks, it is a Balanced Block Design (BBD) (vide Kiefer [1])involving v treatments in k blocks of b plots each;

(b) treating the columns as blocks, it is also a BBD involving v treatments in b blocks of k plots each.

A GYD with parameters v,b and k will be denoted as GYD(v,b,k). The optimality considerations have lead to a classification of the GYDs as regular and nonregular as defined below.

Definition 2. A GYD(v,b,k) is called regular if at least one of b and k is an integral multiple of v; otherwise, it is called nonregular.

Now we present an inequality involving b,v and k in the non-regular set-up. The more familiar inequality ($b,k > v$ necessarily) is based on Fisher's inequality, stated below, for the BBDs.

For a BBD with parameters b,v and k, $b \geq v$ unless k is an integral multiple of v.

We recall that the proof of this inequality itself provides a general method of construction of the BBDs. We will refer to this method as method of reduction as it reduces the problem to one of getting some BIBD, vide Kiefer [4], Proposition 1.

The setting for our inequality would be as follows. It has been developed in the context of optimality considerations for nonregular GYDs (Kiefer [5], Sinha [7]). Suppose there is a nonregular setting with the parameters v,b and k where $b \geq k > v$ Let $\underline{\underline{N}} = \{n : o \leq n \leq bk,\ n$ is an integral multiple of b or $k\}$. If $C,D \in \underline{\underline{N}}$, $C < D$ and no integer between C and D is in $\underline{\underline{N}}$, we call $[C,D]$ an elementary interval. The interval $[C_o,D_o]$ containing the integer $r\ (= bk/v)$ is called the basic interval. Since the setting is nonregular, we have $C_o < r < D_o$. Then our result is stated as follows.

Lemma 1. $k[b/v] \leq \frac{bk}{v} - \frac{k}{k-v}$ if C_o is an integral multiple of k,

$b[k/v] \leq \frac{bk}{v} - \frac{b}{b-v}$ if C_o is an integral multiple of b.

Proof. Suppose $C_o = k\ \delta \geq b\ B$ where $B = [C_o/b]$. Clearly, then, $\delta = [b/v]$. Set $v = v_1v_2$, $b = v_1b'$ and $k = v_2k'$ for some v_1, v_2, b' and k'. Then

$$C_o = k[b/v] = k[b'/v_2] \leq k(\frac{b'-1}{v_2}) = b'k' - k' = \frac{bk}{v} - k'.$$

Again, $k > v \rightarrow k' > v_1$, i.e., $k' \geq v_1 + 1$. Hence, $k = v_2k' \geq v_2(v_1+1) = v + v_2$, so $k - v \geq v_2$. Using this, we get $C_o \leq \frac{bk}{v} - \frac{k}{v_2} \leq \frac{bk}{v} - \frac{k}{k-v}$. The same argument holds for the other inequality as well.

Note 1. These inequalities are of a general nature and apply to any nonregular setting. Since, in such a setting, $v \geq 6$, one might state the inequalities as

$$k[b/v] \leq \frac{bk}{6} - \frac{k}{k-6},\quad b[k/v] < \frac{bk}{6} - \frac{b}{b-6}.$$

3. CONSTRUCTIONAL ASPECTS OF THE NONREGULAR GYDs

We intend to present the essential features of the patchwork methods of Kiefer. While doing so, we will incorporate some new observations and additional results with a view to point out the wide applicability of these methods. We will not discuss the regular case further.

Suppose the parameters are v,b and k with $b = a_2v + c_2$ and $k = a_1v + c_1$. The $(k \times b)$ array can be thought of as a partitioned array

$$G = \left[\begin{array}{c|c} G_{11}(a_1v \times a_2v) & G_{12}(a_1v \times c_2) \\ \hline G_{21}(c_1 \times a_2v) & G_{22}(c_1 \times c_2) \end{array}\right]$$

We now recall the definition of a GYD and the method of reduction (for constructing the BBDs) to conclude as follows.

For the GYD(v,b,k) to exist, it is necessary and sufficient that

(i) $\begin{bmatrix} G_{12} \\ G_{22} \end{bmatrix}$ forms a BIBD(a_1v+c_1,v,c_2) provided $\begin{bmatrix} G_{11} \\ G_{21} \end{bmatrix}$ is already composed of a_2 replications of each of the v symbols in each row, and

(ii) (G_{21} G_{22}) forms another BIBD(a_2v+c_2,v,c_1) provided (G_{11} G_{12}) is already composed of a_1 replications of each of the v symbols in each column.

Combining the two, we may set down a set of sufficient conditions for the GYD to exist (and to be constructed practically) :

(a) $G_{11}(a_1v \times a_2v)$ is a regular GYD(v,a_2v,a_1v),

(b) (G_{21} G_{22}) is a BIBD(a_2v+c_2,v,c_1) with the part G_{22} (consisting of c_2 blocks each of size c_1) containing an equal number of replications of the v symbols,

(c) $\begin{bmatrix} G_{12} \\ G_{22} \end{bmatrix}$ is also a BIBD(a_1v+c_1,v,c_2) with the part G_{22} (this time consisting of c_1 blocks each of size c_2) containing an equal number of replications of the v symbols.

The form of G_{22}, as suggested by (b) and (c), in combination with Proposition 2 in Kiefer [4], assures one of the required forms of G_{12} and G_{21} in (i) and (ii) above.

Looking to the sufficient conditions, we readily see that (a) always holds. Hence, essentially, one has to get hold of the two BIBDs with the 'dual' role of the array G_{22}. From (b) and (c), we have $vt = c_1c_2$ for some integral $t \geq 1$. One gets striking simplicity in case $t = 1$ which we discuss first. This simply means that G_{22} has to form a set of blocks providing single replication of each of the v symbols in each of the two BIBDs concerned. A BIBD with this property has been called by Kiefer [4], partly resolvable (PR). Thus, existence of the two PRBIBDs solves the problem completely as they can always be matched over G_{22}, thereby enabling G_{22} play its dual role. This is the essence of Proposition 5 in Kiefer [4].

Again, G_{22} can be made to satisfy (b) and (c) if the following hold :

(b') the BIBD in (b) is composed of at least c_2 copies of some BIBD,

(c') the BIBD in (c) is composed of at least c_1 copies of some BIBD.

Clearly, (b') and (c') imply respectively the existence of the PRBIBDs involved in the case of $t = 1$ but certainly they are of wide applicability. One might combine (b) with (c') or (c) with (b') to get a solution. This is the result in Proposition 6 of Kiefer [4].

There are many classes of nonregular GYDs that can be covered by this technique. Kiefer [4] himself illustrated some of them. We will present some further general results. But, before that, we would like to discuss Proposition 9 in Kiefer [4] and clarify an observation made by him.

In his Proposition 9, Kiefer [4] established the following: For every $t > 1$, whenever the BIBD($b = 2(4t-1)$, $v = 4t$, $k = 2t$) exists, the GYD($v = 4t$, $b = 2t(4t-1) \times (2J_2-1)$, $k = 2t(4t-1)(2J_1-1)$) exists if <u>either $J_i \geq 2$ and the other is equal to 1.</u>

Following Kiefer, we enter into the following discussion: Suppose $J_2 \geq 2$ and $J_1 = 1$ so that $k = 2t(4t-1)$. Then the BIBD($b = 2t(4t-1)(2J_2-1)$, $v = 4t$, $k = 2t$) can be thought of as $t(2J_2-1) \geq 2t$ copies of the BIBD($b = 2(4t-1)$, $v = 4t$, $k = 2t$). Hence, the columns of G_{22} can be made to form t pairs with each pair containing a complete replication of the v symbols. However, $J_1 = 1$ implies there are only t copies of this BIBD (referred in the statement of the result) available for the formation of the rows of G_{22}. We want to illustrate that even in such a situation, the rows of G_{22} can suitably be formed. This becomes possible upon observing that in the BIBD($b = 2(4t-1)$, $v = 4t$, $k = 2t$), there are at least two blocks with an even number of symbols in common. To be specific, let the common symbols be $(1,2,\ldots,2t')$ for some $t' < t$. We form $2t$ pairs with the $4t$ symbols as $G_i = (2i-1,2i)$, $1 \leq i \leq 2t$. Further, let $(G_i) = (2i,2i-1)$ (interchange of the positions in G_i), $1 \leq i \leq 2t$. We now indicate below a structure of G_{22} meeting the desired objective. In the following, G_i's and (G_i)'s are to be regarded as column vectors.

$$\underline{t' \text{ odd}} \quad G_{22} = (A \,|\, (A_1) \,|\, B \,|\, (B_1)),$$

$$\underline{t' \text{ even}} \quad G_{22} = (A^* \,|\, (A_1^*) \,|\, B^* \,|\, (B_1^*)),$$

where the components are all partitioned matrices (each an incomplete circulant progressing columnwise) formed of the G_i's or (G_i)'s as a whole.

We write down A in detail.

$$
A = \begin{array}{l} 1 \\ 2 \\ 3 \\ 4 \\ \cdot \\ \cdot \\ t'-2 \\ t'-1 \\ t' \\ t'+1 \\ t'+2 \\ t'+3 \\ \cdot \\ \cdot \\ \cdot \\ \cdot \\ t \end{array}
\left[\begin{array}{ccccccc|cccccc}
G_1 & G_t & \cdot & \cdot & \cdot & \cdot & \cdot & \cdot & \cdot & \cdot & \cdot & \cdot & \cdot \\
G_2 & \cdot & \cdot & \cdot & \cdot & \cdot & \cdot & \cdot & \cdot & \cdot & \cdot & \cdot & \cdot \\
\cdot & G_1 & \cdot & \cdot & \cdot & \cdot & \cdot & \cdot & \cdot & \cdot & \cdot & \cdot & \cdot \\
\cdot & G_2 & \cdot & \cdot & \cdot & \cdot & \cdot & \cdot & \cdot & \cdot & \cdot & \cdot & \cdot \\
\cdot & \cdot & \cdot & \cdot & \cdot & \cdot & \cdot & \cdot & \cdot & \cdot & \cdot & \cdot & \cdot \\
\cdot & \cdot & \cdot & \cdot & \cdot & \cdot & \cdot & \cdot & \cdot & \cdot & \cdot & \cdot & \cdot \\
\cdot & \cdot & \cdot & \cdot & \cdot & G_1 & \cdot & \cdot & \cdot & \cdot & \cdot & \cdot & \cdot \\
\cdot & \cdot & \cdot & \cdot & \cdot & G_2 & \cdot & \cdot & \cdot & \cdot & \cdot & \cdot & \cdot \\
\cdot & \cdot & \cdot & \cdot & \cdot & \cdot & G_1 & \cdot & \cdot & \cdot & \cdot & \cdot & \cdot \\
\cdot & \cdot & \cdot & \cdot & \cdot & \cdot & G_2 & \cdot & \cdot & \cdot & \cdot & \cdot & \cdot \\
\cdot & \cdot & \cdot & \cdot & \cdot & \cdot & \cdot & G_1 & \cdot & \cdot & \cdot & \cdot & \cdot \\
\cdot & \cdot & \cdot & \cdot & \cdot & \cdot & \cdot & G_2 & G_1 & \cdot & \cdot & \cdot & \cdot \\
\cdot & \cdot & \cdot & \cdot & \cdot & \cdot & \cdot & \cdot & G_2 & \cdot & \cdot & \cdot & \cdot \\
\cdot & \cdot & \cdot & \cdot & \cdot & \cdot & \cdot & \cdot & \cdot & \cdot & \cdot & \cdot & \cdot \\
\cdot & \cdot & \cdot & \cdot & \cdot & \cdot & \cdot & \cdot & \cdot & \cdot & \cdot & \cdot & \cdot \\
\cdot & \cdot & \cdot & \cdot & \cdot & \cdot & \cdot & \cdot & \cdot & \cdot & \cdot & G_1 & \cdot \\
\cdot & \cdot & \cdot & \cdot & \cdot & \cdot & \cdot & \cdot & \cdot & \cdot & \cdot & G_2 & G_1
\end{array}\right] = [A_1 | A_2] \quad (t' \text{ odd})
$$

Then $(A_1) = A_1$ with G_i replaced by (G_i), $1 \le i \le t$. B involves $G_{t+1}, G_{t+2}, \ldots, G_{2t}$ and is of the same form as A except that the partitioning is made one step before, that is, when G_{t+1} reaches the row numbered $t'-2$. The two partitions are named B_1 and B_2 and then $(B_1) = B_1$ with G_i replaced by (G_i), $t+1 \le i \le 2t$. For t' even, the matrix A^* is written in the same fashion as A and this time proceeds through the rows numbered $1, 3, \ldots$ and stops at $t'-1$ and this yields A_1^* which in its turn produces (A_1^*). A_2^* starts at row number $t'+1$ and reaches the bottom without any discontinuity. This time, B^*, B_1^*, B_2^*, (B_1^*) are of identical forms as $A^*, A_1^*, A_2^*, (A_1^*)$ respectively but with considerations of $(G_{t+1}, \ldots, G_{2t})$ instead of $(G_1, \ldots, G_t)$. This supports an observation made by Kiefer [4].

Next we present some further results under the general case: $vt = c_1c_2$, $b = a_2v+c_2$, $k = a_1v+c_1$, $t \ge 1$. Kiefer [4] has also discussed some of the cases covered under this set-up. The cases include : (a) each c_i-1 is relatively prime to c_1c_2-1, (b) $c_1 = 3$, $c_2 = 2q+1$.

We consider here the case of $v > c_1 = c_2 = c$(say), $vt = c^2$ with $t > 1$. Note that c must be of the form $c = f_1f_2$ with each $f_i > 1$. Take $t = f_1$ and $v = f_1f_2^2$ so that for the GYD, the parameters are $v = f_1f_2^2$, $b = f_1f_2(a_2f_2+1)$, $k = f_1f_2(a_1f_2+1)$ for some $a_1, a_2 \geq 1$. The problem gets reduced to solving for the BIBDs($b = f_1f_2(a_if_2+1)$, $v = f_1f_2^2$, $k = f_1f_2$). This yields $\lambda = f_1(a_if_2+1)\ (f_1f_2-1)/(f_1f_2^2-1)$ and since $f_i > 1$, $f_1(f_1f_2-1)$ is relatively prime to $(f_1f_2^2-1)$ so that we must set $a_if_2+1 = t_i(f_1f_2^2-1)$. The BIBDs now may be formed of $t_i\ f_1$ copies of the BIBD($b = f_2(f_1f_2^2-1)$, $v = f_1f_2^2$, $k = f_1f_2$). Again, in order that the method applies, we set $t_i \geq f_2$. But, then, for a_i to be an integer, it is n.s. that $t_i = J_i(f_2-1)$ for some $J_i \geq 2$. Hence, finally, we have the following result :

Existence of the BIBD($b = f_2(f_1f_2^2-1)$, $v = f_1f_2^2$, $k = f_1f_2$) implies existence of the GYD($v = f_1f_2^2$, $b = f_1f_2(f_1f_2^2-1)\ (f_2J_2-1)$, $k = f_1f_2(f_1f_2^2-1)\ (f_2J_1-1)$) for every $J_i \geq 2$, $i = 1,2$.

In the particular case of $f_1 = f_2 = x$ (say), the BIBD involved has the parameters $b = x(x^3-1)$, $v = x^3$ and $k = x^2$ and this BIBD again can be regarded as $(x-1)$ copies of a different BIBD whose parameters are $b = x(x^2+x+1)$, $v = x^3$, $k = x^2$. These designs can easily be constructed from $EG(3,p^n)$ for $x = p^n$ = prime or a prime power.

Note 2. Kiefer [4] also illustrates uses of union, derivation, residuation and complementation in the construction of GYDs.

ACKNOWLEDGEMENT

The author would like to thank the referee for suggesting this form of presentation in preference to an earlier version which was presented during the conference.

REFERENCES

1. J. Kiefer, On the nonrandomized optimality and randomized nonoptimality of symmetrical designs, Ann. Math. Stat., 29(1958), 675-699.

2. J. Kiefer, Optimum experimental designs, J. Roy. Stat. Soc., Ser.B, 21(1959), 272-319.

3. J. Kiefer, The role of symmetry and approximation in exact design optimality. Statistical Decision Theory and Related Topics, 1971, 109-118. Academic Press, New York.

4. J. Kiefer, Balanced block designs and generalized Youden designs, I. Construction (patchwork), Ann. Stat., 3(1975), 109-118.

5. J. Kiefer, Construction and optimality of generalized Youden designs in A survey of Statistical Design and Linear Models, 1975, 333-353, Ed J.N. Srivastava, North-Holland.

6. F. Ruiz and E. Seiden, On construction of some families of generalized Youden designs, Ann. Stat. 2(1974), 503-519.

7. Bikas Kumar Sinha, Optimal experimental designs, Unpublished Seminar notes, Indian Statistical Institute, Calcutta, 1979.

8. C.A.B. Smith and H.O. Hartley, The construction of Youden squares, Jour. Roy. Stat. Soc., Ser. B, 10(1948), 262-263.

9. W.J. Youden, Use of incomplete block replications in estimating tobacco mosaic virus, Contributions from Boyca Thompson Institute, 9(1937), 317-326.

MAXIMUM DEGREE AMONG VERTICES OF A NON-HAMILTONIAN HOMOGENEOUSLY TRACEABLE GRAPH

ZDZISLAW SKUPIEN
Inst. Mat. AGH. Krakow, Poland
and
Mathematics Department
Kuwait University
P. O. Box 5969, Kuwait

The aim of this note is to present a simple proof of the following

Theorem 1. If G is a non-Hamiltonian homogeneously traceable graph of order $n \geq 3$, then $\Delta(G) \leq n - 4$.

This theorem is proved in [1]. In what follows a simpler proof and some refinements are presented. The theorem itself and the proof given below were found by the present author independently of [1] at the beginning of 1978.

Throughout, "graph G" means a "simple graph G" with vertex set $V(G)$, edge set $E(G)$, and with the minimum degree $\delta(G)$ and the maximum degree $\Delta(G)$. The degree of a vertex x of G is denoted by $d(x)$.

Following Skupien [7], a graph G is called homogeneously traceable if, for each vertex x of G, there is a Hamiltonian path beginning at x. Following Jung [3], the scattering number $s(G)$ of G is defined as follows :

$$s(G) = \max\{k(G-S) - |S| : S \subseteq V(G) \text{ and } K(G-S) \neq 1\}$$

where $k(G-S)$ stands for the number of components of $G-S$.

We shall make use of the following lemma.

Lemma 1. ([6]) If G is a HT graph then

(i) G is 2-connected (whence $\delta(G) \geq 2$) if $K_1 \neq G \neq K_2$,

(ii) its scattering number $s(G) \leq 0$,

(iii) if G is non-trivially non-Hamiltonian (i.e., non-Hamiltonian with $n \geq 3$ vertices) then each vertex of G has at most one neighbour of degreee ≤ 2.

Theorem 1 is an immediate consequence of the part (I) of the following

Lemma 2. (I) If G is a NHHT (non-Hamiltonian homogeneously traceable) graph on $n \geq 3$ vertices then, for each vertex x of G, there is another vertex y of G such that x and y are end-vertices of a Hamiltonian path, say

$$P = [x = v_1, v_2, \ldots, v_n = y], \qquad \ldots \quad (1)$$

$d(y) \geq 3$, the edge $xy \notin E(G)$, and $d(x) + d(y) \leq n-1$.

(II) Moreover, the equality $d(x) + d(y) = n-1$ is equivalent to the condition (a) and implies the condition (b) where

(a) For any integer k such that $3 \leq k \leq n$, if $xv_k \notin E(G)$ then $v_{k-1}y \in E(G)$.

(b) There is $v \in \{v_1, v_n\}$ with $d(v) \geq (n-1)/2$ and there is a triangle in G, which contains both v and an edge (or two edges) belonging to P.

The following result related to the condition (a) is well-known in Hamiltonian graph theory (cf. Ore [5], p.54).

Lemma 3. If $P = [v_1, v_2, \ldots, v_n]$ is a Hamiltonian path of a non-Hamiltonian graph G on $n \geq 3$ vertices then if the edge $v_1v_k \in E(G)$, the edge $v_{k-1}v_n \notin E(G)$ $(k = 2, 3, \ldots, n-1)$.

Proof of Lemma 2. Given any vertex x of G, let $P_1 = [x = x_1, x_2, \ldots, x_n]$ be a Hamiltonian path of G beginning at x. On the strength of Lemma 1 (i), we have $d(x_n) \geq 2$. So there is j, $2 < j < n$, such that edge $x_{j-1}x_n \in E(G)$. Hence $P_2 := P_1 \cup x_{j-1}x_n - x_{j-1}x_j$ is a Hamiltonian path of G with end-vertices x and x_j. Consequently, by Lemma 1 (iii), since both x_j and x_n are adjacent to x_{j-1}, we can choose $y \in \{x_j, x_n\}$ such that $d(y) \geq 3$ and $xy \notin E(G)$. To obtain a Hamiltonian path (1), we can put either $P = P_1$ if $y = x_n$ or $P = P_2$ if $y = x_j$. Now, by Lemma 3, the set

$$\tilde{V} = \{v_{k-1} : v_1v_k \in E(G) \text{ with } 2 \leq k \leq n-1\}$$

contains $d(x)$ vertices each of which is different from and non-adjacent to v_n. Therefore

$$d(v_n) \leq n - 1 - d(v_1) \qquad \ldots \quad (2)$$

whence part (I) of Lemma 2 follows.

Moreover, it is easily seen that the equality in (2) does hold iff each vertex $x \in V(G) - \tilde{V} - \{y = v_n\}$ is adjacent to y, i.e., iff (a) is true.

It remains to prove that the equality in (2) implies the condition (b). To this end, observe that there is $v \in \{v_1, v_n\}$ with $d(v) \geq (n-1)/2$. Now, the supposition that (b) is false, i.e., that any two vertices of G, adjacent to v, are separated in P by a vertex non-adjacent to v, implies the inequality $2d(v) - 1 \leq n-2$. This inequality follows from the fact that $d(v) - 1$ vertices of P are necessary and sufficient to separate $d(v)$ verices from each other in P and that the vertices adjacent to v and those separating them in P are among $n-2$ vertices $v_2, v_3, \ldots, v_{n-1}$. Thus there is no contradiction only if n is odd and $d(v) = d(v_1) = d(v_n) = (n-1)/2$. If it is the case, however, then the vertices non-adjacent to and those adjacent to v_1 alternate while passing along P. The same is true with v_1 replaced by v_n. So the vertices $v_2, v_4, \ldots, v_{n-1}$ are all adjacent to both v_1 and v_n. Furthermore, no two vertices v_i and v_j with odd subscripts are adjacent because otherwise G is Hamiltonian. Consequently, if $S = \{v_2, v_4, \ldots, v_{n-1}\}$, then $s(G) \geq k(G-S) - |S| = 1$, contrary to Lemma 1(ii). This completes the proof.

Remark 1. Lemma 2 both resembles a lemma by Lesniak-Foster [4] and suggests a slight refinement of it.

Part (I) of Lemma 2 can be strengthened so that the following generalization of Theorem 1 may be obtained.

Theorem 2. If G is a NHHT graph on $n \geq 3$ vertices then

$$\Delta(G) + \max_{v} \min_{y} \{d(y) \mid y, v \in V(G),\ vy \notin E(G) \text{ and } d(v) = \Delta(G)\} \leq n-2,$$

whence $\Delta(G) + \delta(G) \leq n-2$.

Theorems 1 and 2 can be used to prove the following result ([1]).

Theorem 3. NHHT n-vertex graphs do exist iff either $n \leq 2$ or $n \geq 9$.

Theorem 4. There are exactly four NHHT graphs on 9 vertices (ordered by inclusion) of which the smallest, Ξ , and the greatest, $\hat{\Xi}$, are presented in Figure 1.

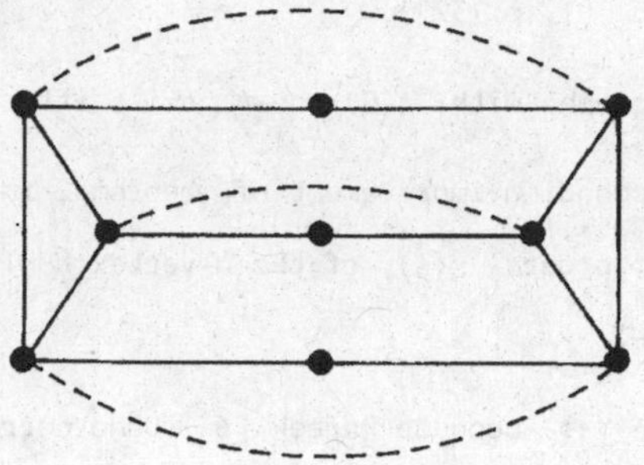

FIGURE 1

Because $\Delta(\hat{\Xi}) = 4$ therefore from Theorems 3 and 4 it follows that the upper bound $n-4$ for $\Delta(G)$ in Theorem 1 can be attained only if $n \geq 10$.

Theorem 5. There are exactly six NHHT graphs on $n = 10$ vertices with $\Delta(G) = n-4$, of which the smallest, Φ, and the greatest, $\hat{\Phi}$, are shown in Figure 2.

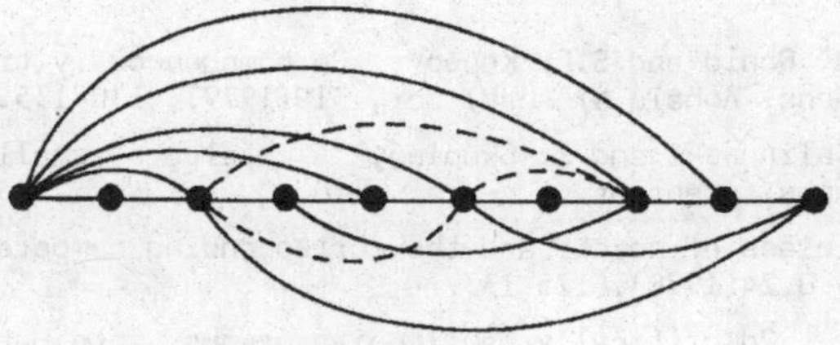

FIGURE 2

Though Theorems 4 and 5 were obtained independently of [2], they follow easily from the following results of [2] : $\hat{\Xi}$ is the only maximal NHHT graph on 9 vertices and among (four) maximal NHHT graphs G on $n = 10$ vertices only $G = \hat{\Phi}$ has $\Delta(G) = n-4$.

The above graph Φ can be generalized. Given non-negative integers r, s, t, and n such that $r + s + t = n-10$, let $G(r,s,t)$ be a graph on n vertices $x_1, x_2, \ldots, x_n$ containing the Hamiltonian path $[x_1, x_2, \ldots, x_n]$ and the edges x_1x_j with $3 \leq j \leq n-1$ and $j \notin \{r+4, r+s+7\}$; x_ix_n with $i \in \{r+3, r+s+6\}$, and the edge $x_{r+4}\, x_{r+s+8}$. Then $G(0,0,0) = \Phi$ and graphs $G = G(r,s,t)$ are (and even from the family of all) NHHT graphs (of minimum possible size $2n-3$) with $\Delta(G) = n-4$. In particular, because $d(x_2) = 2$, the non-Hamiltonicity of $G(r,s,t)$ is clearly reducible to that of Φ. Thus we can prove the following improvement of Theorem 1 (and Theorem 2 in case $\delta(G) = 2$) (see also [1, p.132]).

Theorem 6. NHHT n-vertex graphs with $\Delta(G) = n-4$ exist iff $n \geq 10$.

Problem 1. Find the corresponding improvement of Theorem 2 in case $\delta(G) \geq 3$. In particular, find the minimum order, $\nu(\delta)$, of the n-vertex NHHT graph G with $\delta(G) = \delta(\geq 3)$ and $\Delta(G) + \delta(G) = n-2$.

Remark 2. $\nu(\delta) \leq 4\delta$ if $\delta \geq 3$ because Pareek [6] found corresponding graphs G of each order $n \geq 4\delta$ for each $\delta \geq 3$.

Problem 2. Find or estimate the number of maximal NHHT graphs of order n(cf. [8]).

Notice that the number of NHHT graphs of order n is bounded below by an exponential function in n (cf. [7]).

REFERENCES

1. G. Chartrand, R.J. Gould and S.F. Kapoor, On homogeneously traceable non-hamiltonian graphs, Annals NY Acad. Sci, 319(1979), 130-135.
2. J. Jamrozik, R. Kalinowski and Z. Skupień, A catalog of small maximal non-hamiltonian graphs, preprint.
3. H.A. Jung, On a class of posets and the corresponding comparability graphs, J. Comb. Theory B,24(1978), 125-133.
4. L. Lesniak-Foster, On critically hamiltonian graphs, Acta Math. Acad. Sci. Hungar., 29(1977), 255-258.
5. O. Ore, Theory of Graphs, Providence, R.I. (1962).
6. C.M. Pareek, private communication.
7. Z. Skupień, Homogeneously traceable and Hamiltonian connected graphs, preprint (1976).
8. Z. Skupień, On maximal non-Hamiltonian graphs, Rostock. Math. Kolloq., 11(1979), 97-106.